Microbes for Legume Improvement

Almas Zaidi • Mohammad Saghir Khan
Javed Musarrat
Editors

Microbes for Legume Improvement

Second Edition

Editors
Almas Zaidi
Dept of Agricultural Microbiology
Faculty of Agricultural Sciences
Aligarh Muslim University
Aligarh
India

Mohammad Saghir Khan
Dept of Agricultural Microbiology
Faculty of Agricultural Sciences
Aligarh Muslim University
Aligarh
India

Javed Musarrat
Dept of Agricultural Microbiology
Faculty of Agricultural Sciences
Aligarh Muslim University
Aligarh
India

ISBN 978-3-319-59173-5   ISBN 978-3-319-59174-2   (eBook)
DOI 10.1007/978-3-319-59174-2

Library of Congress Control Number: 2017954922

Printed on acid-free paper

This Springer imprint is published by Springer Nature
The registered company is Springer International Publishing AG
The registered company address is: Gewerbestrasse 11, 6330 Cham, Switzerland

Preface

Globally, farming communities are finding it difficult to fulfill food demands of human populations due largely to declining crop production. The crop production is dwindling due to declining cultivable lands, fluctuating environments, and excessive usage of chemical fertilizers in order to optimize crop yields. Apart from these, the nutrient pool of soil is deteriorating rapidly, which further intensifies agricultural problems. Due to these, there is a pressing need to find solutions to expensive and environmentally disruptive problems. To solve these problems, soil microbiota have been found as inexpensive and environmentally sustainable options as organic fertilizers in providing adequate nutrients to growing crops including legumes. Legumes grown in many countries improve soil quality by increasing soil organic matter and soil structure and porosity, recycling nutrients, decreasing soil pH, diversifying the rhizosphere microbes, and decreasing disease incidence. The application of rhizobial inoculants and other free-living/associative plant growth-promoting rhizobacteria (PGPR) and mycorrhizal fungi in legume production has been found extremely useful.

Microbes for Legume Improvement (second edition) written by qualified teachers and scientists presents exceptional, recent, and wide-ranging information on the use of beneficial soil microbiota in legume production across different production systems. The revised edition presents the current status on the taxonomy of bacteria able to establish nitrogen-fixing legume symbiosis. Recent developments in the active biomolecules involved in rhizobia-legume symbiosis are highlighted. The importance of flavonoids and nod factors in legume-microbe interactions and their role in legume improvement is dealt separately. The advances made in recent times on the role of ethylene and bacterial ACC deaminase in legume-*Rhizobium* interactions are also included in this second edition. The latest developments in the field of some novel rhizobial exopolysaccharides and their role in legume-rhizobia symbiosis and environmental monitoring in legume improvement are discussed separately. The rhizobial diversity for tropical pulses and forage and tree legumes in Brazil is discussed separately. The book further describes the potential of rhizobia as plant growth-promoting rhizobacteria for enhancing the production of legumes in different agronomic regions. The deficiency of phosphorus restricts the legume production severely. To address and resolve such problems, meaningful and extensive information on the role of phosphate-solubilizing bacteria in the improvement of legumes is highlighted. The mycorrhizosphere interactions involving mycorrhizal establishment, mycorrhizal management for improving legume productivity, and interactive influence of mycorrhiza on legume development are described. The role of associative plant growth-promoting rhizobacteria especially *Azospirillum* used either alone or as mixture with other PGPR in increasing the productivity of legumes is highlighted. Also, endophytic microbes affecting legume performance are included in this second edition. This book further provides some novel microbial strategies and proposes alternative solution, which if properly applied could help to boost the overall performance of legumes growing under various stressed environments including salt, drought, and heavy metal-polluted soils. Also, this book gives information on how rhizobia abate metal toxicity and consequently enhance legume production in metal-contaminated soil, when used as metal-tolerant inoculants. The information and strategies described in this second edition are very useful which may serve as an important and updated reference material. This revised edition provides an elaborate overview for persons interested in legume research. This revised edition will, therefore, be of great practical interest to research scientists, postgraduate students, bioscience professionals, decision-makers, and farmers who aim to apply microbes for enhancing legume production. It is also likely to serve as a precious resource for agronomists, soil microbiologists, soil scientists, biologists, and biotechnologists involved in legume research.

We are extremely grateful to our well-qualified and internationally renowned colleague authors from different countries for providing their important, authoritative, and cutting-edge scientific information to upgrade this book. All chapters presented in this revised edition have the latest information with well-placed tables and figures and most recent references. The timely help and generous support extended by our loyal and trusted research scholars in revising this book are commendable. We are indeed very thankful to our family members for their unconditional and constant support, who, in their own ways, motivated us to complete this herculean task. We must also appreciate the honest efforts of the book publishing team in responding to all our queries very promptly and without any delay. Finally, if any one finds any mistake, factual or otherwise, or printing errors in this book, they may inform us so that the same is corrected and improved in subsequent print/edition.

Aligarh, UP, India   Almas Zaidi
Aligarh, UP, India   Mohammad Saghir Khan
Aligarh, UP, India   Javed Musarrat

Contents

Editors and Contributors

About the Editors

Almas Zaidi, M.Sc., Ph.D. received her M.Sc., Ph.D. (agricultural microbiology) from Aligarh Muslim University, Aligarh, India, and is currently serving as assistant professor at the Department of Agricultural Microbiology, Aligarh Muslim University, Aligarh, India. Dr. Zaidi has been teaching microbiology at postgraduate level for more than 12 years and has research experience of 16 years. She has published more than 50 research papers, book chapters, and review articles in journals of national and international repute. Dr. Zaidi has edited seven books published by leading publishers. Her main focus of research is to address problems related with rhizo-microbiology, microbiology, environmental microbiology, and biofertilizer technology.

Mohammad Saghir Khan, Ph.D. is a professor at the Department of Agricultural Microbiology, Aligarh Muslim University, Aligarh, India. Dr. Khan received his M.Sc. from Aligarh Muslim University, Aligarh, India, and Ph.D. (microbiology) from Govind Ballabh Pant University of Agriculture and Technology, Pantnagar, India. He has been teaching microbiology to postgraduate students for the last 20 years and has research experience of 24 years. In addition to his teaching, Dr. Khan is engaged in guiding students for their doctoral degree in microbiology. He has published over 100 scientific papers including original research articles, review articles, and book chapters in various national and international publication media. Dr. Khan has also edited nine books published by the leading publishers. Dr. Khan is deeply involved in research activities focusing mainly on rhizobiology, microbiology, environmental microbiology especially heavy metal-microbe-legume interaction, bioremediation, pesticide-PGPR-plant interaction, biofertilizers, and rhizo-immunology.

Javed Musarrat, M.Sc., Ph.D. (Biochem.) former chairman of the Department of Agricultural Microbiology and ex-dean of the Faculty of Agricultural Sciences, Aligarh Muslim University, Aligarh, India, is presently working as a vice-chancellor of the Baba Ghulam Shah Badshah University, Rajouri, Jammu and Kashmir, India. He has been teaching biochemistry, microbiology, and molecular biology to postgraduate students for the last 27 years and has research experience of about 33 years. He has published more than 175 national and international scientific publications. Dr. Musarrat has edited five books published by the leading publishers. He is associated with several scientific bodies such as DBT, CSIR, UGC, ICAR, UPCST, and CCRUM in various capacities. His major area of interest includes the molecular microbiology, microbial ecology, nanotechnology, and genetic toxicology.

Contributors

Bilal Ahmad Department of Agricultural Microbiology, Faculty of Agricultural Sciences, Aligarh Muslim University, Aligarh, UP, India

Zulfiqar Ahmad Department of Environmental Sciences, PMAS Arid Agriculture University, Rawalpindi, Pakistan

Ademir Sérgio Ferreira Araújo Centro de Ciências Agrárias, Universidade Federal do Piauí, Campus da Soccopo, Teresina, PI, Brazil

Rosario Azcón Departamento de Microbiología del Suelo y Sistemas Simbióticos, Estación, Experimental del Zaidín, Granada, Spain

Concepción Azcón-Aguilar Departamento de Microbiología del Suelo y Sistemas Simbióticos, Estación, Experimental del Zaidín, Granada, Spain

Muhammad Zahir Aziz Institute of Soil and Environmental Sciences, University of Agriculture, Faisalabad, Pakistan

José-Miguel Barea Departamento de Microbiología del Suelo y Sistemas Simbióticos, Estación, Experimental del Zaidín, Granada, Spain

Anelise Beneduzi Department of Genetics, Institute of Biosciences, Universidade Federal do Rio Grande do Sul, Porto Alegre, Brazil

Raktim Bhattacharya Microbiology Laboratory, Department of Life Sciences, Presidency University, Kolkata, West Bengal, India

Rabindranath Bhattacharyya Microbiology Laboratory, Department of Life Sciences, Presidency University, Kolkata, West Bengal, India

Madhurima Chatterjee Microbiology Laboratory, Department of Life Sciences, Presidency University, Kolkata, West Bengal, India

Sandip Das Microbiology Laboratory, Department of Life Sciences, Presidency University, Kolkata, West Bengal, India

Abhijit Dey Microbiology Laboratory, Department of Life Sciences, Presidency University, Kolkata, West Bengal, India

Erdal Elkoca Department of Agronomy, Faculty of Agriculture, Ataturk University, Erzurum, Turkey

Ahmet Eşitken Horticulture and Viticulture Department, Faculty of Agriculture, Selcuk University, Konya, Turkey

Felipe José Cury Fracetto Departamento de Agronomia, Dom Manoel de Medeiros, Universidade Federal Rural de Pernambuco, Recife, PE, Brazil

Giselle Gomes Monteiro Fracetto Departamento de Agronomia, Dom Manoel de Medeiros, Universidade Federal Rural de Pernambuco, Recife, PE, Brazil

Adeneide Candido Galdino Departamento de Agronomia, Universidade Federal Rural de Pernambuco, Recife, PE, Brazil

Paula García-Fraile Microbiology Institute, Academy of Science of the Czech Republic, Prague, Czech Republic

Adriana Giongo Institute of Petroleum and Natural Resources, Pontifical Catholic University of Rio Grande do Sul, Porto Alegre, RS, Brazil

Adem Güneş Agricultural Faculty, Soil and Plant Nutrition Science, Erciyes University, Kayseri, Turkey

M. Baqir Hussain Institute of Soil and Environmental Sciences, University of Agriculture, Faisalabad, Pakistan

Department of Soil Science, Faculty of Agricultural Sciences and Technology, Bahauddin Zakariya University, Multan, Pakistan

Muhammad Imran Department of Soil Science, Muhammad Nawaz Shareef University of Agriculture, Multan, Pakistan

M.Rüştü Karaman Vocational High School of Medical and Aromatic Plants, Afyon Kocatepe University, Afyonkarahisar, Turkey

Azeem Khalid Department of Environmental Sciences, PMAS Arid Agriculture University, Rawalpindi, Pakistan

Mohammad Saghir Khan Department of Agricultural Microbiology, Faculty of Agricultural Sciences, Aligarh Muslim University, Aligarh, UP, India

Dominika Kidaj Department of Genetics and Microbiology, Maria Curie-Skłodowska University, Lublin, Poland

Nurgül Kitir Engineering Faculty, Genetics and Bioengineering Department, Yeditepe University, Istanbul, Turkey

Mario Andrade Lira Junior Departamento de Agronomia, Dom Manoel de Medeiros, Universidade Federal Rural de Pernambuco, Recife, PE, Brazil

Bruno Brito Lisboa Laboratory of Agricultural Chemistry, Fundação Estadual de Pesquisa Agropecuária, Porto Alegre, Brazil

Shahid Mahmood Department of Environmental Sciences, PMAS Arid Agriculture University, Rawalpindi, Pakistan

Tariq Mahmood Department of Environmental Sciences, PMAS Arid Agriculture University, Rawalpindi, Pakistan

Pilar Martínez-Hidalgo Departamento de Microbiología y Genética and Instituto Hispano-Luso de Investigaciones Agrarias (CIALE), Universidad de Salamanca, Salamanca, Spain

Department of Molecular, Cell and Developmental Biology, University of California, Los Angeles, CA, USA

Eustoquio Martínez-Molina Departamento de Microbiología y Genética, Edificio Departamental de Biología, Doctores de la Reina s/n, Universidad de Salamanca, Salamanca, Spain

Pedro F. Mateos Departamento de Microbiología y Genética and Instituto Hispano-Luso de Investigaciones Agrarias (CIALE), Universidad de Salamanca, Salamanca, Spain

Ijaz Mehboob District Fertility Lab, Kasur, Pakistan

Esther Menéndez Departamento de Microbiología y Genética and Instituto Hispano-Luso de Investigaciones Agrarias (CIALE), Universidad de Salamanca, Salamanca, Spain

Negar Ebrahim Pour Mokhtari Islahiye Vocational School, Organic Farming Department, Gaziantep University, Gaziantep, Turkey

Diriba Muleta Institute of Biotechnology, College of Natural Sciences, Addis Ababa University, Addis Ababa, Ethiopia

Muhammad Naveed Institute of Soil and Environmental Sciences, University of Agriculture, Faisalabad, Pakistan

Emrah Nikerel Engineering Faculty, Genetics and Bioengineering Department, Yeditepe University, Istanbul, Turkey

Rafaela Simão Abrahão Nóbrega Universidade Federal do Recôncavo da Bahia, Centro de Ciências Agrárias, Ambientais e Biológicas, Cruz das Almas, BA, Brazil

Luciane Maria Pereira Passaglia Department of Genetics, Institute of Biosciences, Universidade Federal do Rio Grande do Sul, Porto Alegre, Brazil

Alvaro Peix Departamento de Desarrollo Sostenible de Sistemas Agroforestales y Ganaderos, Instituto de Recursos Naturales y Agrobiología (IRNASA-CSIC), Salamanca, Spain

Martha-Helena Ramírez-Bahena Departamento de Microbiología y Genética, Edificio Departamental de Biología, Universidad de Salamanca, Salamanca, Spain

Raúl Rivas Departamento de Microbiología y Genética, Edificio Departamental de Biología, Universidad de Salamanca, Salamanca, Spain

Asfa Rizvi Department of Agricultural Microbiology, Faculty of Agricultural Sciences, Aligarh Muslim University, Aligarh, UP, India

Saima Saif Department of Agricultural Microbiology, Faculty of Agricultural Sciences, Aligarh Muslim University, Aligarh, UP, India

Mohammad Shahid Department of Agricultural Microbiology, Faculty of Agricultural Sciences, Aligarh Muslim University, Aligarh, UP, India

Luis R. Silva Health Sciences Research Center (CICS), Faculty of Health Sciences—University of Beira Interior, Covilhã, Portugal

Krisle Silva Empresa Brasileira de Pesquisa Agropecuária, Embrapa Roraima, Boa Vista, RR, Brazil

Anna Skorupska Maria Curie-Skłodowska University, Department of Genetics and Microbiology, Lublin, Poland

Bahar Soğutmaz Özdemir Engineering Faculty, Genetics and Bioengineering Department, Yeditepe University, Istanbul, Turkey

Leyla Tarhan Engineering Faculty, Genetics and Bioengineering Department, Yeditepe University, Istanbul, Turkey

Şefik Tüfenkçi Department of Agricultural Faculty, Biosystems Engineering, Yüzüncüyıl University, Van, Turkey

Metin Turan Engineering Faculty, Genetics and Bioengineering Department, Yeditepe University, Istanbul, Turkey

Ceren Ünek Engineering Faculty, Genetics and Bioengineering Department, Yeditepe University, Istanbul, Turkey

Deniz Uras Engineering Faculty, Genetics and Bioengineering Department, Yeditepe University, Istanbul, Turkey

Luciano Kayser Vargas Laboratory of Agricultural Chemistry, Fundação Estadual de Pesquisa Agropecuária, Porto Alegre, Brazil

Encarna Velázquez Departamento de Microbiología y Genética, Edificio Departamental de Biología, Doctores de la Reina s/n, Universidad de Salamanca, Salamanca, Spain

Camila Gazolla Volpiano Department of Genetics, Institute of Biosciences, Universidade Federal do Rio Grande do Sul, Porto Alegre, Brazil

Jerzy Wielbo Department of Genetics and Microbiology, Maria Curie-Skłodowska University, Lublin, Poland

Muhammad Yaseen Institute of Soil and Environmental Sciences, University of Agriculture, Faisalabad, Pakistan

Ertan Yildirim Faculty of Agriculture, Department of Horticulture, Ataturk University, Erzurum, Turkey

Zahir Ahmad Zahir Institute of Soil and Environmental Sciences, University of Agriculture, Faisalabad, Pakistan

Hamdi H. Zahran Faculty of Science, Department of Botany, University of Beni-Suef, Beni-Suef, Egypt

Almas Zaidi Department of Agricultural Microbiology, Faculty of Agricultural Sciences, Aligarh Muslim University, Aligarh, UP, India

Current Status of the Taxonomy of Bacteria Able to Establish Nitrogen-Fixing Legume Symbiosis

Encarna Velázquez, Paula García-Fraile,
Martha-Helena Ramírez-Bahena, Raúl Rivas,
and Eustoquio Martínez-Molina

Abstract

Bacteria forming nitrogen-fixing symbiosis with legumes, classically named *rhizobia*, currently include more than 100 species distributed in the old genera *Allorhizobium, Azorhizobium, Bradyrhizobium, Ensifer* (formerly *Sinorhizobium*), *Mesorhizobium* and *Rhizobium* and in the new genera *Neorhizobium* and *Pararhizobium*. In addition, several new *rhizobia* have been described in the twenty-first century belonging, as the classical rhizobia, to the alpha *Proteobacteria* genera *Aminobacter, Devosia, Methylobacterium, Microvirga, Ochrobactrum, Phyllobacterium* and *Shinella* and to the beta *Proteobacteria Burkholderia, Paraburkholderia* (formerly *Burkholderia*) and *Cupriavidus*. These species carry symbiotic genes encoding for nodulation and nitrogen fixation which are located on plasmids or symbiotic islands. These genes determine the host range and confer rhizobia the ability to fix nitrogen in the legume nodules. Depending on the harboured nodulation genes, several symbiovars have recently been described in the classical *rhizobia* genera. In this chapter, we review the different groups of bacteria able of forming symbiosis with legumes and their classification based on core genes (genera and species) as well as on auxiliary ones (symbiovars).

E. Velázquez (✉) • M.-H. Ramírez-Bahena • R. Rivas • E. Martínez-Molina
Departamento de Microbiología y Genética and Instituto Hispanoluso de Investigaciones
Agrarias (CIALE), Edificio Departamental de Biología, Doctores de la Reina s/n, Universidad
de Salamanca, 37007 Salamanca, Spain
e-mail: evp@usal.es

P. García-Fraile
Microbiology Institute, Academy of Science of the Czech Republic,
Vídeňská 1083, Praha 4-Krč, 14220 Prague, Czech Republic

© Springer International Publishing AG 2017
A. Zaidi et al. (eds.), *Microbes for Legume Improvement*,
DOI 10.1007/978-3-319-59174-2_1

1.1   Introduction

Bacteria able to induce the formation of nitrogen-fixing nodules on the root systems of legumes were discovered during nineteenth century. The first isolate of this bacterial group was obtained from a nodule suspension of *Vicia* and was named as *Bacillus radicicola* (Beijerinck 1888). Later on, it was renamed as *Rhizobium leguminosarum* (Frank 1889). Since then, the bacteria able to produce nodules on legume roots were called *rhizobia* and have been included in the genus *Rhizobium*, which initially contained only six species; all of them were able to nodulate legumes. These species were defined primarily on the basis of the legumes they nodulated and the cross-inoculation concepts given by Baldwin and Fred (1929). Physiological characteristics were not used with taxonomic purposes in rhizobia until year 1982 when the genus *Bradyrhizobium* was created to include the species from genus *Rhizobium* that presented slow growth in culture media (Jordan 1982). Nevertheless, the symbiotic criteria continued to be used in both genera for species definition during several years.

The taxonomic changes in this group of bacteria started when Woese placed the nodule bacteria into the alpha subdivision of *Proteobacteria* using 16S rRNA gene sequence analysis (Woese et al. 1984). Following 16S rRNA gene sequencing, the number of rhizobial genera and species began to increase which further led to several relevant changes in the rhizobial classification, most of them were, however, not accepted by the rhizobiologists, because they implied a change in the previous rhizobial concept. For example, the reclassification of *Agrobacterium* and *Sinorhizobium* in *Rhizobium* and *Ensifer*, respectively, (Young et al. 2001; Young 2003) implied that some pathogenic bacteria were included into genus *Rhizobium*, which until year 2000 only contained legume endosymbionts, and *Sinorhizobium*, also a classical rhizobial genus, disappeared after its inclusion into the non-rhizobial genus *Ensifer*. The discomfort of rhizobiologists with this situation was evident, and recently, Mousavi et al. (2015) recovered the old names *Agrobacterium* and *Allorhizobium*. Nevertheless, they could avoid to include pathogenic strains in classic rhizobial genera, since the former species *Agrobacterium vitis* (Ophel and Kerr 1990) belonged to the genus *Allorhizobium*. In the case of *Ensifer*, the genus name has been imposed in taxonomic journals for the recent descriptions of new species, but some old species have been recently reclassified, such as *Ensifer morelensis* and *Ensifer americanus* (Wang et al. 2013c), and one species, *Sinorhizobium chiapanecum*, is still pending for reclassification into the genus *Ensifer* (Rincón-Rosales et al. 2009). Nevertheless, this species should be reclassified into genus *Ensifer* since it is phylogenetically related to *Ensifer mexicanus* (Rincón-Rosales et al. 2009).

Recent advances in the gene sequencing have further allowed the analysis of several genes and even complete genomes. There are evidences that classic rhizobial families now encompass genera that have not been isolated from legume nodules, such as *Ciceribacter* (Kathiravan et al. 2013) and *Pseudorhizobium* (Kimes et al. 2015) from family *Rhizobiaceae* and *Tardiphaga* (de Meyer et al. 2012) and *Metalliresistens* (Noisangiam et al. 2010) from family *Bradyrhizobiaceae* (*Nitrobacteriaceae*). It is however, possible that some of these genera, in the future, will include nodulating species, as happened with the genus *Shinella*, initially found in environmental sources (An et al. 2006). Currently, *Shinella* contains the species *Shinella kummerowiae* able to form nodules in the legume *Kummerowia stipulata* (Lin et al. 2008). Conversely, classic rhizobial genera now also contain many species isolated from sources other than legumes. Some of them coming from non-rhizobial genera, such as *Blastobacter denitrificans* and *Agromonas oligotrophica* that were transferred to genus *Bradyrhizobium* (van Berkum et al. 2006; Ramírez-Bahena et al. 2013a) and *Blastobacter aggregatus* that was first transferred to genus *Rhizobium* (Kaur et al. 2011) and later to the new genus *Pararhizobium* (Mousavi et al. 2015). Nevertheless, the most relevant advancement in the present century was the finding that bacteria that do not belong to classic rhizobial genera are able to induce nodules in legumes. Most of these bacteria belong to genera from the alpha subdivision of Proteobacteria, while some of them belong to the beta subdivision of *Proteobacteria* (Velázquez et al. 2010a; Peix et al. 2015b). The first beta *Proteobacteria* isolated from legume nodules were included into the genera, *Burkholderia* and *Ralstonia*, able to nodulate *Mimosa* (Moulin et al. 2001; Chen et al. 2003a), but they are currently included in the genera *Paraburkholderia* (Sawana et al. 2014) and *Cupriavidus* (Vandamme and Coenye 2004). Although some studies have shown that in some American and Asian countries *Mimosa* is preferably nodulated by alpha *Proteobacteria* (Gehlot et al. 2013; Bontemps et al. 2016), beta *Proteobacteria* are widespread in nodules of legumes from several tribes in different countries and continents (Barrett and Parker 2005; Lemaire et al. 2015). For this reason, the number of new species of this Class of bacteria that are able to nodulate legumes has considerably increased in the last years, although only in the genus *Burkholderia* (*Paraburkholderia*).

In parallel, the development of sequencing techniques also allowed the analysis of symbiotic genes that have been used to define biovars, which are currently named symbiovars (Rogel et al. 2011). The old biovars were defined on the basis of nodulation assays in different legumes (Jordan 1984), but currently the symbiovars are defined on the basis of the nodulation gene analysis and particularly on that of *nodC* gene (Rogel et al. 2011; Peix et al. 2015b). As occurred in the case of genera and species, the number of symbiovars is increasing, but many lineages found in classical rhizobial genera are still undescribed. Taking also into account that symbiovars have not been described to date in non-rhizobial genera and that there are many legumes whose endosymbionts are still poorly studied, it is predicted that there could be a significant increase in the number of symbiovars in the near future. In this chapter, an attempt is made to present a recent update on species in the classic and new genera of rhizobia and symbiovars within the classic rhizobial genera.

1.2   The Classic Rhizobia

In the beginning, the classification of rhizobial strains in different species was based on the legume they effectively nodulated. Following this criterion, in the same year, three fast-growing species, *Rhizobium phaseoli*, *R. trifolii* and *R. meliloti*, nodulating *Phaseolus*, *Trifolium* and *Melilotus*, respectively, were described by Dangeard (1926), and the slow-growing species *Rhizobium japonicum*, nodulating *Glycine*, was described by Buchanan (1926). Five years later, a second slow-growing species *Rhizobium lupini*, nodulating *Lupinus*, was added to the genus *Rhizobium* (Eckhardt et al. 1931). These species were recorded in *Bergey's Manual of Determinative Bacteriology* published in year 1974 (Jordan and Allen 1974) and were later included in the validation lists of Skerman et al. (1980) published in the International Journal of Systematic Bacteriology (IJSB), which is the official journal for prokaryotes systematic, currently named International Journal of Systematic and Evolutionary Microbiology (IJSEM). Therefore, in 1980, the genus *Rhizobium* contained five species, and it was included in the family *Rhizobiaceae*, which presently also contains several old genera (Conn 1938). One of these genera was *Alcaligenes*, which in 1938 had several species, later reclassified as *Agrobacterium* (Conn 1942). When the Bergey's Manual was published in 1984, the Family *Rhizobiaceae* contained the two old genera *Rhizobium* and *Agrobacterium* and two new ones named *Bradyrhizobium* and *Phyllobacterium* (Jordan 1984). The genus *Phyllobacterium* was isolated from leaves of plants (Knösel 1984) and was reported first time in the Bergey's Manual in 1984, being validated in the same year in IJSB. The genus *Bradyrhizobium* was described by Jordan 2 years before the Bergey's Manual publication to include the slow-growing rhizobia which was previously placed into genus *Rhizobium*. Nevertheless, of the two species, *R. japonicum* and *R. lupini*, only *R. japonicum* was reclassified into the genus *Bradyrhizobium* and the first species with the name *Bradyrhizobium japonicum*, because it was considered that *R. lupini* mainly differed from this species in its ability to nodulate *Lupinus* and *Ornithopus* (Jordan 1982). Although the proposal of the new genus *Bradyrhizobium* was considered in the IJSB, and since the species *R. lupini* was not formally rejected in this publication, the name *R. lupini* remained valid. In addition to the description of the new genus *Bradyrhizobium*, Jordan (1984) revised the taxonomic status of the species *R. trifolii* and *R. phaseoli*, which were reclassified into the species *R. leguminosarum*. Therefore, in the Bergey's Manual from year 1984, the genus *Rhizobium* contained only four species, two of them already included in the version from year 1974, *R. leguminosarum*, *R. meliloti*, and a third species described in IJSB and named *R. loti* (Jarvis et al. 1982). Nevertheless, since the reclassification of *R. trifolii* and *R. phaseoli* was made outside the IJSB, these two names remained valid as occurred in the case of *R. lupini*.

Despite variation in physiological traits, the differences in growth rate were used to differentiate the genera *Rhizobium* and *Bradyrhizobium*. However, the ability to nodulate different legumes remained the main criterion for differentiating the species within these genera. In this way the *Bradyrhizobium* strains that were not isolated from soybean nodules were named *Bradyrhizobium* sp. placing in parentheses the name of the species they nodulated. At this time in other bacterial groups, the numerical taxonomy developed by Sokal and Sneath (1963) had already been applied for the classification of their members at genus and species levels, including *Agrobacterium*, which belongs to the same family than rhizobial genera (Kersters and de Ley 1984). The scarce importance that rhizobiologists paid to the phenotypic characteristics, the low number of legumes studied and the erroneous link between the species concept and the ability to nodulate legumes were the probable causes of the underestimation of rhizobial biodiversity.

This situation began to change in 1980s when it was discovered that soybean may also be nodulated by fast-growing rhizobia belonging to the new species *R. fredii* (Scholla and Elkan 1984). The description of this species thus revealed that a single legume may be infected by different rhizobial genera. This led to the reclassification of *Rhizobium fredii* into a new genus named *Sinorhizobium* (Chen et al. 1988). Moreover, from this date, the number of phenotypic characteristics included in the description of new taxa was higher, as occurred in the case of *Rhizobium galegae* isolated from *Galega* nodules (Lindström 1989) and *Rhizobium huakuii* isolated from *Astragalus* nodules (Chen et al. 1991). Also, a high number of phenotypic characteristics and several molecular approaches, such as the rRNA-DNA hybridization, were used in the description of a new genus named *Azorhizobium* whose type species *Azorhizobium caulinodans* forms stem nodules on the legume *Sesbania rostrata* (Dreyfus et al. 1988). This study showed that not only the nodules formed in legume roots are induced by bacteria but also those formed in the stems of tropical legumes.

The most relevant change for bacterial taxonomy was Woese's proposal of a new classification of prokaryotes based on their 16S ribosomal gene sequences (Woese and Fox 1977). The findings of Woese converted this gene in an essential tool for bacterial classification and identification, although initially only partial sequences were obtained. According to the further analysis of the 16S rRNA gene, the rhizobia were placed within the alpha subdivision of *Proteobacteria* (Woese et al. 1984). However, the sequencing of this gene was not included in the rhizobial species descriptions until year 1991, when the minimal standards for species description of new rhizobia and *Agrobacterium* were published (Graham et al. 1991). The first species described on the basis of partial sequencing of the 16S rRNA gene was *Rhizobium tropici* (Martínez-Romero et al. 1991). From 1991 onwards, the 16S rRNA gene sequences were included in all descriptions or reclassifications of the different taxa within family *Rhizobiaceae*, and the reclassification of *R. fredii* into genus *Sinorhizobium* was confirmed by the analysis of this gene (Jarvis et al. 1992). Several years later, the existence of two phylogenetic groups within genus *Rhizobium* was evidenced, and a new genus named *Mesorhizobium* with an intermediate growth rate between *Rhizobium* and *Bradyrhizobium* was described (Jarvis et al. 1997). A year later, the new genus *Allorhizobium* was described with a single species named *Allorhizobium undicola* isolated in Senegal from nodules of *Neptunia natans* (de Lajudie et al. 1992). In addition to these new genera, in the 1990s several new rhizobial species were described (Table 1.1), and some species were changed from the old to the new described genera as occurred in the case of the species *R. meliloti*, which was reclassified into the new genus *Sinorhizobium* (de Lajudie et al. 1994) and the species *Rhizobium loti*, *Rhizobium huakuii*, *Rhizobium ciceri*, *Rhizobium mediterraneum* and *Rhizobium tianshanense* that were transferred to the new genus *Mesorhizobium* (Jarvis et al. 1997).

These reclassifications were based on the 16S rRNA gene sequences whose phylogenetic analyses also led Young et al. (2001) to reclassify the genera *Agrobacterium* and *Allorhizobium* into genus *Rhizobium*, published in the International Journal of Systematic and Evolutionary Microbiology (IJSEM). Many researchers did not

Table 1.1 Species of 'classical' rhizobia isolated from different sources

Species	Isolation source	References
Family *Rhizobiaceae*		
Genus *Rhizobium*		
Species isolated from legume nodules		
R. acidisoli	*Phaseolus vulgaris*	Román-Ponce et al. (2016)
R. aegyptiacum	*Trifolium alexandrinum*	Shamseldin et al. (2016)
R. altiplani	*Mimosa pudica*	Baraúna et al. (2016)
R. anhuiense	*Vicia faba, Pisum sativum*	Zhang et al. (2015b)
R. bangladeshense	*Lens culinaris*	Rashid et al. (2015)
R. binae	*Lens culinaris*	Rashid et al. (2015)
R. calliandrae	*Calliandra grandiflora*	Rincón-Rosales et al. (2013)
R. cauense	*Kummerowia stipulacea*	Liu et al. (2012a)
R. ecuadorense	*Phaseolus vulgaris*	Ribeiro et al. (2015)
R. etli	*Phaseolus vulgaris*	Segovia et al. (1993)
R. fabae	*Vicia faba*	Tian et al. (2008)
R. favelukesii	*Medicago sativa*	Torres Tejerizo et al. (2016)
R. freirei	*Phaseolus vulgaris*	Dall'Agnol et al. (2013)
R. gallicum	*Phaseolus vulgaris*	Amarger et al. (1997)
R. grahamii	*Dalea leporina, Leucaena leucocephala, Clitoria ternatea*	López-López et al. (2012)
R. helanshanense	*Spaherophysa salsula*	Qin et al. (2012)
R. hidalgonense	*Phaseolus vulgaris*	Yan et al. (2017)
R. jaguaris	*Calliandra grandiflora*	Rincón-Rosales et al. (2013)
R. hainanense	*Desmodium* spp., *Stylosanthes guianensis, Centrosema pubescens, Tephrosia candida, Acacia sinuata, Arachis hypogaea, Zornia diphylla, Uraria crinita, Macroptilium lathyroides*	Chen et al. (1997)
R. indigoferae	*Indigofera* spp.	Wei et al. (2002)
R. laguerreae	*Vicia* spp.	Saïdi et al. (2014)
R. leguminosarum	*Pisum sativum*	Frank (1889), Ramírez-Bahena et al. (2008)
R. lentis	*Lens culinaris*	Rashid et al. (2015)
R. leucaenae	*Leucaena leucocephala, Phaseolus vulgaris*	Ribeiro et al. (2012)
R. loessense	*Astragalus* spp.	Wei et al. (2003)
R. lusitanum	*Phaseolus vulgaris*	Valverde et al. (2006)
R. mayense	*Calliandra grandiflora*	Rincón-Rosales et al. (2013)

Table 1.1 (continued)

Species	Isolation source	References
R. mesoamericanum	Phaseolus vulgaris, Macroptilium atropurpureum, Vigna unguiculata, Mimosa púdica	López-López et al. (2012)
R. mesosinicum	Albizia julibrissin, Kummerowia spp., Dalbergia spp.	Lin et al. (2009)
R. miluonense	Lespedeza chinensis	Gu et al. (2008)
R. mongolense	Medicago ruthenica	van Berkum et al. (1998)
R. multihospitium	Lotus spp., Alhagi spp., Astragalus spp., Halimodendron halodendron, Oxytropis spp., Robinia pseudoacacia, Sophora alopecuroides, Caragana jubata, Lathyrus odoratus, Vicia hirsuta	Han et al. (2008b)
R. pakistanensis	Arachis hypogaea	Khalid et al. (2015)
R. paranaense	Phaseolus vulgaris	Dall'Agnol et al. (2014)
R. phaseoli	Phaseolus vulgaris	Dangeard (1926), Ramírez-Bahena et al. (2008)
R. pisi	Pisum sativum	Ramírez-Bahena et al. (2008)
R. pongamiae	Pongamia pinnata	Kesari et al. (2013)
R. puerariae	Pueraria candollei	Boonsnongcheep et al. (2015)
R. sophorae	Sophora flavescens	Jiao et al. (2015b)
R. sophoriradicis	Sophora flavescens	Jiao et al. (2015b)
R. sullae	Hedysarum coronarium	Squartini et al. (2002)
R. tibeticum	Trigonella archiducis-nicolai	Hou et al. (2009)
R. tropici	Phaseolus vulgaris, Leucaena leucocephala	Martínez-Romero et al. (1991)
R. tubonense	Oxytropis glabra	Zhang et al. (2011a)
R. vallis	Phaseolus vulgaris, Mimosa pudica, Indigofera spicata	Wang et al. (2011)
R. vignae[c]	Vigna radiata, Desmodium microphyllum, Astragalus spp.	Ren et al. (2011a)
R. yanglingense	Coronilla varia, Gueldenstaedtia multiflora, Amphicarpaea trisperma	Tan et al. (2001)
Species isolated from other sources		
R. alamii	Rhizosphere of Arabidopsis thaliana	Berge et al. (2009)
R. albus	Lake water	Li et al. (2017)
R. alvei	Freshwater river	Sheu et al. (2015c)
R. azooxidifex	Soil	Behrendt et al. (2016)
R. capsici	Root tumour on Capsicum annuum plants	Lin et al. (2015)

(continued)

Table 1.1 (continued)

Species	Isolation source	References
R. cellulosilyticum	Sawdust of Populus alba	García-Fraile et al. (2007)
R. daejeonense	Cyanide treatment bioreactor	Quan et al. (2005)
R. endolithicum	Beach sand	Parag et al. (2013)
R. endophyticum	Seeds of Phaseolus vulgaris	López-López et al. (2010)
R. flavum	Soil	Gu et al. (2014)
R. gei	Geum aleppicum stem	Shi et al. (2016)
R. halophytocola	Roots of Rosa rugosa	Bibi et al. (2012)
R. halotolerans	Chloroethylene-contaminated soil	Diange and Lee (2013)
R. helianthi	Rhizosphere of sunflower	Wei et al. (2015)
R. ipomoeae	Water convolvulus field	Sheu et al. (2016)
R. lemnae	Lemna aequinoctialis	Kittiwongwattana and Thawai (2014)
R. marinum	Seawater	Liu et al. (2015)
R. metallidurans	Seedlings of Anthyllis vulneraria	Grison et al. (2015)
R. naphtalenivorans	Sediment of a polychlorinated-dioxin-transforming microcosm	Kaiya et al. (2012)
R. oryzicola	Rice roots	Zhang et al. (2015a)
R. oryziradicis	Rice roots	Zhao et al. (2016)
R. petrolearium[a]	Oil-contaminated soil	Zhang et al. (2012c)
R. phenanthrenilyticum[a]	Petroleum residue treatment system	Wen et al. (2011)
R. populi	Populus euphratica	Rozahon et al. (2014)
R. pusense	Rhizosphere of Cicer arietinum	Panday et al. (2011)
R. qilianshanense	Oxytropis ochrocephala	Xu et al. (2013)
R. rhizoryzae	Oryza sativa roots	Zhang et al. (2014a)
R. rosettiformans	Hexachlorocyclohexane dumpsite	Kaur et al. (2011)
R. selenitireducens	Bioreactor	Hunter et al. (2007)
R. smilacinae	Smilacina japonica	Zhang et al. (2014c)
R. soli	Soil	Yoon et al. (2010)
R. straminoryzae	Rice straw	Lin et al. (2014)
R. subbaraonis	Beach sand	Ramana et al. (2013)
R. tarimense	Soil	Turdahon et al. (2013)
R. yantingense	Mineral water	Chen et al. (2015)
Genus Ensifer (formerly Sinorhizobium)		
Species isolated from legume nodules		
E. alkalisoli	Sesbania cannabina	Li et al. (2016b)
E. americanus	Acacia spp.	Toledo et al. (2003), Wang et al. (2013c), Oren and Garrity (2015a)

Table 1.1 (continued)

Species	Isolation source	References
E. arboris	Acacia senegal, Prosopis chilensis	Nick et al. (1999), Young (2003)
E. fredii	Glycine max	Scholla and Elkan (1984), Jarvis et al. (1992), Young (2003)
E. garamanticus	Argyrolobium uniflorum, Medicago sativa	Merabet et al. (2010)
E. glycinis	Glycine max, Astragalus mongholicus	Yan et al. (2016)
E. kostiensis	Acacia senegal, Prosopis chilensis	Nick et al. (1999), Young (2003)
E. kummerowiae	Kummerowia stipulacea	Wei et al. (2002), Young (2003)
E. medicae	Medicago truncatula	Rome et al. (1996), Young (2003)
E. meliloti	Medicago sativa	Dangeard (1926), de Lajudie et al. (1994), Young (2003)
E. mexicanus	Acacia angustissima	Lloret et al. (2007)
E. numidicus	Argyrolobium uniflorum, Lotus creticus	Merabet et al. (2010)
E. psoraleae	Psoralea corylifolia, Sesbania cannabina	Wang et al. (2013c)
E. saheli	Sesbania spp.	de Lajudie et al. (1994), Young (2003)
E. sesbaniae	Sesbania cannabina, Medicago lupulina	Wang et al. (2013c)
E. sojae	Glycine max	Li et al. (2011)
E. terangae	Acacia spp., Sesbania spp.	de Lajudie et al. (1994), Young (2003)
Species isolated from other sources		
E. adhaerens	Soil	Casida (1982)
E. morelensis	Associated to nodules Leucaena leucocephala	Wang et al. (2002), Wang et al. (2013c), Oren and Garrity (2015a)
Genus *Neorhizobium*		
Species isolated from legume nodules		
N. alkalisoli	Caragana intermedia	Lu et al. (2009a), Mousavi et al. (2014)
N. galegae	Galega officinalis	Lindström (1989), Mousavi et al. (2014)
N. huautlense	Sesbania herbacea	Wang et al. (1998), Mousavi et al. (2014)

(continued)

Table 1.1 (continued)

Species	Isolation source	References
Genus *Allorhizobium*		
Species isolated from legume nodules		
A. taibaishanense	Kummerowia striata	Yao et al. (2012), Mousavi et al. (2015)
A. undicola	Neptunia natans	de Lajudie et al. (1998), Young et al. (2001), Mousavi et al. (2015)
Species isolated from other sources		
A. borbori	Activated sludge	Zhang et al. (2011b), Mousavi et al. (2015)
A. oryzae	Roots of Oryza alta	Peng et al. (2008), Mousavi et al. (2015)
A. paknamense	Lemna aequinoctialis tissues	Kittiwongwattana and Thawai (2013), Mousavi et al. (2015)
A. pseudoryzae	Rhizosphere of Oryza sativa	Zhang et al. (2011c), Mousavi et al. (2015)
A. vitis	tumours on Vitis vinífera	Ophel and Kerr (1990), Young et al. (2001), Mousavi et al. (2015)
Genus *Pararhizobium*		
Species isolated from legume nodules		
P. giardinii	Phaseolus vulgaris	Amarger et al. (1997), Mousavi et al. (2015)
P. herbae	Astragalus membranaceus, Oxytropis cashemiriana	Ren et al. (2011b), Mousavi et al. (2015)
P. sphaerophysae	Sphaerophysa salsula	Xu et al. (2011), Mousavi et al. (2015)
Species isolated from other sources		
P. capsulatum	Surface lake water	Kaur et al. (2011), Mousavi et al. (2015)
P. polonicum	Galls of Prunus rootstocks	Puławska et al. (2016)
Genus *Sinorhizobium*[d]		
Species isolated from legume nodules		
S. abri	Abrus precatorius	Ogasawara et al. (2003)
S. chiapanecum	Acaciella angustissima	Rincón-Rosales et al. (2009)
S. indiaense	Sesbania rostrata	Ogasawara et al. (2003)
Family *Phyllobacteriaceae*		
Genus *Mesorhizobium*		
Species isolated from legume nodules		
M. abyssinicae	Acacia abyssinica	Degefu et al. (2013)
M. acaciae	Acacia melanoxylon	Zhu et al. (2015)

Table 1.1   (continued)

Species	Isolation source	References
M. albiziae	*Albizia*	Wang et al. (2007)
M. alhagi	*Alhagi sparsifolia*	Chen et al. (2010)
M. amorphae	*Amorpha fruticosa*	Wang et al. (1999b)
M. australicum	*Biserrula pelecinus*	Nandasena et al. (2009)
M. calcicola	*Sophora* spp.	De Meyer et al. (2015)
M. camelthorni	*Alhagi sparsifolia*	Chen et al. (2011)
M. cantuariense	*Sophora microphylla*	De Meyer et al. (2015)
M. caraganae	*Caragana*	Guan et al. (2008)
M. chacoense	*Prosopis*	Velázquez et al. (2001a)
M. ciceri	*Cicer arietinum*	Nour et al. (1994), Jarvis et al. (1997)
M. erdmanii	*Lotus* spp.	Martínez-Hidalgo et al. (2015b)
M. gobiense	*Astragalus filicaulis*, *Lotus* spp., *Oxytropis glabra*	Han et al. (2008a)
M. hawassense	*Sesbania sesban*	Degefu et al. (2013)
M. huakuii	*Astragalus sinicus*	Chen et al. (1991), Jarvis et al. (1997)
M. japonicum	*Lotus corniculatus*	Martínez-Hidalgo et al. (2016)
M. jarvisii	*Lotus corniculatus*	Martínez-Hidalgo et al. (2015b)
M. kowhaii	*Sophora* spp.	De Meyer et al. (2016)
M. loti	*Lotus corniculatus*	Jarvis et al. (1997), Martínez-Hidalgo et al. (2015b)
M. mediterraneum	*Cicer arietinum*	Nour et al. (1995), Jarvis et al. (1997)
M. metallidurans	*Anthyllis vulneraria*	Vidal et al. (2009)
M. muleiense	*Cicer arietinum*	Zhang et al. (2012a)
M. newzealandense	*Sophora* spp.	De Meyer et al. (2016)
M. olivaresii	*Lotus corniculatus*	Lorite et al. (2016)
M. opportunistum	*Biserrula pelecinus*	Nandasena et al. (2009)
M. plurifarium	*Acacia* spp., *Prosopis juliflora*, *Chamaecrista ensiformis*, *Leucaena* spp.	de Lajudie et al. (1998)
M. qingshengii	*Astragalus sinicus*	Zheng et al. (2013)
M. robiniae	*Robinia pseudoacacia*	Zhou et al. (2010)
M. sangaii	*Astragalus* spp.	Zhou et al. (2013)
M. septentrionale	*Astragalus adsurgens*	Gao et al. (2004)
M. shangrilense	*Caragana* spp.	Lu et al. (2009b)
M. shonense	*Acacia abyssinica*	Degefu et al. (2013)

(continued)

Table 1.1 (continued)

Species	Isolation source	References
M. silamurunense	Astragalus spp.	Zhao et al. (2012)
M. sophorae	Sophora spp.	De Meyer et al. (2016)
M. tamadayense	Anagyris latifolia, Lotus berthelotii	Ramírez-Bahena et al. (2012)
M. tarimense	Lotus frondosus	Han et al. (2008a)
M. temperatum	Astragalus adsurgens	Gao et al. (2004)
M. tianshanense	Glycyrrhiza spp., Sophora alopecuroides, Halimodendron holodendron, Caragana polourensis, Swainsona salsula, Glycine spp.	Chen et al. (1995), Jarvis et al. (1997)
M. waimense	Sophora longicarinata	De Meyer et al. (2015)
M. waitakense	Sophora spp.	De Meyer et al. (2016)
Species isolated from other sources		
M. sediminum	Deep-sea sediment	Yuan et al. (2016)
M. soli	Rhizosphere of Robinia pseudoacacia	Nguyen et al. (2015)
M. thiogangeticum	Rhizosphere of Clitoria ternatea	Ghosh and Roy (2006)
Family Nitrobacteriaceae (Bradyrhizobiaceae)		
Genus Bradyrhizobium		
Species isolated from legume nodules		
B. americanum	Centrosema macrocarpum	Ramírez Bahena et al. (2016)
B. arachidis	Arachis hypogaea	Wang et al. (2013b)
B. canariense	Chamaecytisus proliferus	Vinuesa et al. (2005)
B. centrosemae	Centrosema molle	Ramírez Bahena et al. (2016)
B. cytisi	Cytisus villosus	Chahboune et al. (2011)
B. daqingense	Glycine max	Wang et al. (2013a)
B. diazoefficiens	Glycine max	Delamuta et al. (2013)
B. elkanii	Glycine max	Kuykendall et al. (1992)
B. embrapense	Neonotonia wightii, Desmodium heterocarpon	Delamuta et al. (2015)
B. erythrophlei	Erythrophleum fordii	Yao et al. (2015)
B. ferriligni	Erythrophleum fordii	Yao et al. (2015)
B. ganzhouense	Acacia melanoxylon	Lu et al. (2014)
B. guangdongense	Arachis hypogaea	Li et al. (2015)
B. guangxiense	Arachis hypogaea	Li et al. (2015)
B. huanghuaihaiense	Glycine max	Zhang et al. (2012a)
B. icense	Phaseolus lunatus	Durán et al. (2014a)
B. ingae	Inga laurina	da Silva et al. (2014)
B. japonicum	Glycine max	Buchanan (1926), Jordan (1982)

Table 1.1 (continued)

Species	Isolation source	References
B. jicamae	Pachyrhizus erosus	Ramírez-Bahena et al. (2009)
B. kavangense	Vigna subterranea, Arachis hypogaea	Grönemeyer et al. (2015b)
B. lablabi	Lablab purpureus, Arachis hypogaea	Chang et al. (2011)
B. lupini	Lupinus spp.	Eckhardt et al. (1931), Peix et al. (2015a)
B. liaoningense	Glycine max	Xu et al. (1995)
B. manausense	Vigna unguiculata	Silva et al. (2014)
B. neotropicale	Centrolobium paraense	Zilli et al. (2014)
B. ottawaense	Glycine max	Yu et al. (2014)
B. pachyrhizi	Pachyrhizus erosus	Ramírez-Bahena et al. (2009)
B. paxllaeri	Phaseolus lunatus	Durán et al. (2014a)
B. retamae	Retama spp.	Guerrouj et al. (2013)
B. rifense	Cytisus villosus	Chahboune et al. (2012)
B. subterraneum	Vigna subterranea, Arachis hypogaea	Grönemeyer et al. (2015a)
B. stylosanthis	Stylosanthes	Delamuta et al. (2016)
B. tropiciagri	Neonotonia wightii, Desmodium heterocarpon	Delamuta et al. (2015)
B. valentinum	Lupinus mariae-josephae	Durán et al. (2014b)
B. vignae	Vigna subterranea, Arachis hypogaea	Grönemeyer et al. (2016)
B. viridifuturi	Centrosema pubescens	Helene et al. (2015)
B. yuanmingense	Lespedeza cuneata	Yao et al. (2002)
Species isolated from other sources		
B. betae	Root tumours on Beta vulgaris	Rivas et al. (2004)
B. denitrificans[b]	Water	Hirsch and Müller (1985), van Berkum et al. (2006)
B. iriomotense	Root tumours on Entada koshunensis	Islam et al. (2008)
B. oligotrophicum	Rice paddy soil	Ohta and Hattori (1983), Ramírez-Bahena et al. (2013a)
Family *Hyphomicrobiaceae*		
Genus *Azorhizobium*		
Species isolated from legume nodules		
A. dobereinerae	Sesbania virgata	Souza Moreira et al. (2006)
A. caulinodans	Sesbania rostrata	Dreyfus et al. (1988)

(continued)

Table 1.1 (continued)

Species	Isolation source	References
Species isolated from other sources		
A. oxalatiphilum	Macerated petioles of *Rumex* sp.	Lang et al. (2013)

[a]These species are synonyms since they have identical core gene sequences

[b]This species is able to fix nitrogen in *Aeschynomene* nodules according to van Berkum and Eardly (2002)

[c]This species was not officially rejected by Mousavi et al. (2014)

[d]The genus *Sinorhizobium* has been reclassified into genus *Ensifer*, but the species *S. abri*, *S. chiapanecum* and *S. indiaense* have still been not reclassified

accept the reclassification of the genus *Agrobacterium* into genus *Rhizobium* (Farrand et al. 2003), which was only justified in the case of the species *Agrobacterium rhizogenes* closely related to the species *R. tropici* (Yanagi and Yamasato 1993; Young et al. 2001; Velázquez et al. 2005, 2010b). Also, the reclassification of *Sinorhizobium*, a classical rhizobial genus, into genus *Ensifer* (Casida 1982) was performed according to the decision of the Judicial Commission of the International Committee on Systematic of Prokaryotes (2008) after several requests for an opinion sent to IJSEM (Willems et al. 2003; Young 2003). Despite that these two reclassifications were controversial, they were finally accepted because they were officially published and were based on the 16S rRNA gene analysis, which until year 2001 was the only gene analysed in rhizobia.

The taxonomic relevance of the 16S rRNA gene was also pointed out after the publication of the Bergey's Manual published in year 2005 in which the genera from the family *Rhizobiaceae* were dispersed into several new families within the new order *Rhizobiales* (Kuykendall et al. 2005) whose name is illegitimate because the order *Hyphomicrobiales* has preference (this order includes the family *Hyphomicrobiaceae* encompassing the rhizobial genus *Azorhizobium*). The rhizobia were included in the old families *Rhizobiaceae* and *Hyphomicrobiaceae* and in two new ones named *Bradyrhizobiaceae* and *Phyllobacteriaceae*. In the Bergey's Manual of 2005, the family *Rhizobiaceae* included the rhizobial genera *Rhizobium*, *Allorhizobium* and *Sinorhizobium*, because the reclassification of this last genus into genus *Ensifer* was not considered. The old family *Hyphomicrobiaceae* included the genus *Azorhizobium*, the only legume nodulating genus within this family. The new family *Phyllobacteriaceae* was proposed in Bergey's Manual (Mergaert and Swings 2005) and later validated (Validation list No. 107 2006) and contained the rhizobial genera *Phyllobacterium* and *Mesorhizobium*. The genus *Bradyrhizobium* was included in the new family *Bradyrhizobiaceae* (Garrity et al. 2005) whose name is also illegitimate since *Nitrobacteraceae*, a previously described family, includes the genus *Nitrobacter*, closely related to *Bradyrhizobium*. After the sequencing of the 16S rRNA gene, which currently remain as the basic for the classification and identification of rhizobia, the most important contribution to the taxonomy of this group of bacteria was the sequencing of several housekeeping genes. Two of these genes, *recA* and *atpD*, were firstly analysed in rhizobia by Gaunt et al.

(2001), and currently they are the most analysed housekeeping genes in rhizobia. Nevertheless, other genes such as *glnII*, *rpoB*, *dnaK* or *gyrB* have commonly been analysed in the recently described species of rhizobia, and the analysis of multilocus sequences (MLSA or MLST) including three or more housekeeping genes has supported the description of new genera and species, the reclassification of several genera and the recovery of old genera. For example, the new genera *Neorhizobium* (Mousavi et al. 2014) and *Pararhizobium* (Mousavi et al. 2015) have been described, and the old genera *Agrobacterium* and *Allorhizobium* have been recovered (Mousavi et al. 2015). The analysis of several housekeeping genes had also allowed the confirmation of *R. phaseoli* as a valid species (Ramírez-Bahena et al. 2008) and the reclassification of the old valid species *R. lupini* into the genus *Bradyrhizobium* as *Bradyrhizobium lupini* (Peix et al. 2015a). Moreover, since at least three housekeeping genes are commonly analysed, the descriptions of all recently described species within the classic rhizobial genera are based in multilocus sequence analysis.

The reliability of the analysis of housekeeping genes exceeds that of the 16S rRNA gene for rhizobial species differentiation which was pointed out during the description of *Rhizobium lusitanum*, the first species in which the phylogenetic analyses of two housekeeping genes, *recA* and *atpD*, were included (Valverde et al. 2006). Even several new rhizobial species have identical 16S rRNA genes and only differ in their housekeeping genes. This has been found in *Bradyrhizobium icense* and *Bradyrhizobium paxllaeri* which had identical 16S rRNA genes (Durán et al. 2014a) and in *Mesorhizobium acaciae* which had 16S rRNA gene identical to that of *Mesorhizobium plurifarium* (Zhu et al. 2015). Nevertheless, the higher number of valid species which have identical 16S rRNA genes belongs to genus *Rhizobium* and particularly to the *R. leguminosarum-Rhizobium etli* phylogenetic group. Species such as *R. laguerreae* (Saïdi et al. 2014), *R. sophorae* (Jiao et al. 2015b), *R. anhuiense* (Zhang et al. 2015b), *R. acidisoli* (Román-Ponce et al. 2016) and *R. leguminosarum* (Ramírez-Bahena et al. 2008) have identical 16S rRNA genes. Also, the 16S rRNA genes of *R. binae* and *R. bangladeshense* are identical (Rashid et al. 2015), while *R. ecuadorense* (Ribeiro et al. 2015) has a 16S rRNA gene identical to that of *R. pisi* (Ramírez-Bahena et al. 2008).

The most recent innovation in rhizobial taxonomy has been the sequencing of complete genomes. The complete genome sequence is currently available for the type strains of several old and recent rhizobial species and has already been used to describe new rhizobial species isolated from lentil, such as *R. lentis*, *R. binae* and *R. bangladeshense* (Rashid et al. 2015), and from soybean, such as *Ensifer glycinis* (Yan et al. 2016). Moreover, many strains isolated from nodules of different legumes in different ecosystems are in project for genome analysis which will lead to important changes in the taxonomy of nodule bacteria. Currently, the classic rhizobial species are included in the genera *Allorhizobium*, *Azorhizobium*, *Bradyrhizobium*, *Ensifer* (formerly *Sinorhizobium*), *Mesorhizobium*, *Neorhizobium*, *Pararhizobium* and *Rhizobium* (Table 1.1). Nevertheless, the complete list of valid species of rhizobia is constantly updated and recorded in the List of Prokaryotic Names with Standing in Nomenclature (http:// www.bacterio.cict.fr).

1.3   Symbiovars and Legume Promiscuity in Rhizobia

Since the first decade of the past century, several studies on root nodule bacteria to establish nitrogen-fixing symbiosis with legumes were conducted. The findings of such studies resulted in the definition of cross-inoculation groups (Baldwin and Fred 1929), and both legumes and endosymbionts were divided in restrictive and promiscuous (Wilson 1939). In the 1970s, it was discovered that nodulation determinants are codified in plasmids, and in many cases it was autoconjugative, for example, in the genus *Rhizobium* (Zurkowski and Lorkiewic 1979). In the 1980s, symbiosis-specific genes were found located in megaplasmids in *R. meliloti* (Rosenberg et al. 1981), and the nodulation (*nod*) and nitrogen fixation (*nif*) genes were sequenced in several species of rhizobia (Fuhrmann and Hennecke 1984; Schofield and Watson 1986; Debellé and Sharma 1986; Norel and Elmerich 1987; Goethals et al. 1989). From the 1990s onwards, the functions of the symbiotic genes were widely revealed, and it was established that the *nodABC* genes are determinants of the host range (Relic et al. 1994; Roche et al. 1996; Perret et al. 2000). Finally, in the first years of twenty-first century, it has been reported that in *Bradyrhizobium* and several *Mesorhizobium* species these genes are integrated in the chromosome (Göttfert et al. 2001; Sullivan et al. 2002) and that in some photosynthetic bradyrhizobia, the nodulation genes are absent (Giraud et al. 2007). The symbiotic genes, also named 'auxiliary' or 'accessory' genes, particularly the *nodA*, *nodC* and *nifH* genes, have been included in MLST analyses comparing their phylogenies with those obtained after the 'core' gene analyses (Wei et al. 2009; Diouf et al. 2010; Mierzwa et al. 2010; Wdowiak-Wróbel and Małek 2010; Degefu et al. 2011; Lorite et al. 2012; Bakhoum et al. 2015; Gnat et al. 2015; Mousavi et al. 2016; Bontemps et al. 2016). The symbiotic genes are also very useful for biogeography studies (Stepkowski et al. 2007; Steenkamp et al. 2008; Lu et al. 2009c; Wei et al. 2009; Zhang et al. 2014a; Ji et al. 2015; Li et al. 2016a, b), and the *nodC* gene is the most used gene in recent studies (Ramírez-Bahena et al. 2013b; Bejarano et al. 2014; Díaz-Alcántara et al. 2014; Horn et al. 2014; Verástegui-Valdés et al. 2014; Ji et al. 2015; Wang et al. 2016). Also, some symbiotic genes are still included in most descriptions of rhizobial species, although they are not useful in taxonomy because of their ability to be transferred in nature (Finan 2002) from plasmids to islands (Nakatsukasa et al. 2008), from bacteria to plants (Broothaerts et al. 2005) and among bacteria (Rogel et al. 2001). Nevertheless, they are essential for definition of symbiotic biovarieties, initially named biovars (Jordan 1984) and currently named symbiovars (Rogel et al. 2011).

The first biovars proposed in rhizobia were those from the species *R. leguminosarum* with the names viciae, trifolii and phaseoli, defined on the basis of their host specificity (Jordan 1984). According to this proposal, the two former species *R. trifolii* and *R. phaseoli* were included within *R. leguminosarum* as biovars. This is a clear example of an erroneous use of the symbiotic abilities in taxonomy, because a revision of the taxonomic status of these three old species concluded that *R. trifolii* is a later subjective synonym of *R. leguminosarum*, but *R. phaseoli* is a valid species (Ramírez-Bahena et al. 2008). The reiteration of *nifH* genes was later proposed to identify the biovar phaseoli of the species *R. etli* (Aguilar et al. 1998), and this feature was used to differentiate this biovar from biovars gallicum and giardinii described in both *Rhizobium gallicum* and *Rhizobium giardinii* (Amarger et al. 1997). The *nifH* gene analyses and the ability to nodulate *Leucaena leucocephala* allowed the description of a new biovar, mimosae, in the species *R. etli* that also contains strains from the symbiovar etli able to nodulate *P. vulgaris* but not *L. leucocephala* (Wang et al. 1999a). In the same period, other authors used the ability to nodulate different legumes together with the production of different nod factors to define two biovars, acaciae and sesbaniae, within *Sinorhizobium* (*Ensifer*) *terangae* (Lortet et al. 1996). Later, nodulation of different hosts and hybridization with different symbiotic gene probes were used for the definition of two symbiovars within *R. galegae*, orientalis and officinalis (Radeva et al. 2001). In the same year, Laguerre et al. (2001) pointed out that the biovars phaseoli, gallicum and giardinii nodulating *P. vulgaris* may be differentiated by their *nodC* gene sequences, but this was not used to define symbiovars until the year 2005 when two symbiovars, glycinearum and genistearum, were described within the genus *Bradyrhizobium* (Vinuesa et al. 2005). From this date ahead, although some biovars (symbiovars) have been described on the basis of their ability to establish symbiosis with specific legumes (Gubry-Rangin et al. 2013) and the gene analysis of *nodA* (Villegas et al. 2006; Nandasena et al. 2007) and *nifH* (Rincón-Rosales et al. 2013), most of them have been described on the basis of the *nodC* gene (Table 1.2). This analysis helps us to clarify the proposal of Jordan (1984), since *R. leguminosarum* contains three symbiovars, viciae, trifolii and phaseoli, clearly differentiated by their *nodC* gene sequences, but to date the species *R. phaseoli* only harbours the symbiovar phaseoli (García-Fraile et al. 2010). The *nodC* gene has also been related with the promiscuity degree of legumes (Roche et al. 1996), since those considered highly promiscuous hosts such as *Macroptilium atropurpureum*, *Phaseolus vulgaris*, *Leucaena leucocephala* or *Vigna unguiculata* (Perret et al. 2000) are nodulated by strains belonging to different *nodC* groups or lineages (symbiovars), whereas restrictive hosts from tribes trifoliae, viceae and cicereae considered as restrictive hosts (Perret et al. 2000) are nodulated by strains with closely related *nodC* genes from the same symbiovar (Laguerre et al. 2001; Rivas et al. 2007; Iglesias et al. 2007; Zurdo-Piñeiro et al. 2009; Ramírez-Bahena et al. 2013b; Bejarano et al. 2014; Jiao et al. 2015a; Martínez-Hidalgo et al. 2015a). Concerning to the endosymbionts, it has been reported that strains from *R. leguminosarum* bv. *trifolii* only can nodulate *Trifolium*, whereas the strain *Ensifer* sp. NGR234 (formerly *Rhizobium* sp.) nodulates over 100 legumes as well as the nonlegume *Parasponia* (Pueppke and Broughton 1999). Nevertheless, this is not true in all cases, because in a recent work we showed that strains of the symbiovar trifolii, theoretically restricted to nodulate clover, are able to nodulate *Cicer canariense*, which belong to the restrictive tribe cicereae (Martínez-Hidalgo et al. 2015a). Therefore, the promiscuity is related with the number of symbiovars, not with the number of species that are able to induce nodules in a legume. This has been shown for *Cicer arietinum*, a very restrictive host that can be nodulated by several species of *Mesorhizobium*, but all of them carry nearly identical *nodC* genes from the symbiovar ciceri (Rivas et al. 2007;

Table 1.2 Symbiovars (formerly biovars) of rhizobial species

Species	Symbiovar	Isolation legume	References
Genus *Rhizobium*			
R. aegyptiacum	trifolii	*Trifolium alexandrinum*	Shamseldin et al. (2016)
R. bangladeshense	trifolii	*Trifolium alexandrinum*	Shamseldin et al. (2016)
	viciae	*Lens culinaris*	Rashid et al. (2015)
R. binae	viciae	*Lens culinaris*	Rashid et al. (2015)
R. calliandrae	calliandrae	*Calliandra grandiflora*	Rincón-Rosales et al. (2013)
R. etli	mimosae	*Mimosa affinis*	Wang et al. (1999a)
	phaseoli	*Phaseolus vulgaris*	Segovia et al. (1993)
R. gallicum	gallicum	*Phaseolus vulgaris*	Amarger et al. (1997)
	phaseoli	*Phaseolus vulgaris*	Amarger et al. (1997)
R. jaguaris	calliandrae	*Calliandra grandiflora*	Rincón-Rosales et al. (2013)
R. laguerreae	viciae	*Vicia* spp.	Saïdi et al. (2014)
R. leguminosarum	phaseoli	*Phaseolus vulgaris*	Jordan (1984); García-Fraile et al. (2010)
	trifolii	*Trifolium* spp.	Jordan (1984); García-Fraile et al. (2010)
	viciae	*Pisum sativum, Vicia* spp.	Jordan (1984); García-Fraile et al. (2010)
R. lentis	viciae	*Lens culinaris*	Rashid et al. (2015)
R. mayense	calliandrae	*Calliandra grandiflora*	Rincón-Rosales et al. (2013)
R. pisi	trifolii	*Trifolium* spp.	Marek-Kozaczuk et al. (2013)
	viciae	*Pisum sativum, Vicia* spp.	Robledo et al. (2011)
R. tropici	tropici	*Phaseolus vulgaris*	Ormeño-Orrillo et al. (2012)
Genus *Ensifer*			
E. americanus	mediterranense	*Phaseolus vulgaris*	Mnasri et al. (2012)
E. mexicanus	acaciellae	*Acaciella angustissima*	Rogel et al. (2011)
E. fredii	fredii	*Glycine max*	Mnasri et al. (2007)
	mediterranense	*Phaseolus vulgaris*	Mnasri et al. (2007)
E. meliloti	lancerottense	*Lotus* spp.	León-Barrios et al. (2009)
	medicaginis	*Medicago laciniata*	Villegas et al. (2006)
	mediterranense	*Phaseolus vulgaris*	Mnasri et al. (2007)
	meliloti	*Medicago sativa*	Villegas et al. (2006)
	rigiduloides	*Medicago rigiduloides*	Gubry-Rangin *et al.* (2013)
E. saheli	acacieae	*Acacia*	Lortet et al. (1996)
	sesbaniae	*Sesbania*	Lortet et al. (1996)

Table 1.2 (continued)

Species	Symbiovar	Isolation legume	References
E. terangae	acacieae	*Acacia*	Lortet et al. (1996)
	sesbaniae	*Sesbania*	Lortet et al. (1996)
Genus *Neorhizobium*			
N. galegae	officinalis	*Galega officinalis*	Radeva et al. (2001)
	orientalis	*Galega orientalis*	Radeva et al. (2001)
Genus *Pararhizobium*			
P. giardinii	giardinii	*Phaseolus vulgaris*	Amarger et al. (1997)
	phaseoli	*Phaseolus vulgaris*	Amarger et al. (1997)
Genus *Sinorhizobium*[a]			
S. chiapanecum	acaciellae	*Acaciella angustissima*	Rogel et al. (2011)
Genus *Mesorhizobium*			
M. amorphae	ciceri	*Cicer arietinum*	Rivas et al. (2007)
M. ciceri	biserrulae	*Biserrula pelecinus*	Nandasena et al. (2007)
	ciceri	*Cicer arietinum*	Rivas et al. (2007)
M. loti	loti	*Lotus* spp.	Martínez-Hidalgo et al. (2015b)
M. mediterraneum	ciceri	*Cicer arietinum*	Rivas et al. (2007)
M. tamadayense	loti	*Lotus berthelotii*	Ramírez-Bahena et al. (2012)
M. tianshanense	ciceri	*Cicer arietinum*	Rivas et al. (2007)
Genus *Bradyrhizobium*			
B. americanum	phaseolarum	*Centrosema macrocarpum*	Ramírez Bahena et al. (2016)
B. canariense	genistearum	Genisteae legumes	Vinuesa et al. (2005)
B. centrosemae	centrosemae	*Centrosema molle*	Ramírez Bahena et al. (2016)
B. cytisi	genistearum	*Cytisus villosus*	Chahboune et al. (2011)
B. embrapense	tropici	*Desmodium heterocarpon*	Ramírez Bahena et al. (2016)
B. japonicum	genistearum	Genisteae legumes	Vinuesa et al. (2005)
	glycinearum	*Glycine max*	Vinuesa et al. (2005)
B. lupini	genistearum	*Lupinus* spp.	Peix et al. (2015a)
B. retamae	retamae	*Retama* spp.	Guerrouj et al. (2013)
B. rifense	genistearum	*Cytisus villosus*	Chahboune et al. (2012)
B. tropiciagri	tropici	*Neonotonia wightii*	Ramírez Bahena et al. (2016)
B. viridifuturi	tropici	*Centrosema* spp.	Ramírez Bahena et al. (2016)
Bradyrhizobium sp.	sierranevadense	*Genista versicolor*	Cobo-Díaz et al. (2014)
Bradyrhizobium sp.	vignae	*Vigna unguiculata*	Bejarano et al. (2014)

[a]The genus *Sinorhizobium* has been reclassified into genus *Ensifer*, but the species *Sinorhizobium chiapanecum* has still been not reclassified

Laranjo et al. 2008). Conversely, there are strains of *Ensifer meliloti* (formerly *Sinorhizobium meliloti*) that are not able to nodulate *Medicago sativa* because they harbour *nodC* genes phylogenetically divergent to those of strains nodulating this host. These strains were included in a new symbiovar named mediterranense that has been found to date in Africa (Mnasri et al. 2007) and Canary Islands in *P. vulgaris* nodules (Zurdo-Piñeiro et al. 2009). Also, some symbiovars are restricted to nodulate a promiscuous legume, as occurs in the case or the symbiovar phaseoli which nodulates only *P. vulgaris* (Amarger et al. 1997; Valverde et al. 2006), a promiscuous legume which establish symbiosis with strains that belong to different symbiovars (Peix et al. 2015b).

The number of symbiovars is continuously increasing, and, although in most of species only a symbiovar has been found to date (Table 1.2), several rhizobial species contain more than two symbiovars, most of them differentiated by the nodulation gene analyses, as occurs in the already mentioned *R. leguminosarum*, which contains three symbiovars named viciae, phaseoli and trifolii (Jordan 1984; Laguerre et al. 2001; García-Fraile et al. 2010). The species *R. etli* contains two biovars, phaseoli and mimosae, which differs from biovar mimosae to nodulate *Leucaena* (Wang et al. 1999a). Other species such as *R. gallicum* and *R. giardinii* also contain strains from symbiovar phaseoli and another ones named gallicum and giardinii, respectively, all of them able to nodulate *P. vulgaris* (Amarger et al. 1997). The species *R. galegae* contains two symbiovars, officinalis and orientalis, able to nodulate *Galega* (Radeva et al. 2001). The species *R. pisi* contains the symbiovars viciae and trifolii, nodulating *Vicia* and *Trifolium*, respectively (Robledo et al. 2011; Marek-Kozaczuk et al. 2013), and recently the symbiovar trifolii has been described in the species *R. bangladeshense* already containing the symbiovar viciae (Shamseldin et al. 2016). In the genus *Ensifer*, the species *E. meliloti* contains the symbiovars meliloti, medicaginis and rigiduloides able to nodulate *Medicago* (Villegas et al. 2006; Gubry-Rangin et al. 2013), mediterranense able to nodulate *P. vulgaris* but not *Medicago* (Mnasri et al. 2007) and lancerottense able to nodulate *Lotus* (León-Barrios et al. 2009). The species *E. terangae* contains two symbiovars named acaciae and sesbaniae able to nodulate *Acacia* and *Sesbania*, respectively (Lortet et al. 1996). The species *E. fredii* contains two symbiovars named fredii and mediterranense able to nodulate *Glycine max* and *P. vulgaris*, respectively (Mnasri et al. 2007). In the genus *Mesorhizobium*, the species *M. ciceri* contains the symbiovar ciceri able to nodulate *Cicer* (Rivas et al. 2007) and the symbiovar biserrulae nodulating *Biserrula pelecinus* (Nandasena et al. 2007). In the slow-growing genus *Bradyrhizobium*, the species *B. japonicum* contains two biovars named glycinearum and genistearum nodulating *Glycine max* and Genisteae legumes, respectively (Vinuesa et al. 2005). The species *B. retamae* contains the symbiovar genistearum nodulating Genisteae and retamae nodulating *Retama* sp. and *Lablab purpureus* (Guerrouj et al. 2013). To date, no symbiovars have been described within the classic rhizobial genera *Allorhizobium* and *Azorhizobium* and neither in the new rhizobial genera from alpha and beta *Proteobacteria*. Therefore, a significant increase in the number of symbiovars after the analysis of the nodulation genes of more legume endosymbionts is predictable.

1.4   The New Rhizobia

For more than one century, rhizobia were considered the unique bacteria able to induce nodules in legumes. However, in 2001, the report of two atypical bacteria nodulating legumes showed the legume nodulation by 'non-rhizobial' bacteria, one of them belonging to the alpha *Proteobacteria* (Sy et al. 2001) and the other to the beta *Proteobacteria* (Moulin et al. 2001). Some 'non-rhizobial' bacteria from to the alpha *Proteobacteria* belong to families that did not encompass rhizobial genera (Table 1.3). This was the case of *Methylobacterium*, the first alpha *Proteobacteria* described as legume endosymbiont, which belongs to the family *Methylobacteriaceae* (Sy et al. 2001). The species *Methylobacterium nodulans* was isolated from nodules of *Crotalaria* and harbours the common nodulation *nodABC* genes and the *nifH* gene (Sy et al. 2001; Jourand et al. 2004). More recently, the nodulation of legumes by several species from the genus *Microvirga*, which also belongs to the family *Methylobacteriaceae*, has been reported. The species *Microvirga lupini*, *M. lotononidis* and *M. zambiensis* carry *nodA* genes which are phylogenetically related to those from different rhizobial genera (Ardley et al. 2012). *Microvirga vignae* carries *nifH* genes which are closely related to *Rhizobium* and *Mesorhizobium* (Radl et al. 2014). Nevertheless, the most surprising finding was the nodulation ability of species belonging to the family *Brucellaceae* which contains several human pathogens (Garrity et al. 2005). The species *Ochrobactrum lupini* and *Ochrobactrum cytisi* were isolated from nodules of *Lupinus* and *Cytisus*, respectively, and had *nifH* and *nodD* genes which are phylogenetically related with those of rhizobial species (Trujillo et al. 2005; Zurdo-Piñeiro et al. 2007). Other non-rhizobial alpha *Proteobacteria* able to nodulate legumes belong to families that also contain classic rhizobial genera (Table 1.3), as occurs with the genus *Devosia* belonging to the family *Hyphomicrobiaceae*, which also encompasses the genus *Azorhizobium*. The species *Devosia neptuniae* forms nodules in the aquatic legume *Neptunia natans* and carry *nodD* and *nifH* genes closely related to those of *Rhizobium tropici* (Rivas et al. 2002). The genus *Shinella*, which belongs to the family *Rhizobiaceae*, contains several species isolated from different sources (An et al. 2006) and includes the species *Shinella kummerowiae* isolated from *Kummerowia stipulata* nodules, which carries *nodD*, *nodC* and *nifH* nodules related to those of *R. tropici* (Lin et al. 2008). In the case of the family *Phyllobacteriaceae*, which contains the classic rhizobial genus *Mesorhizobium*, two genera, *Phyllobacterium* and *Aminobacter*, contain species able to nodulate legumes. In the genus *Phyllobacterium*, the first reported species was *P. trifolii*, which forms nodules on *Trifolium repens* and carries *nod* and *nif* genes related to *Rhizobium* species (Valverde et al. 2005), and recently the species *P. sophorae*, which forms nodules on *Sophora japonica* and carries *nodC* and *nifH* genes related to those of *Mesorhizobium* (Jiao et al. 2015c). Finally, in the genus *Aminobacter*, the recently described species *Aminobacter anthyllidis* is able to nodulate *Anthyllis vulneraria* and carries *nodA* genes closely related to those of *Mesorhizobium loti* (Maynaud et al. 2012).

Although the nodulation of legumes by non-rhizobial genera from alpha *Proteobacteria* constituted a relevant change in the concept of rhizobia, the most

Table 1.3 Species of new rhizobia able to nodulate legumes

Species	Nodulated legume	Reference
Alpha *Proteobacteria*		
Aminobacter anthyllidis	*Anthyllis vulneraria*	Maynaud et al. (2012)
Devosia neptuniae	*Neptunia natans*	Rivas et al. (2003)
Methylobacterium nodulans	*Crotalaria* spp.	Jourand et al. (2004)
Microvirga lotononidis	*Listia angolensis*	Ardley et al. (2012)
Microvirga lupini	*Lupinus texensis*	Ardley et al. (2012)
Microvirga vignae	*Vigna unguiculata*	Radl et al. (2014)
Microvirga zambiensis	*Listia angolensis*	Ardley et al. (2012)
Ochrobactrum lupini	*Lupinus* spp.	Trujillo et al. (2005)
Ochrobactrum cytisi	*Cytisus scoparius*	Zurdo-Piñeiro et al. (2007)
Phyllobacterium sophorae	*Sophora flavescens*	Jiao et al. (2015c)
Phyllobacterium trifolii	*Trifolium pratense*	Valverde et al. (2005)
Shinella kummerowiae	*Kummerowia stipulata*	Lin et al. (2008)
Beta *Proteobacteria*		
Cupriavidus necator	*Phaseolus vulgaris, Leucaena leucocephala, Mimosa caesalpiniaefolia, Macroptilium atropurpureum, Vigna unguiculata*	da Silva et al. (2012)
Cupriavidus taiwanensis	*Mimosa* spp.	Chen et al. (2001), Vandamme and Coenye (2004), Barrett and Parker (2006), Andam et al. (2007)
Burkholderia cepacia	*Dalbergia* spp.	Rasolomampianina et al. (2005)
Burkholderia dipogonis[a]	*Dipogon lignosus*	Sheu et al. (2015a)
Paraburkholderia caballeronis	*Phaseolus vulgaris*	Martínez-Aguilar et al. (2013), Sawana et al. (2014)
Paraburkholderia caribensis	*Mimosa* spp.	Liu et al. (2011), Sawana et al. (2014), Oren and Garrity (2015b)
Paraburkholderia diazotrophica	*Mimosa* spp.	Sheu et al. (2015b), Sawana et al. (2014), Oren and Garrity (2015b)
Paraburkholderia dilworthii	*Lebeckia ambigua*	de Meyer et al. (2014)
Paraburkholderia kirstenboschensis	*Hypocalyptus* spp., *Virgilia oroboides*	Steenkamp et al. (2015), Dobritsa and Samadpour (2016)
Paraburkholderia mimosarum	*Mimosa* spp.	Chen et al. (2006), Sawana et al. (2014), Oren and Garrity (2015b)
Paraburkholderia nodosa	*Mimosa* spp.	Chen et al. (2007), Sawana et al. (2014), Oren and Garrity (2015b)
Paraburkholderia phymatum	*Machaerium lunatum, Mimosa* spp., *Phaseolus vulgaris, Acacia, Prosopis*	Moulin et al. (2001), Vandamme et al. (2002), Elliott et al. (2007b), Talbi et al. (2010), Sawana et al. (2014), Oren and Garrity (2015b)

Table 1.3 (continued)

Species	Nodulated legume	Reference
Paraburkholderia piptadeniae	*Piptadenia gonoacantha*	Bournaud et al. (2016)
Paraburkholderia ribeironis	*Piptadenia gonoacantha*	Bournaud et al. (2016)
Paraburkholderia sabiae	*Mimosa caesalpiniifolia*	Chen et al. (2008), Sawana et al. (2014), Oren and Garrity (2015b)
Paraburkholderia rhynchosiae	*Rhynchosia ferulifolia*	de Meyer et al. (2013b), Sawana et al. (2014)
Paraburkholderia sprentiae	*Lebeckia ambigua*	de Meyer et al. (2013a), Sawana et al. (2014), Oren and Garrity (2015c)
Paraburkholderia symbiotica	*Mimosa* spp.	Sheu et al. (2012), Sawana et al. (2014), Oren and Garrity (2015b)
Paraburkholderia tuberum	*Aspalathus* spp., *Cyclopia* spp., *Macroptilium atropurpureum*, *Mimosa* spp.	Vandamme et al. (2002), Elliott et al. (2007a), Barrett and Parker (2006), Sawana et al. (2014), Oren and Garrity (2015b)

[a]Due to its phylogenetically closeness with the species *Paraburkholderia phytofirmans* and *P. caledonica*, this species probably belongs to the genus *Paraburkholderia*

surprisingly noticed was the ability to nodulate legumes by a beta *Proteobacteria*, and it was published in year 2001 in the journal Nature (Moulin et al. 2001). The strains analysed by these authors belong to the genus *Burkholderia* and nodulate *Mimosa*, but only 2 years later it was described that the nodulation of this legume also occurs by members of genus *Ralstonia* (Chen et al. 2003a). The species nodulating legumes from the genus *Ralstonia* was soon reclassified into a new genus named *Cupriavidus* (Vandamme and Coenye 2004), and recently most of them from the genus *Burkholderia* have also been reclassified into the new genus *Paraburkholderia* (Sawana et al. 2014; Dobritsa and Samadpour 2016), which has been recently validated in IJSEM although not all species reclassified by Sawana et al. (2014) have been included in the validation lists 164 and 165 (Oren and Garrity 2015a, b). Both genera *Burkholderia* (now *Paraburkholderia*) and *Cupriavidus* have been found in nodules of several legumes grown in different countries such as Taiwan (Chen et al. 2005), Madagascar (Rasolomampianina et al. 2005), Costa Rica (Barrett and Parker 2006), Japan (Shiraishi et al. 2010) and China (Liu et al. 2011; Liu et al. 2012b), and their *nod* and *nif* genes have been analysed (Chen et al. 2005; Andam et al. 2007; Amadou et al. 2008; Andrus et al. 2012; Taulé et al. 2012; Platero et al. 2016). After many studies carried out in the first two decades of the present century, it was concluded that beta *Proteobacteria* are widespread in nodules of legumes from several tribes (Barrett and Parker 2005; Chen et al. 2003b; Lemaire et al. 2015), and many new species of *Burkholderia* (now *Paraburkholderia*) able to nodulate legumes have been described (Table 1.3). Although the species of *Cupriavidus* identified to date in legume nodules are restricted to *Cupriavidus taiwanensis* (Chen et al. 2003a) and *Cupriavidus necator* (da Silva et al. 2012), some recent species showed that putative new species of this genus are endosymbionts of *Mimosa* in Uruguay (Platero et al. 2016). Although a strain of the genus *Pseudomonas* belonging to the gamma *Proteobacteria* has been reported as endosymbiont of *Robinia pseudoacacia* (Shiraishi et al. 2010), some authors have reported nodulation of legumes by Gram-positive sporulating strains (Ampomah and Huss-Danell 2011; Latif et al. 2013). In order to prevent an erroneous attribution of the ability to nodulate legumes to a nodule endophyte, it is necessary to fulfil Koch's postulates in a reliable way, such as those proposed in the work in which the nodulation of *Cicer canariense* by *R. leguminosarum* symbiovar trifolii was proposed (Martínez-Hidalgo et al. 2015a). In this work the rhizobial strain was marked with GFP and followed during the infection process, and also we demonstrated by metagenomic techniques that in the nodules the *nodC* genes of *Mesorhizobium*, the common endosymbiont of *Cicer canariense* (Armas-Capote et al. 2014), are not detected by using specific primers (Martínez-Hidalgo et al. 2015a). The use of these or other similar techniques is necessary to avoid confusion between a nodule endophyte and the endosymbiont responsible for the nodule formation.

Conclusion

The taxonomy of bacteria inducing legume nodules has dramatically been changed in the recent years with the rearrangement of species and genera in different families, the reclassification of several species in new genera and the recovering of old genera and species. In addition, the description of new species and symbiovars is continuously increasing with new emerging legumes. These studies further demand the comprehensive analysis of both chromosomal and symbiotic genes so that the rhizobia could be identified at species and symbiovars levels. The application of new techniques, particularly the genome sequence analysis, is likely to contribute to a better understanding of the legume-rhizobia interactions and to modify/change some taxonomic criteria currently adopted for description of species and genera. Also, a considerable increase in the number of species and symbiovars is expected in years to come which will help to further identify nitrogen-fixing rhizobia which ultimately could be used as biofertilizers in order to limit the use of chemical fertilizers in agronomic practices.

Acknowledgements The authors would like to thank our numerous collaborators and students involved in this research over the years. Funding was provided by Ministerio de Economía, Industria y Competitividad (MINECO) and Junta de Castilla y León.

References

Aguilar OM, Grasso DH, Riccillo PM, López MV, Szafer E (1998) Rapid identification of bean *Rhizobium* isolates by a *nifH* gene-pcr assay. Soil Biol Biochem 30:1655–1661

Amadou C, Pascal G, Mangenot S, Glew M, Bontemps C, Capela D, Carrere S, Cruveiller S, Dossat C, Lajus A, Marchetti M, Poinsot V, Rouy Z, Servin B, Saad M, Schenowitz C, Barbe V, Batut J, Medigue C, Masson-Boivin C (2008) Genome sequence of the beta-rhizobium *Cupriavidus taiwanensis* and comparative genomics of rhizobia. Genome Res 18:1472–1483

Amarger N, Macheret V, Laguerre G (1997) *Rhizobium gallicum* sp. nov. and *Rhizobium giardinii* sp. nov., from *Phaseolus vulgaris* nodules. Int J Syst Bacteriol 47:996–1006

Ampomah OY, Huss-Danell K (2011) Genetic diversity of root nodule bacteria nodulating *Lotus corniculatus* and *Anthyllis vulneraria* in Sweden. Syst Appl Microbiol 34:267–275

An DS, Im WT, Yang HC, Lee ST (2006) *Shinella granuli* gen. nov., sp. nov., and proposal of the reclassification of *Zoogloea ramigera* ATCC 19623 as *Shinella zoogloeoides* sp. nov. Int J Syst Evol Microbiol 56:443–448

Andam CP, Mondo SJ, Parker MA (2007) Monophyly of *nodA* and *nifH* genes across Texan and Costa Rican populations of *Cupriavidus* nodule symbionts. Appl Environ Microbiol 73:4686–4690

Andrus AD, Andam C, Parker MA (2012) American origin of *Cupriavidus* bacteria associated with invasive *Mimosa* legumes in the Philippines. FEMS Microbiol Ecol 80:747–750

Ardley JK, Parker MA, De Meyer SE, Trengove RD, O'Hara GW, Reeve WG, Yates RJ, Dilworth MJ, Willems A, Howieson JG (2012) *Microvirga lupini* sp. nov., *Microvirga lotononidis* sp. nov. and *Microvirga zambiensis* sp. nov. are alphaproteobacterial root-nodule bacteria that specifically nodulate and fix nitrogen with geographically and taxonomically separate legume hosts. Int J Syst Evol Microbiol 62:2579–2588

Armas-Capote N, Pérez-Yépez J, Martínez-Hidalgo P, Garzón-Machado V, Del Arco-Aguilar M, Velázquez E, León-Barrios M (2014) Core and symbiotic genes reveal nine *Mesorhizobium* genospecies and three symbiotic lineages among the rhizobia nodulating *Cicer canariense* in its natural habitat (La Palma, Canary Islands). Syst Appl Microbiol 37:140–148

Bakhoum N, Galiana A, Le Roux C, Kane A, Duponnois R, Ndoye F, Fall D, Noba K, Sylla SN, Diouf D (2015) Phylogeny of nodulation genes and symbiotic diversity of *Acacia senegal* (L.) Willd. and *A. seyal* (Del.) mesorhizobium strains from different regions of Senegal. Microb Ecol 69:641–651

Baldwin IL, Fred EB (1929) Nomenclature of the root nodule bacteria of the Leguminosae. J Bacteriol 17:141–150

Baraúna AC, Rouws LF, Simoes-Araujo JL, Dos Reis Junior FB, Iannetta PP, Maluk M, Goi SR, Reis VM, James EK, Zilli JE (2016) *Rhizobium altiplani* sp. nov., isolated from effective nodules on *Mimosa pudica* growing in untypically alkaline soil in central Brazil. Int J Syst Evol Microbiol 66:4118–4124

Barrett CF, Parker MA (2005) Prevalence of *Burkholderia* sp. nodule symbionts on four mimosoid legumes from Barro Colorado Island, Panama. Syst Appl Microbiol 28:57–65

Barrett CF, Parker MA (2006) Coexistence of *Burkholderia*, *Cupriavidus*, and *Rhizobium* sp. nodule bacteria on two *Mimosa* spp. in Costa Rica. Appl Environ Microbiol 72:1198–1206

Behrendt U, Kämpfer P, Glaeser SP, Augustin J, Ulrich A (2016) Characterisation of the N$_2$O producing soil bacterium *Rhizobium azooxidifex* sp. nov. Int J Syst Evol Microbiol. doi:10.1099/ijsem.0.001036

Beijerinck MW (1888) Cultur des *Bacillus radicicola* aus den Knöllchen. Bot Ztg 46:740–750

Bejarano A, Ramírez-Bahena MH, Velázquez E, Peix A (2014) *Vigna unguiculata* is nodulated in Spain by endosymbionts of Genisteae legumes and by a new symbiovar (vignae) of the genus *Bradyrhizobium*. Syst Appl Microbiol 37:533–540

Berge O, Lodhi A, Brandelet G, Santaella C, Roncato MA, Christen R, Heulin T, Achouak W (2009) *Rhizobium alamii* sp. nov., an exopolysaccharide-producing species isolated from legume and non-legume rhizospheres. Int J Syst Evol Microbiol 59:367–372

van Berkum P, Eardly BD (2002) The aquatic budding bacterium *Blastobacter denitrificans* is a nitrogen-fixing symbiont of *Aeschynomene indica*. Appl Environ Microbiol 68:1132–1136

van Berkum P, Beyene D, Bao G, Campbell TA, Eardly BD (1998) *Rhizobium mongolense* sp. nov. is one of three rhizobial genotypes identified which nodulate and form nitrogen-fixing symbioses with *Medicago ruthenica* [(L.) Ledebour]. Int J Syst Bacteriol 48:13–22

van Berkum P, Leibold JM, Eardly BD (2006) Proposal for combining *Bradyrhizobium* spp. (*Aeschynomene indica*) with *Blastobacter denitrificans* and to transfer *Blastobacter denitrificans* (Hirsch and Muller, 1985) to the genus *Bradyrhizobium* as *Bradyrhizobium denitrificans* (comb. nov.) Syst Appl Microbiol 29:207–215

Bibi F, Chung EJ, Khan A, Jeon CO, Chung YR (2012) *Rhizobium halophytocola* sp. nov., isolated from the root of a coastal dune plant. Int J Syst Evol Microbiol 62:1997–2003

Bontemps C, Rogel MA, Wiechmann A, Mussabekova A, Moody S, Simon MF, Moulin L, Elliott GN, Lacercat-Didier L, Dasilva C, Grether R, Camargo-Ricalde SL, Chen W, Sprent JI, Martínez-Romero E, Young JP, James EK (2016) Endemic *Mimosa* species from Mexico prefer alphaproteobacterial rhizobial symbionts. New Phytol 209:319–333

Boonsnongcheep P, Prathanturarug S, Takahashi Y, Matsumoto A (2015) *Rhizobium puerariae* sp. nov., an endophytic bacterium from the root nodules of medicinal plant *Pueraria candollei* var. candollei. Int J Syst Evol Microbiol. doi:10.1099/ijsem.0.000863

Bournaud C, Moulin L, Cnockaert M, de Faria SM, Prin Y, Severac D, Vandamme P (2016) *Paraburkholderia piptadeniae* sp. nov. and *Paraburkholderia ribeironis* sp. nov., two root-nodulating symbiotic species of *Piptadenia gonoacantha* in Brazil. Int J Syst Evol Microbiol. doi:10.1099/ijsem.0.001648

Broothaerts W, Mitchell HJ, Weir B, Kaines S, Smith LM, Yang W, Mayer JE, Roa-Rodríguez C, Jefferson RA (2005) Gene transfer to plants by diverse species of bacteria. Nature 433:629–633

Buchanan RE (1926) What names should be used for the organisms producing nodules on the roots of leguminous plants? Proc Iowa Acad Sci 33:81–90

Casida LE (1982) *Ensifer adhaerens* gen. nov., sp. nov.: a bacterial predator of bacteria in soil. Int J Syst Bacteriol 32:339–345

Chahboune R, Carro L, Peix A, Barrijal S, Velázquez E, Bedmar EJ (2011) *Bradyrhizobium cytisi* sp. nov., isolated from effective nodules of *Cytisus villosus*. Int J Syst Evol Microbiol 61:2922–2927

Chahboune R, Carro L, Peix A, Ramírez-Bahena MH, Barrijal S, Velázquez E, Bedmar EJ (2012) *Bradyrhizobium rifense* sp. nov. isolated from effective nodules of *Cytisus villosus* grown in the Moroccan Rif. Syst Appl Microbiol 35:302–305

Chang YL, Wang JY, Wang ET, Liu HC, Sui XH, Chen WX (2011) *Bradyrhizobium lablabi* sp. nov., isolated from effective nodules of *Lablab purpureus* and *Arachis hypogaea*. Int J Syst Evol Microbiol 61:2496–5202

Chen WX, Yan GH, Li JL (1988) Numerical taxonomy study of fast-growing soybean rhizobia and a proposal that *Rhizobium fredii* be assigned to *Sinorhizobium* gen. nov. Int J Syst Bacteriol 38:392–397

Chen WX, Li GS, Qi YL, Wang ET, Yuan HL, Li JL (1991) *Rhizobium huakuii* sp. nov., isolated from the root nodules of *Astragalus sinicus*. Int J Syst Bacteriol 41:275–280

Chen WX, Wang E, Wang S, Li Y, Chen X, Li Y (1995) Characteristics of *Rhizobium tian-shanense* sp. nov., a moderately and slowly growing root nodule bacterium isolated from an arid saline environment in Xinjiang, People's Republic of China. Int J Syst Bacteriol 45:153–159

Chen WX, Tan ZY, Gao JL, Li Y, Wang ET (1997) *Rhizobium hainanense* sp. nov., isolated from tropical legumes. Int J Syst Bacteriol 47:870–873

Chen WM, Laevens S, Lee TM, Coenye T, De Vos P, Mergeay M, Vandamme P (2001) *Ralstonia taiwanensis* sp. nov., isolated from root nodules of *Mimosa* species and sputum of a cystic fibrosis patient. Int J Syst Evol Microbiol 51:1729–1735

Chen WM, James EK, Prescott AR, Kierans M, Sprent JI (2003a) Nodulation of *Mimosa* spp. by the beta-proteobacterium *Ralstonia taiwanensis*. Mol Plant Microbe Interact 16:1051–1061

Chen WM, Moulin L, Bontemps C, Vandamme P, Béna G, Boivin-Masson C (2003b) Legume symbiotic nitrogen fixation by beta-proteobacteria is widespread in nature. J Bacteriol 185:7266–7272

Chen WM, James EK, Chou JH, Sheu SY, Yang SZ, Sprent JI (2005) Beta-rhizobia from *Mimosa pigra*, a newly discovered invasive plant in Taiwan. New Phytol 168:661–675

Chen WM, James EK, Coenye T, Chou JH, Barrios E, de Faria SM, Elliott GN, Sheu SY, Sprent JI, Vandamme P (2006) *Burkholderia mimosarum* sp. nov., isolated from root nodules of *Mimosa* spp. from Taiwan and South America. Int J Syst Evol Microbiol 56:1847–1851

Chen WM, de Faria SM, James EK, Elliott GN, Lin KY, Chou JH, Sheu SY, Cnockaert M, Sprent JI, Vandamme P (2007) *Burkholderia nodosa* sp. nov., isolated from root nodules of the woody Brazilian legumes *Mimosa bimucronata* and *Mimosa scabrella*. Int J Syst Evol Microbiol 57:1055–1059

Chen WM, de Faria SM, Chou JH, James EK, Elliott GN, Sprent JI, Bontemps C, Young JP, Vandamme P (2008) *Burkholderia sabiae* sp. nov., isolated from root nodules of *Mimosa caesalpiniifolia*. Int J Syst Evol Microbiol 58:2174–2179

Chen WM, Zhu WF, Bontemps C, Young JP, Wei GH (2010) *Mesorhizobium alhagi* sp. nov., isolated from wild *Alhagi sparsifolia* in north-western China. Int J Syst Evol Microbiol 60:958–962

Chen WM, Zhu WF, Bontemps C, Young JP, Wei GH (2011) *Mesorhizobium camelthorni* sp. nov., isolated from *Alhagi sparsifolia*. Int J Syst Evol Microbiol 61:574–579

Chen W, Sheng XF, He LY, Huang Z (2015) *Rhizobium yantingense* sp. nov., a mineral-weathering bacterium. Int J Syst Evol Microbiol 65:412–417

Cobo-Díaz JF, Martínez-Hidalgo P, Fernández-González AJ, Martínez-Molina E, Toro N, Velázquez E, Fernández-López M (2014) The endemic *Genista versicolor* from Sierra Nevada National Park in Spain is nodulated by putative new *Bradyrhizobium* species and a novel symbiovar (sierranevadense). Syst Appl Microbiol 37:177–185

Conn HJ (1938) Taxonomic relationships of certain non-sporeforming rods in soil. J Bacteriol 36:320–321

Conn HJ (1942) Validity of the genus *Alcaligenes*. J Bacteriol 44:353–360

Dall'Agnol RF, Ribeiro RA, Ormeño-Orrillo E, Rogel MA, Delamuta JR, Andrade DS, Martínez-Romero E, Hungria M (2013) *Rhizobium freirei* sp. nov., a symbiont of *Phaseolus vulgaris* that is very effective at fixing nitrogen. Int J Syst Evol Microbiol 63:4167–4173

Dall'Agnol RF, Ribeiro RA, Delamuta JR, Ormeño-Orrillo E, Rogel MA, Andrade DS, Martínez-Romero E, Hungria M (2014) *Rhizobium paranaense* sp. nov., an effective N2-fixing symbiont of common bean (*Phaseolus vulgaris* L.) with broad geographical distribution in Brazil. Int J Syst Evol Microbiol 64:3222–3229

Dangeard PA (1926) Recherches sur les tubercules radicaux des Légumineuses. Botaniste (Paris) 16:1–275

De Meyer SE, Coorevits A, Willems A (2012) *Tardiphaga robiniae* gen. nov, sp. nov., a new genus in the family *Bradyrhizobiaceae* isolated from *Robinia pseudoacacia* in Flanders (Belgium). Syst Appl Microbiol 35:205–214

De Meyer SE, Cnockaert M, Ardley JK, Maker G, Yates R, Howieson JG, Vandamme P (2013a) *Burkholderia sprentiae* sp. nov., isolated from *Lebeckia ambigua* root nodules. Int J Syst Evol Microbiol 63:3950–3957

De Meyer SE, Cnockaert M, Ardley JK, Trengove RD, Garau G, Howieson JG, Vandamme P (2013b) *Burkholderia rhynchosiae* sp. nov., isolated from *Rhynchosia ferulifolia* root nodules. Int J Syst Evol Microbiol 63:3944 -3949

De Meyer SE, Cnockaert M, Ardley JK, Van Wyk BE, Vandamme PA, Howieson JG (2014) *Burkholderia dilworthii* sp. nov., isolated from *Lebeckia ambigua* root nodules. Int J Syst Evol Microbiol 64:1090–1095

De Meyer SE, Tan HW, Heenan PB, Andrews M, Willems A (2015) *Mesorhizobium waimense* sp. nov. isolated from *Sophora longicarinata* root nodules and *Mesorhizobium cantuariense* sp. nov. isolated from *Sophora microphylla* root nodules. Int J Syst Evol Microbiol 65:3419–3426

De Meyer SE, Tan HW, Andrews M, Heenan PB, Willems A (2016) *Mesorhizobium calcicola* sp. nov., *Mesorhizobium waitakense* sp. nov., *Mesorhizobium sophorae* sp. nov., *Mesorhizobium newzealandense* sp. nov. and *Mesorhizobium kowhaii* sp. nov. isolated from *Sophora* root nodules in New Zealand. Int J Syst Evol Microbiol. doi:10.1099/ijsem.0.000796

Debellé F, Sharma SB (1986) Nucleotide sequence of *Rhizobium meliloti* RCR2011 genes involved in host specificity of nodulation. Nucleic Acids Res 14:7453–7472

Degefu T, Wolde-Meskel E, Frostegård Å (2011) Multilocus sequence analyses reveal several unnamed *Mesorhizobium* genospecies nodulating *Acacia* species and *Sesbania* sesban trees in Southern regions of Ethiopia. Syst Appl Microbiol 34:216–226

Degefu T, Wolde-Meskel E, Liu B, Cleenwerck I, Willems A, Frostegård Å (2013) *Mesorhizobium shonense* sp. nov., *Mesorhizobium hawassense* sp. nov. and *Mesorhizobium abyssinicae* sp. nov., isolated from root nodules of different agroforestry legume trees. Int J Syst Evol Microbiol 63:1746–1753

Delamuta JR, Ribeiro RA, Ormeño-Orrillo E, Melo IS, Martínez-Romero E, Hungria M (2013) Polyphasic evidence supporting the reclassification of *Bradyrhizobium japonicum* group Ia strains as *Bradyrhizobium diazoefficiens* sp. nov. Int J Syst Evol Microbiol 63:3342–3351

Delamuta JR, Ribeiro RA, Ormeño-Orrillo E, Parma MM, Melo IS, Martínez-Romero E, Hungria M (2015) *Bradyrhizobium tropiciagri* sp. nov. and *Bradyrhizobium embrapense* sp. nov., nitrogen-fixing symbionts of tropical forage legumes. Int J Syst Evol Microbiol 65:4424–4433

Delamuta JR, Ribeiro RA, Araújo JL, Rouws LF, Zilli JÉ, Parma MM, Melo IS, Hungria M (2016) *Bradyrhizobium stylosanthis* sp. nov., comprising nitrogen-fixing symbionts isolated from nodules of the tropical forage legume *Stylosanthes* spp. Int J Syst Evol Microbiol. doi:10.1099/ijsem.0.001148

Diange EA, Lee SS (2013) *Rhizobium halotolerans* sp. nov., isolated from chloroethylenes contaminated soil. Curr Microbiol 66:599–605

Díaz-Alcántara CA, Ramírez-Bahena MH, Mulas D, García-Fraile P, Gómez-Moriano A, Peix A, Velázquez E, González-Andrés F (2014) Analysis of rhizobial strains nodulating *Phaseolus vulgaris* from Hispaniola Island, a geographic bridge between Meso and South America and the first historical link with Europe. Syst Appl Microbiol 37:149–156

Diouf D, Fall D, Chaintreuil C, Ba AT, Dreyfus B, Neyra M, Ndoye I, Moulin L (2010) Phylogenetic analyses of symbiotic genes and characterization of functional traits of *Mesorhizobium* spp. strains associated with the promiscuous species *Acacia seyal* Del. J Appl Microbiol 108:818–830

Dobritsa AP, Samadpour M (2016) Transfer of eleven *Burkholderia* species to the genus *Paraburkholderia* and proposal of *Caballeronia* gen. nov., a new genus to accommodate twelve species of *Burkholderia* and *Paraburkholderia*. Int J Syst Evol Microbiol. doi:10.1099/ijsem.0.001065

Dreyfus B, Garcia JL, Gillis M (1988) Characterization of *Azorhizobium caulinodans* gen. nov., sp. nov., a stem-nodulating nitrogen-fixing bacterium isolated from *Sesbania rostrata*. Int J Syst Bacteriol 38:89–98

Durán D, Rey L, Mayo J, Zúñiga-Dávila D, Imperial J, Ruiz-Argüeso T, Martínez-Romero E, Ormeño-Orrillo E (2014a) *Bradyrhizobium paxllaeri* sp. nov. and *Bradyrhizobium icense* sp. nov., nitrogen-fixing rhizobial symbionts of Lima bean (*Phaseolus lunatus* L.) in Peru. Int J Syst Evol Microbiol 64:2072–2078

Durán D, Rey L, Navarro A, Busquets A, Imperial J, Ruiz-Argüeso T (2014b) *Bradyrhizobium valentinum* sp. nov., isolated from effective nodules of *Lupinus mariae-josephae*, a lupine endemic of basic-lime soils in Eastern Spain. Syst Appl Microbiol 37:336–341

Eckhardt MM, Baldwin IR, Fred EB (1931) Studies on the root-nodule bacteria of *Lupinus*. J Bacteriol 21:273–285

Elliott GN, Chen WM, Bontemps C, Chou JH, Young JP, Sprent JI, James EK (2007a) Nodulation of *Cyclopia* spp. (Leguminosae, Papilionoideae) by *Burkholderia tuberum*. Ann Bot 100:1403–1411

Elliott GN, Chen WM, Chou JH, Wang HC, Sheu SY, Perin L, Reis VM, Moulin L, Simon MF, Bontemps C, Sutherland JM, Bessi R, de Faria SM, Trinick MJ, Prescott AR, Sprent JI, James EK (2007b) *Burkholderia phymatum* is a highly effective nitrogen-fixing symbiont of *Mimosa* spp. and fixes nitrogen ex planta. New Phytol 173:168–180

Farrand SK, van Berkum PB, Oger P (2003) *Agrobacterium* is a definable genus of the family *Rhizobiaceae*. Int J Syst Evol Microbiol 53:1681–1687

Finan TM (2002) Evolving insights: symbiosis islands and horizontal gene transfer. J Bacteriol 184:2855–2856

Frank B (1889) Ueber die Pilzsymbiose der Leguminosen. Bet Dtsch Bot Ges 7:332–346

Fuhrmann M, Hennecke H (1984) *Rhizobium japonicum* nitrogenase Fe protein gene (*nifH*). J Bacteriol 158:1005–1011

Gao JL, Turner SL, Kan FL, Wang ET, Tan ZY, Qiu YH, Gu J, Terefework Z, Young JP, Lindström K, Chen WX (2004) *Mesorhizobium septentrionale* sp. nov. and *Mesorhizobium temperatum* sp. nov., isolated from *Astragalus adsurgens* growing in the northern regions of China. Int J Syst Evol Microbiol 54:2003–20012

García-Fraile P, Rivas R, Willems A, Peix A, Martens M, Martínez-Molina E, Mateos PF, Velázquez E (2007) *Rhizobium cellulosilyticum* sp. nov., isolated from sawdust of *Populus alba*. Int J Syst Evol Microbiol 57:844–848

García-Fraile P, Mulas-García D, Peix A, Rivas R, González-Andrés F, Velázquez E (2010) *Phaseolus vulgaris* is nodulated in northern Spain by *Rhizobium leguminosarum* strains harboring two *nodC* alleles present in American *Rhizobium etli* strains: biogeographical and evolutionary implications. Can J Microbiol 56:657–666

Garrity GM, Bell JA, Lilburn T (2005) Brucellaceae. Bergey's manual of systematics of archaea and bacteria. John Wiley & Sons, Inc., New York

Gaunt MW, Turner SL, Rigottier-Gois L, Lloyd-Macgilp SA, Young JPW (2001) Phylogenies of *atpD* and *recA* support the small subunit rRNA-based classification of rhizobia. Int J Syst Evol Microbiol 51:2037–2048

Gehlot HS, Tak N, Kaushik M, Mitra S, Chen WM, Poweleit N, Panwar D, Poonar N, Parihar R, Tak A, Sankhla IS, Ojha A, Rao SR, Simon MF, Reis Junior FB, Perigolo N, Tripathi AK, Sprent JI, Young JP, James EK, Gyaneshwar P (2013) An invasive *Mimosa* in India does not adopt the symbionts of its native relatives. Ann Bot 112:179–196

Ghosh W, Roy P (2006) *Mesorhizobium thiogangeticum* sp. nov., a novel sulfur-oxidizing chemolithoautotroph from rhizosphere soil of an Indian tropical leguminous plant. Int J Syst Evol Microbiol 56:91–97

Giraud E, Moulin L, Vallenet D, Barbe V, Cytryn E, Avarre JC, Jaubert M, Simon D, Cartieaux F, Prin Y, Bena G, Hannibal L, Fardoux J, Kojadinovic M, Vuillet L, Lajus A, Cruveiller S, Rouy Z, Mangenot S, Segurens B, Dossat C, Franck WL, Chang WS, Saunders E, Bruce D, Richardson P, Normand P, Dreyfus B, Pignol D, Stacey G, Emerich D, Verméglio A, Médigue C, Sadowsky M (2007) Legumes symbioses: absence of *nod* genes in photosynthetic bradyrhizobia. Science 316:1307–1312

Gnat S, Małek W, Oleńska E, Wdowiak-Wróbel S, Kalita M, Łotocka B, Wójcik M (2015) Phylogeny of symbiotic genes and the symbiotic properties of rhizobia specific to *Astragalus glycyphyllos* L. PLoS One 23:e0141504

Goethals K, Gao M, Tomekpe K, Van Montagu M, Holsters M (1989) Common *nodABC* genes in Nod locus 1 of *Azorhizobium caulinodans*: Nucleotide sequence and plant-inducible expression. Mol Gen Genetics 219:289–298

Göttfert M, Röthlisberger S, Kündig C, Beck C, Marty R, Hennecke H (2001) Potential symbiosis-specific genes uncovered by sequencing of a 410-kilobase DNA region of the *Bradyrhizobium japonicum* chromosome. J Bacteriol 183:1405-1412

Graham PH, Sadowsky MJ, Keyser HH, Barnet YM, Bradley RS, Cooper JE, De Ley J, Jarvis BDW, Roslycky EB, Strijdom BW, Young JPW (1991) Proposed minimal standards for the description of new genera and species of root- and stem-nodulation bacteria. Int J Syst Bacteriol 41:582–587

Grison CM, Jackson S, Merlot S, Dobson A, Grison C (2015) *Rhizobium metallidurans* sp. nov., a symbiotic heavy metal resistant bacterium isolated from the *Anthyllis vulneraria* Zn-hyperaccumulator. Int J Syst Evol Microbiol 65:1525–1530

Grönemeyer JL, Chimwamurombe P, Reinhold-Hurek B (2015a) *Bradyrhizobium subterraneum* sp. nov., a symbiotic nitrogen-fixing bacterium from root nodules of groundnuts. Int J Syst Evol Microbiol 65:3241–3247

Grönemeyer JL, Hurek T, Reinhold-Hurek B (2015b) *Bradyrhizobium kavangense* sp. nov., a symbiotic nitrogen-fixing bacterium from root nodules of traditional Namibian pulses. Int J Syst Evol Microbiol 65:4886–4894

Grönemeyer JL, Hurek T, Bünger W, Reinhold-Hurek B (2016) *Bradyrhizobium vignae* sp. nov., a nitrogen-fixing symbiont isolated from effective nodules of *Vigna* and *Arachis*. Int J Syst Evol Microbiol 66:62–69

Gu CT, Wang ET, Tian CF, Han TX, Chen WF, Sui XH, Chen WX (2008) *Rhizobium miluonense* sp. nov., a symbiotic bacterium isolated from *Lespedeza* root nodules. Int J Syst Evol Microbiol 58:1364–1368

Gu T, Sun LN, Zhang J, Sui XH, Li SP (2014) *Rhizobium flavum* sp. nov., a triazophos-degrading bacterium isolated from soil under the long-term application of triazophos. Int J Syst Evol Microbiol 64:2017–2022

Guan SH, Chen WF, Wang ET, Lu YL, Yan XR, Zhang XX, Chen WX (2008) *Mesorhizobium caraganae* sp. nov., a novel rhizobial species nodulated with *Caragana* spp. in China. Int J Syst Evol Microbiol 58:2646–2653

Gubry-Rangin C, Béna G, Cleyet-Marel JC, Brunel B (2013) Definition and evolution of a new symbiovar, sv. riguloides, among *Ensifer meliloti* efficiently nodulating *Medicago* species. Syst Appl Microbiol 36:490–496

Guerrouj K, Ruíz-Díez B, Chahboune R, Ramírez-Bahena MH, Abdelmoumen H, Quiñones MA, El Idrissi MM, Velázquez E, Fernández-Pascual M, Bedmar EJ, Peix A (2013) Definition of a novel symbiovar (sv. retamae) within *Bradyrhizobium retamae* sp. nov., nodulating *Retama sphaerocarpa* and *Retama monosperma*. Syst Appl Microbiol 36:218–223

Han TX, Han LL, Wu LJ, Chen WF, Sui XH, Gu JG, Wang ET, Chen WX (2008a) *Mesorhizobium gobiense* sp. nov. and *Mesorhizobium tarimense* sp. nov., isolated from wild legumes growing in desert soils of Xinjiang, China. Int J Syst Evol Microbiol 58:2610–2618

Han TX, Wang ET, Wu LJ, Chen WF, Gu JG, Gu CT, Tian CF, Chen WX (2008b) *Rhizobium multihospitium* sp. nov., isolated from multiple legume species native of Xinjiang, China. Int J Syst Evol Microbiol 58:1693–1699

Helene LC, Marçon Delamuta JR, Augusto Ribeiro R, Ormeño-Orrillo E, Antonio Rogel M, Martínez-Romero E, Hungria M (2015) *Bradyrhizobium viridifuturi* sp. nov., encompassing nitrogen-fixing symbionts of legumes used for green manure and environmental services. Int J Syst Evol Microbiol 65:4441–4448

Hirsch P, Müller M (1985) *Blastobacter aggregatus* sp. nov., *Blastobacter capsulatus* sp. nov., and *Blastobacter denitrificans* sp. nov., new budding bacteria from freshwater habitats. Syst Appl Microbiol 6:281–286

Horn K, Parker IM, Malek W, Rodríguez-Echeverría S, Parker MA (2014) Disparate origins of *Bradyrhizobium* symbionts for invasive populations of *Cytisus scoparius* (Leguminosae) in North America. FEMS Microbiol Ecol 89:89–98

Hou BC, Wang ET, Li Y, Jia RZ, Chen WF, Gao Y, Dong R, Chen WX (2009) *Rhizobium tibeticum* sp. nov., a symbiotic bacterium isolated *from Medicago archiducis-nicolai Vassilcz*. Int J Syst Evol Microbiol 59:3051–3057

Hunter WJ, Kuykendall LD, Manter DK (2007) *Rhizobium selenireducens* sp. nov.: a selenite-reducing alpha-Proteobacteria isolated from a bioreactor. Curr Microbiol 55:455–460

Iglesias O, Rivas R, García-Fraile P, Abril A, Mateos PF, Martinez-Molina E, Velázquez E (2007) Genetic characterization of fast-growing rhizobia able to nodulate *Prosopis alba* in North Spain. FEMS Microbiol Lett 277:210–216

Islam MS, Kawasaki H, Muramatsu Y, Nakagawa Y, Seki T (2008) *Bradyrhizobium iriomotense* sp. nov., isolated from a tumor-like root of the legume *Entada koshunensis* from Iriomote Island in Japan. Biosci Biotechnol Biochem 72:1416–1429

Jarvis BDW, Pankhurst CE, Patel JJ (1982) *Rhizobium loti*, a new species of legume root nodule bacteria. Int J Syst Bacteriol 32:378–380

Jarvis BDW, Downer HL, Young JPW (1992) Phylogeny of fast-growing soybean-nodulating rhizobia supports synonymy of *Sinorhizobium* and *Rhizobium* and assignment to *Rhizobium fredii*. Int J Syst Bacteriol 42:93–96

Jarvis BDW, van Berkum P, Chen WX, Nour SM, Fernandez MP, Cleyet-Marel JC, Gillis M (1997) Transfer of *Rhizobium loti*, *Rhizobium huakuii*, *Rhizobium ciceri*, *Rhizobium mediterraneum*, and *Rhizobium tianshanense* to *Mesorhizobium* gen. nov. Int J Syst Bacteriol 47:895–898

Ji Z, Yan H, Cui Q, Wang E, Chen W, Chen W (2015) Genetic divergence and gene flow among *Mesorhizobium* strains nodulating the shrub legume *Caragana*. Syst Appl Microbiol 38:176–183

Jiao YS, Liu YH, Yan H, Wang ET, Tian CF, Chen WX, Guo BL, Chen WF (2015a) Rhizobial diversity and nodulation characteristics of the extremely promiscuous legume *Sophora flavescens*. Mol Plant Microbe Interact 28:1338–1352

Jiao YS, Yan H, Ji ZJ, Liu YH, Sui XH, Wang ET, Guo BL, Chen WX, Chen WF (2015b) *Rhizobium sophorae* sp. nov. and *Rhizobium sophoriradicis* sp. nov., nitrogen-fixing rhizobial symbionts of the medicinal legume *Sophora flavescens*. Int J Syst Evol Microbiol 65:497–503

Jiao YS, Yan H, Ji ZJ, Liu YH, Sui XH, Zhang XX, Wang ET, Chen WX, Chen WF (2015c) *Phyllobacterium sophorae* sp. nov., a symbiotic bacterium isolated from root nodules of *Sophora flavescens*. Int J Syst Evol Microbiol 65:399–406

Jordan DC (1982) Transfer of *Rhizobium japonicum* Buchanan 1980 to *Bradyrhizobium* gen. nov., a genus of slow-growing, root nodule bacteria from leguminous plants. Int J Syst Bacteriol 32:136–139

Jordan DC (1984) Family III *Rhizobiaceae*. In: Krieg NR, Holt JG (eds) Bergey's manual of systematic bacteriology, vol I. Williams and Wilkins Co., Baltimore, pp 234–242

Jordan DC, Allen ON (1974) Family 111. *Rhizobiaceae* Conn, 1938. In: Buchanan RE, Gibbons NE (eds) Bergey's manual of determinative bacteriology, 8th edn. Williams & Wilkins Co., Baltimore, pp 261–264

Jourand P, Giraud E, Béna G, Sy A, Willems A, Gillis M, Dreyfus B, de Lajudie P (2004) *Methylobacterium nodulans* sp. nov., for a group of aerobic, facultatively methylotrophic, legume root-nodule-forming and nitrogen-fixing bacteria. Int J Syst Evol Microbiol 54:2269–2273

Judicial Commission of the International Committee on Systematics of Prokaryotes (2008) The genus name *Sinorhizobium* Chen et al. 1988 is a later synonym of *Ensifer* Casida 1982 and is not conserved over the latter genus name, and the species name 'Sinorhizobium adhaerens' is not validly published. Opinion 84. Int J Syst Evol Microbiol 58:1973

Kaiya S, Rubaba O, Yoshida N, Yamada T, Hiraishi A (2012) Characterization of *Rhizobium naphthalenivorans* sp. nov. with special emphasis on aromatic compound degradation and multilocus sequence analysis of housekeeping genes. J Gen Appl Microbiol 58:211–224

Kathiravan R, Jegan S, Ganga V, Prabavathy VR, Tushar L, Sasikala C, Ramana CV (2013) *Ciceribacter lividus* gen. nov., sp. nov., isolated from rhizosphere soil of chick pea (*Cicer arietinum* L.) Int J Syst Evol Microbiol 63:4484–4488

Kaur J, Verma M, Lal R (2011) *Rhizobium rosettiformans* sp. nov., isolated from a hexachlorocyclohexane dump site, and reclassification of *Blastobacter aggregatus* Hirsch and Muller 1986 as *Rhizobium aggregatum* comb. nov. Int J Syst Evol Microbiol 61:1218–1225

Kersters K, de Ley J (1984) Genus III *Agrobacterium*. In: Krieg NR, Holt JG (eds) Bergey's manual of systematic bacteriology, vol I. Williams and Wilkins Co., Baltimore, pp 244–254

Kesari V, Ramesh AM, Rangan L (2013) *Rhizobium pongamiae* sp. nov. from root nodules of *Pongamia pinnata*. Biomed Res Int 2013:165198

Khalid R, Zhang YJ, Ali S, Sui XH, Zhang XX, Amara U, Chen WX, Hayat R (2015) *Rhizobium pakistanensis* sp. nov., isolated from groundnut (*Arachis hypogaea*) nodules grown in rainfed Pothwar, Pakistan. Antonie van Leeuwenhoek 107:281–290

Kimes NE, López-Pérez M, Flores-Félix JD, Ramírez-Bahena MH, Igual JM, Peix A, Rodriguez-Valera F, Velázquez E (2015) *Pseudorhizobium pelagicum* gen. nov, sp. nov. isolated from a pelagic Mediterranean zone. Syst Appl Microbiol 38:293–299

Kittiwongwattana C, Thawai C (2013) *Rhizobium paknamense* sp. nov., isolated from lesser duckweeds (*Lemna aequinoctialis*). Int J Syst Evol Microbiol 63:3823–3828

Kittiwongwattana C, Thawai C (2014) *Rhizobium lemnae* sp. nov., a bacterial endophyte of *Lemna aequinoctialis*. Int J Syst Evol Microbiol 64:2455–2460

Knösel DH (1984) Genus IV. *Phyllobacterium* nom. rev. In: Krieg NR, Holt JG (eds) Bergey's manual of systematic bacteriology, vol 1. Williams & Wilkins Co., Baltimore, pp 254–256

Kuykendall LD, Saxena B, Devine TE, Udell SE (1992) Genetic diversity in *Bradyrhizobium japonicum* Jordan 1982 and a proposal for *Bradyrhizobium elkanii* sp. nov. Can J Microbiol 38:501–505

Kuykendall LD, Young JM, Martínez-Romero E, Kerr A, Sawada H (2005) Order *Rhizobiales* (new) Family *Rhizobiaceae* Genus *Rhizobium*. In: Brenner DJ, Krieg NR, Staley JT, Garrity GM (eds) The alpha-, beta-, delta- and epsilonproteobacteria, the proteobacteria: Part C. Bergey's manual of systematic bacteriology, vol 2, 2nd edn. Springer, New York, pp 324–340

Laguerre G, Nour SM, Macheret V, Sanjuan J, Drouin P, Amarger N (2001) Classification of rhizobia based on *nodC* and *nifH* gene analysis reveals a close phylogenetic relationship among *Phaseolus vulgaris* symbionts. Microbiology 147:981–993

de Lajudie P, Laurent-Fulele E, Willems A, Torck U, Coopman R, Collins MD, Kersters K, Dreyfus B, Gillis M (1992) *Allorhizobium undicola* gen. nov., sp. nov., nitrogen-fixing bacteria that efficiently nodulate *Neptunia natans* in Senegal. Int J Syst Bacteriol 42:93–96

de Lajudie P, Willems A, Pot B, Dewettinck D, Maestrojuan G, Neyra M, Collins MD, Dreyfus B, Kersters K, Gillis M (1994) Polyphasic taxonomy of Rhizobia: emendation of the genus *Sinorhizobium* and description of *Sinorhizobium meliloti* comb. nov., *Sinorhizobium saheli* sp. nov., and *Sinorhizobium teranga* sp. nov. Int J Syst Bacteriol 44:715–733

de Lajudie P, Willems A, Nick G, Moreira F, Molouba F, Hoste B, Torck U, Neyra M, Collins MD, Lindström K, Dreyfus B, Gillis M (1998) Characterization of tropical tree rhizobia and description of *Mesorhizobium plurifarium* sp. nov. Int J Syst Bacteriol 48:369–382

Lang E, Schumann P, Adler S, Spröer C, Sahin N (2013) *Azorhizobium oxalatiphilum* sp. nov., and emended description of the genus *Azorhizobium*. Int J Syst Evol Microbiol 63:1505–1511

Laranjo M, Alexandre A, Rivas R, Velázquez E, Young JP, Oliveira S (2008) Chickpea rhizobia symbiosis genes are highly conserved across multiple *Mesorhizobium* species. FEMS Microbiol Ecol 66:391–400

Latif S, Khan S, Naveed M, Mustafa G, Bashir T, Mumtaz AS (2013) The diversity of Rhizobia, Sinorhizobia and novel non-rhizobial *Paenibacillus* nodulating wild herbaceous legumes. Arch Microbiol 195:647–653

Lemaire B, Dlodlo O, Chimphango S, Stirton C, Schrire B, Boatwright JS, Honnay O, Smets E, Sprent J, James EK, Muasya AM (2015) Symbiotic diversity, specificity and distribution of rhizobia in native legumes of the Core Cape subregion (South Africa). FEMS Microbiol Ecol 91:1–17

León-Barrios M, Lorite MJ, Donate-Correa J, Sanjuán J (2009) *Ensifer meliloti* bv. lancerottense establishes nitrogen-fixing symbiosis with *Lotus* endemic to the Canary Islands and shows distinctive symbiotic genotypes and host range. Syst Appl Microbiol 32:413–420

Li QQ, Wang ET, Chang YL, Zhang YZ, Zhang YM, Sui XH, Chen WF, Chen WX (2011) *Ensifer sojae* sp. nov., isolated from root nodules of *Glycine max* grown in saline-alkaline soils. Int J Syst Evol Microbiol 61:1981–1988

Li YH, Wang R, Zhang XX, Young JP, Wang ET, Sui XH, Chen WX (2015) *Bradyrhizobium guangdongense* sp. nov. and *Bradyrhizobium guangxiense* sp. nov., isolated from effective nodules of peanut. Int J Syst Evol Microbiol 65:4655–4661

Li Y, Li X, Liu Y, Wang ET, Ren C, Liu W, Xu H, Wu H, Jiang N, Li Y, Zhang X, Xie Z (2016a) Genetic diversity and community structure of rhizobia nodulating *Sesbania cannabina* in saline-alkaline soils. Syst Appl Microbiol 39:195–202

Li Y, Yan J, Yu B, Wang ET, Li X, Yan H, Liu W, Xie Z (2016b) *Ensifer alkalisoli* sp. nov., isolated from root nodules of *Sesbania cannabina* grown in saline-alkaline soils. Int J Syst Evol Microbiol. doi:10.1099/ijsem.0.001510

Li Y, Lei X, Xu Y, Zhu H, Xu M, Fu L, Zheng W, Zhang J, Zheng T (2017) *Rhizobium albus* sp. nov., isolated from lake water in Xiamen, Fujian province of China. Curr Microbiol 74:42–48

Lin DX, Wang ET, Tang H, Han TX, He YR, Guan SH, Chen WX (2008) *Shinella kummerowiae* sp. nov., a symbiotic bacterium isolated from root nodules of the herbal legume *Kummerowia stipulacea*. Int J Syst Evol Microbiol 58:1409–1413

Lin DX, Chen WF, Wang FQ, Hu D, Wang ET, Sui XH, Chen WX (2009) *Rhizobium mesosinicum* sp. nov., isolated from root nodules of three different legumes. Int J Syst Evol Microbiol 59:1919–1923

Lin SY, Hsu YH, Liu YC, Hung MH, Hameed A, Lai WA, Yen WS, Young CC (2014) *Rhizobium straminoryzae* sp. nov., isolated from the surface of rice straw. Int J Syst Evol Microbiol 64:2962–2968

Lin SY, Hung MH, Hameed A, Liu YC, Hsu YH, Wen CZ, Arun AB, Busse HJ, Glaeser SP, Kämpfer P, Young CC (2015) *Rhizobium capsici* sp. nov., isolated from root tumor of a green bell pepper (*Capsicum annuum* var. grossum) plant. Antonie van Leeuwenhoek 107:773–784

Lindström K (1989) *Rhizobium galegae*, a new species of legume root nodule bacteria. Int J Syst Bacteriol 39:365–367

Liu XY, Wu W, Wang ET, Zhang B, Macdermott J, Chen WX (2011) Phylogenetic relationships and diversity of β-rhizobia associated with *Mimosa* species grown in Sishuangbanna, China. Int J Syst Evol Microbiol 61:334–342

Liu TY, Li Y Jr, Liu XX, Sui XH, Zhang XX, Wang ET, Chen WX, Chen WF, Puławska J (2012a) *Rhizobium cauense* sp. nov., isolated from root nodules of the herbaceous legume *Kummerowia stipulacea* grown in campus lawn soil. Syst Appl Microbiol 35:415–420

Liu X, Wei S, Wang F, James EK, Guo X, Zagar C, Xia LG, Dong X, Wang YP (2012b) *Burkholderia* and *Cupriavidus* spp. are the preferred symbionts of *Mimosa* spp. in southern China. FEMS Microbiol Ecol 80:417–426

Liu Y, Wang RP, Ren C, Lai QL, Zeng RY (2015) *Rhizobium marinum* sp. nov., a malachite-green-tolerant bacterium isolated from seawater. Int J Syst Evol Microbiol 65:4449–4454

Lloret L, Ormeño-Orrillo E, Rincón R, Martínez-Romero J, Rogel-Hernández MA, Martínez-Romero E (2007) *Ensifer mexicanus* sp. nov. a new species nodulating *Acacia angustissima* (Mill.) Kuntze in Mexico. Syst Appl Microbiol 30:280–290

López-López A, Rogel MA, Ormeño-Orrillo E, Martínez-Romero J, Martínez-Romero E (2010) *Phaseolus vulgaris* seed-borne endophytic community with novel bacterial species such as *Rhizobium endophyticum* sp. nov. Syst Appl Microbiol 33:322–327

López-López A, Rogel-Hernández MA, Barois I, Ortiz Ceballos AI, Martínez J, Ormeño-Orrillo E, Martínez-Romero E (2012) *Rhizobium grahamii* sp. nov., from nodules of *Dalea leporina*, Leucaena leucocephala and *Clitoria ternatea*, and *Rhizobium mesoamericanum* sp. nov., from nodules of *Phaseolus vulgaris*, siratro, cowpea and *Mimosa pudica*. Int J Syst Evol Microbiol 62:2264–2271

Lorite MJ, Videira e Castro I, Muñoz S, Sanjuán J (2012) Phylogenetic relationship of *Lotus uliginosus* symbionts with bradyrhizobia nodulating genistoid legumes. FEMS Microbiol Ecol 79:454–464

Lorite MJ, Flores-Félix JD, Peix Á, Sanjuán J, Velázquez E (2016) *Mesorhizobium olivaresii* sp. nov. isolated from *Lotus corniculatus* nodules. Syst Appl Microbiol 39:557–561

Lortet G, Mear N, Lorquin J, Dreyfus B, de Lajudie P, Rosenberg C, Boivin C (1996) Nod factor thin-layer chromatography profiling as a tool to characterize symbiotic specificity of rhizobial strains: application to *Sinorhizobium saheli*, *S. teranga*, and *Rhizobium* sp. strains isolated from *Acacia* and *Sesbania*. Mol Plant Microbe Interact 9:736–747

Lu YL, Chen WF, Han LL, Wang ET, Chen WX (2009a) *Rhizobium alkalisoli* sp. nov., isolated from the legume *Caragana intermedia* growing in saline-alkaline soils. Int J Syst Evol Microbiol 59:3006–3011

Lu YL, Chen WF, Han LL, Wang ET, Zhang XX, Chen WX, Han SZ (2009b) *Mesorhizobium shangrilense* sp. nov., isolated from root nodules of *Caragana* spp. Int J Syst Evol Microbiol 59:3012–3018

Lu YL, Chen WF, Wang ET, Guan SH, Yan XR, Chen WX (2009c) Genetic diversity and biogeography of rhizobia associated with *Caragana* species in three ecological regions of China. Syst Appl Microbiol 32:351–361

Lu JK, Dou YJ, Zhu YJ, Wang SK, Sui XH, Kang LH (2014) *Bradyrhizobium ganzhouense* sp. nov., an effective symbiotic bacterium isolated from *Acacia melanoxylon* R. Br. nodules. Int J Syst Evol Microbiol 64:1900–1905

Marek-Kozaczuk M, Leszcz A, Wielbo J, Wdowiak-Wróbel S, Skorupska A (2013) *Rhizobium pisi* sv. trifolii K3.22 harboring nod genes of the *Rhizobium leguminosarum* sv. trifolii cluster. Syst Appl Microbiol 36:252–258

Martínez-Aguilar L, Salazar-Salazar C, Méndez RD, Caballero-Mellado J, Hirsch AM, Vásquez-Murrieta MS, Estrada-de los Santos P (2013) *Burkholderia caballeronis* sp. nov., a nitrogen fixing species isolated from tomato (*Lycopersicon esculentum*) with the ability to effectively nodulate *Phaseolus vulgaris*. Antonie van Leeuwenhoek 104:1063–1071

Martínez-Hidalgo P, Flores-Félix JD, Menéndez E, Rivas R, Carro L, Mateos PF, Martínez-Molina E, León-Barrios M, Velázquez E (2015a) *Cicer canariense*, an endemic legume to the Canary Islands, is nodulated in mainland Spain by fast-growing strains from symbiovar trifolii phylogenetically related to *Rhizobium leguminosarum*. Syst Appl Microbiol 38:346–350

Martínez-Hidalgo P, Ramírez-Bahena MH, Flores-Félix JD, Rivas R, Igual JM, Mateos PF, Martínez-Molina E, León-Barrios M, Peix Á, Velázquez E (2015b) Revision of the taxonomic status of type strains of *Mesorhizobium loti* and reclassification of strain USDA 3471^T as the type strain of *Mesorhizobium erdmanii* sp. nov. and ATCC 33669^T as the type strain of *Mesorhizobium jarvisii* sp. nov. Int J Syst Evol Microbiol 65:1703–1708

Martínez-Hidalgo P, Ramírez-Bahena MH, Flores-Félix JD, Igual JM, Sanjuán J, León-Barrios M, Peix A, Velázquez E (2016) Reclassification of strains MAFF 303099^T and R7A into the new species *Mesorhizobium japonicum* sp. nov. Int J Syst Evol Microbiol. doi:10.1099/ijsem.0.001448

Martínez-Romero E, Segovia L, Mercante FM, Franco AA, Graham P, Pardo MA (1991) *Rhizobium tropici*: a novel species nodulating *Phaseolus vulgaris* L. beans and *Leucaena* sp. trees. Int J Syst Bacteriol 41:417–426

Maynaud G, Willems A, Soussou S, Vidal C, Mauré L, Moulin L, Cleyet-Marel JC, Brunel B (2012) Molecular and phenotypic characterization of strains nodulating *Anthyllis vulneraria* in mine tailings, and proposal of *Aminobacter anthyllidis* sp. nov., the first definition of *Aminobacter* as legume-nodulating bacteria. Syst Appl Microbiol 35:65–72

Merabet C, Martens M, Mahdhi M, Zakhia F, Sy A, Le Roux C, Domergue O, Coopman R, Bekki A, Mars M, Willems A, de Lajudie P (2010) Multilocus sequence analysis of root nodule isolates from *Lotus arabicus* (Senegal), *Lotus creticus*, *Argyrolobium uniflorum* and *Medicago sativa* (Tunisia) and description of *Ensifer numidicus* sp. nov. and *Ensifer garamanticus* sp. nov. Int J Syst Evol Microbiol 60:664–674

Mergaert J, Swings J (2005) Genus I. *Phyllobacterium* (ex Knösel 1962) Knösel 1984, 356VP (Effective publication: Knösel 1984, 254). In: Brenner DJ, Krieg NR, Staley JT, Garrity GM (eds) Bergey's manual of systematic bacteriology: Part C, vol 2, 2nd edn. Springer, New York, pp 394–396

Mierzwa B, Łotocka B, Wdowiak-Wróbel S, Kalita M, Gnat S, Małek W (2010) Insight into the evolutionary history of symbiotic genes of *Robinia pseudoacacia* rhizobia deriving from Poland and Japan. Arch Microbiol 192:341–350

Mnasri B, Mrabet M, Laguerre G, Aouani ME, Mhamdi R (2007) Salt-tolerant rhizobia isolated from a Tunisian oasis that are highly effective for symbiotic N2-fixation with *Phaseolus vulgaris* constitute a novel biovar (bv. mediterranense) of *Sinorhizobium meliloti*. Arch Microbiol 187:79–85

Mnasri B, Saïdi S, Chihaoui SA, Mhamdi R (2012) *Sinorhizobium americanum* symbiovar mediterranense is a predominant symbiont that nodulates and fixes nitrogen with common bean (*Phaseolus vulgaris* L.) in a Northern Tunisian field. Syst Appl Microbiol 35:263–269

Moulin L, Munive A, Dreyfus B, Boivin-Masson C (2001) Nodulation of legumes by members of the beta-subclass of Proteobacteria. Nature 411:948–950. Erratum in: Nature 412:926

Mousavi SA, Österman J, Wahlberg N, Nesme X, Lavire C, Vial L, Paulin L, de Lajudie P, Lindström K (2014) Phylogeny of the *Rhizobium-Allorhizobium-Agrobacterium* clade supports the delineation of *Neorhizobium* gen. nov. Syst Appl Microbiol 37:208–215

Mousavi SA, Willems A, Nesme X, de Lajudie P, Lindström K (2015) Revised phylogeny of Rhizobiaceae: proposal of the delineation of *Pararhizobium* gen. nov., and 13 new species combinations. Syst Appl Microbiol 38:84–90

Mousavi SA, Li L, Wei G, Räsänen L, Lindström K (2016) Evolution and taxonomy of native mesorhizobia nodulating medicinal *Glycyrrhiza* species in China. Syst Appl Microbiol. doi:10.1016/j.syapm.2016.03.009

Nakatsukasa H, Uchiumi T, Kucho K, Suzuki A, Higashi S, Abe M (2008) Transposon mediation allows a symbiotic plasmid of *Rhizobium leguminosarum* bv. trifolii to become a symbiosis island in *Agrobacterium* and *Rhizobium*. J Gen Appl Microbiol 54:107–118

Nandasena KG, O'Hara GW, Tiwari RP, Willlems A, Howieson JG (2007) *Mesorhizobium ciceri* biovar biserrulae, a novel biovar nodulating the pasture legume *Biserrula pelecinus* L. Int J Syst Evol Microbiol 57:1041–1045

Nandasena KG, O'Hara GW, Tiwari RP, Willems A, Howieson JG (2009) *Mesorhizobium australicum* sp. nov. and *Mesorhizobium opportunistum* sp. nov. isolated from *Biserrula pelecinus* L. growing in Australia. Int J Syst Evol Microbiol 59:2140–2147

Nguyen TM, Pham VH, Kim J (2015) *Mesorhizobium soli* sp. nov., a novel species isolated from the rhizosphere of *Robinia pseudoacacia* L. in South Korea by using a modified culture method. Antonie van Leeuwenhoek 108:301–310

Nick G, de Lajudie P, Eardly BD, Suomalainen S, Paulin L, Zhang X, Gillis M, Lindström K (1999) *Sinorhizobium arboris* sp. nov. and *Sinorhizobium kostiense* sp. nov., isolated from leguminous trees in Sudan and Kenya. Int J Syst Bacteriol 49:1359–1368

Noisangiam R, Nuntagij A, Pongsilp N, Boonkerd N, Denduangboripant J, Ronson C, Teaumroong N (2010) Heavy metal tolerant *Metalliresistens boonkerdii* gen. nov, sp. nov., a new genus in the family *Bradyrhizobiaceae* isolated from soil in Thailand. Syst Appl Microbiol 33:374–382. Erratum in: Syst Appl Microbiol 34:166–168

Norel FF, Elmerich C (1987) Nucleotide sequence and functional analysis of the two nifH copies of Rhizobium ORS571. Microbiology 133:1563–1576

Nour SM, Fernandez MP, Normand P, Cleyet-Marel JC (1994) *Rhizobium ciceri* sp. nov., consisting of strains that nodulate chickpeas (*Cicer arietinum* L.) Int J Syst Bacteriol 44:511–522

Nour SM, Cleyet-Marel JC, Normand P, Fernandez MP (1995) Genomic heterogeneity of strains nodulating chickpeas (*Cicer arietinum* L.) and description of *Rhizobium mediterraneum* sp. nov. Int J Syst Bacteriol 45:640–648

Ogasawara M, Suzuki T, Mutoh I, Annapurna K, Arora NK, Nishimura Y, Maheshwari DK (2003) *Sinorhizobium indiaense* sp. nov. and *Sinorhizobium abri* sp. nov. isolated from tropical legumes, *Sesbania rostrata* and *Abrus precatorius*, respectively. Symbiosis 34:53–68

Ohta H, Hattori T (1983) *Agromonas oligotrophica* gen. nov., sp. nov., a nitrogen-fixing oligotrophic bacterium. Antonie van Leeuwenhoek 49:429–446

Ophel K, Kerr A (1990) *Agrobacterium vitis* sp. nov. for strains of *Agrobacterium* biovar 3 from grapevines. Int J Syst Bacteriol 40:236–241

Oren A, Garrity GM (2015a) List of new names and new combinations previously effectively, but not validly, published. Int J Syst Evol Microbiol 65:741–744

Oren A, Garrity GM (2015b) List of new names and new combinations previously effectively, but not validly, published. Int J Syst Evol Microbiol 65:2017–2025

Oren A, Garrity GM (2015c) List of new names and new combinations previously effectively, but not validly, published. Int J Syst Evol Microbiol 65:2777–2783

Ormeño-Orrillo E, Menna P, Almeida LG, Ollero FJ, Nicolás MF, Pains Rodrigues E, Shigueyoshi Nakatani A, Silva Batista JS, Oliveira Chueire LM, Souza RC, Ribeiro Vasconcelos AT, Megías M, Hungria M, Martínez-Romero E (2012) Genomic basis of broad host range and environmental adaptability of *Rhizobium tropici* CIAT 899 and *Rhizobium* sp. PRF 81 which are used in inoculants for common bean (*Phaseolus vulgaris* L.) BMC Genomics 13:735

Panday D, Schumann P, Das SK (2011) *Rhizobium pusense* sp. nov., isolated from the rhizosphere of chickpea (*Cicer arietinum* L.) Int J Syst Evol Microbiol 61:2632–2639

Parag B, Sasikala C, Ramana CV (2013) Molecular and culture dependent characterization of endolithic bacteria in two beach sand samples and description of *Rhizobium endolithicum* sp. nov. Antonie van Leeuwenhoek 104:1235–1244

Peix A, Ramírez-Bahena MH, Flores-Félix JD, Alonso de la Vega P, Rivas R, Mateos PF, Igual JM, Martínez-Molina E, Trujillo ME, Velázquez E (2015a) Revision of the taxonomic status of the species *Rhizobium lupini* and reclassification as *Bradyrhizobium lupini* comb. nov. Int J Syst Evol Microbiol 65:1213–1219

Peix A, Ramírez-Bahena MH, Velázquez E, Bedmard EJ (2015b) Bacterial associations with legumes. Crit Rev Plant Sci 34:17–42

Peng G, Yuan Q, Li H, Zhang W, Tan Z (2008) *Rhizobium oryzae* sp. nov., isolated from the wild rice *Oryza alta*. Int J Syst Evol Microbiol 58:2158–2163

Perret X, Staehelin C, Broughton WJ (2000) Molecular basis of symbiotic promiscuity. Microbiol Mol Biol Rev 64:180–201

Platero R, James EK, Rios C, Iriarte A, Sandes L, Zabaleta M, Battistoni F, Fabiano E (2016) Novel *Cupriavidus* strains isolated from root nodules of native Uruguayan *Mimosa* species. Appl Environ Microbiol. pii: AEM.04142-15

Pueppke SG, Broughton WJ (1999) *Rhizobium* sp. strain NGR234 and R. fredii USDA257 share exceptionally broad, nested host ranges. Mol Plant Microbe Interact 12:293–318

Puławska J, Kuzmanović N, Willems A, Pothier JF (2016) *Pararhizobium polonicum* sp. nov. isolated from tumors on stone fruit rootstocks. Syst Appl Microbiol 39:164–169

Qin W, Deng ZS, Xu L, Wang NN, Wei GH (2012) *Rhizobium helanshanense* sp. nov., a bacterium that nodulates *Sphaerophysa salsula* (Pall.) DC. in China. Arch Microbiol 194:371–378

Quan ZX, Bae HS, Baek JH, Chen WF, Im WT, Lee ST (2005) *Rhizobium daejeonense* sp. nov. isolated from a cyanide treatment bioreactor. Int J Syst Evol Microbiol 55:2543–2549

Radeva G, Jurgens G, Niemi M, Nick G, Suominen L, Lindström K (2001) Description of two biovars in the *Rhizobium galegae* species: Biovar orientalis and biovar officinalis. Syst Appl Microbiol 24:192–205

Radl V, Simões-Araújo JL, Leite J, Passos SR, Martins LM, Xavier GR, Rumjanek NG, Baldani JI, Zilli JE (2014) *Microvirga vignae* sp. nov., a root nodule symbiotic bacterium isolated from cowpea grown in semi-arid Brazil. Int J Syst Evol Microbiol 64:725–730

Ramana CV, Parag B, Girija KR, Ram BR, Ramana VV, Sasikala C (2013) *Rhizobium subbaraonis* sp. nov., an endolithic bacterium isolated from beach sand. Int J Syst Evol Microbiol 63:581–585

Ramírez Bahena MH, Flores Félix JD, Chahboune R, Toro M, Velázquez E, Peix A (2016) *Bradyrhizobium centrosemae* (symbiovar centrosemae) sp. nov., *Bradyrhizobium americanum* (symbiovar phaseolarum) sp. nov. and a new symbiovar (tropici) of *Bradyrhizobium viridifuturi* establish symbiosis with *Centrosema* species native to America. Syst Appl Microbiol 39:378–383

Ramírez-Bahena MH, García-Fraile P, Peix A, Valverde A, Rivas R, Igual JM, Mateos PF, Martínez-Molina E, Velázquez E (2008) Revision of the taxonomic status of the species *Rhizobium leguminosarum* (Frank 1879) Frank 1889AL, *Rhizobium phaseoli* Dangeard 1926AL and *Rhizobium trifolii* Dangeard 1926AL. *R. trifolii* is a later synonym of *R. leguminosarum*. Reclassification of the strain R. leguminosarum DSM 30132 (=NCIMB 11478) as *Rhizobium pisi* sp. nov. Int J Syst Evol Microbiol 58:2484–2490

Ramírez-Bahena MH, Peix A, Rivas R, Camacho M, Rodríguez-Navarro DN, Mateos PF, Martínez-Molina E, Willems A, Velázquez E (2009) *Bradyrhizobium pachyrhizi* sp. nov. and *Bradyrhizobium jicamae* sp. nov., isolated from effective nodules of *Pachyrhizus erosus*. Int J Syst Evol Microbiol 59:1929–1934

Ramírez-Bahena MH, Hernández M, Peix A, Velázquez E, León-Barrios M (2012) Mesorhizobial strains nodulating *Anagyris latifolia* and *Lotus berthelotii* in Tamadaya ravine (Tenerife, Canary Islands) are two symbiovars of the same species, *Mesorhizobium tamadayense* sp. nov. Syst Appl Microbiol 35:334–341

Ramírez-Bahena MH, Chahboune R, Peix A, Velázquez E (2013a) Reclassification of *Agromonas oligotrophica* into the genus *Bradyrhizobium* as *Bradyrhizobium oligotrophicum* comb. nov. Int J Syst Evol Microbiol 63:1013–1016

Ramírez-Bahena MH, Chahboune R, Velázquez E, Gómez-Moriano A, Mora E, Peix A, Toro M (2013b) *Centrosema* is a promiscuous legume nodulated by several new putative species and symbiovars of *Bradyrhizobium* in various American countries. Syst Appl Microbiol 36:392–400

Rashid MH, Young JP, Everall I, Clercx P, Willems A, Santhosh Braun M, Wink M (2015) Average nucleotide identity of genome sequences supports the description of *Rhizobium lentis* sp. nov., *Rhizobium bangladeshense* sp. nov. and *Rhizobium binae* sp. nov. from lentil (*Lens culinaris*) nodules. Int J Syst Evol Microbiol 65:3037–3045

Rasolomampianina R, Bailly X, Fetiarison R, Rabevohitra R, Béna G, Ramaroson L, Raherimandimby M, Moulin L, De Lajudie P, Dreyfus B, Avarre JC (2005) Nitrogen-fixing nodules from rose wood legume trees (*Dalbergia* spp.) endemic to Madagascar host seven different genera belonging to alpha- and beta-Proteobacteria. Mol Ecol 14:4135–4146

Relic B, Perret X, Estrada-García MT, Kopcinska J, Golinowski W, Krishnan HB, Pueppke SG, Broughton WJ (1994) Nod factors of *Rhizobium* are a key to the legume door. Mol Microbiol 13:171–178

Ren d W, Chen WF, Sui XH, Wang ET, Chen WX (2011a) *Rhizobium vignae* sp. nov., a symbiotic bacterium isolated from multiple legume species. Int J Syst Evol Microbiol 61:580–586

Ren d W, Wang ET, Chen WF, Sui XH, Zhang XX, Liu HC, Chen WX (2011b) *Rhizobium herbae* sp. nov. and *Rhizobium giardinii*-related bacteria, minor microsymbionts of various wild legumes in China. Int J Syst Evol Microbiol 61:1912–1920

Ribeiro RA, Rogel MA, López-López A, Ormeño-Orrillo E, Barcellos FG, Martínez J, Thompson FL, Martínez-Romero E, Hungria M (2012) Reclassification of *Rhizobium tropici* type A strains as *Rhizobium leucaenae* sp. nov. Int J Syst Evol Microbiol 62:1179–1184

Ribeiro RA, Martins TB, Ormeño-Orrillo E, Marçon Delamuta JR, Rogel MA, Martínez-Romero E, Hungria M (2015) *Rhizobium ecuadorense* sp. nov., an indigenous N2-fixing symbiont of the Ecuadorian common bean (*Phaseolus vulgaris* L.) genetic pool. Int J Syst Evol Microbiol 65:3162–3169

Rincón-Rosales R, Lloret L, Ponce E, Martínez-Romero E (2009) Rhizobia with different symbiotic efficiencies nodulate *Acaciella angustissima* in Mexico, including *Sinorhizobium chiapanecum* sp. nov. which has common symbiotic genes with *Sinorhizobium mexicanum*. FEMS Microbiol Ecol 67:103–117

Rincón-Rosales R, Villalobos-Escobedo JM, Rogel MA, Martinez J, Ormeño-Orrillo E, Martínez-Romero E (2013) *Rhizobium calliandrae* sp. nov., *Rhizobium mayense* sp. nov. and *Rhizobium jaguaris* sp. nov. rhizobial species nodulating the medicinal legume *Calliandra grandiflora*. Int J Syst Evol Microbiol 63:3423–3429

Rivas R, Velázquez E, Willems A, Vizcaíno N, Subba-Rao NS, Mateos PF, Gillis M, Dazzo FB, Martínez-Molina E (2002) A new species of *Devosia* that forms a unique nitrogen-fixing root-nodule symbiosis with the aquatic legume *Neptunia natans* (L.f.) druce. Appl Environ Microbiol 68:5217–5222

Rivas R, Willems A, Subba-Rao NS, Mateos PF, Dazzo FB, Kroppenstedt RM, Martínez-Molina E, Gillis M, Velázquez E (2003) Description of *Devosia neptuniae* sp. nov. that nodulates and fixes nitrogen in symbiosis with *Neptunia natans*, an aquatic legume from India. Syst Appl Microbiol 26:47–53

Rivas R, Willems A, Palomo JL, García-Benavides P, Mateos PF, Martínez-Molina E, Gillis M, Velázquez E (2004) *Bradyrhizobium betae* sp. nov., isolated from roots of *Beta vulgaris* affected by tumour-like deformations. Int J Syst Evol Microbiol 54:1271–1275

Rivas R, Laranjo M, Mateos PF, Oliveira S, Martínez-Molina E, Velázquez E (2007) Strains of *Mesorhizobium amorphae* and *Mesorhizobium tianshanense*, carrying symbiotic genes of common chickpea endosymbiotic species, constitute a novel biovar (ciceri) capable of nodulating *Cicer arietinum*. Lett Appl Microbiol 44:412–418

Robledo M, Velázquez E, Ramírez-Bahena MH, García-Fraile P, Pérez-Alonso A, Rivas R, Martínez-Molina E, Mateos PF (2011) The *celC* gene, a new phylogenetic marker useful for taxonomic studies in *Rhizobium*. Syst Appl Microbiol 34:393–399

Roche P, Maillet F, Plazanet C, Debelle F, Ferro M, Truchet G, Prome JC, Denarié J (1996) The common *nodabc* genes of *Rhizobium meliloti* are host-range determinants. Proc Natl Acad Sci U S A 93:15305–15310

Rogel MA, Hernández-Lucas I, Kuykendall LD, Balkwill DL, Martínez-Romero E (2001) Nitrogen-fixing nodules with *Ensifer adhaerens* harboring *Rhizobium tropici* symbiotic plasmids. Appl Environ Microbiol 67:3264–3268

Rogel MA, Ormeño-Orrillo E, Martinez Romero E (2011) Symbiovars in rhizobia reflect bacterial adaptation to legumes. Syst Appl Microbiol 34:96–104

Román-Ponce B, Jing Zhang Y, Soledad Vásquez-Murrieta M, Hua Sui X, Feng Chen W, Carlos Alberto Padilla J, Wu Guo X, Lian Gao J, Yan J, Hong Wei G, Tao Wang E (2016) *Rhizobium acidisoli* sp. nov., isolated from root nodules of *Phaseolus vulgaris* in acid soils. Int J Syst Evol Microbiol 66:398–406

Rome S, Fernandez MP, Brunel B, Normand P, Cleyet-Marel JC (1996) *Sinorhizobium medicae* sp. nov., isolated from annual *Medicago* spp. Int J Syst Bacteriol 46:972–980

Rosenberg C, Boistard P, Dénarié J, Casse-Delbart F (1981) Genes controlling early and late functions in symbiosis are located on a megaplasmid in *Rhizobium meliloti*. Mol Gen Genetics 184:326–333

Rozahon M, Ismayil N, Hamood B, Erkin R, Abdurahman M, Mamtimin H, Abdukerim M, Lal R, Rahman E (2014) *Rhizobium populi* sp. nov., an endophytic bacterium isolated from *Populus euphratica*. Int J Syst Evol Microbiol 64:3215–3221

Saïdi S, Ramírez-Bahena MH, Santillana N, Zúñiga D, Álvarez-Martínez E, Peix A, Mhamdi R, Velázquez E (2014) *Rhizobium laguerreae* sp. nov. nodulates *Vicia faba* on several continents. Int J Syst Evol Microbiol 64:242–247

Sawana A, Adeolu M, Gupta RS (2014) Molecular signatures and phylogenomic analysis of the genus *Burkholderia*: proposal for division of this genus into the emended genus *Burkholderia* containing pathogenic organisms and a new genus *Paraburkholderia* gen. nov. harboring environmental species. Front Genet 5:429

Schofield PR, Watson JM (1986) DNA sequence of *Rhizobium trifolii* nodulation genes reveals a reiterated and potentially regulatory sequence preceding *nodABC* and *nodFE*. Nucl Acids Res 14:2891–2903

Scholla MH, Elkan GH (1984) *Rhizobium fredii* sp. nov., a fast-growing species that effectively nodulates soybeans. Int J Syst Bacteriol 34:484–486

Segovia L, Young JP, Martínez-Romero E (1993) Reclassification of American *Rhizobium leguminosarum* biovar phaseoli type I strains as *Rhizobium etli* sp. nov. Int J Syst Bacteriol 43:374–377

Shamseldin A, Carro L, Peix A, Velázquez E, Moawad H, Sadowsky MJ (2016) The symbiovar trifolii of *Rhizobium bangladeshense* and *Rhizobium aegyptiacum* sp. nov. nodulate *Trifolium alexandrinum* in Egypt. Syst Appl Microbiol. doi:10.1016/j.syapm.2016.05.002

Sheu SY, Chou JH, Bontemps C, Elliott GN, Gross E, James EK, Sprent JI, Young JP, Chen WM (2012) *Burkholderia symbiotica* sp. nov., isolated from root nodules of *Mimosa* spp. native to north-east Brazil. Int J Syst Evol Microbiol 62:2272–2278

Sheu SY, Chen MH, Liu WY, Andrews M, James EK, Ardley JK, De Meyer SE, James TK, Howieson JG, Coutinho BG, Chen WM (2015a) *Burkholderia dipogonis* sp. nov., isolated from root nodules of *Dipogon lignosus* in New Zealand and Western Australia. Int J Syst Evol Microbiol 65:4716–4723

Sheu SY, Chou JH, Bontemps C, Elliott GN, Gross E, dos Reis Junior FB, Melkonian R, Moulin L, James EK, Sprent JI, Young JP, Chen WM (2015b) *Burkholderia diazotrophica* sp. nov., isolated from root nodules of *Mimosa* spp. Int J Syst Evol Microbiol 63:435–441

Sheu SY, Huang HW, Young CC, Chen WM (2015c) *Rhizobium alvei* sp. nov., isolated from a freshwater river. Int J Syst Evol Microbiol 65:472–478

Sheu SY, Chen ZH, Young CC, Chen WM (2016) *Rhizobium ipomoeae* sp. nov., isolated from a water convolvulus field. Int J Syst Evol Microbiol 66:1633–1640

Shi X, Li C, Zhao L, Si M, Zhu L, Xin K, Chen C, Wang Y, Shen X, Zhang L (2016) *Rhizobium gei* sp. nov., a bacterial endophyte of *Geum aleppicum*. Int J Syst Evol Microbiol 66:4282–4288

Shiraishi A, Matsushita N, Hougetsu T (2010) Nodulation in black locust by the Gammaproteobacteria *Pseudomonas* sp. and the Betaproteobacteria *Burkholderia* sp. Syst Appl Microbiol 33:269–274 da Silva K, Florentino LA, Barroso da Silva KB, de Brandt E, Vandamme P, de Souza Moreira FM (2012) *Cupriavidus necator* isolates are able to fix nitrogen in symbiosis with different legume species. Syst Appl Microbiol 35:175–182

da Silva K, De Meyer SE, Rouws LF, Farias EN, dos Santos MA, O'Hara G, Ardley JK, Willems A, Pitard RM, Zilli JE (2014) *Bradyrhizobium ingae* sp. nov., isolated from effective nodules of *Inga laurina* grown in Cerrado soil. Int J Syst Evol Microbiol 64:3395–3401

Silva FV, De Meyer SE, Simões-Araújo JL, Barbé Tda C, Xavier GR, O'Hara G, Ardley JK, Rumjanek NG, Willems A, Zilli JE (2014) *Bradyrhizobium manausense* sp. nov., isolated from effective nodules of *Vigna unguiculata* grown in Brazilian Amazonian rainforest soils. Int J Syst Evol Microbiol 64:2358–2363

Skerman VBD, McGowan V, Sneath PHA (1980) Approved lists of bacterial names. Int J Syst Bacteriol 30:225–420

Sokal RR, Sneath P (1963) Principles of numerical taxonomy. WH Freeman, San Francisco

Souza Moreira MF, Cruz L, Miana de Faria S, Marsh T, Martínez-Romero E, de Oliveira Pedrosa F, Pitard MR, Young PWJ (2006) *Azorhizobium doebereinerae* sp. nov. microsymbiont of *Sesbania virgata* (Caz.) Pers. Syst Appl Microbiol 29:197–206

Squartini A, Struffi P, Döring H, Selenska-Pobell S, Tola E, Giacomini A, Vendramin E, Velázquez E, Mateos PF, Martínez-Molina E, Dazzo FB, Casella S, Nuti MP (2002) *Rhizobium sullae* sp. nov. (formerly '*Rhizobium hedysari*'), the root-nodule microsymbiont of *Hedysarum coronarium* L. Int J Syst Evol Microbiol 52:1267–1276

Steenkamp ET, Stepkowski T, Przymusiak A, Botha WJ, Law IJ (2008) Cowpea and peanut in southern Africa are nodulated by diverse *Bradyrhizobium* strains harboring nodulation genes that belong to the large pantropical clade common in Africa. Mol Phylogenet Evol 48:1131–1144

Steenkamp ET, van Zyl E, Beukes CW, Avontuur JR, Chan WY, Palmer M, Mthombeni LS, Phalane FL, Sereme TK, Venter SN (2015) *Burkholderia kirstenboschensis* sp. nov. nodulates papilionoid legumes indigenous to South Africa. Syst Appl Microbiol 38:545–554

Stepkowski T, Hughes CE, Law IJ, Markiewicz L, Gurda D, Chlebicka A, Moulin L (2007) Diversification of lupine *Bradyrhizobium* strains: evidence from nodulation gene trees. Appl Environ Microbiol 73:3254–3264

Sullivan JT, Trzebiatowski JR, Cruickshank RW, Gouzy J, Brown SD, Elliot RM, Fleetwood DJ, McCallum NG, Rossbach U, Stuart GS, Weaver JE, Webby RJ, De Bruijn FJ, Ronson CW (2002) Comparative sequence analysis of the symbiosis island of *Mesorhizobium loti* strain R7A. J Bacteriol 184:3086–3095

Sy A, Giraud E, Jourand P, Garcia N, Willems A, de Lajudie P, Prin Y, Neyra M, Gillis M, Boivin-Masson C, Dreyfus B (2001) Methylotrophic *Methylobacterium* bacteria nodulate and fix nitrogen in symbiosis with legumes. J Bacteriol 183:214–220

Talbi C, Delgado MJ, Girard L, Ramírez-Trujillo A, Caballero-Mellado J, Bedmar EJ (2010) *Burkholderia phymatum* strains capable of nodulating *Phaseolus vulgaris* are present in Moroccan soils. Appl Environ Microbiol 76:4587–4591

Tan ZY, Kan FL, Peng GX, Wang ET, Reinhold-Hurek B, Chen WX (2001) *Rhizobium yanglingense* sp. nov., isolated from arid and semi-arid regions in China. Int J Syst Evol Microbiol 51:909–914

Taulé C, Zabaleta M, Mareque C, Platero R, Sanjurjo L, Sicardi M, Frioni L, Battistoni F, Fabiano E (2012) New betaproteobacterial *Rhizobium* strains able to efficiently nodulate *Parapiptadenia rigida* (Benth.) Brenan. Appl Environ Microbiol 78:1692–1700

Tian CF, Wang ET, Wu LJ, Han TX, Chen WF, Gu CT, Gu JG, Chen WX (2008) *Rhizobium fabae* sp. nov., a bacterium that nodulates *Vicia faba*. Int J Syst Evol Microbiol 58:2871–2875

Toledo I, Lloret L, Martínez-Romero E (2003) *Sinorhizobium americanus* sp. nov., a new *Sinorhizobium* species nodulating native Acacia spp. in Mexico. Syst Appl Microbiol 26:54–64

Torres Tejerizo G, Rogel MA, Ormeño-Orrillo E, Althabegoiti MJ, Nilsson JF, Niehaus K, Schlüter A, Pühler A, Del Papa MF, Lagares A, Martínez-Romero E, Pistorio M (2016) *Rhizobium favelukesii* sp. nov., isolated from the root nodules of alfalfa (*Medicago sativa* L.) Int J Syst Evol Microbiol 66:4451–4457

Trujillo ME, Willems A, Abril A, Planchuelo AM, Rivas R, Ludeña D, Mateos PF, Martínez-Molina E, Velázquez E (2005) Nodulation of *Lupinus albus* by strains of *Ochrobactrum lupini* sp. nov. Appl Environ Microbiol 71:1318–1327

Turdahon M, Osman G, Hamdun M, Yusuf K, Abdurehim Z, Abaydulla G, Abdukerim M, Fang C, Rahman E (2013) *Rhizobium tarimense* sp. nov., isolated from soil in the ancient Khiyik River. Int J Syst Evol Microbiol 63:2424–2429

Validation List no. 107 (2006) List of new names and new combinations previously effectively, but not validly, published. Int J Syst Evol Microbiol 56:1–6

Valverde A, Velázquez E, Fernández-Santos F, Vizcaíno N, Rivas R, Mateos PF, Martínez-Molina E, Igual JM, Willems A (2005) *Phyllobacterium trifolii* sp. nov., nodulating *Trifolium* and *Lupinus* in Spanish soils. Int J Syst Evol Microbiol 55:1985–1989

Valverde A, Igual JM, Peix A, Cervantes E, Velázquez E (2006) *Rhizobium lusitanum* sp. nov. a bacterium that nodulates *Phaseolus vulgaris*. Int J Syst Evol Microbiol 56:2631–2637

Vandamme P, Coenye T (2004) Taxonomy of the genus *Cupriavidus:* a tale of lost and found. Int J Syst Evol Microbiol 54:2285–2289

Vandamme P, Goris J, Chen WM, de Vos P, Willems A (2002) *Burkholderia tuberum* sp. nov. and *Burkholderia phymatum* sp. nov., nodulate the roots of tropical legumes. Syst Appl Microbiol 25:507–512

Velázquez E, Igual JM, Willems A, Fernández MP, Muñoz E, Mateos PF, Abril A, Toro N, Normand P, Cervantes E, Gillis M, Martínez-Molina E (2001) *Mesorhizobium chacoense* sp. nov., a novel species that nodulates *Prosopis alba* in the Chaco Arido region (Argentina). Int J Syst Evol Microbiol 51:1011–1021

Velázquez E, Peix A, Zurdo-Piñeiro JL, Palomo JL, Mateos PF, Rivas R, Muñoz-Adelantado E, Toro N, García-Benavides P, Martínez-Molina E (2005) The coexistence of symbiosis and pathogenicity-determining genes in *Rhizobium rhizogenes* strains enables them to induce nodules and tumours or hairy roots in plants. Mol Plant Microbe Interact 18:1325–1332

Velázquez E, García-Fraile P, Ramírez-Bahena MH, Rivas R, Martínez-Molina E (2010a) Bacteria involved in nitrogen-fixing legume symbiosis: current taxonomic perspective. In: Khan MS, Zaidi A, Mussarrat J (eds) Microbes for legume improvement. Springer, Germany, pp 1–25

Velázquez E, Palomo JL, Rivas R, Guerra H, Peix A, Trujillo ME, García-Benavides P, Mateos PF, Wabiko H, Martínez-Molina E (2010b) Analysis of core genes supports the reclassification of strains *Agrobacterium radiobacter* K84 and *Agrobacterium tumefaciens* AKE10 into the species *Rhizobium rhizogenes*. Syst Appl Microbiol 33:247–251

Verástegui-Valdés MM, Zhang YJ, Rivera-Orduña FN, Cheng HP, Sui XH, Wang ET (2014) Microsymbionts of *Phaseolus vulgaris* in acid and alkaline soils of Mexico. Syst Appl Microbiol 37:605–612

Vidal C, Chantreuil C, Berge O, Mauré L, Escarré J, Béna G, Brunel B, Cleyet-Marel JC (2009) *Mesorhizobium metallidurans* sp. nov., a metal-resistant symbiont of *Anthyllis vulneraria* growing on metallicolous soil in Languedoc, France. Int J Syst Evol Microbiol 59:850–855

Villegas MC, Rome S, Mauré L, Domergue O, Gardan L, Bailly X, Cleyet-Marel JC, Brunel B (2006) Nitrogen-fixing sinorhizobia with *Medicago laciniata* constitute a novel biovar (bv. medicaginis) of *S. meliloti*. Syst Appl Microbiol 29:526–538

Vinuesa P, León-Barrios M, Silva C, Willems A, Jarabo-Lorenzo A, Pérez-Galdona R, Werner D, Martínez-Romero E (2005) *Bradyrhizobium canariense* sp. nov., an acid-tolerant endosymbiont that nodulates endemic genistoid legumes (*Papilionoideae: Genisteae*) from the Canary Islands, along with *Bradyrhizobium japonicum* bv. *genistearum*, *Bradyrhizobium* genospecies alpha and *Bradyrhizobium genospecies* beta. Int J Syst Evol Microbiol 55:569–575

Wang ET, van Berkum P, Beyene D, Sui XH, Dorado O, Chen WX, Martínez-Romero E (1998) *Rhizobium huautlense* sp. nov., a symbiont of *Sesbania herbacea* that has a close phylogenetic relationship with *Rhizobium galegae*. Int J Syst Bacteriol 48:687–699

Wang ET, Rogel MA, García-de los Santos A, Martínez-Romero J, Cevallos MA, Martínez-Romero E (1999a) *Rhizobium etli* bv. mimosae, a novel biovar isolated from *Mimosa affinis*. Int J Syst Bacteriol 49:1479–1491

Wang ET, van Berkum P, Sui XH, Beyene D, Chen WX, Martínez-Romero E (1999b) Diversity of rhizobia associated with *Amorpha fruticosa* isolated from Chinese soils and description of *Mesorhizobium amorphae* sp. nov. Int J Syst Bacteriol 49:51–65

Wang ET, Tan ZY, Willems A, Fernández-López M, Reinhold-Hurek B, Martínez-Romero E (2002) *Sinorhizobium morelense* sp. nov., a *Leucaena leucocephala*-associated bacterium that is highly resistant to multiple antibiotics. Int J Syst Evol Microbiol 52:1687–1693

Wang FQ, Wang ET, Liu J, Chen Q, Sui XH, Chen WF, Chen WX (2007) *Mesorhizobium albiziae* sp. nov., a novel bacterium that nodulates *Albizia kalkora* in a subtropical region of China. Int J Syst Evol Microbiol 57:1192–1199

Wang F, Wang ET, Wu LJ, Sui XH, Li Y Jr, Chen WX (2011) *Rhizobium vallis* sp. nov., isolated from nodules of three leguminous species. Int J Syst Evol Microbiol 61:2582–2588

Wang JY, Wang R, Zhang YM, Liu HC, Chen WF, Wang ET, Sui XH, Chen WX (2013a) *Bradyrhizobium daqingense* sp. nov., isolated from soybean nodules. Int J Syst Evol Microbiol 63:616–624

Wang R, Chang YL, Zheng WT, Zhang D, Zhang XX, Sui XH, Wang ET, Hu JQ, Zhang LY, Chen WX (2013b) *Bradyrhizobium arachidis* sp. nov., isolated from effective nodules of *Arachis hypogaea* grown in China. Syst Appl Microbiol 36:101–105

Wang YC, Wang F, Hou BC, Wang ET, Chen WF, Sui XH, Chen WX, Li Y, Zhang YB (2013c) Proposal of *Ensifer psoraleae* sp. nov, *Ensifer sesbaniae* sp. nov., *Ensifer morelense* comb. nov. and *Ensifer americanum* comb. nov. Syst Appl Microbiol 36:467–473

Wang L, Cao Y, Wang ET, Qiao YJ, Jiao S, Liu ZS, Zhao L, Wei GH (2016) Biodiversity and biogeography of rhizobia associated with common bean (*Phaseolus vulgaris* L.) in Shaanxi Province. Syst Appl Microbiol 39:211–219

Wdowiak-Wróbel S, Małek W (2010) Following phylogenetic tracks of *Astragalus cicer* microsymbionts. Antonie van Leeuwenhoek 97:21–34

Wei GH, Wang ET, Tan ZY, Zhu ME, Chen WX (2002) *Rhizobium indigoferae* sp. nov. and *Sinorhizobium kummerowiae* sp. nov., respectively isolated from *Indigofera* spp. and *Kummerowia stipulacea*. Int J Syst Evol Microbiol 52:2231–2239

Wei GH, Tan ZY, Zhu ME, Wang ET, Han SZ, Chen WX (2003) Characterization of rhizobia isolated from legume species within the genera *Astragalus* and *Lespedeza* grown in the Loess Plateau of China and description of *Rhizobium loessense* sp. nov. Int J Syst Evol Microbiol 53:1575–1583

Wei G, Chen W, Zhu W, Chen C, Young JP, Bontemps C (2009) Invasive *Robinia pseudoacacia* in China is nodulated by *Mesorhizobium* and *Sinorhizobium* species that share similar nodulation genes with native American symbionts. FEMS Microbiol Ecol 68:320–328

Wei X, Yan S, Li D, Pang H, Li Y, Zhang J (2015) *Rhizobium helianthi* sp. nov., isolated from the rhizosphere of sunflower. Int J Syst Evol Microbiol 65:4455–4460

Wen Y, Zhang J, Yan Q, Li S, Hong Q (2011) *Rhizobium phenanthrenilyticum* sp. nov., a novel phenanthrene-degrading bacterium isolated from a petroleum residue treatment system. J Gen Appl Microbiol 57:319–329

Willems A, Fernández-López M, Muñoz-Adelantado E, Goris J, De Vos P, Martínez-Romero E, Toro N, Gillis M (2003) Description of new *Ensifer* strains from nodules and proposal to transfer *Ensifer adhaerens* Casida 1982 to *Sinorhizobium* as *Sinorhizobium adhaerens* comb. nov. Request for an opinion. Int J Syst Evol Microbiol 53:1207–1217

Wilson JK (1939) Leguminous plants and their associated organisms. Cornell University Press, NY

Woese CR, Fox GE (1977) Phylogenetic structure of the prokaryotic domain: the primary kingdoms. Proc Natl Acad Sci U S A 74:5088–5090

Woese CR, Stackebrandt E, Weisburg WG, Paster BJ, Madigan MT, Fowler VJ, Hahn CM, Blanz P, Gupta R, Nealson KH, Fox GE (1984) The phylogeny of purple bacteria: the alpha subdivision. Syst Appl Microbiol 5:315–326

Xu LM, Ge C, Cui Z, Li J, Fan H (1995) *Bradyrhizobium liaoningense* sp. nov., isolated from the root nodules of soybeans. Int J Syst Bacteriol 45:706–711

Xu L, Shi JF, Zhao P, Chen WM, Qin W, Tang M, Wei GH (2011) *Rhizobium sphaerophysae* sp. nov., a novel species isolated from root nodules of *Sphaerophysa salsula* in China. Antonie van Leeuwenhoek 99:845–854

Xu L, Zhang Y, Deng ZS, Zhao L, Wei XL, Wei GH (2013) *Rhizobium qilianshanense* sp. nov., a novel species isolated from root nodule of *Oxytropis ochrocephala* Bunge in China. Antonie van Leeuwenhoek 103:559–565

Yan H, Yan J, Sui XH, Wang ET, Chen WX, Zhang XX, Chen WF (2016) *Ensifer glycinis* sp. nov., an novel rhizobial species associated with *Glycine* spp. Int J Syst Evol Microbiol. doi:10.1099/ijsem.0.001120

Yan J, Yan H, Liu LX, Chen WF, Zhang XX, Verástegui-Valdés MM, Wang ET, Han XZ (2017) *Rhizobium hidalgonense* sp. nov., a nodule endophytic bacterium of *Phaseolus vulgaris* in acid soil. Arch Microbiol 199:97–104

Yanagi M, Yamasato K (1993) Phylogenetic analysis of the family *Rhizobiaceae* and related bacteria by sequencing of 16S rRNA gene using PCR and DNA sequencer. FEMS Microbiol Lett 107:115–120

Yao ZY, Kan FL, Wang ET, Wei GH, Chen WX (2002) Characterization of rhizobia that nodulate legume species of the genus *Lespedeza* and description of *Bradyrhizobium yuanmingense* sp. nov. Int J Syst Evol Microbiol 52:2219–2230

Yao LJ, Shen YY, Zhan JP, Xu W, Cui GL, Wei GH (2012) *Rhizobium taibaishanense* sp. nov., isolated from a root nodule of *Kummerowia striata*. Int J Syst Evol Microbiol 62:335–341

Yao Y, Sui XH, Zhang XX, Wang ET, Chen WX (2015) *Bradyrhizobium erythrophlei* sp. nov. and *Bradyrhizobium ferriligni* sp. nov., isolated from effective nodules of *Erythrophleum fordii*. Int J Syst Evol Microbiol 65:1831–1837

Yoon JH, Kang SJ, Yi HS, Oh TK, Ryu CM (2010) *Rhizobium soli* sp. nov., isolated from soil. Int J Syst Evol Microbiol 60:1387–1393

Young JM (2003) The genus name Ensifer Casida 1982 takes priority over Sinorhizobium Chen et al. 1988, and *Sinorhizobium morelense* Wang et al. 2002 is a later synonym of *Ensifer adhaerens* Casida 1982. Is the combination "Sinorhizobium adhaerens" Casida 1982 Willems et al. 2003 legitimate? Request for an opinion. Int J Syst Evol Microbiol 53:2107–2110

Young JM, Kuykendall LD, Martínez -Romero E, Kerr A, Sawada H (2001) A revision of *Rhizobium* Frank 1889, with an emended description of the genus, and the inclusion of all species of *Agrobacterium* Conn 1942 and *Allorhizobium undicola* de Lajudie et al. 1998 as new combinations: *Rhizobium radiobacter, R. rhizogenes, R. rubi, R. undicola* and *R. vitis*. Int J Syst Evol Microbiol 51:89–103

Yu X, Cloutier S, Tambong JT, Bromfield ES (2014) *Bradyrhizobium ottawaense* sp. nov., a symbiotic nitrogen fixing bacterium from root nodules of soybeans in Canada. Int J Syst Evol Microbiol 64:3202–3207

Yuan CG, Jiang Z, Xiao M, Zhou EM, Kim CJ, Hozzein WN, Park DJ, Zhi XY, Li WJ (2016) *Mesorhizobium sediminum* sp. nov., isolated from deep-sea sediment. Int J Syst Evol Microbiol 66:4797–4802

Zhang RJ, Hou BC, Wang ET, Li Y Jr, Zhang XX, Chen WX (2011a) *Rhizobium tubonense* sp. nov., isolated from root nodules of *Oxytropis glabra*. Int J Syst Evol Microbiol 61:512–517

Zhang GX, Ren SZ, Xu MY, Zeng GQ, Luo HD, Chen JL, Tan ZY, Sun GP (2011b) *Rhizobium borbori* sp. nov., aniline-degrading bacteria isolated from activated sludge. Int J Syst Evol Microbiol 61:816–822

Zhang X, Sun L, Ma X, Sui XH, Jiang R (2011c) *Rhizobium pseudoryzae* sp. nov., isolated from the rhizosphere of rice. Int J Syst Evol Microbiol 61:2425–2429

Zhang YM, Li Y Jr, Chen WF, Wang ET, Sui XH, Li QQ, Zhang YZ, Zhou YG, Chen WX (2012a) *Bradyrhizobium huanghuaihaiense* sp. nov., an effective symbiotic bacterium isolated from soybean (*Glycine max* L.) nodules. Int J Syst Evol Microbiol 62:1951–1957

Zhang JJ, Liu TY, Chen WF, Wang ET, Sui XH, Zhang XX, Li Y, Li Y, Chen WX (2012b) *Mesorhizobium muleiense* sp. nov., nodulating with *Cicer arietinum* L. Int J Syst Evol Microbiol 62:2737–2742

Zhang X, Li B, Wang H, Sui X, Ma X, Hong Q, Jiang R (2012c) *Rhizobium petrolearium* sp. nov., isolated from oil-contaminated soil. Int J Syst Evol Microbiol 62:1871–1876

Zhang XX, Tang X, Sheirdil RA, Sun L, Ma XT (2014a) *Rhizobium rhizoryzae* sp. nov., isolated from rice roots. Int J Syst Evol Microbiol 64:1373–1377

Zhang JJ, Yu T, Lou K, Mao PH, Wang ET, Chen WF, Chen WX (2014b) Genotypic alteration and competitive nodulation of *Mesorhizobium muleiense* against exotic chickpea rhizobia in alkaline soils. Syst Appl Microbiol 37:520–524

Zhang L, Shi X, Si M, Li C, Zhu L, Zhao L, Shen X, Wang Y (2014c) *Rhizobium smilacinae* sp. nov., an endophytic bacterium isolated from the leaf of *Smilacina japonica*. Antonie van Leeuwenhoek 106:715–723

Zhang XX, Gao JS, Cao YH, Sheirdil RA, Wang XC, Zhang L (2015a) *Rhizobium oryzicola* sp. nov., potential plant-growth-promoting endophytic bacteria isolated from rice roots. Int J Syst Evol Microbiol 65:2931–2936

Zhang YJ, Zheng WT, Everall I, Young JP, Zhang XX, Tian CF, Sui XH, Wang ET, Chen WX (2015b) *Rhizobium anhuiense* sp. nov., isolated from effective nodules of *Vicia faba* and *Pisum sativum*. Int J Syst Evol Microbiol 65:2960–2967

Zhao CT, Wang ET, Zhang YM, Chen WF, Sui XH, Chen WX, Liu HC, Zhang XX (2012) *Mesorhizobium silamurunense* sp. nov., isolated from root nodules of *Astragalus* species. Int J Syst Evol Microbiol 62:2180–2186

Zhao JJ, Zhang J, Sun L, Zhang RJ, Zhang CW, Yin HQ, Zhang XX (2016) *Rhizobium oryziradicis* sp. nov., isolated from the root of rice. Int J Syst Evol Microbiol. doi:10.1099/ijsem.0.001724

Zheng WT, Li Y Jr, Wang R, Sui XH, Zhang XX, Zhang JJ, Wang ET, Chen WX (2013) *Mesorhizobium qingshengii* sp. nov., isolated from effective nodules of *Astragalus sinicus*. Int J Syst Evol Microbiol 63:2002–2007

Zhou PF, Chen WM, Wei GH (2010) *Mesorhizobium robiniae* sp. nov., isolated from root nodules of *Robinia pseudoacacia*. Int J Syst Evol Microbiol 60:2552–2556

Zhou S, Li Q, Jiang H, Lindström K, Zhang X (2013) *Mesorhizobium sangaii* sp. nov., isolated from the root nodules of *Astragalus luteolus* and *Astragalus ernestii*. Int J Syst Evol Microbiol 63:2794–2799

Zhu YJ, Kun J, Chen YL, Wang SK, Sui XH, Kang LH (2015) *Mesorhizobium acaciae* sp. nov., isolated from root nodules of *Acacia melanoxylon* R. Br. Int J Syst Evol Microbiol 65:3558–3563

Zilli JE, Baraúna AC, da Silva K, De Meyer SE, Farias EN, Kaminski PE, da Costa IB, Ardley JK, Willems A, Camacho NN, Dourado Fdos S, O'Hara G (2014) *Bradyrhizobium neotropicale* sp. nov, isolated from effective nodules of *Centrolobium paraense*. Int J Syst Evol Microbiol 64:3950–3957

Zurdo-Piñeiro JL, Rivas R, Trujillo ME, Vizcaíno N, Carrasco JA, Chamber M, Palomares A, Mateos PF, Martínez-Molina E, Velázquez E (2007) *Ochrobactrum cytisi* sp. nov., isolated from nodules of *Cytisus scoparius* in Spain. Int J Syst Evol Microbiol 57:784–788

Zurdo-Piñeiro JL, García-Fraile P, Rivas R, Peix A, León-Barrios M, Willems A, Mateos PF, Martínez-Molina E, Velázquez E, van Berkum P (2009) Rhizobia from Lanzarote, the Canary Islands, that nodulate *Phaseolus vulgaris* have characteristics in common with *Sinorhizobium meliloti* from mainland Spain. Appl Environ Microbiol 75:2354–2359

Zurkowski W, Lorkiewicz Z (1979) Plasmid-mediated control of nodulation in *Rhizobium trifolii*. Arch Microbiol 123:195–201

Recent Advances in the Active Biomolecules Involved in Rhizobia-Legume Symbiosis

Esther Menéndez, Pilar Martínez-Hidalgo, Luis R. Silva, Encarna Velázquez, Pedro F. Mateos, and Alvaro Peix

Abstract

The mutualistic interactions between nodule-forming rhizobia and specific legume host plants involve a series of signalling molecules leading to the establishment of a strong and functional symbiosis between the two partners. The competitive ability and legume host specificity of rhizobia together with the ability of both rhizobia and legumes to release functionally divergent active molecules determines the success of symbiotic relationships. Here, recent developments in the key active biomolecules affecting legume-rhizobia symbiosis are surveyed and discussed.

E. Menéndez • E. Velázquez • P.F. Mateos (✉)
Departamento de Microbiología y Genética and Instituto Hispano-Luso de Investigaciones Agrarias (CIALE), Universidad de Salamanca, 37007 Salamanca, Spain
e-mail: pfmg@usal.es

P. Martínez-Hidalgo
Departamento de Microbiología y Genética and Instituto Hispano-Luso de Investigaciones Agrarias (CIALE), Universidad de Salamanca, 37007 Salamanca, Spain

Department of Molecular, Cell and Developmental Biology, University of California, Los Angeles, Los Angeles, CA 90095-1606, USA

L.R. Silva
Health Sciences Research Center (CICS), Faculty of Health Sciences—University of Beira Interior, 6201-506 Covilhã, Portugal

A. Peix
Departamento de Desarrollo Sostenible de Sistemas Agroforestales y Ganaderos, Instituto de Recursos Naturales y Agrobiología, IRNASA-CSIC, Salamanca, Spain

© Springer International Publishing AG 2017
A. Zaidi et al. (eds.), *Microbes for Legume Improvement*,
DOI 10.1007/978-3-319-59174-2_2

2.1   Introduction

Nitroge-fixing bacteria (often called rhizobia) form symbiotic relationships with legumes under different environmental conditions. Initially, rhizobia infect host legumes, which requires a full coordination between the two partners. The successful establishment of rhizobia-legume symbiosis requires rhizobia to be competitive enough to infect specifically and effectively the legume plants (Fig. 2.1). This interaction between both symbionts is mediated by certain signalling molecules released during this process by the bacterium, the host legumes, or both interacting partners (Table 2.1). Signalling during the first stages is critical where flavonoids, Nod factors, and other molecules play an important role in the establishment of symbiosis. This molecular dialog continues with a complex exchange of molecules between plant and microorganism (Janczarek et al. 2015a). The host plant produces flavonoids, kinases, lectins, and other molecules, while the bacterium is responsible for the production of Nod factors that trigger the organogenesis of the nodule in the plant (Broghammer et al. 2012), cellulose, and exopolysaccharides, important for infection and release of bacterium inside host (Kelly et al. 2013). Quorum sensing-related compounds that play some important roles in the symbiotic process (Cubo et al. 1992; Loh and Stacey 2003; Marketon et al. 2003) modulate the concentration of bacteria in and around the plant roots and nodules. As a consequence, rhizobial

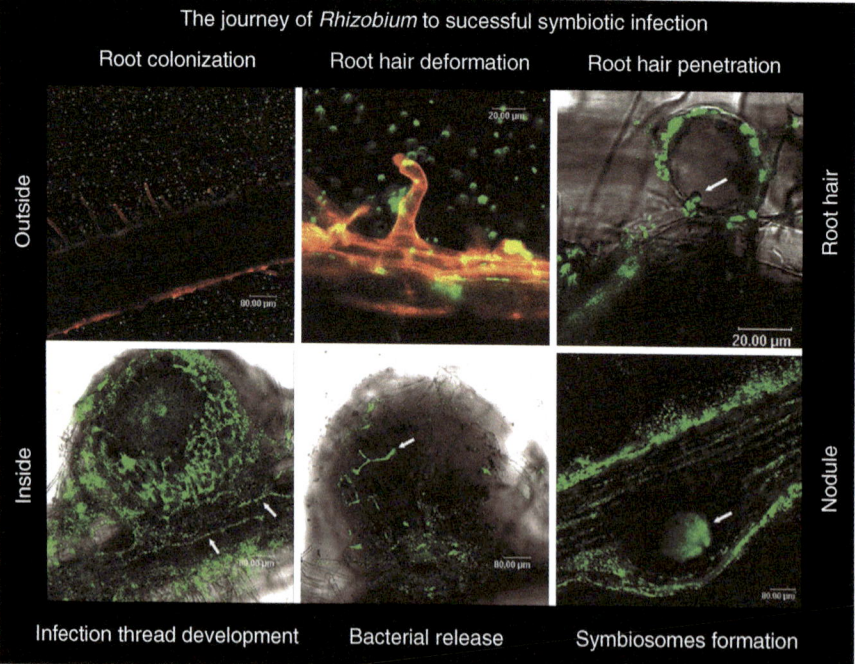

Fig. 2.1 The journey of *Rhizobium* to successful symbiotic infection. Cascade of events outside and inside the plant root rhizobia-legume symbiosis establishment

Table 2.1 Examples of some active signalling biomolecules affecting rhizobia-legume symbiosis

Signalling molecules	Functions in the rhizobia-legume symbiosis	Target stage of the symbiotic process
Produced by plant		
Flavonoids flavonones, isoflavones, flavanones, flavonols, anthocyanins, chalcones	Attraction of rhizobia by the plant	Initial stages of the symbiotic process
Non-flavonoids Jasmonates, strigolactones, simple phenolic compounds	Modulates the level of rhizobial Nod factor production	Initial stages of the symbiotic process
Nod receptors Plant receptor kinases	Rhizobial recognition	Initial stages of the symbiotic process
Remorins and flotillins	Regulation of rhizobial infection	Preinfection, infection, and bacterial release into nodule cell stages (nodulation)
Lectins	Rhizobial attachment to plant roots	Early steps of plant infection
Plant hydrolytic enzymes—polygalacturonases/pectins	Softening of root hair cell wall	Infection thread initiation and elongation stages
Nodulins (early and late)	Initiation and maintenance of nodulation	Infection/invasion proccess. Cortex cell differentiation (nodule initiation; *noi*). Initiation of N_2 fixation activity
Nodule-specific cysteine-rich (NCR) peptides	Promotion of rhizobial differentiation to bacteroid form	Later stages of nodulation process
Produced by rhizobia		
Exopolysaccharides (EPS, LPS, KPS, GPS, NP (glucomannan), cyclic ß-glucan, cellulose microfibrils)	Protection against stresses, attachment to surfaces, and nutrient gathering/enhancement of contact of rhizobia to root surfaces	Early steps of plant infection. Later stages of nodulation process (penetration of infection threads, N_2-fixing phenotype)
Nod factors (lipochitinoligosaccharides)	Activation of plant receptor kinases; signalling	Initial stages of the symbiotic process
Rhizobial hydrolytic enzymes (cellulases and pectinases/polygalacturonases)	Erosion of root hair tips (cell walls at the tip)/softening of root hair cell walls	Infection and rhizobial release stages
Rhizobial IAA	Establishment of the symbiotic proccess	Early stages of symbiotic process/contribution to hormonal balance

(continued)

Table 2.1 (continued)

Signalling molecules	Functions in the rhizobia-legume symbiosis	Target stage of the symbiotic process
Rhizobial AHLs	Establishment of the symbiotic proccess)	Early stages of symbiotic process
(NO and H_2O_2)	Establishment of the symbiotic proccess)/nodule senescence	Early stages of symbiotic proccess and/or further nodulation stages
Nitrogenase	Reduction of N_2 to ammonia	Atmospheric N_2 fixation in bacteroid forms
Hydrogenase	Hydrogen recycling	Atmospheric N_2 fixation in bacteroid forms

cell density is increased (González and Marketon 2003). The quorum sensing in rhizobia has been recently reviewed (Rinaudi-Marron and González 2013; Bogino et al. 2015) and reported in many rhizobia, for example, *R. leguminosarum* bv. viciae (Cantero et al. 2006; McAnulla et al. 2007; Edwards et al. 2009), *Rhizobium etli* (Daniels et al. 2006; Braeken et al. 2008), *Ensifer (Sinorhizobium) meliloti* (Pellock et al. 2002), and *Mesorhizobium huakuii* (Wang et al. 2004), *Mesorhizobium loti* (Yang et al. 2009), or *Mesorhizobium tianshanense* (Cao et al. 2009). It involves the use of acylated homoserine lactones (AHLs) as signal molecules, which is a general mechanism found in Gram-negative bacteria (González and Marketon 2003). Here, a sincere attempt is made to highlight the role of signalling molecules in *Rhizobium*-legume symbiosis.

2.2 Flavonoids and Nod Factors

Two molecules are mainly involved during the early stages of infection, flavonoids, which are synthesized by the plant and nodulation factors (NFs), produced by the microsymbiont. The host releases flavonoids in a signalling process, which are perceived by its rhizobial symbionts. These flavonoids are derived mostly from p-coumaroyl-CoA and malonyl-CoA, belonging to a secondary plant metabolites family synthesized via phenylpropanoid and acetate-malonate pathways (Gibson et al. 2008; Hassan and Mathesius 2012; Sugiyama and Yazaki 2014). Flavonoids, synthesized and secreted by legume roots, act as signalling molecules during the early stages of rhizobia-legume interactions, attract rhizobia toward host plants, and induce the expression of nodulation (*nod*) genes of rhizobia (Subramanian et al. 2007; Wei et al. 2008; el Zahar Haichar et al. 2014; Janczarek et al. 2015a). Nodulation genes express Nod factors (lipochitooligosaccharides), which act as receptors for the plant flavonoid signal, being the interaction flavonoids-NodD an essential point in the early symbiotic interaction (Oldroyd et al. 2011; Wang et al. 2012; Oldroyd 2013). The *nodD* gene inducers include different types of flavonoids, such as flavones, isoflavones, flavanones, flavonols, chalcones, and anthocyanins (reviewed by Sugiyama and Yazaki 2014; Janczarek et al. 2015a). Of these molecules, flavones and flavonols have been found to play vital roles during nodulation of *Medicago truncatula* inoculated by *Ensifer (Sinorhizobium) meliloti* (Zhang et al. 2009). Some other new functions have also been assigned to flavonoids (Buer et al. 2010; Weston and Mathesius 2013). For example, some flavonoids are anti-inducers of *nod* gene expression, mainly by modulation of flavone-dependent Nod factor synthesis (Zuanazzi et al. 1998; Zhang et al. 2009; Hassan and Mathesius 2012). In other studies naringenin has been reported to induce the expression of *nod* genes in *R. leguminosarum-Pisum sativum* interaction, whereas quercetin inhibited the nodulation (Novák et al. 2002). Moreover, naringenin regulates the expression of genes involved in cell wall synthesis in other species of nitrogen-fixing bacteria (Tadra-Sfeir et al. 2011). The isoflavone daidzein can induce *nod* gene expression in *B. japonicum* (nodulating soybean) inhibiting by contrast the expression in *R. leguminosarum* strains nodulating clover or pea, thus contributing to host specificity (Andersen and Markham 2006). In the case of *E. meliloti-Medicago sativa* symbiosis, in the presence of luteolin, NodD1 exhibited increased binding to *nod* gene promoters compared to binding in the absence of luteolin. The flavonoids naringenin, eriodictyol, and daidzein did not stimulate *nod* gene expression in *E. meliloti*, but stimulated an increase in the DNA binding affinity of NodD1 to *nod* gene promoters. In vivo competition assays demonstrated that these noninducing flavonoids act as competitive inhibitors of luteolin, suggesting that both inducing and noninducing flavonoids are able to directly bind to NodD1 and mediate conformational changes at *nod* gene promoters, but only luteolin is capable of promoting the downstream changes necessary for *nod* gene induction (Peck et al. 2006). Moreover, flavonoids influence competitiveness of rhizobia and their symbiotic activity, since competitive strains of *R. leguminosarum* showed a better response to a wide range of synthetic flavonoids and seed exudates, increasing some plant growth parameters (Maj et al. 2010).

Indeed, the *nod* gene expression may sufficiently be induced by exogenous isoflavones, but endogenous isoflavones have also been found to be indispensable for the development of symbiosis, for example, between soybean and *B. japonicum*. Expression of isoflavone synthase (IFS), a key enzyme involved in the synthesis of isoflavones, is explicitly induced by *B. japonicum*. The silencing of IFS in soybean plants, however, severely declined nodulation. Also, inoculation of *B. japonicum* or treatment of soybean roots with isoflavones, daidzein, or genistein failed to restore normal nodulation process (Subramanian et al. 2006). Similarly, RNAi silencing assays in *Medicago truncatula* suggested that flavonols and flavones play different roles in nodulation by *E. meliloti,* which depends on the rhizobia-legume system used (Zhang et al. 2009). Genistein extracted from soybean not only induces *nod* genes but functions differently, and about 100 genes were induced in *B. japonicum* (Lang et al. 2008). Various flavonoids such as apigenin, daidzein, genistein, hesperetin, kaempferol, luteolin, naringenin and rutin, hesperetin, and naringenin have been found as most effective molecules for plant-to-bacteria interaction in the case of *R. leguminosarum* symbiosis with *Pisum sativum* and *Lens culinaris* (Begum et al. 2001). Moreover, the combined application of hesperetin and apigenin significantly enhanced growth, nodulation, and nitrogen fixation of *R. tibeticum* inoculated *Trigonella foenum-graecum* (fenugreek) plants (Abd-Alla et al. 2014). In contrast, an unidentified chalcone was found as the better inducer of *nodD1* gene during *R. galegae-Galega orientalis* symbiosis, while apigenin or luteolin showed a moderate induction potential (Suominen et al. 2003). Peck et al. (2013) character- ized several strains with full or partial defect in the NodD1 protein of *Ensifer meli- loti*, presenting altered responses to luteolin, during the activation of *nod* gene expression.

Flavonoids bind bacterial NodD proteins and activate these proteins to induce the transcription of rhizobial genes (Perret et al. 2000; Barnett and Fisher 2006). The expression of these genes produces the Nod factors, which triggers the earliest stages of nodule development, including root hair deformation and curling, and cortical cell divisions of their legume host (Lerouge et al. 1990; Spaink et al. 1991; Cullimore et al. 2001). The backbone of Nod factors is ß-1,4-*N*-acetyl-D-glucosamine residues, which differs in number both among bacterial species and also within the repertoire of a single species. These Nod factors are identified by specific LysM domain receptor kinases, known as NFP protein in *M. truncatula* (Limpens et al. 2003; Amor et al. 2003). Apart from the LysM kinases, numerous other compo- nents, for instance, putative cations channel DMI1 (Ané et al. 2004), the leucine- rich-repeat-containing receptor kinase DMI2 (Limpens et al. 2005), and the calcium calmodulin-dependent kinase DMI3 (Lévy et al. 2004) of Nod factor-induced sig- nalling cascade have been identified and reported. In general, the Nod factor recep- tors trigger a signal transduction cascade which in turn induces early stages of symbiotic process like root hair deformation, preinfection thread formation, and the induction of cell division in the root cortex leading to the formation of nodule pri- mordium (Oldroyd 2013; Gough and Jacquet 2013; Libault 2015). Despite such activities, the addition of purified compatible Nod factors to plant roots has not been found enough to cause the formation of tightly curled root hairs (shepherd's crooks), a complete differentiation of the infection thread and mature nodules. This indicates that Nod factors are not the only required effector produced by these symbionts to enter plant tissues and colonize plant cells, but there are other factors that affect the symbiotic process (Gage 2002; Jones et al. 2007; Gibson et al. 2008). Some recent studies have, however, reported the discovery of ionic channels that maintains Ca^{2+} spiking (Charpentier et al. 2016) or new proteins and/or transcription factors involved in the rhizobia-legume recognition (Sinharoy et al. 2016; del Cerro et al. 2016; Fliegmann et al. 2016), which can help to better understand the symbiotic signalling pathways. Also, these findings are likely to answer many unresolved questions about the specificity and signalling between rhizobia and legumes (Remigi et al. 2016).

Even though, the Nod factors are considered the integral component of rhizobia- legume symbiosis, yet there are cases where nodulation has occurred even in the absence of Nod factors. As an example, the recent genome sequencing of some photosynthetic *Bradyrhizobium* strains, capable of forming nodules on the roots and stems of an aquatic host, *Aeschynomene sensitiva*, revealed that the common *nod- ABC* genes were absent in these species (Giraud et al. 2007). Moreover, other species from the genus *Aeschynomene*, *A. evenia*, was also described as a model for the study of NF-independent rhizobia-legume symbiosis (Arrighi et al. 2012; Fabre et al. 2015). Thus, the host initiated nodule development in a NF-independent manner and responded well to the bacterial purine derivatives having cytokinin-like activity, suggesting that the hormones secreted by legume host plants played an important role in nodule formation (el Zahar Haichar et al. 2014).

2.3   Cellulases and Polygalacturonases (Pectinases)

Infection of legumes is the first step in the development of *Rhizobium*-legume symbiosis, which requires that rhizobia must cross the root hair wall of host plants. For this, cell wall degradation process must be delicate so that the rhizobia enter inside the root cells without any damage to the root hairs and subsequent abortion of the infection process.

To understand the mechanistic basis of infection process, several hypotheses have been proposed:

1. Production of polygalacturonases by the rhizobia induces host plants—polygalacturonases soften the root hair wall at the infection sites and hence allow the bacteria to penetrate between microfibrils onto the cell membrane and initiate the infection thread formation (Ljunggren and Fahraeus 1961).
2. Rhizobia redirect the growth of root hair wall starting from the tip to the localized infection sites, resulting in invagination rather than the penetration through the root hair wall (Nutman et al. 1973).
3. Production of cell wall-degrading enzymes cause a localized degradation that completely crosses the root hair wall and allows bacteria to penetrate directly (Hubbell 1981).

Of all the three hypotheses, the role of hydrolytic enzymes in the infection process is widely accepted. Interestingly, a recent study showed that pectate lyases, related to polygalacturonases in terms of cell wall pectin hydrolysis, have a role in the infection process of *Lotus japonicus–Mesorhizobium loti* symbiotic system (Xie et al. 2012). Defective mutants in the so-called *L. japonicus nodulation pectate lyase* (LjNPL) gene showed that rhizobial infection is affected; however, a reduced number of infection threads were still present, suggesting its combined action with other enzymes of plant or rhizobial origin. Given the role of hydrolytic enzymes in the active penetration of plant cell walls, McCoy (1932) was the first who described the participation of hydrolytic enzymes in the rhizobial infection of legumes. However, author failed to pin point the contribution of these enzymes in the rhizobial interaction due largely to the unavailability of sensitive procedures to detect smallest concentration of cell wall-degrading enzymes. Since then, pectinolytic, cellulolytic, and hemicellulolytic activities in rhizobial pure cultures have been detected in numerous studies (Hubbell et al. 1978; Morales et al. 1984; Martínez-Molina et al. 1979). Generally, the activities of these enzymes are very low. Using improved and reliable increased sensitivity assays (Mateos et al. 1992), detected cellulases in all of the official type strains of rhizobia (Jiménez-Zurdo et al. 1996a; Robledo et al. 2008). The genes codifying for cellulases (*celC* genes) are located close to genes involved in cellulose biosynthesis, which suggest the hypothesis of a possible involvement of rhizobial cellulases in the colonization of legume roots. Also, molecular analyses show that *celC*-encoded protein has homologies to other rhizobial endoglucanases (*R. leguminosarum*, *R. etli*, and *E. medicae*, among others) (Robledo et al. 2012). In contrast, polygalacturonases are less common in rhizobia (Jiménez-Zurdo et al. 1996a). The model wild-type strain of clover-nodulating rhizobia, *R. leguminosarum* bv. trifolii ANU843, produces at least two cell-bound cellulase isozymes, CelC1 and CelC2 (Mateos et al. 1992). Engineering pSym-plasmid-cured and *nod*-recombinant derivatives of strain ANU843, cellulase CelC1 gene locus was localized on the symbiotic plasmid (pSym), outside the *nod* region, whereas the cellulase CelC2 gene locus was not located on the pSym (Jiménez-Zurdo et al. 1996b). Using a combination of phase contrast/polarized light microscopy and enzymology, we found that only cellulase CelC2 can completely erode the root hair wall at a highly localized site on the isotropic, noncrystalline apex of the root hair tip, forming the so-called HoT phenotype (*H*ole *O*n the *T*ip), and can extensively degrade clover root hair walls when grown in the presence rather than in the absence of clover rhizobia Nod factors (Mateos et al. 2001). These and other associated data suggest a complementary role of rhizobial cellulases and Nod factors in promoting root hair infectivity at strategic sites during primary host infection.

Rhizobial CelC2 is a 1,4-ß-D-endoglucanase (EC 3.2.1.4) with high substrate specificity for noncrystalline (amorphous) cellulose. This endoglucanase has an approximate molecular mass of 33.2 kDa, an optimal pH of 5, optimal temperature of 40 °C, and an apparent Km of 84.4 mg/ml for CMC as substrate (Mateos et al. 1992; Robledo et al. 2008). These biochemical characteristics restrict (and hence tightly control) the symbiotically relevant activity of CelC2 cellulase during primary host infection. The cell-bound location (rather than largely excreted), the high Km value, and the relatively low activity of CelC2 cellulase are features that would restrain its degradative action to root hairs, thereby minimizing indiscriminate host cell lyses and death. These features provide further opportunities to restrict its short-range action based on physical positioning of the bacterium at the host cell wall interface. The specificity of CelC2 cellulase for noncrystalline cellulose significantly restricts its in vivo erosion site to the highly localized root hair infection site that lacks crystalline cell wall architecture. The optimum pH (5) of CelC2 cellulase is consistent with the slightly acidic pH at the external surface of white clover root hairs. Finally, the host plant specificity, exhibited by the HoT biological activity of CelC2 cellulase which includes the compatible white clover legume but excludes the heterologous, nonhost legume alfalfa, is consistent with the host specificity of infection thread formation in legume root hairs (Robledo et al. 2008). Moreover, the application of purified CelC2 induced a complete erosion of the highly localized noncrystalline tip of the host root hair, forming a hole whose geometry and location match the entry point into white clover (primary host infection). *celC*-defective strains are unable to breach the host wall at the root hair tip, form infection threads within the host root hair, and induce effective nodules, indicating that this bacterial enzyme is absolutely required for development of the nitrogen-fixing *R. leguminosarum* bv. trifolii-*Trifolium repens* symbiosis. *celC*-complemented strain was able to revert this phenotype, providing further evidences of the requirement of this enzyme in successful development of the canonical *Rhizobium*-white clover symbiosis (Robledo et al. 2008). In more recent studies, Robledo et al. (2011) obtained a *celC*+ derivative, which constitutively overproduces the CelC2 enzyme. This derivative leads to the formation of aberrant phenotypes in clover root hair apex as a result of an extensive hydrolysis of the noncrystalline cellulose located at the root hair tip and also at infection thread ends. Moreover, this CelC2-overproducing strain produces an altered phenotype in nodules, which are not efficient in nitrogen fixation, and also elicits plant defense responses at the infection sites. Taken all together, these data suggest that there is a tight regulation over the CelC2 cellulase for a successful *R. leguminosarum* bv. trifolii-*T. repens* symbiotic interaction, and this particular enzyme, CelC2 cellulase, is required during rhizobia-legume interactions. Further studies are required to elucidate its role in other symbiotic systems as well as to determine host specificity. Moreover, the strategic location of *celC* genes, near putative cellulose synthase genes confined to a region of the chromosome (*celABC*), confirmed its involvement in bacterial cellulose biosynthesis (Robledo et al. 2012). This finding raises the possibility of new infection strategies based on cellulases associated with cellulose biosynthesis with a view to avoid the elicitation of plant defenses.

2.4 Lectins and Polysaccharides

Lectins are proteins that reversibly and non-enzymatically bind specific carbohydrates. They have been recently redefined by de Hoff et al. (2009) and are classified into different families according to their carbohydrate recognition domains (van Damme et al. 2004). Legume lectins constitute one of these families and bind different sugar residues, for example, concanavalin A, a lectin from *Canavalia ensiformis*, binds glucose/mannose residues, the *Glycine max* agglutinin binds N-acetyl-D-galactosamine/galactose, and *Ulex europaeus* lectin binds L-fucose (de Hoff et al. 2009). The legume lectins are the most extensively studied group (van Damme et al. 2004) and generally consist of two or four 25–30 kDa subunits that may present identical chains as for phytohemagglutinin (PHA; *Phaseolus vulgaris* L. lectin) or distinct, as occurs in the case of *Pisum sativum* L. lectin (Lioi et al. 2006).

The successful bacterial attachment to root hairs through plant lectins facilitates infection thread formation, a prerequisite for effective root nodule development (de Hoff et al. 2009). Several experiments with transgenic plants supported that lectins facilitate rather than direct the symbiosis (van Rhijn et al. 2001; Sreevidya et al. 2005). According to these experiments, the rhizobia preferentially get attached to root hair tips, a location where legume lectins are typically localized, and the authors hypothesized that recognition of lectin and enhanced attachment by rhizobia led to structural modifications of the cell wall, similar to a model proposed by Kijne et al.

(1997). It is now known that lectins in fact enhance this attachment, and nodulation can be inhibited when lectins are not present (Sytnikov 2013; Roberts et al. 2013). Additionally, it has been recently discovered that lectins purified from legume seeds can enhance the growth of *R. tropici* (de Vasconcelos et al. 2013).

A second phase in the attachment process implies a significantly increased force of adhesion of attached rhizobial cells concurrent with the formation of extracellular microfibrils that enhance the degree of contact of bacteria to the root hair surface (Dazzo et al. 1984). The extracellular microfibrils are cellulosic in nature and have been found in *R. leguminosarum* bv. trifolii cells colonizing white clover root epidermis (Mateos et al. 1995). Besides lipochitooligosaccharides (Nod factors) and cellulose, rhizobia produce surface polysaccharides, which protect rhizobia from environmental stress (Jaszek et al. 2014), protect against plant antimicrobial compounds (Becker and Pühler 1998; Fraysse et al. 2003), help rhizobia to attach to surfaces or nutrient gathering, and are also crucial for establishment of successful symbiosis (Fraysse et al. 2003; Skorupska et al. 2006). Involvement of polysaccharides in bacterial adherence to the tip of root hairs and the subsequent development of the infective process seems to be critical, because besides their role in competitiveness of strains, it has been demonstrated that mutants defective in their biosynthesis are characterized by low infectivity, a low capacity for nodulation, and in some cases changes in the host range (Gibson et al. 2008).

Polysaccharides produced by rhizobia may be of different types, such as lipopolysaccharides (LPS), capsular polysaccharides (CPS or K-antigens), acidic EPS, and neutral beta-1,2-glucans. EPSs have been reported to have a role in bacterial attachment and biofilm formation in *Rhizobium* (Janczarek et al. 2015b). Thus, they seem to be essential for the early infection process and may be involved in nodule ontogenia (Gray et al. 1991), which suggest a key role for the establishment of nitrogen-fixing symbiosis on legume developing indeterminate nodules, such as *E. meliloti*-alfalfa, *R. leguminosarum* bv. viciae/*Vicia sativa* and bv. trifolii/*Trifolium*, and ssp. and strain NGR234/*Leucaena* (Djordjevic et al. 1987a). This is not the case for associations leading to determinate nodules, such as *E. fredii-Glycine max* and *R. etli*/*Phaseolus* ssp. (Diebold and Noel 1989; Kim et al. 1989). In these cases other polysaccharides such as LPS could complement the EPS defficiency in the determinate nodule formation. The cyclic beta-(1,2)-glucans are predominantly localized in the periplasmic compartment (Breedveld and Miller 1994) and consist of a neutral homopolymer of about 20 beta-(1,2)-linked glucose residues—often substituted by phosphoglycerol, phosphocholin, or succinyls—and probably play a passive role in the bacterial cell adaptation to hypoosmotic conditions in its surroundings (Chen et al. 1985). Nevertheless, the possible involvement in some aspects of the symbiotic interactions has been shown in several works (Breedveld and Miller 1994, 1998; Skorupska et al. 2006). Capsular polysaccharides (KPSs) surround the bacterium and constitute a hydrated matrix, which confers bacterial resistance to bacteriophages and to the dry conditions often encountered in the rhizosphere environments (Fraysse et al. 2003). Basically, all known KPSs have been described from *Ensifer (Sinorhizobium)* species (Forsberg and Reuhs 1997; Forsberg and Carlson 1998) and mediated the contact between legume and rhizobia (Fraysse et al. 2003; Skorupska et al. 2006). Interestingly, KPS have been also found to be able to substitute EPS in the *Ensifer (Sinorhizobium)-Medicago* symbiosis, allowing nodulation (Downie 2010)

Lipopolysaccharide affects the later stages of the nodulation process such as penetration of the infection thread into the cortical cells or the setting up of the nitrogen-fixing phenotype (Fraysse et al. 2003). Dazzo and coworkers demonstrated that *R. trifolii* LPS plays an important role by modulating infection thread development in white clover root hairs (Dazzo et al. 1991). Although LPS is a constitutive component of the bacterial membrane, it could be found in very low concentrations in growth media. Consequently, a putative role from a distance or in the early steps of symbiosis could be attributed to rhizobial LPS. The importance of LPS in the rhizobia-legume symbiosis has been widely explored (Fraysse et al. 2003; Becker et al. 2005), and the sequencing of complete genomes of different rhizobia is likely to facilitate the identification of genes involved in the synthesis of EPS and the understanding of their function during the establishment of nitrogen-fixing legume symbiosis.

2.5   Nodulins and Leghemoglobins

Root nodule formation in leguminous plants is mediated by differential expression of nodulin genes: early and late nodulin genes. Those nodulins are found mainly in the root nodules, but they have also been found in other organs of leguminous plants. Of these nodulin genes, the early nodulin genes are expressed prior to the onset of nitrogen fixation, and early nodulins are involved in the formation of nodule structure and infection process. Late nodulins are expressed around the onset of nitrogen fixation and facilitate the establishment and maintenance of suitable conditions inside nodule so that a proper nitrogen fixation occurs leading to the assimilation of ammonia inside nodule. Early nodulin (ENOD) genes have been found to be related with cytoskeletal remodeling, cell wall deposition, cell growth, and division (Downie 2010). Several early nodulins are involved in signalling during the early stages of rhizobia-legume symbiosis, such as those codified by the plant genes *enod40*, *enod2*, and *enod12* (Bahyrycz and Konopinska 2007; Laporte et al. 2007; Hashimoto et al. 2008); functions of some of these nodulins are unknown, but they are essential for nodule development (van Hameren et al. 2013). The expression of the early nodulin genes is elicited by the Nod factors produced by rhizobia (Stacey et al. 2006; Mathesius 2009). The presence of nodulin genes encoding proteins and nodulins affecting plant development and nodule organogenesis has been reported in numerous leguminous plants, for example, *Glycine* (Matvienko et al. 1994; Girard et al. 2003), *Pisum* (Matvienko et al. 1994), *Medicago* (Campalans et al. 2004; Wan et al. 2007), *Vicia* (Vijn et al. 1995), *Phaseolus* (Papadopoulou et al. 1996), *Lotus* (Kumagai et al. 2006; Gronlund et al. 2005), *Trifolium* (Varkonyi-Gasic and White 2002), or *Lupinus* (Podkowinski et al. 2009). The nodulin codified by *enod40* gene is not exclusive of legumes (Kouchi et al. 1999; Vleghels et al. 2003) but is one of the earliest nodulins to be expressed upon *Rhizobium* inoculation, whose role as "riboregulator" of the *enod40* genes during plant development was proposed by Crespi et al. (1994). Later, Charon et al. (1997) showed that the early nodulin gene *enod40*, which encodes a small peptide comprising 12 or 13 amino acids, induces dedifferentiation and division of root cortical cells in *Medicago*. Therefore, it was proposed that *enod40* gene is involved in the initiation of root nodule organogenesis in legumes. A year later, the involvement of *enod40* gene in the nodule development on stems of *S. rostrata* inoculated with *A. caulinodans* was also reported (Corich et al. 1998). It has been reported that RNAi knockdown of *enod40*s leads to significant suppression of nodule formation in the model legume *Lotus japonicus* (Kumagai et al. 2006). Based on the data from Rohrig et al. (2002), soybean *enod40* encodes two peptides that bind to sucrose synthase and may be involved in the control of sucrose use in nitrogen-fixing nodules.

There are two types of late nodulins: (1) metabolic nodulins and (2) symbiosome membrane nodulins. Metabolic nodulins include leghemoglobin and uricase (a key enzyme of uriede biosynthesis), glutamine synthetase (catalyzes the first step in ammonium assimilation), and sucrose synthase (catalyzes the breakdown of sucrose leading to the generation of energy). In addition, several enzymes have been detected in root nodules of legumes that differ in their physical, kinetic, and immunological properties from the corresponding root enzymes: phosphoenolpyruvate carboxykinase, choline kinase, xanthine dehydrogenase, purine nucleosidase, and malate dehydrogenase. These enzymes may not be true nodulins but rather posttranslational modifications of the root enzymes. Symbiosome membrane nodulins originate from the plasma membrane with modifications due to coalescence with Golgi vesicles. The symbiosome membrane serves as the interface between eukaryotic and prokaryotic symbionts and thus is expected to possess transporters for nutrient exchange. Examples of soybean symbiosome membrane nodulins include nodulin 24 and 26. Nodulin 24 is synthesized as a lumenal protein in the endoplasmic reticulum and posttranslationally attached to the membranes en route to the symbiosome membrane (Cheon et al. 1994). Nodulin 26 is an aquaporin channel with a modest water transport rate (Guenther et al. 2003). Phosphorylation of nodulin 26 on Ser-262, which is catalyzed by a symbiosome membrane-associated calcium-dependent protein kinase, stimulates its intrinsic water transport rate.

Hemoglobins (Hbs) are hemoproteins that reversibly bind O_2 (Kundu et al. 2003; Garrocho-Villegas et al. 2007; Hoy and Hargrove 2008), and three types of Hbs have been identified in plants: symbiotic (sHb), nonsymbiotic (nsHb), and truncated (2/2) Hbs (tHbs) (Ross et al. 2002). nsHbs are found at nanomolar to micromolar concentrations in many plant tissues and can be divided into two groups: Class 1 is involved in NO metabolism and in the maintenance of ATP production in low oxygen concentrarions, and Class 2, whose function in nodulation is still unknown. The role of tHbs in nodulation has not also been conclusively demonstrated yet, but their involvement in the suppression of the defense response during symbiosis was recently suggested (Becana et al. 2015). The first identification of Hbs in plant was reported in legume nodules by Kubo (1939), who concluded that the physiological role of the hemoprotein in nodules is to stimulate the assimilation, and transport of O_2. Kubo's plant Hb was named as leghemoglobin (Lhb) by Virtanen and Laine (1946), and it is also known as plant symbiotic Hb. Further, analyses showed that Kubo's hemoprotein is a plant Hb with similar (i.e., structural) properties to animal Hbs. For many years Hbs were identified in N_2-fixing legumes and were called leghemoglobins (Lhb) because they were firstly discovered in legumes. The role of Lhb in N_2-fixing nodules was elucidated by Wittenberg et al. (1974). The function of Lhb in nodules is to facilitate the diffusion of O_2 to bacteroids at an internal concentration low enough to avoid the inhibition or destruction of their O_2-sensitive nitrogenase. This concept of high O_2 flux at low free O_2 in the bacteroid vicinity is now generally accepted. Although sHb were initially found in legumes, sHbs have been also found in the root nodules of nonlegumes such as, *Parasponia andersonii* in symbiosis with *Rhizobium*, and dicotyledoneous plants, such as *Casuarina glauca* in symbiosis with the actinobacteria *Frankia* (Appleby et al. 1998). Tjepkema (1983) detected high concentrations of Hb-like proteins in nodule extracts of actinorhizal plants (which are nodulated by *Frankia*), such as *C. cunninghamiana* and *Myrica gale*, and low concentrations in nodules of *Comptonia peregrina*, *Alnus rubra*, and *Elaeagnus angustifolia*. The spectral properties of actinorhizal Hbs are similar to those of Lhbs and non-plant Hbs. Although Hbs are abundant in nodules playing a crucial role in nitrogen fixation (Ott et al. 2005), nsHbs have been found in many nonlegume plants such as barley, rice, maize, and wheat (Taylor et al. 1994), in *Arabidopsis* where a knockout of a class 2 hemoglobin causes seedlings to die at a very early stage (Hebelstrup et al. 2006) of actinorhizal plants *Causarina* (Jacobsen-Lyon et al. 1995) and leguminous plants such as soybean, clover, alfalfa, and pea (Andersson et al. 1996). This confirmed their existence in monocot and dicot plants including legumes and strengthened the theory that Hbs have an ancient origin and are ubiquitous in the plant kingdom through vertical evolution.

NsHbs differ from sHbs and mammalian Hb and myoglobins (Mb) in that they are generally "hexacoordinate" in both the ferric and ferrous states due to a histidine in the distal pocket that reversibly binds the sixth coordination site of the hemo iron (Arredondo-Peter et al. 1997; Qu et al. 2005). Two classes (classes 1 and 2) of nsHbs have been distinguished using phylogenetic analysis and shown to differ in their patterns of expression (Trevaskis et al. 1997). Despite hexacoordination, class 1 nsHbs have high oxygen affinities and low oxygen dissociation rate constants (Arredondo-Peter et al. 1997; Hoy et al. 2007) due to stabilization between the distal histidine and the bound ligand akin to that in Mb (Arredondo-Peter et al. 1997; Das et al. 1999). Class 2 nsHbs have lower oxygen affinities and greater similarity to sHbs than nsHbs, consistent with the observation that most sHbs evolved from class 2 nsHbs (Trevaskis et al. 1997). In addition to Lbs and nsHbs, Hb sequences that are similar to those of microbial truncated (2/2) Hbs (Pesce et al. 2000; Wittenberg et al. 2002) are detected in primitive and evolved plants. However, the function of 2/2-like Hbs in plant organs is not yet known, although kinetic properties of a recombinant *Arabidopsis* 2/2-like Hb suggest that these proteins may function as O_2 carriers (Watts et al. 2001).

2.6   Nitrogenase and Hydrogenase

Nitrogenase is an enzyme complex that reduces atmospheric N to ammonium in the microaerophillic environment of the nodule (Appleby 1992; Haag et al. 2013). Nitrogenase catalyzes the MgATP-dependent reduction of N_2 to ammonia:

$$N_2 + 8e^- + 16\,MgATP + 8H^+ \rightarrow 2NH_3 + H_2 + 16\,MgADP + P_i$$

The most studied nitrogenase contains two metallocomponents, dinitrogenase [molybdenum-iron (MoFe) protein] and dinitrogenase reductase (Fe protein). Nitrogenase requires an electron donor and a minimum of 16 moles of ATP per mole of N_2; the energy needed for the complete nitrogen fixation process is around 40 moles of ATP per mole of N_2, or in terms of carbon, 6 g of carbon for every gram of N reduced (Vance and Heichel 1991). Electrons are generated in vivo either oxidatively or photosynthetically, depending on the organism. These electrons are transferred to flavodoxin or ferredoxin, a (4Fe-4S)-containing electron carrier that transfers an electron to the Fe protein of nitrogenase, beginning a series of oxidoreduction cycles. Two molecules of MgATP bind to the reduced Fe protein and are hydrolyzed to drive an electron from the Fe protein to the MoFe protein. The actual reduction of N_2 occurs in the MoFe protein in a multistep reaction. Electron transfer must occur six times per each fixed N_2 molecule. Therefore, a total of 12 ATPs are required to fix one N_2 molecule, but as nitrogenase also reduces protons to H_2 consuming two electrons, the total cost of N_2 reduction is eight electrons transferred and 16 MgATPs (Cheng 2008). In nitrogen-fixing bacteria, nitrogenase is encoded by a set of operons that includes regulatory genes (such as *nifLA*), structural genes (such as *nifHDK*), and other supplementary genes. The free-living diazotrophic bacterium, *Klebsiella pneumoniae*, has been the most extensively analyzed and provides a model for studies of nitrogenase regulation, synthesis, and assembly. A 24 kb DNA region contains the entire *K. pneumoniae nif* cluster, which includes 20 genes (Dos Santos et al. 2004). Since then, other complete genomes have been sequenced for free-living and symbiotic nitrogen fixers alike (de Oliveira Cunha et al. 2012; Martínez-Abarca et al. 2013; Halim et al. 2016). Although there are several different types of nitrogenases, rhizobia possess only the molybdenum-containing type. The Mo nitrogenases are composed of two proteins: a MoFe protein and a Fe protein. The MoFe protein is a 220–240 kDa tetramer, a $(\alpha\beta)_2$ complex, of the *nif*D (α-subunit) and *nif*K (β-subunit) gene products each of which contains complex metalloclusters. Each tetramer of two $\alpha\beta$ pairs contains two P-clusters [Fe_8S_7] and two FeMo cofactors. The FeMo cofactor, located within the α-subunit, consists of a $MoFe_3$-S_3 cluster bridged to a Fe_4-S_3 cluster by three sulfur ligands with a homocitrate coordinated to the molybdenum. The Fe protein is ~60 kDa dimer of the *nif*H gene with a single 4Fe-4S cluster located between the subunits. A MgATP binding site is located on each subunit (Howard and Rees 2006). During catalysis, electrons are delivered one at a time from the Fe protein to the MoFe protein in a reaction coupled to the hydrolysis of two MgATP for each electron transferred (Rees and Howard 2000). The P-clusters are thought to mediate electron transfer from the Fe protein to the FeMo cofactor of the MoFe protein, the site for substrate binding and reduction.

Hydrogen evolution is an inherent step of the catalytic mechanism of nitrogenase (Simpson and Burris 1984) being the nitrogen fixation process as one of the most relevant biogenic hydrogen sources (see review by Palacios et al. 2005). Generally in proteobacteria, genetic determinants for hydrogenase synthesis can be found in large clusters that can encode between 15 to 18 proteins related to the process. Hydrogenase genes are usually conserved in different proteobacteria, suggesting a conserved mechanism for their synthesis (Vignais and Billoud 2007). For the *Rhizobium*-legume symbiosis, over one million tonnes of hydrogen/year was evolved from root nodules into the air (Evans et al. 1987). A number of strains of the slow-growing rhizobia possess hydrogenases that recycle the hydrogen, thereby recapturing some of the lost energy (Baginsky et al. 2002). Until now the hydrogenase system has been characterized in a limited number of rhizobia, such as *Rhizobium leguminosarum* (Rey et al. 1993), *Bradyrhizobium* (Baginsky et al. 2005), and *Azorhizobium caulinodans* (Baginsky et al. 2004). Although the hydrogen-uptake system of *Rhizobium leguminosarum* bv. viciae UPM791 has been analyzed in detail (Ruiz-Argüeso et al. 2000), only few strains have this system (Fernández et al. 2005). However sequencing of some regions of Hup cluster showed that they are conserved in these strains (Fernández et al. 2005). In *Rhizobium leguminosarum* bv. viciae UPM791, the hydrogen-uptake system is based on a membrane-bound, heterodimeric [NiFe] hydrogenase (Hidalgo et al. 1990; Brito et al. 1994) similar to those described in *Bradyrhizobium japonicum* and other aerobic bacteria. In these bacteria, electron flow from hydrogen to oxygen results in the generation of ATP, although for *R. leguminosarum*, the degree of coupling is variable for the different strains (Ruiz-Argüeso et al. 2000). *R. leguminosarum* hydrogenase contains two subunits: a large subunit (HupL) of approx. 60 kDa carrying the heterometallic FeNi active center and a small subunit (HupS) of approx. 30 kDa harboring three FeS clusters. As in other bacteria, the synthesis of this enzyme is a complex process that occurs in the cytoplasm through the concerted action of over 15 proteins (*hup* and *hyp* gene products). Functions ascribed to the proteins involved in hydrogenase synthesis and activity include electron transport (HupC), processing of large subunit (HupD), nickel provision (HypAB), and synthesis of NiFe cofactor (HypFCDE). Albareda et al. (2012) found evidence of HupF having a dual role during hydrogenase biosynthesis: it is involved in the processing of the hydrogenase large subunit, and it also stabilizes HupL as a chaperone when the hydrogenase is synthesized in aerobic conditions. In *R. leguminosarum* bv. viciae, hydrogenase genes (*hupSLCDEFGHIJK* and *hypABFCDEX*) are clustered in a 20 kb DNA region of the symbiotic plasmid (Leyva et al. 1990). This plasmid also contains genes for nodulation and nitrogen fixation. The location of hydrogenase genes in the symbiotic plasmid is a general trait for hydrogenase-positive strains of *R. leguminosarum* (Leyva et al. 1987), suggesting an adaptation of hydrogen recycling to the symbiotic lifestyle in this bacterial species. The environment surrounding the bacteroid affects the expression of *R. leguminosarum* hydrogenase activity, and permissive (*Pisum, Vicia*) and nonpermissive (*Lens*) hosts for hydrogenase activity in bacteroids have been described (López et al. 1983).

2.7 Other Molecules Affecting Rhizobia-Legume Symbiosis

Apart from the key molecules discussed earlier, there are other active biomolecules that greatly influence the *Rhizobium*-legume interactions leading to the formation of symbiotic relationships. In the following section, focus is given on various kinds of such molecules affecting symbiosis.

2.7.1 Phytohormones

Plant hormones (phytohormones) regulate plant development and elicit different responses in plant organs, tissues, or individual cells. Some of the phytohormones, such as strigolactones, jasmonates, auxins, and cytokinins, have been found to play significant role in plant-microbe interactions.

2.7.1.1 Strigolactones

Strigolactones (SL) are plant hormones, which promotes shoot branching (Dun et al. 2009). These phytohormones are also described as seed germination stimulants for parasitic plants and act as signals for the pre-symbiotic growth of AM fungi (Awad et al. 2006; Besserer et al. 2006). Multiple functions including their role in plant defense against stresses have been reported (de Saint-Germain et al. 2013; Al-Babili and Bouwmeester 2015; Pandey et al. 2016). Also, SLs play a role as signalling molecules in rhizobia-legume symbiotic associations (Oldroyd 2013; Breakspear et al. 2014, 2015). The presence of the strigolactone analogue GR24 has been found to promote nodulation in different rhizobia-legume symbioses, such as the *Medicago sativa-Ensifer (Sinorhizobium) meliloti* 1021 (Soto et al. 2010) or the *Pisum sativum-Rhizobium leguminosarum* bv. viciae RLV248 symbiotic interaction (Foo and Davies 2011). However, the results further suggest that even though SLs are involved in symbiotic association, they are not essentially required for nodulation. In a recent investigation, strigolactones have been observed to act as stimulators of the establishment of the rhizobia-legume interactions and promoting bacterial swarming motility in a *Medicago sativa-Ensifer (Sinorhizobium) meliloti* model system (Peláez-Vico et al. 2016).

2.7.1.2 Jasmonic Acid

Jasmonic acid (JA) and its methylated derivatives (MeJA) play an important role in plant defense against biotic and abiotic stresses. During *Rhizobium*-legume symbiosis, JA and its derivatives stimulate the expression of nod genes as reported in *R. leguminosarum* (Rosas et al. 1998). Other studies have also confirmed the JA-derived stimulation of nod genes in *Bradyrhizobium japonicum* and nodulation was enhanced when *Rhizobium* was preincubated/inoculated with JA and a specific flavonoid (Mabood et al. 2006; Poustini et al. 2010). On the contrary, low concentrations of this phytohormone decreased the number of nodules in the *E. meliloti-M. truncatula* symbiotic system (Sun et al. 2006). However, other studies suggest that jasmonates are not involved in nodule development and functionality (Zdyb et al. 2011). Although JA may affect the autoregulation of signalling and nodulation process, results are still controversial (Hause and Schaarschmidt 2009; Ferguson et al. 2010).

2.7.1.3 Auxins

Auxins affect the early steps of the rhizobia-legume interactions, which have a cross talk with cytokinins. The role of these two phytohormones in the regulation of infection and nodulation was reviewed by Suzaki et al. (2013). In a recent study, Breakspear et al. (2014) obtained transcription profiles of *Medicago truncatula* root hairs during the establishment of rhizobial infection, revealing changes in production of phytohormones. SLs and auxins showed the most notable expression changes in infection. The regulation of auxin signalling is completely necessary for the initiation of the infection process (Laplaze et al. 2015). Miri et al. (2015) reviewed the role of cytokinins as nodule primordium inducer and discussed their implication in the regulation of bacterial root colonization by nitrogen-fixing bacteria. Plants containing mutations in cytokinin receptors showed altered infection phenotypes. Indole-3-acetic acid (IAA) is a phytohormone commonly synthesized by rhizobial strains (Spaepen and Vanderleyden 2011). Spaepen et al. (2009) engineered a miniTn5 transposon mutant library in *R. etli* CNPAF512, performing a search for up- or downregulated genes related with IAA. In silico analysis to identify affected genes predicted the involvement of IAA in the early stages and in nodulation of symbiotic interactions. It seems very clear the role of IAA in the rhizobia-legume symbiosis; however, the studies referring to its role in nodulation are controversial. IAA-deficient *Bradyrhizobium elkanii* mutants induced a reduction in soybean nodule number (Fukuhara et al. 1994). On the contrary, a low IAA-producing mutant of *R.* sp NGR234 induced similar number of nodules in *Vigna unguiculata* and *Tephrosia vogelii* (Theunis et al. 2004).

2.7.2 Simple Phenolic Compounds

Phenolic compounds are the main group of polyphenols present in plants, which have a role as signalling molecules in the initial stages of root nodule symbiosis, among other functionalities (Mandal et al. 2010; Cheynier et al. 2013). Flavonoids are the most known phenolic compounds involved in symbiosis; however, simple phenolic compounds (phenolic compounds with a C6 skeleton) are also involved in some stages. Vanillin and isovanillin are inducers of nod gene expression in the broad range strain *Ensifer.* sp. NGR234 (Le Strange et al. 1990). Xanthones have similar effects inducing Nod factors in *B. japonicum* (Yuen et al. 1995). Some of these compounds (ferulic and coumaric acid) can also be used by rhizobia as carbon sources (Prinsen et al. 1991; Van Rossum et al. 1995).

2.7.3   Remorins and Flotillins

Remorins, named after *Remora*, a shark "parasitic" fish genus, are proteins recently identified in *M. truncatula* and *L. japonicus*, associated to lipid rafts (Lefebvre et al. 2010; Tóth et al. 2012; Janczarek et al. 2015a). In *M. truncatula*, symbiotic remorin 1 (MtSYMREM1) gene encodes a remorin, which is nodule specific, and interacts with three receptor-like kinases (LYK3, NFP, and SYMRK), playing an important role in rhizobial infection (Lefebvre et al. 2010; Murray 2011). Moreover, the over-expression of the symbiotic remorin 1 ortolog in *L. japonicus* (LjSYMREM1) increases nodulation on *L. japonicus* transgenic roots (Tóth et al. 2012). Plant flotil-lins are also lipid raft-associated proteins, which are located in membranes. Haney and Long (2010) reported that the silencing of flotillins results in a decrease in the number of infection threads in *M. truncatula-E. meliloti* symbiosis. Moreover, genes codifying flotillins are induced by rhizobial infection. The combination of flotillins and remorins, at least in this model, might be important to facilitate the interaction ligand receptor in the recognition rhizobia-legume (Oldroyd 2013)

2.7.4   Nodule-Specific Cysteine-Rich (NCR) Peptides

Antimicrobial peptides are a group of natural "antibiotics," acting as effectors in the immunity of several organisms, such as animals or plants (Maróti et al. 2011). Nodule-specific cysteine-rich peptides (NCRs), a group of these antimicrobial pep-tides, are encoded by a large family of genes, the *ncr* gene family, which are expressed specifically in the nodule cells. There are more than 300 identified NCR peptides, all of them with similar features, such as their small size and their con-served Cys motifs. However, these genes have different patterns of expression dur-ing nodule organogenesis (Mergaert et al. 2003). NCR peptides are toxic for rhizobia, but in the interaction *E. meliloti-M. truncatula*, rhizobia is protected by its BacA protein, which is reported to partially protect the bacteroid from the NCR peptide toxicity, increasing bacteroid persistence into nodule cells, hence allowing nitrogen fixation (Haag et al. 2011). In fact, NCR peptides prevent the use of plant resources by rhizobia, such as carbon sources, by compelling them to the differen-tiation in the bacteroid form (van de Velde et al. 2010). Recently, Montiel et al. (2015) reported the existence of NCR-encoding genes in all IRLC (inverted repeat-lacking clade) legume species with available sequence data. These findings suggest that NCR peptides could indeed be considered as key molecules in the rhizobia-legume symbiosis, and possibly some new roles have still to be undiscovered.

2.7.5   Other Rhizobial Compounds Involved in Rhizobia-Legume Symbiosis

2.7.5.1   *N*-Acyl Homoserine Lactones

N-Acyl homoserine lactones (AHLs) are considered signalling molecules produced by rhizobia, among other bacteria, involved in quorum-sensing. AHLs can be very different among rhizobial species, but their basic structure consist in a homoserine lactone linked to an acyl chain from 4 to 18 C. Substitutions in the acyl chain provide the specificity in quorum-sensing communication mechanisms with both, plants and other rhizospheric bacteria. A modification in the composition or in the levels of AHLs affects rhizobia-legume symbiosis (Giordano 2015). AHLs enhance the motility of *Bradyrhizobium* sp. and help in finding and reaching its host plant, peanut (Nievas et al. 2012). Bogino et al. (2015) reviewed the role of AHLs and quorum sensing, mostly in *Bradyrhizobium* species, and proposed a model, which should be confirmed in other symbiotic interactions.

2.7.5.2 Hydrogen Peroxide and Nitric Oxide

Hydrogen peroxide (H_2O_2) and nitric oxide (NO) act as signalling molecules in the early stages of nodulation during rhizobia-legume interactions. Several authors reported that H_2O_2 may control somehow infection process and bacterial differentiation, while NO is required for an optimal establishment of symbiosis and appears to be a key molecule in nodule senescence, inhibiting nitrogenase and hence nitrogen fixation. Balance between both molecules is however strictly necessary for a correct interaction between rhizobia and host legumes (Puppo et al. 2013; Damiani et al. 2016).

Conclusion

In the last decade, the number of investigations and publications concerning the role of different active biomolecules in symbiotic interactions has increased considerably due to its greater agronomic importance. The outcomes of these studies are providing some incredibly valuable information, which could be extremely useful in understanding the mechanistic basis of symbiotic process. Apart from the molecules surveyed and identified in this chapter, there might be several other molecules, which could be handy in legume root nodule formation; however, such molecules need to be identified. Further investigations will probably lead to the discovery of many more other molecules essential for the regulation of an effective nitrogen-fixing symbiosis. The knowledge of these molecules will undoubtly contribute to the better exploitation of this fascinating ecological event and therefore to the comprehensive management of the symbiosis for enhancing the production of legumes and protecting and preserving the environment in a more sustainable manner.

References

Abd-Alla MH, El-enany AWE, Bagy MK, Bashandy SR (2014) Alleviating the inhibitory effect of salinity stress on nod gene expression in *Rhizobium tibeticum*–fenugreek (*Trigonella foenum graecum*) symbiosis by isoflavonoids treatment. J Plant Interact 9:275–284

Al-Babili S, Bouwmeester HJ (2015) Strigolactones, a novel carotenoid-derived plant hormone. Annu Rev Plant Biol 66:161–186

Albareda M, Manyani H, Imperial J, Brito B, Ruiz-Argüeso T, Böck A, Palacios JM (2012) Dual role of HupF in the biosynthesis of [NiFe] hydrogenase in *Rhizobium leguminosarum*. BMC Microbiol 12:256

Amor BB, Shaw SL, Oldroyd GED, Maillet F, Penmetsa RV, Cook D, Long SR, Dénarié J, Gough C (2003) The NFP locus of *Medicago truncatula* controls an early step of Nod factor signal transduction upstream of a rapid calcium flux and root hair deformation. Plant J 34: 495–506

Andersen OM, Markham KR (2006) Flavonoids: chemistry, biochemistry and applications. CRC Press, Boca Raton

Andersson CR, Jensen EO, Llewellyn DJ, Dennis ES, Peacock WJ (1996) A new hemoglobin gene from soybean: a role for hemoglobin in all plants. Proc Natl Acad Sci U S A 93:5682–5687

Ané JM, Kiss GB, Riely BK, Penmetsa RV, Oldroyd GED, Ayax C, Lévy J, Debellé F, Baek JM, Kalo P, Rosenberg C, Roe BA, Long SR, Dénarié J, Cook DR (2004) *Medicago truncatula* DMI1 required for bacterial and fungal symbioses in legumes. Science 303:1364–1367

Appleby CA (1992) The origin and functions of haemoglobin in plants. Sci Prog 76:365–398

Appleby CA, Bogusz D, Dennis ES, Peacock WJ (1998) A role for haemoglobin in all plant roots? Plant Cell Environ 11:359–367

Arredondo-Peter R, Hargrove MS, Sarath G, Moran JF, Lohrman J, Olson JS, Klucas RV (1997) Rice hemoglobins. Gene cloning, analysis, and O_2-binding kinetics of a recombinant protein synthesized in *Escherichia coli*. Plant Physiol 115:1259–1266

Arrighi JF, Cartieaux F, Brown SC, Rodier-Goud M, Boursot M, Fardoux J, Patrel D, Gully D, Fabre S, Chaintreuil C, Giraud E (2012) *Aeschynomene evenia*, a model plant for studying the molecular genetics of the Nod-independent rhizobium-legume symbiosis. Mol Plant Microbe Interact 25:851–861

Awad AA, Sato D, Kusumoto D, Kamioka H, Takeuchi Y, Yoneyama K (2006) Characterization of strigolactones, germination stimulants for the root parasitic plants *Striga* and *Orobanche*, produced by maize, millet and sorghum. Plant Growth Regul 48:221–227

Baginsky C, Brito B, Imperial J, Palacios JM, Ruiz-Argüeso T (2002) Diversity and evolution of hydrogenase systems in rhizobia. Appl Environ Microbiol 68:4915–4924

Baginsky C, Brito B, Imperial J, Ruiz-Argüeso T, Palacios JM (2005) Symbiotic hydrogenase activity in *Bradyrhizobium* sp. (vigna) increases nitrogen content in *Vigna unguiculata* plants. Appl Environ Microbiol 71:7536–7538

Baginsky C, Palacios JM, Imperial J, Ruiz-Argüeso T, Brito B (2004) Molecular and functional characterization of the *Azorhizobium caulinodans* ORS571 hydrogenase gene cluster. FEMS Microbiol Lett 237:399–405

Bahyrycz A, Konopinska D (2007) Plant signalling peptides: some recent developments. J Pept Sci 13:787–797

Barnett MJ, Fisher RF (2006) Global gene expression in the rhizobial–legume symbiosis. Symbiosis 42:1–24

Becana M, Navascués J, Pérez-Rontomé C, Walker A, Desbois A, Abian J (2015) Leghemoglobins with nitrated hemes in legume root nodules. In: de Bruijn F (ed) Biological nitrogen fixation, vol 2. John Wiley & Sons, New York, pp 705–713

Becker A, Pühler A (1998) Production of exopolysaccharides. In: Spaink HP, Kondorosi A, Hooykaas JJ (eds) The rhizobiaceae. Kluwer Academic Publishers, Dordrecht, The Netherlands, pp 97–118

Becker A, Fraysse N, Sharypova L (2005) Recent advances in studies on structure and symbiosis-related function of rhizobial K-antigens and lipopolysaccharides. Mol Plant Microbe Interact 18:899–905

Begum AA, Leibovitch S, Migner P, Zhang F (2001) Specific flavonoids induced nod gene expression and pre-activated nod genes of *Rhizobium leguminosarum* increased pea (*Pisum sativum* L.) and lentil (*Lens culinaris* L.) nodulation in controlled growth chamber environments. J Exp Bot 52:1537–1543

Besserer A, Puech-Pagès V, Kiefer P, Gomez-Roldan V, Jauneau A, Roy S, Séjalon-Delmas N (2006) Strigolactones stimulate arbuscular mycorrhizal fungi by activating mitochondria. PLoS Biol 4(7):e226

Bogino PC, Nievas FL, Giordano W (2015) A review: Quorum sensing in *Bradyrhizobium*. Appl Soil Ecol 94:49–58

Braeken K, Daniels R, Vos K, Fauvart M, Bachaspatimayum D, Vanderleyden J, Michiels J (2008) Genetic determinants of swarming in *Rhizobium etli*. Microb Ecol 55:54–64

Breakspear A, Liu C, Roy S, Stacey N, Rogers C, Trick M, Morieri G, Mysore KS, Wen J, Oldroyd GED, Downie JA, Murray JD (2014) The root hair "infectome" of *Medicago truncatula* uncovers changes in cell cycle genes and reveals a requirement for auxin signaling in rhizobial infection. Plant Cell 26:4680–4701

Breakspear A, Liu C, Cousins DR, Roy S, Guan D, Murray JD (2015) The role of hormones in rhizobial infection. In: de Bruijn F (ed) Biological nitrogen fixation, vol 2. John Wiley & Sons, New York, pp 555–566

Breedveld MW, Miller KJ (1994) Cyclic b-glucans of members of the family *Rhizobiaceae*. Microbiol Rev 58:145–161

Breedveld MW, Miller KJ (1998) Cell surface b-glucans. In: Spaink HP, Kondorosi A, Hooykaas JJ (eds.) The Rhizobiaceae Kluwer Academic Publishers, Dordrecht, The Netherlands, pp. 81-96

Brito B, Palacios JM, Hidalgo E, Imperial J, Ruiz-Argüeso T (1994) Nickel availability to pea (*Pisum sativum* L.) plants limit hydrogenase activity of *Rhizobium leguminosarum* bv. viciae bacteroids by affecting the processing of the hydrogenase structural subunits. J Bacteriol 176:5297–5303

Broghammer A, Krusell L, Blaise M, Sauer J, Sullivan JT, Maolanon N, Vinther M, Lorentzen A, Madsen EB, Jensen KJ, Roepstorff P, Thirup S, Ronson CW, Thygesen MB, Stougaard J (2012) Legume receptors perceive the rhizobial lipochitin oligosaccharide signal molecules by direct binding. Proc Natl Acad Sci U S A 109:13859–13864

Buer CS, Imin N, Djordjevic MA (2010) Flavonoids: new roles for old molecules. J Integr Plant Biol 52:98–111

Campalans A, Kondorosi A, Crespi M (2004) Enod40, a short open reading frame-containing mRNA, induces cytoplasmic localization of a nuclear RNA binding protein in *Medicago truncatula*. Plant Cell 16:1047–1059

Cantero L, Palacios JM, Ruiz-Argüeso T, Imperial J (2006) Proteomic analysis of quorum sensing in *Rhizobium leguminosarum* bv. viciae UPM791. Proteomics 6:S97–106

Cao H, Yang M, Zheng H, Zhang J, Zhong Z, Zhu J (2009) Complex quorum-sensing regulatory systems regulate bacterial growth and symbiotic nodulation in *Mesorhizobium tianshanense*. Arch Microbiol 191:283–289

Charon C, Johansson C, Kondorosi E, Kondorosi A, Crespi M (1997) enod40 induces dedifferentiation and division of root cortical cells in legumes. Proc Natl Acad Sci U S A 94: 8901–8906

Charpentier M, Sun J, Martins TV, Radhakrishnan GV, Findlay K, Soumpourou E, Thouin J, Véry AA, Sanders D, Morris RJ, Oldroyd GED (2016) Nuclear-localized cyclic nucleotide-gated channels mediate symbiotic calcium oscillations. Science 352(6289):1102–1105

Chen H, Batley M, Redmond J, Rolfe BG (1985) Alteration of effective nodulation properties of a fast-growing broad host range *Rhizobium* due to changes in exopolysaccharide synthesis. J Plant Physiol 120:331–349

Cheng Q (2008) Perspectives in biological nitrogen fixation research. J Integr Plant Biol 50:786–798

Cheon CI, Hong Z, Verma DP (1994) Nodulin-24 follows a novel pathway for integration into the peribacteroid membrane in soybean root nodules. J Biol Chem 269:6598–6602

Cheynier V, Comte G, Davies KM, Lattanzio V, Martens S (2013) Plant phenolics: recent advances on their biosynthesis, genetics, and ecophysiology. Plant Physiol Biochem 72:1–20

Corich V, Goormachtig S, Lievens S, Van Montagu M, Holsters M (1998) Patterns of ENOD40 gene expression in stem-borne nodules of *Sesbania rostrata*. Plant Mol Biol 37:67–76

Crespi MD, Jurkevitch E, Poiret M, d'Aubenton-Carafa Y, Petrovics G, Kondorosi E, Kondorosi A (1994) enod40, a gene expressed during nodule organogenesis, codes for a non-translatable RNA involved in plant growth. EMBO J 13:5099–5112

Cullimore JV, Ranjeva R, Bono JJ (2001) Perception of lipo-chitooligosaccharidic Nod factors in legumes. Trends Plant Sci 6:24–30

Cubo MT, Economou A, Murphy G, Johnston AW, Downie JA (1992) Molecular characterization and regulation of the rhizosphere-expressed genes *rhiABCR* that can influence nodulation by *Rhizobium leguminosarum* biovar viciae. J Bacteriol 174:4026–4035

Damiani I, Pauly N, Puppo A, Brouquisse R, Boscari A (2016) Reactive oxygen species and nitric oxide control early steps of the legume–*Rhizobium* symbiotic interaction. Front Plant Sci 7:454

Daniels R, Reynaert S, Hoekstra H, Verreth C, Janssens J, Braeken K, Fauvart M, Beullens S, Heusdens C, Lambrichts I, De Vos DE, Vanderleyden J, Vermant J, Michiels J (2006) Quorum signal molecules as biosurfactants affecting swarming in *Rhizobium etli*. Proc Natl Acad Sci U S A 103:14965–14970

Das TK, Lee HC, Duff SM, Hill RD, Peisach J, Rousseau DL, Wittenberg BA, Wittenberg JB (1999) The heme environment in barley hemoglobin. J Biol Chem 274:4207–4212

Dazzo F, Truchet G, Sherwood J, Hrabak E, Abe M, Pankratz HS (1984) Specific phases of root hair attachment in the *Rhizobium trifolii*-clover symbiosis. Appl Environ Microbiol 48:1140–1150

Dazzo FB, Truchet GL, Hollingsworth RI, Hrabak EM, Pankratz HS, Philip-Hollingsworth S, Salzwedel JL, Chapman K, Appenzeller L, Squartini A, Gerhold D, Orgambide G (1991) *Rhizobium* LPS modulates infection thread development in white clover root hairs. J Bacteriol 173:5371–5384

del Cerro P, Rolla-Santos AA, Valderrama-Fernández R, Gil-Serrano A, Bellogín RA, Gomes DF, Pérez-Montaño F, Megías M, Hungría M, Ollero FJ (2016) NrcR, a new transcriptional regulator of *Rhizobium tropici* CIAT 899 involved in the legume root-nodule symbiosis. PLoS One 11(4):e0154029

de Hoff D, Brill LM, Hirsch AM (2009) Plant lectins: the ties that bind in root symbiosis and plant defense. Mol Genet 282:1–15

de Oliveira Cunha C, Goda Zuleta LF, Paula de Almeida LG, Prioli Ciapina L, Lustrino Borges W, Pitard RM, Ivo Baldani J, Straliotto R, Miana de Faria S, Hungria M, Sousa Cavadae B, Martins Mercante F, Ribeiro de Vasconcelos AT (2012) Complete genome sequence of *Burkholderia phenoliruptrix* BR3459a (CLA1), a heat-tolerant, nitrogen-fixing symbiont of *Mimosa flocculosa*. J Bacteriol 194:6675–6676

de Saint-Germain A, Bonhomme S, Boyer FD, Rameau C (2013) Novel insights into strigolactone distribution and signalling. Curr Opin Plant Biol 16:583–589

de Vasconcelos M, Oliveira Cunha C, Sousa Arruda FV, Alves Carneiro V, Mesquita Bastos R, Martins Mercante F, Santiago do Nascimento K, Sousa Cavada B, Pires dos Santos R, Holanda Teixeira E (2013) Effect of leguminous lectins on the growth of *Rhizobium tropici* CIAT899. Molecules 18:5792–5803

Diebold R, Noel KD (1989) *Rhizobium leguminosarum* exopolysaccharide mutants: biochemical and genetic analyses and symbiotic behaviour on three hosts. J Bacteriol 171:4821–4827

Djordjevic SP, Chen H, Batley M, Redmond JW, Rolfe BG (1987a) Nitrogen fixation ability of exopolysaccharide synthesis mutants of *Rhizobium* sp. strain NGR234 and *Rhizobium trifolii* is restored by the addition of homologous exopolysaccharides. J Bacteriol 169:53–60

Dos Santos PC, Dean DR, Hu Y, Ribbe MW (2004) Formation and insertion of the nitrogenase iron-molybdenum cofactor. Chem Rev 104:1159–1173

Downie JA (2010) The roles of extracellular proteins, polysaccharides and signals in the interactions of rhizobia with legume roots. FEMS Microbiol Rev 34:150–170

Dun EA, Brewer PB, Beveridge CA (2009) Strigolactones: discovery of the elusive shoot branching hormone. Trends Plant Sci 14:364–372

Edwards A, Frederix M, Wisniewski-Dyé F, Jones J, Zorreguieta A, Downie JA (2009) The cin and rai quorum-sensing regulatory systems in *Rhizobium leguminosarum* are coordinated by ExpR and CinS, a small regulatory protein coexpressed with CinI. J Bacteriol 191:3059–3067

el Zahar Haichar F, Santaella C, Heulin T, Achouak W (2014) Root exudates mediated interactions belowground. Soil Biol Biochem 77:69–80

Evans HJ, Harker AR, Papen H, Russell SA, Hanus FJ, Zuber M (1987) Physiology, biochemistry and genetics of the uptake hydrogenase in *Rhizobium*. Annu Rev Microbiol 41:335–361

Fabre S, Gully D, Poitout A, Patrel D, Arrighi JF, Giraud E, Czernic P, Cartieaux F (2015) Nod factor-independent nodulation in *Aeschynomene evenia* required the common plant-microbe symbiotic toolkit. Plant Physiol 169:2654–2664

Ferguson BJ, Indrasumunar A, Hayashi S, Lin MH, Lin YH, Reid DE, Gresshoff PM (2010) Molecular analysis of legume nodule development and autoregulation. J Integr Plant Biol 52:61–76

Fernández D, Toffanin A, Palacios JM, Ruiz-Argüeso T, Imperial J (2005) Hydrogenase genes are uncommon and highly conserved in *Rhizobium leguminosarum* bv. viciae. FEMS Microbiol Lett 253:83–88

Fliegmann J, Jauneau A, Pichereaux C, Rosenberg C, Gasciolli V, Timmers ACJ, Burlet-Schiltz O, Cullimore J, Bono JJ (2016) LYR3, a high-affinity LCO-binding protein of *Medicago truncatula*, interacts with LYK3, a key symbiotic receptor. FEBS Lett 590:1477–1487

Foo E, Davies NW (2011) Strigolactones promote nodulation in pea. Planta 234:1073–1081

Forsberg LS, Reuhs B (1997) Structural characterization of the K antigens from *Rhizobium fredii* USDA257: evidence for a common structural motif, with strain-specific variation, in the capsular polysaccharides of *Rhizobium* spp. J Bacteriol 179:5366–5371

Forsberg LS, Carlson RW (1998) The structures of the lipopolysaccharides from *Rhizobium etli* strains CE358 and CE359 – the complete structure of the core region of *R. etli* lipopolysaccharides. J Biol Chem 273:2747–2757

Fraysse N, Couderc F, Poinsot V (2003) Surface polysaccharide involvement in establishing the *Rhizobium*-legume symbiosis. Eur J Biochem 270:1365–1380

Fukuhara H, Minakawa Y, Akao S, Minamisawa K (1994) The involvement of indole-3-acetic acid produced by *Bradyrhizobium elkanii* in nodule formation. Plant Cell Physiol 35:1261–1265

Gage DJ (2002) Analysis of infection thread development using Gfp- and DsRed-expressing *Sinorhizobium meliloti*. J Bacteriol 184:7042–7046

Garrocho-Villegas V, Gopalasubramaniam SK, Arredondo-Peter R (2007) Plant hemoglobins: what we know six decades after their discovery. Gene 398:78–85

Gibson KE, Kobayashi H, Walker GC (2008) Molecular determinants of a symbiotic chronic infection. Annu Rev Genet 42:413–441

Giordano W (2015) Rhizobial extracellular signaling molecules and their functions in symbiotic interactions with legumes. In: Quorum sensing vs quorum quenching: a battle with no end in sight. Springer, India, pp 123–132

Girard G, Roussis A, Gultyaev AP, Pleij CW, Spaink HP (2003) Structural motifs in the RNA encoded by the early nodulation gene enod40 of soybean. Nucleic Acids Res 31:5003–5015

Giraud E, Moulin L, Vallenet D, Barbe V, Cytryn E, Avarre JC, Jaubert M, Simon D, Cartieaux F, Prin Y, Bena G, Hannibal L, Fardoux J, Kojadinovic M, Vuillet L, Lajus A, Cruveiller S, Rouy Z, Mangenot S, Segurens B, Dossat C, Franck WL, Chang WS, Saunders E, Bruce D, Richardson P, Normand P, Dreyfus B, Pignol D, Stacey G, Emerich D, Verméglio A, Médigue C, Sadowsky M (2007) Legumes symbioses: absence of Nod genes in photosynthetic bradyrhizobia. Science 316:1307–1312

González JE, Marketon MM (2003) Quorum sensing in nitrogen-fixing rhizobia. Microbiol Mol Biol Rev 67:574–592

Gough C, Jacquet C (2013) Nod factor perception protein carries weight in biotic interactions. Trends Plant Sci 18(10):566–574

Gray XJ, Zhan H, Levery SB, Battisti L, Rolfe BG, Leigh JA (1991) Heterologous exopolysaccharide production in *Rhizobium* sp. strain NGR234 and consequences for nodule development. J Bacteriol 173:3066–3077

Gronlund M, Roussis A, Flemetakis E, Quaedvlieg NE, Schlaman HR, Umehara Y, Katinakis P, Stougaard J, Spaink HP (2005) Analysis of promoter activity of the early nodulin Enod40 in *Lotus japonicus*. Mol Plant Microbe Interact 18:414–427

Guenther JF, Chanmanivone N, Galetovic MP, Wallace IS, Cobb JA, Roberts DM (2003) Phosphorylation of soybean nodulin 26 on serine 262 enhances water permeability and is regulated developmentally and by osmotic signals. Plant Cell 15:981–991

Haag AF, Baloban M, Sani M, Kerscher B, Pierre O, Farkas A, Longhi R, Boncompagni E, Hérouart D, Dall'Angelo S, Kondorosi E, Zanda M, Mergaert P, Ferguson GP (2011) Protection of *Sinorhizobium* against host cysteine-rich antimicrobial peptides is critical for symbiosis. PLoS Biol 9(10):e1001169

Haag AF, Arnold MFF, Myka KK, Kerscher B, Dall'Angelo S, Zanda M, Mergaert P, Ferguson GP (2013) Molecular insights into bacteroid development during Rhizobium–legume symbiosis. FEMS Microbiol Rev 37:364–383

Halim MA, Rahman AY, Sim K-S, Yam H-C, Rahim AA, Ghazali AHA, Najimudin N (2016) Genome sequence of a Gram-positive diazotroph, *Paenibacillus durus* type strain ATCC 35681. Genome Announc 4:e00005–e00016

Haney CH, Long SR (2010) Plant flotillins are required for infection by nitrogen-fixing bacteria. Proc Natl Acad Sci U S A 107:478–483

Hashimoto Y, Kondo T, Kageyama Y (2008) Lilliputians get into the limelight: novel class of small peptide genes in morphogenesis. Dev Growth Differ 50(Suppl 1):S269–S276

Hassan S, Mathesius U (2012) The role of flavonoids in root–rhizosphere signalling: opportunities and challenges for improving plant–microbe interactions. J Exp Bot 63(9):3429–3444

Hause B, Schaarschmidt S (2009) The role of jasmonates in mutualistic symbioses between plants and soil-born microorganisms. Phytochemistry 70:1589–1599

Hebelstrup KH, Hunt P, Dennis E, Jensen SB, Jensen EO (2006) Hemoglobin is essential for normal growth of *Arabidopsis* organs. Physiol Plant 127:157–166

Hidalgo E, Leyva A, Ruiz-Argüeso T (1990) Nucleotide sequence of the hydrogenase structural genes from *Rhizobium leguminosarum*. Plant Mol Biol 15:367–370

Howard JB, Rees DC (2006) Nitrogen fixation special feature: how many metals does it take to fix N_2? A mechanistic overview of biological nitrogen fixation. Proc Natl Acad Sci U S A 103:17088–17093

Hoy JA, Robinson JT, Kakar S, Smagghe BJ, Hargrove MS (2007) Plant hemoglobins: a molecular fossil record for the evolution of oxygen transport. J Mol Biol 371:168–179

Hoy JA, Hargrove MS (2008) The structure and function of plant hemoglobins. Plant Physiol Biochem 46:371–379

Hubbell DH, Morales VM, Umali-García M (1978) Pectolytic enzymes in *Rhizobium*. Appl Environ Microbiol 35:210–213

Hubbell D (1981) Legume infection by *Rhizobium*: a conceptual approach. BioScience 31:832–837

Jacobsen-Lyon K, Jensen EO, Jorgensen JE, Marcker KA, Peacock WJ, Dennis ES (1995) Symbiotic and nonsymbiotic hemoglobin genes of *Casuarina glauca*. Plant Cell 7:213–223

Janczarek M, Rachwał K, Marzec A, Grządziel J, Palusińska-Szysz M (2015a) Signal molecules and cell-surface components involved in early stages of the legume–rhizobium interactions. Appl Soil Ecol 84:94–113

Janczarek M, Rachwał K, Cieśla J, Ginalska G, Bieganowski A (2015b) Production of exopolysaccharide by *Rhizobium leguminosarum* bv. trifolii and its role in bacterial attachment and surface properties. Plant and Soil 388:211–227

Jaszek M, Janczarek M, Kuczyński K, Piersiak T, Grzywnowicz K (2014) The response of the *Rhizobium leguminosarum* bv. trifolii wild-type and exopolysaccharide-deficient mutants to oxidative stress. Plant Soil 376:75–94

Jiménez-Zurdo JI, Mateos PF, Dazzo FB, Martínez-Molina E (1996a) Cell-bound cellulase and polygalacturonase production by *Rhizobium* and *Bradyrhizobium* species. Soil Biol Biochem 28:917–921

Jiménez-Zurdo JI, Mateos PF, Dazzo FB, Martínez-Molina E (1996b) Influence of the symbiotic plasmid (pSym) on cellulase production by *Rhizobium leguminosarum* bv. trifolii ANU843. Soil Biol Biochem 28:131–133

Jones KM, Kobayashi H, Davies BW, Taga ME, Walker GH (2007) How rhizobial symbionts invade plants: the *Sinorhizobium-Medicago* model. Nat Rev Microbiol 5:619–633

Kelly SJ, Muszynski A, Kawaharada Y, Hubber AM, Sullivan JT, Sandal N, Carlson RW, Stougaard J, Ronson CW (2013) Conditional requirement for exopolysaccharide in the *Mesorhizobium-Lotus* symbiosis. Mol Plant Microbe Interact 26:319–329

Kijne JW, Bauchrowitz MA, Díaz CL (1997) Root lectins and rhizobia. Plant Physiol 115:869–873

Kim CH, Tully RE, Keister DL (1989) Exopolysaccharide-deficient mutants of *Rhizobium fredii* HH303 which are symbiotically effective. Appl Environ Microbiol 55:1852–1859

Kouchi H, Takane K, So RB, Ladha JK, Reddy PM (1999) Rice ENOD40: isolation and expression analysis in rice and transgenic soybean root nodules. Plant J 18:121–129

Kubo H (1939) Über hämoprotein aus den wurzelknöllchen von leguminosen. Acta Phytochim (Tokyo) 11:195–200

Kumagai H, Kinoshita E, Ridge RW, Kouchi H (2006) RNAi knock-down of ENOD40s leads to significant suppression of nodule formation in *Lotus japonicus*. Plant Cell Physiol 47:1102–1111

Kundu S, Trent JT III, Hargrove MS (2003) Plants, humans and hemoglobins. Trends Plant Sci 8:387–393

Lang K, Lindemann A, Hauser F, Göttfert M (2008) The genistein stimulon of *Bradyrhizobium japonicum*. Mol Genet Genomics 279:203–211

Laplaze L, Lucas M, Champion A (2015) Rhizobial root hair infection requires auxin signaling. Trends Plant Sci 20(6):332–334

Laporte P, Merchan F, Amor BB, Wirth S, Crespi M (2007) Riboregulators in plant development. Biochem Soc Trans 35:1638–1642

Lefebvre B, Timmers T, Mbengue M, Moreau S, Hervé C, Tóth K, Bittencourt-Silvestre J, Klaus D, Deslandes L, Godiard L, Murray JD, Udvardi MK, Raffaele S, Mongrand S, Cullimore J, Gamas P, Niebel A, Ott T (2010) A remorin protein interacts with symbiotic receptors and regulates bacterial infection. Proc Natl Acad Sci U S A 107:2343–2348

Lerouge P, Roché P, Faucher C, Maillet F, Truchet G, Promé J-C, Dénarié J (1990) Symbiotic host-specificity of *Rhizobium meliloti* is determined by a sulphated and acylated glucosamine oligosaccharide signal. Nature 344:781–784

Le Strange KK, Bender GL, Djordjevic MA, Rolfe BG, Redmond JW (1990) The *Rhizobium* strain NGR234 nodD1 gene product responds to activation by the simple phenolic compounds vanillin and isovanillin present in wheat seedling extracts. Mol Plant Microbe Interact 3:214–220

Lévy J, Bres C, Geurts R, Chalhoub B, Kulikova O, Duc G, Journet EP, Ané JM, Lauber E, Bisseling T, Dénarié J, Rosenberg C, Debellé F (2004) A putative Ca^{2+} and calmodulin-dependent protein kinase required for bacterial and fungal symbioses. Science 303(5662):1361–1364

Leyva A, Palacios JM, Ruiz-Argüeso T (1987) Cloning and characterization of hydrogen uptake genes from *Rhizobium leguminosarum*. Appl Environ Microbiol 53:2539–2543

Leyva A, Palacios JM, Murillo J, Ruiz-Argüeso T (1990) Genetic organization of the hydrogen uptake (hup) cluster from *Rhizobium leguminosarum*. J Bacteriol 172:1647–1655

Libault M (2015) The root hair: a single cell model for systems biology. In: de Bruijn F (ed) Biological nitrogen fixation, vol 2. John Wiley & Sons, New York, pp 417–424

Limpens E, Franken C, Smit P, Willemse J, Bisseling T, Geurts R (2003) LysM domain receptor kinases regulating rhizobial Nod factor-induced infection. Science 302(5645):630–633

Limpens E, Mirabella R, Fedorova E, Franken C, Franssen H, Bisseling T, Geurts R (2005) Formation of organelle-like N_2-fixing symbiosomes in legume root nodules is controlled by DMI2. Proc Natl Acad Sci U S A 102:10375–10380

Lioi L, Galasso I, Santantonio M, Lanave C, Bollini R, Sparvoli F (2006) Lectin gene sequences and species relationships among cultivated legumes. Genet Resour Crop Evol 53:1615–1623

Ljunggren H, Fahraeus G (1961) The role of polygalacturonase in root-hair invasion by nodule bacteria. J Gen Microbiol 26:521–528

Loh J, Stacey G (2003) Nodulation gene regulation in *Bradyrhizobium japonicum*: a unique integration of global regulatory circuits. Appl Environ Microbiol 69:10–17

López M, Carbonero V, Cabrera E, Ruiz-Argüeso T (1983) Effects of host on the expression of the H_2-uptake hydrogenase of *Rhizobium* in legume nodules. Plant Sci Lett 29:191–199

Mabood F, Souleimanov A, Khan W, Smith DL (2006) Jasmonates induce Nod factor production by *Bradyrhizobium japonicum*. Plant Physiol Biochem 44:759–765

Maj D, Wielbo J, Marek-Kozaczuk M, Skorupska A (2010) Response to flavonoids as a factor influencing competitiveness and symbiotic activity of *Rhizobium leguminosarum*. Microbiol Res 165:50–60

Mandal SM, Chakraborty D, Dey S (2010) Phenolic acids act as signaling molecules in plant-microbe symbioses. Plant Signal Behav 5:359–368

Mateos PF, Jiménez-Zurdo JI, Chen J, Squartini A, Haack S, Martínez-Molina E, Hubbell DH, Dazzo F (1992) Cell-associated pectinolytic and cellulolytic enzymes in *Rhizobium leguminosarum* biovar trifolii. Appl Environ Microbiol 58:1816–1822

Mateos PF, Baker DL, Philip-Hollingsworth S, Squartini A, Paruffo AD, Nuti MP, Dazzo FB (1995) Direct *in situ* identification of cellulose microfibrils associated with *Rhizobium leguminosarum* biovar trifolii attached to the root epidermis of white clover. Can J Microbiol 41:202–207

Mateos PF, Baker D, Petersen M, Velázquez E, Jiménez-Zurdo JI, Martínez-Molina E, Squartini A, Orgambide G, Hubbell D, Dazzo FB (2001) Erosion of root epidermal cell walls by *Rhizobium* polysaccharide-degrading enzymes as related to primary host infection in the *Rhizobium*-legume symbiosis. Can J Microbiol 47:475–487

Mathesius U (2009) Comparative proteomic studies of root-microbe interactions. J Proteomics 72:353–366

Matvienko M, van de Sande K, Yang WC, van Kammen A, Bisseling T, Franssen H (1994) Comparison of soybean and pea ENOD40 cDNA clones representing genes expressed during both early and late stages of nodule development. Plant Mol Biol 26:487–493

Marketon MM, Glenn SA, Eberhard A, González JA (2003) Quorum sensing controls exopolysaccharide production in *Sinorhizobium meliloti*. J Bacteriol 185:325–331

Maróti G, Kereszt A, Kondorosi E, Mergaert P (2011) Natural roles of antimicrobial peptides in microbes, plants and animals. Res Microbiol 162:363–374

Martínez-Abarca F, Martínez-Rodríguez L, López-Contreras JA, Jiménez-Zurdo JI, Toro N (2013) Complete genome sequence of the alfalfa symbiont *Sinorhizobium/Ensifer meliloti* strain GR4. Genome Announc 1:e00174–e00112

Martínez-Molina E, Morales VM, Hubbell DH (1979) Hydrolytic enzyme production by *Rhizobium*. Appl Environ Microbiol 38:1186–1188

McAnulla C, Edwards A, Sanchez-Contreras M, Sawers RG, Downie JA (2007) Quorum-sensing-regulated transcriptional initiation of plasmid transfer and replication genes in *Rhizobium leguminosarum* biovar viciae. Microbiology 153:2074–2082

McCoy E (1932) Infection by *Bact. radicicola* in relation to the microchemistry of the host's cell walls. Proc R Soc B 110:514–533

Mergaert P, Nikovics K, Kelemen Z, Maunoury N, Vaubert D, Kondorosi A, Kondorosi E (2003) A novel family in *Medicago truncatula* consisting of more than 300 nodule-specific genes coding for small, secreted polypeptides with conserved cysteine motifs. Plant Physiol 132:161–173

Miri M, Janakirama P, Held M, Ross L, Szczyglowski K (2015) Into the root: how cytokinin controls rhizobial infection. Trends Plant Sci 21:178–186

Montiel J, Szűcs A, Boboescu IZ, Gherman VD, Kondorosi É, Kereszt A (2015) Terminal bacteroid differentiation is associated with variable morphological changes in legume species belonging to the inverted repeat-lacking clade. Mol Plant Microbe Interact 29(3):210–219

Morales V, Martínez-Molina E, Hubbell D (1984) Cellulase production by *Rhizobium*. Plant Soil 80:407–415

Murray JD (2011) Invasion by invitation: rhizobial infection in legumes. Mol Plant Microbe Interact 24(6):631–639

Nievas F, Bogino P, Sorroche F, Giordano W (2012) Detection, characterization and biological effect of quorum-sensing signaling molecules in peanut-nodulating Bradyrhizobia. Sensors 12:2851–2873

Novák K, Chovanec P, Skrdleta V, Kropáčová M, Lisá L, Nemcová M (2002) Effect of exogenous flavonoids on nodulation of pea (*Pisum sativum* L.) J Exp Bot 53:1735–1745

Nutman P, Doncaster C, Dart P (1973) Infection of clover by root-nodule bacteria. British Film Institute, London

Oldroyd GED, Murray JD, Poole PS, Downie JA (2011) The rules of engagement in the legume-rhizobial symbiosis. Annu Rev Genet 45:119–144

Oldroyd GED (2013) Speak, friend, and enter: signalling systems that promote beneficial symbiotic associations in plants. Nat Rev Microbiol 11:252–263

Ott T, van Dongen JT, Günther C, Krusell L, Desbrosses G, Vigeolas H, Bock V, Czechowski T, Geigenberger P, Udvardi MK (2005) Symbiotic leghemoglobins are crucial for nitrogen fixation in legume root nodules but not for general plant growth and development. Curr Biol 15:531–535

Palacios JM, Manyani H, Martínez M, Ureta AC, Brito B, Básscones E, Rey L, Imperial J, Ruiz-Argüeso T (2005) Genetics and biotechnology of the H₂-uptake [NiFe] hydrogenase from *Rhizobium leguminosarum* bv. viciae, a legume endosymbiotic bacterium. Biochem Soc Trans 33:94–96

Pandey A, Sharma M, Pandey GK (2016) Emerging roles of strigolactones in plant responses to stress and development. Front Plant Sci 7:434

Papadopoulou K, Roussis A, Katinakis P (1996) *Phaseolus* ENOD40 is involved in symbiotic and non-symbiotic organogenetic processes: expression during nodule and lateral root development. Plant Mol Biol 30:403–417

Peck MC, Fisher RF, Long SR (2006) Diverse flavonoids stimulate NodD1 binding to nod gene promoters in *Sinorhizobium meliloti*. J Bacteriol 188:5417–5427

Peck MC, Fisher RF, Bliss R, Long SR (2013) Isolation and characterization of mutant *Sinorhizobium meliloti* NodD1 proteins with altered responses to luteolin. J Bacteriol 195:3714–3723

Peláez-Vico MA, Bernabéu-Roda L, Kohlen W, Soto MJ, López-Ráez JA (2016) Strigolactones in the *Rhizobium*-legume symbiosis: stimulatory effect on bacterial surface motility and down-regulation of their levels in nodulated plants. Plant Sci 245:119–127

Pellock BJ, Teplitski M, Boinay RP, Bauer WD, Walker GC (2002) A LuxR homolog controls production of symbiotically active extracellular polysaccharide II by *Sinorhizobium meliloti*. J Bacteriol 184(18):5067–5076

Perret X, Staehelin C, Broughton WJ (2000) Molecular basis of symbiotic promiscuity. Microbiol Mol Biol Rev 64:180–201

Pesce A, Couture M, Dewilde S, Guertin M, Yamauchi K, Ascenzi P, Moens L, Bolognesi M (2000) A novel two-over-two alpha-helical sandwich fold is characteristic of the truncated hemoglobin family. EMBO J 19:2424–2434

Podkowinski J, Zmienko A, Florek B, Wojciechowski P, Rybarczyk A, Wrzesinski J, Ciesiolka J, Blazewicz J, Kondorosi A, Crespi M, Legocki A (2009) Translational and structural analysis of the shortest legume ENOD40 gene in *Lupinus luteus*. Acta Biochim Pol 56:89–102

Poustini K, Mabood F, Smith DL (2010) Preincubation of *Rhizobium leguminosarum* bv. phaseoli with jasmonate and genistein signal molecules increases bean (*Phaseolus vulgaris* L.) nodulation, nitrogen fixation and biomass production. J Agric Sci Technol 9:107–117

Prinsen E, Chauvaux N, Schmidt J, John M, Wieneke U, Greef JD, Schell J, Onckelen HV (1991) Stimulation of indole-3-acetic acid production in *Rhizobium* by flavonoids. FEBS Lett 282:53–55

Puppo A, Pauly N, Boscari A, Mandon K, Brouquisse R (2013) Hydrogen peroxide and nitric oxide: key regulators of the legume—rhizobium and mycorrhizal symbioses. Antioxid Redox Signal 18:2202–2219

Qu ZL, Wang HY, Xia GX (2005) GhHb1: a nonsymbiotic hemoglobin gene of cotton responsive to infection by *Verticillium dahliae*. Biochim Biophys Acta 1730:103–113

Rees DC, Howard JB (2000) Nitrogenase: standing at the crossroads. Curr Opin Chem Biol 4:559–566

Remigi P, Zhu J, Young JPW, Masson-Boivin C (2016) Symbiosis within symbiosis: evolving nitrogen-fixing legume symbionts. Trends Microbiol 24:63–75

Rey L, Murillo J, Hernando Y, Hidalgo E, Cabrera E, Imperial J, Ruiz-Argüeso T (1993) Molecular analysis of a microaerobically induced operon required for hydrogenase synthesis in *Rhizobium leguminosarum* bv. viciae. Mol Microbiol 8:471–481

Rinaudi-Marron LV, González JE (2013) Role of quorum sensing in the Sinorhizobium Meliloti–Alfalfa Symbiosis. In: de Bruijn FJ (ed) Molecular microbial ecology of the rhizosphere, vol 1–2. John Wiley & Sons, Inc., Hoboken, NJ, pp 535–540

Roberts NJ, Morieri G, Kalsi G, Rose A, Stiller J, Edwards A, Xie F, Gresshoff PM, Oldroyd GED, Downie JA, Etzler ME (2013) Rhizobial and mycorrhizal symbioses in *Lotus japonicus* require lectin nucleotide phosphohydrolase, which acts upstream of calcium signaling. Plant Physiol 161:556–567

Robledo M, Jiménez-Zurdo JI, Velázquez E, Trujillo ME, Zurdo-Piñeiro JL, Ramírez-Bahena MH, Ramos B, Díaz-Mínguez JM, Dazzo F, Martínez-Molina E, Mateos PF (2008) *Rhizobium* cellulase CelC2 is essential for primary symbiotic infection of legume host roots. Proc Natl Acad Sci U S A 105:7064–7069

Robledo M, Jiménez-Zurdo JI, Soto MJ, Velázquez E, Dazzo F, Martínez-Molina E, Mateos PF (2011) Development of functional symbiotic white clover root hairs and nodules requires tightly regulated production of rhizobial cellulase CelC2. Mol Plant Microbe Interact 24:798–807

Robledo M, Rivera L, Jiménez-Zurdo JI, Rivas R, Dazzo F, Velázquez E, Martinez-Molina E, Hirsch AM, Mateos PF (2012) Role of *Rhizobium* endoglucanase CelC2 in cellulose biosynthesis and biofilm formation on plant roots and abiotic surfaces. Microb Cell Fact 11:125

Rohrig H, Schmidt J, Miklashevichs E, Schell J, John M (2002) Soybean ENOD40 encodes two peptides that bind to sucrose synthase. Proc Natl Acad Sci U S A 99:1915–1920

Rosas S, Soria R, Correa N, Abdala G (1998) Jasmonic acid stimulates the expression of nod genes in *Rhizobium*. Plant Mol Biol 38:1161–1168

Ross EJH, Lira-Ruan V, Arredondo-Peter R, Klucas RV, Sarath G (2002) Recent insights into plant hemoglobins. Rev Plant Biochem Biotechnol 1:173–189

Ruiz-Argüeso T, Imperial J, Palacios JM (2000) In: Triplett EW (ed) Prokaryotic nitrogen fixation: a model system for analysis of a biological process. Horizon Scientific Press, Wymondham, pp 489–507

Sinharoy S, Liu C, Breakspear A, Guan D, Shailes S, Nakashima J, Zhang D, Wen J, Torres-Jerez I, Oldroyd GED, Murray JD, Udvardi MK (2016) A *Medicago truncatula* cystathionine-β-synthase-like domain-containing protein is required for rhizobial infection and symbiotic nitrogen fixation. Plant Physiol 170:2204–2217

Simpson FB, Burris RH (1984) A nitrogen pressure of 50 atmospheres does not prevent evolution of hydrogen by nitrogenase. Science 224:1095–1097

Skorupska A, Janczarek M, Marczak M, Mazur A, Król J (2006) Rhizobial exopolysaccharides: genetic control and symbiotic functions. Microb Cell Fact 16:5–7

Soto MJ, Fernández-Aparicio M, Castellanos-Morales V, García-Garrido JM, Ocampo JA, Delgado MJ, Vierheilig H (2010) First indications for the involvement of strigolactones on nodule formation in alfalfa (*Medicago sativa*). Soil Biol Biochem 42:383–385

Spaepen S, Das F, Luyten E, Michiels J, Vanderleyden J (2009) Indole-3-acetic acid-regulated genes in *Rhizobium etli* CNPAF512. FEMS Microbiol Lett 291(2):195–200

Spaepen S, Vanderleyden J (2011) Auxin and plant-microbe interactions. Cold Spring Harb Perspect Biol 3:185–190

Spaink HP, Sheeley DM, van Brussel AAN, Glushka J, York WS, Tak T, Geiger O, Kennedy EP, Reinhold VN, Lugtenberg BJJ (1991) A novel highly saturated fatty acid moiety of lipo-oligosaccharide signals determines host specificity of *Rhizobium*. Nature 354:125–130

Sreevidya VS, Hernandez-Oane RJ, So RB, Sullia SB, Stacey G, Ladha JK, Reddy PM (2005) Expression of the legume symbiotic lectin genes psl and gs52 promotes rhizobial colonization of roots in rice. Plant Sci 169:726–736

Stacey G, Libault M, Brechenmacher L, Wan J, May GD (2006) Genetics and functional genomics of legume nodulation. Curr Opin Plant Biol 9:110–121

Subramanian S, Stacey G, Yu O (2006) Endogenous isoflavones are essential for the establishment of symbiosis between soybean and *Bradyrhizobium japonicum*. Plant J 48:261–273

Subramanian S, Stacey G, Yu O (2007) Distinct, crucial roles of flavonoids during legume nodulation. Trends Plant Sci 12:282–285

Sugiyama A, Yazaki K (2014) Flavonoids in plant rhizospheres: secretion, fate and their effects on biological communication. Plant Biotechnol 31:431–443

Sun J, Cardoza V, Mitchell DM, Bright L, Oldroyd GED, Harris JM (2006) Crosstalk between jasmonic acid, ethylene and Nod factor signaling allows integration of diverse inputs for regulation of nodulation. Plant J 46(6):961–970

Suominen L, Luukkainen R, Roos C, Lindström K (2003) Activation of the nodA promoter by the *nodD* genes of *Rhizobium galegae* induced by synthetic flavonoids or *Galega orientalis* root exudate. FEMS Microbiol Lett 219:225–232

Suzaki T, Ito M, Kawaguchi M (2013) Genetic basis of cytokinin and auxin functions during root nodule development. Front Plant Sci 4:42

Sytnikov DM (2013) How to increase the productivity of the soybean–rhizobial symbiosis. In: James EB (ed) A comprehensive survey of international soybean research – genetics, physiology, agronomy and nitrogen relationships. InTech, Rijeka, Croatia, pp 51–82

Tadra-Sfeir MZ, Souza EM, Faoro H, Müller-Santos M, Baura VA, Tuleski TR, Rigo LU, Yates MG, Wassem R, Pedrosa FO, Monteiro RA (2011) Naringenin regulates expression of genes involved in cell wall synthesis in *Herbaspirillum seropedicae*. Appl Environ Microbiol 77:2180–2183

Taylor ER, Nie XZ, MacGregor AW, Hill RD (1994) A cereal haemoglobin gene is expressed in seed and root tissues under anaerobic conditions. Plant Mol Biol 24:853–862

Theunis M, Kobayashi H, Broughton WJ, Prinsen E (2004) Flavonoids, NodD1, NodD2, and nod-box NB15 modulate expression of the y4wEFG locus that is required for indole-3-acetic acid synthesis in *Rhizobium* sp. strain NGR234. Mol Plant Microbe Interact 17: 1153–1161

Tjepkema JD (1983) Hemoglobins in the nitrogen-fixing root nodules of actinorhizal plants. Can J Bot 61:2924–2929

Tóth K, Stratil TF, Madsen EB, Ye J, Popp C, Antolín-Llovera M, Grossman C, Jensen ON, Schüßler A, Parniske M, Ott T (2012) Functional domain analysis of the remorin protein LjSYMREM1 in *Lotus japonicus*. PLoS One 7(1):e30817

Trevaskis B, Watts RA, Andersson CR, Llewellyn DJ, Hargrove MS, Olson JS, Dennis ES, Peacock WJ (1997) Two hemoglobin genes in *Arabidopsis thaliana*: the evolutionary origins of leghemoglobins. Proc Natl Acad Sci U S A 94:12230–12234

Vance CP, Heichel G (1991) Carbon in N_2 fixation: limitation and exquisite adaptation. Annu Rev Plant Physiol 42:373–392

van Damme EJM, Barre A, Rougé P, Peumans WJ (2004) Cytoplasmic/nuclear plant lectins: a new story. Trends Plant Sci 9:484–489

Van de Velde W, Zehirov G, Szatmari A, Debreczeny M, Ishihara H, Kevei Z, Farkas A, Mikulass K, Nagy A, Tiricz H, Satiat-Jeunemaître B, Alunni B, Bourge M, Kucho K, Abe M, Kereszt A, Maroti G, Uchiumi T, Kondorosi E, Mergaert P (2010) Plant peptides govern terminal differentiation of bacteria in symbiosis. Science 327:1122–1126

van Hameren B, Hayashi S, Gresshoff PM, Ferguson BJ (2013) Advances in the identification of novel factors required for soybean nodulation, a process critical to sustainable agriculture and food security. J Plant Biol Soil Health 1:6

van Rhijn P, Fujishige NA, Lim P-O, Hirsch AM (2001) Sugar-binding activity of pea (*Pisum sativum*) lectin is essential for heterologous infection of transgenic alfalfa (*Medicago sativa* L.) plants by *Rhizobium leguminosarum* biovar viciae. Plant Physiol 125:133–144

Van Rossum D, Schuurmans FP, Gillis M, Muyotcha A, Van Verseveld HW, Stouthamer AH, Boogerd FC (1995) Genetic and phenetic analyses of *Bradyrhizobium* strains nodulating peanut (*Arachis hypogaea* L.) roots. Appl Environ Microbiol 61:1599–1609

Varkonyi-Gasic E, White DW (2002) The white clover enod40 gene family: expression patterns of two types of genes indicate a role in vascular function. Plant Physiol 129:1107–1118. Erratum in: Plant Physiol 2002;130:514

Vignais PM, Billoud B (2007) Occurrence, classification, and biological function of hydrogenases: an overview. Chem Rev 107:4206–4272

Vijn I, Yang WC, Pallisgård N, Ostergaard Jensen E, van Kammen A, Bisseling T (1995) VsENOD5, VsENOD12 and VsENOD40 expression during *Rhizobium*-induced nodule formation on *Vicia sativa* roots. Plant Mol Biol 28:1111–1119

Virtanen AI, Laine TR (1946) Brown and green pigments in leguminous root nodules. Nature 1157:25–26

Vleghels I, Hontelez J, Ribeiro A, Fransz P, Bisseling T, Franssen H (2003) Expression of ENOD40 during tomato plant development. Planta 218:42–49

Yang M, Sun K, Zhou L, Yang R, Zhong Z, Zhu J (2009) Functional analysis of three AHL auto-inducer synthase genes in *Mesorhizobium loti* reveals the important role of quorum sensing in symbiotic nodulation. Can J Microbiol 55:210–214

Yuen JPY, Cassini ST, DeOliveira TT, Nagem TJ, Stacey G (1995) Xanthone induction of nod gene expression in *Bradyrhizobium japonicum*. Symbiosis 19:131–140

Wan X, Hontelez J, Lillo A, Guarnerio C, van de Peut D, Fedorova E, Bisseling T, Franssen H (2007) *Medicago truncatula* ENOD40-1 and ENOD40-2 are both involved in nodule initiation and bacteroid development. J Exp Bot 58:2033–2041

Wang H, Zhong Z, Cai T, Li S, Zhu J (2004) Heterologous overexpression of quorum-sensing regulators to study cell-density-dependent phenotypes in a symbiotic plant bacterium *Mesorhizobium huakuii*. Arch Microbiol 182:520–525

Wang D, Yang S, Tang F, Zhu H (2012) Symbiosis specificity in the legume–rhizobial mutualism. Cell Microbiol 14:334–342

Watts RA, Hunt PW, Hvitved AN, Hargrove MS, Peacock WJ, Dennis ES (2001) A hemoglobin from plants homologous to truncated hemoglobins of microorganisms. Proc Natl Acad Sci U S A 98:10119–10124

Wei M, Yokoyama T, Minamisawa K, Mitsui H, Itakura M, Kaneko T, Tabata S, Saeki K, Omori H, Tajima S, Uchiumi T, Abe M, Ohwada T (2008) Soybean seed extracts preferentially express genomic loci of *Bradyrhizobium japonicum* in the initial interaction with soybean, *Glycine max* (L.) Merr. DNA Res 15:201–214

Weston LA, Mathesius U (2013) Flavonoids: their structure, biosynthesis and role in the rhizo-sphere, including allelopathy. J Chem Ecol 39:283–297

Wittenberg JB, Bergersen FJ, Appleby CA, Turner GL (1974) Facilitated oxygen diffusion: the role of leghemoglobin in nitrogen fixation by bacteroids isolated from soybean root nodules. J Biol Chem 249:4057–4066

Wittenberg JB, Bolognesi M, Wittenberg BA, Guertin M (2002) Truncated hemoglobins: a new family of hemoglobins widely distributed in bacteria, unicellular eukaryotes, and plants. J Biol Chem 277:871–874

Xie F, Murray JD, Kim J, Heckmann AB, Edwards A, Oldroyd GED, Downie JA (2012) Legume pectate lyase required for root infection by rhizobia. Proc Natl Acad Sci U S A 109(2):633–638

Zdyb A, Demchenko K, Heumann J, Mrosk C, Grzeganek P, Göbel C, Feussner I, Pawlowski K, Hause B (2011) Jasmonate biosynthesis in legume and actinorhizal nodules. New Phytol 189:568–579

Zhang J, Subramanian S, Stacey G, Yu O (2009) Flavones and flavonols play distinct critical roles during nodulation of *Medicago truncatula* by *Sinorhizobium meliloti*. Plant J 57:171–183

Zuanazzi JAS, Clergeot PH, Quirion JC, Husson HP, Kondorosi A, Ratet P (1998) Production of *Sinorhizobium meliloti* nod gene activator and repressor flavonoids from *Medicago sativa* roots. Mol Plant Microbe Interact 11:784–794

Flavonoids and Nod Factors: Importance in Legume-Microbe Interactions and Legume Improvement

3

Anna Skorupska, Dominika Kidaj, and Jerzy Wielbo

Abstract

Biological nitrogen fixation is one of the most important physiological processes in which atmospheric nitrogen is reduced to ammonia by symbiotic bacteria called rhizobia belonging to α- and β-*Proteobacteria*. Legume plants (Fabaceae) enter into mutualistic symbiosis with nitrogen-fixing rhizobia which enable them to grow in nitrogen-limited agricultural soils. Infection of legumes by rhizobia involves a series of sequential steps in which plant flavonoids and rhizobial Nod factors activate plant transmission signaling and initiate nodule development. Inside the nodule, rhizobia multiply and differentiate into nitrogen-fixing bacteroids. Here, besides an overview of symbiosis, the role of signal molecules, flavonoids, and Nod factors in legume growth and yield enhancement is highlighted. Recent progress in the understanding of the functions of the symbiotic signaling factors in initiation and development of symbiosis is likely to facilitate successful application thereof in sustainable agriculture to promote growth and nodulation of legume plants.

3.1 Introduction

Nitrogen (N) is essential for the growth of all organisms, and although atmospheric nitrogen is the most abundant gas in Earth's atmosphere, this molecular form cannot be used by most organisms. Among variously distributed life forms, only some archaea and bacteria in free-living or symbiotic interactions are capable of transforming gaseous N_2 into ammonia, which can be assimilated by plants. The most

A. Skorupska (✉) • D. Kidaj • J. Wielbo
Department of Genetics and Microbiology, Maria Curie-Skłodowska University,
Akademicka 19, 20-033 Lublin, Poland
e-mail: anna.skorupska@poczta.umcs.lublin.pl

© Springer International Publishing AG 2017
A. Zaidi et al. (eds.), *Microbes for Legume Improvement*,
DOI 10.1007/978-3-319-59174-2_3

75

important N_2-fixing systems in agriculture are the symbiotic associations between crops and forage legumes and rhizobia (Herridge et al. 2008). Leguminous plants and compatible soil bacteria of the order *Rhizobiales* comprising the genera *Agrobacterium, Allorhizobium, Azorhizobium, Bradyrhizobium, Devosia, Mesorhizobium, Methylobacterium, Ochrobactrum, Phyllobacterium, Rhizobium,* and *Sinorhizobium,* commonly referred to as rhizobia, are taxonomically diverse members of the α- and β-subclasses of *Proteobacteria* and are capable of establishing a strong and functionally effective symbiotic interaction in which rhizobia fix nitrogen. Global biological nitrogen fixation (BNF) was estimated at 122 Tg of N per year, in which from 33 to 43 Tg occurs through legume-rhizobia symbiosis (Herridge et al. 2008; Peoples et al. 2009). Therefore, legumes are agriculturally and ecologically very important crop both for the productivity of ecosystems and in agriculture and account for 25% of the world's primary crop production (Peoples et al. 2009; Ferguson et al. 2010). Rhizobial symbioses with 18,000 legume species (Masson-Boivin et al. 2009), including more than 100 agriculturally important legumes in all geographical regions, contribute nearly half of the annual quantity of biological nitrogen fixation (BNF) in soil ecosystems (Graham and Vance 2003). Improving legume inoculation efficiency is extremely important considering the economic and environmental costs associated with the use of chemical nitrogen fertilizers in agriculture. In this chapter, we present an overview of *Rhizobium*-legume symbiotic interactions highlighting plant responses to different signaling molecules such as flavonoids and Nod factors, which mediate the beneficial legume plant-*Rhizobium* symbiosis. The recent research concerning the molecular mechanisms underlying the symbiotic specificity and function of signaling factors can be helpful in the application of Nod factors or flavonoids for promoting plant growth and enhancing crop yields.

3.2   *Rhizobium*-Legume Symbiosis

Establishment of symbiosis between leguminous plants and rhizobia requires a mutual recognition of both partners, which starts when flavonoids (secondary plant metabolites) are recognized as specific inducers of nodulation genes (*nod*) in rhizobia (Peters et al. 1986; Subramanian et al. 2007). Rhizobial infection initiates when the compatible bacteria attach to the tips of growing plant root hairs and a characteristic curl structure ("shepherd's crook") forms at the root hair tip. Within infected root hairs, a bacterial microcolony is established, forming a so-called infection chamber (pocket) from which a tip-growing extension emanates from the infection threadlike compartment created within the infection chamber (Oldroyd and Downie 2008; Fournier et al. 2008; Murray 2011). Rhizobial *nod* genes encoding enzymatic Nod proteins are responsible for the synthesis of species or strain-specific lipochitin oligosaccharides called Nod factors (NF, LCO), which cause root hair curling. At this point, the infection thread (IT), a tubular structure, is filled with growing rhizobia and elongates inside root hairs and, after reaching the root cortex, releases bacteria into plant cells in the nodule primordia (Gage 2004; Fournier et al. 2008).

In the nodule primordia, rhizobia are endocytosed into compartments termed symbiosomes surrounded by host plant-derived peribacteroid membranes (Perret et al. 2000; Gibson et al. 2008; Oldroyd et al. 2011). Inside the symbiosomes, bacteria differentiate into bacteroids, which begin to fix nitrogen, and a nodule is formed (Perret et al. 2000; Jones et al. 2007). Nodules are a specialized niche for intracellular nitrogen-fixing rhizobia in which an efficient metabolic exchange between the symbiotic partners occurs: in return for reduced nitrogen provided to the plant, the symbiotic bacteria are protected from environmental stresses and are supplied with carbon sources within the plant cells (Prell and Poole 2006; Gibson et al. 2008). In the initiation of symbiosis, formation of infection threads and host specificity, several classes of surface polysaccharides such as exopolysaccharides (EPS), capsular polysaccharide (KPS), cyclic glucans, and lipopolysaccharides (LPS) are also essential. Bacterial mutants defective in the synthesis of these cell compounds may produce nodules, but they are not able to fix nitrogen (Fraysse et al. 2003; Skorupska et al. 2006; Jones et al. 2007). These classes of polysaccharides play an important role in plants forming an indeterminate type of nodules with a persistent meristem, such as *Vicia*, *Medicago*, *Pisum*, or *Trifolium* (Skorupska et al. 1995; Becker et al. 2005). It has recently been shown that legume host plants monitor the rhizobial exopolysaccharides (EPS) structure at the initiation of the infection by transmembrane LysM serine/threonine receptor kinase EPR3, which acts positively in response to compatible EPS and negatively in response to incompatible EPS (Kawaharada et al. 2015). It has been suggested that EPS function by amplifying ongoing Nod factor-mediated signaling and/or suppressing plant defense responses. Thus, the plant-bacterial compatibility and bacterial access to legume roots are regulated by a two-stage mechanism involving sequential receptor-mediated recognition of the Nod factor and EPS signals (Kawaharada et al. 2015).

Following successful symbiosis, two major types of nodules are formed: determinate and indeterminate (Vasse et al. 1990; Maroti and Kondorosi 2014). Both types of nodules differ in the activity of the nodule meristem. In the determinate nodules formed by phaseolid legumes, the meristems function until the formation of the nodule primordium and the infected cells start nitrogen fixation. Bacteroids in determinate nodules have the morphology similar to those of cultured cells and can revert to the free-living form (Maroti and Kondorosi 2014). In the indeterminate nodules, the meristem remains active during nodule development, and new generations of cells infected by rhizobia form a developmental gradient. Indeterminate nodules have different nodule zones: (1) the apical meristem, (2) the invasion zone into which infection threads release rhizobia, (2–3) the interzone, (3) the nitrogen-fixing zone, (4) the senescence zone, and (5) the saprophytic zone in older nodules (Timmers et al. 1999). The bacteroids in the indeterminate nodules are terminally differentiated because they are irreversibly transformed to polyploid and cannot reverse to a viable form (Vasse et al. 1990; Maroti and Kondorosi 2014). Terminal bacteroid differentiation is host controlled and dependent on the presence of nodule-specific cysteine-rich (NCR) peptides, which are similar to defensins described in alfalfa, pea, and vetch species. These legumes belong to a so-called inverted-repeat-lacking clade (IRLC), known as the galegoid clade of the *Papilionoideae* subfamily (Mergaert et al. 2006; Van de Velde et al. 2010; Haag et al. 2013). This differentiation involves an alteration of the bacterial cell cycle, cell size, and membrane permeability. The action of NCR peptides at molecular level is not resolved; however, it has been suggested that NCR peptides may act by altering the transcriptional profiles of key cell cycle regulators and remodeling the transcriptome (Penterman et al. 2014). In this type of nodules, cells undergo several rounds of endoreduplication, and resulting cells have a ploidy level up to 32–64C and are ca. 80-fold larger than the meristematic 2C and 4C cells during infection (Cebolla et al. 1999). There is evidence that terminally differentiated bacteroids of IRLC plants are more efficient at fixing nitrogen than nonterminal-differentiated bacteroids of phaseolid plants (Oono et al. 2010). It has been suggested that different NCR peptides interfere with many aspects of the bacteroid metabolism to allow the efficiency of the nitrogen fixation process to be optimized (Van de Velde et al. 2010).

Inside a nodule, the microaerophilic environment is maintained under the legume control of the permeability of nodule cells to the oxygen diffusion barrier (Gibson et al. 2008; Haag et al. 2013); high-level expression of leghemoglobin in the infected cells,

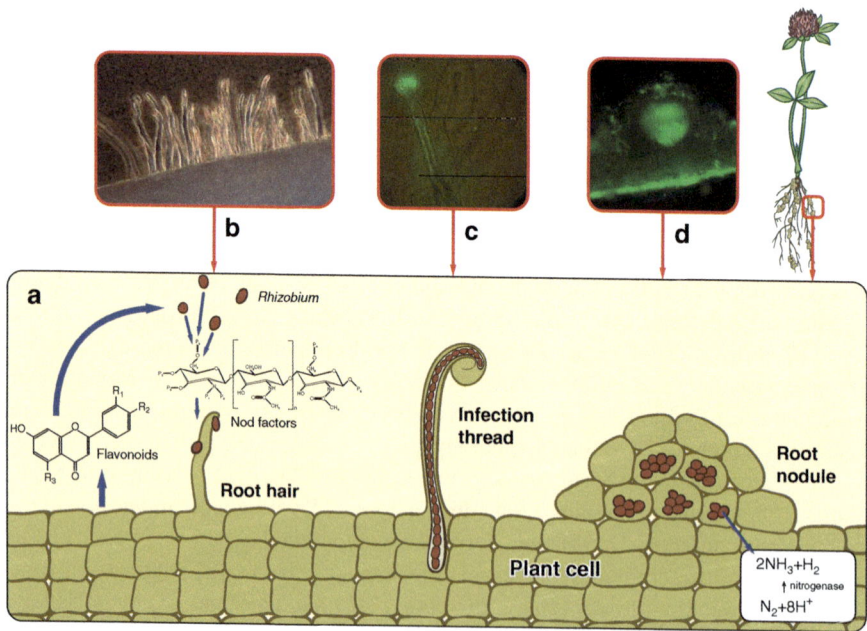

Fig. 3.1 *Rhizobium*-leguminous plant symbiosis: (**a**) plant flavonoid signals activate bacterial NodD protein, which induces *nod* genes expression and Nod factors synthesis; (**b**) Nod factor signals are recognized by specific plant receptors and initiate changes in the root hairs' growth, characteristic root hair curling, and infection thread formation; (**c**) infection threads, after reaching the root cortex, release bacteria into plant cells in the nodule primordia and initiate nodule development; (**d**) *Sinorhizobium meliloti* tagged with *gfp* are seen inside infection thread and young nodule

which reduces the O_2 concentration in the nodule; and the bacterial two-component regulatory system FixJ-FixL (Dixon and Kahn 2004; Haag et al. 2013) (Fig. 3.1).

Biological nitrogen fixation is a highly energy-dependent process, which requires at least 16 molecules of ATP for every two molecules of NH_3 synthesized. Rhizobia possess several *nif* genes encoding proteins involved in nitrogenase synthesis, and these function as regulatory proteins. NifH is a dinitrogenase reductase, also designated Fe-protein. NifD and NifK are α- and β-subunits of dinitrogenase, respectively, which form functional complexes with the FeMo cofactor. *nifB*, *nifE*, and *nifN* genes encode a molecular scaffold for the assembly of the FeMo cofactor. Prosthetic groups containing 4Fe–4S clusters are covalently bound to the MoFe protein bridging the α- and β-subunits. The 4Fe–4S group is also linked to the Fe-protein (Dixon and Kahn 2004; Newton 2007). The NifA performs a regulatory role in the expression of *nif* genes, and its synthesis and activity are regulated by oxygen. In rhizobia, besides *nif* genes with significant homology to *Klebsiella pneumoniae nif* genes (Ruvkun and Ausubel 1980), several other *fix* genes involved in nitrogen fixation have been described, and high plasticity in the composition and regulation of *nif-fix* genes has been observed among rhizobia (Masson-Boivin et al. 2009).

3.3 Role of Flavonoids in Symbiosis

Flavonoids are low molecular weight secondary metabolites produced by plants, and over 10,000 structural variants of flavonoids have been described (Cesco et al. 2010; Hassan and Mathesius 2012; Weston and Mathesius 2013). Flavonoids secreted both as aglycones and glycosides constitute a large part of root exudates. The flavonoid backbone is synthesized by condensation of 4-coumaryl-CoA provided by chalcone synthase, but several modifications of this structure yield different classes of flavonoids: flavanones, flavones, isoflavonoids, chalcones, and anthocyanidines (Harborne and Williams 2000). These molecules are typically accumulated at the tip of the main and lateral roots. High amounts of flavonoids are also secreted by the root hair zone, where infection of legumes by rhizobia occurs. Flavonoids and isoflavonoids play multiple roles at different stages of nodulation. They have natural regulatory roles at different stages in the plant rhizobial infection and signaling during root or nodule development (Peters et al. 1986; Redmond et al. 1986). The flavonoids are thought to serve as signal molecules in the rhizosphere to concentrate compatible rhizobia and induce Nod factor biosynthesis. They are chemoattractants for rhizobia (Caetano-Anollés et al. 1988; Cooper 2007), inducers of *nod* gene expression, determinants of host specificity, and developmental regulators by a role in auxin transport in roots during nodule formation (Wasson et al. 2006; Zhang et al. 2009; Hassan and Mathesius 2012). The most important role of flavonoids in symbiosis is the induction of *nod* genes initiating Nod factor synthesis. Flavonoids bound by the regulatory NodD protein (a transcription factor of the LysR family) induce expression of *nodABC* common nodulation genes encoding the NF core structure and other host-specific genes encoding enzymes that modify NF. In *Bradyrhizobium japonicum* isoflavonoids of soybean, daidzein and genistein, induce NF gene expression. However, daidzein prevents NF production in the noncompatible *S. meliloti*, which responds positively to the flavone luteolin (Cooper 2004).

Nod factors function at low nanomolar or micromolar concentrations ranging from 10^{-6} to 10^{-12} M (Maj et al. 2009; Skorupska et al. 2010). The presence of appropriate flavonoids induces changes in DNA topology at NodD binding site in the promoter region and allows RNA polymerase to initiate gene transcription (Chen et al. 2005). In recent research, the molecular model of NodD1 was generated based on the mutants in *S. meliloti nodD1*, and the domains important for DNA binding, luteolin inducer binding, multimerization, and interaction with RNA polymerase (RNAP) were predicted (Peck et al. 2013). Until now, attempts to crystallize the purified NodD protein have been unsuccessful probably because of the instability of NodD1 (Yeh et al. 2002).

Although the specific function of flavonoids as *nod* gene inducers has been described (Redmond et al. 1986; Subramanian et al. 2007), a mixture of single flavonoids and seeds or root exudates were more effective in the induction of *nod* genes and promotion of nodulation, indicating a possibility of synergistic effects between several *nod*-gene-activating substances (Zaat et al. 1988, 1989; Begum et al. 2001; Maj et al. 2010). These do not exclude the specific interaction of the NodD protein with individual flavonoids to increase binding of this protein to *nod* promoters; however, how NodD proteins from different symbiotic bacteria interact with the distinct flavonoid signals produced by their respective host plants is still unresolved (Peck et al. 2006, 2013).

Successful induction of *nod* genes in the presence of flavonoids has been evidenced by silencing of the isoflavonoid pathway in soybean by RNAi that caused an inhibition of nodulation, which was restored by inoculating plants with a flavonoid hypersensitive *Bradyrhizobium* strain or purified Nod factors (Subramanian et al. 2006, 2007). Pretreatment of rhizobial cells with an appropriate *nod* gene-inducing flavonoid did not restore nodulation in flavonoid-silenced roots, suggesting that *nod* gene induction in the rhizosphere is not sufficient for nodulation. These studies have shown that flavonoids, however, do play a crucial role in nodulation (Subramanian et al. 2007).

Some flavonoids show *nod* gene repressor activity for rhizobia. For example, the isoflavonoids medicarpin and cumesterol have been reported to negatively control Nod factor synthesis by *S. meliloti* (Zuanazzi et al. 1998). Presumably, an optimal level of *nod* inducers and Nod factor synthesis can prevent defense responses by the plant. The antioxidant and antibacterial properties of flavonoids also suggest that some endogenous flavonoids can act to protect the nitrogen-fixing rhizobia from oxidative stresses during rhizobial infection and plant colonization. Stimulation of *nod* genes with multiple flavonoid inducers may be considered as an advantageous feature of the symbiotic interaction and could be exploited in the pre-activation of rhizobia used as inoculants (biofertilizers) (Cooper 2004). Pre-activation of rhizobial strains with plant exudates or single flavonoids was described to increase rhizobial competitiveness in the soil environment (Begum et al. 2001; Maj et al. 2010; Skorupska et al. 2010).

Expression of flavonoid biosynthesis encoding genes and flavonoid accumulation in the host plant is induced by the Nod factor. Recently, a significant role of flavonoids in initiation of infection was demonstrated in the transcriptome analysis of *M. truncatula* root hairs shortly after *S. meliloti* infection or Nod factor treatment (Breakspear et al. 2014). A significant increase in the expression of genes encoding enzymes of the phenylpropanoid pathway producing inducers of *nod* genes such as dihydroxyflavone and methoxychalcone and isoflavones has been shown (Breakspear et al. 2014). In the same study, induction of genes encoding flavanones described earlier as *S. meliloti nod* inducers (Zhang et al. 2009) was not detected. In the first stage of infection, both *S. meliloti* and the purified Nod factor induced several other plant genes responsible for the production of phytoalexins and isoflavones (Breakspear et al. 2014). There is evidence that some flavonoids might be involved in the accumulation of auxins within cortical cells, which is necessary for further steps in nodule organogenesis (Wasson et al. 2006; Subramanian et al. 2007).

3.4 Function of Nod Factors in Symbiosis

In response to plant flavonoids, rhizobia produce Nod factors (NFs), which induce root hair curling, IF formation, initial response in root cortical cell division, and formation of nodule primordia. Nod factors are structurally diverse, and a single rhizobial strain can produce a range of these metabolites. They are produced by rhizobia belonging to α- and majority of β-*Proteobacteria* (Spaink 2000; Moulin et al. 2001). Nod factors are modified lipo-chitooligosaccharides (LCOs) containing two to six β-1,4-linked *N*-acetyl-D-glucosamine residues with a fatty acid replacing the *N*-acetyl group at the terminal nonreducing end and rhizobial species-specific residues at the terminal reducing end (Long 1996; Promé 1996; D'Haeze and Holsters 2002). The core structure of NFs is synthesized by conservative common *nodABC* genes, which are essential for nodulation, and mutations in these genes lead to the Nod⁻ phenotype (Jacobs et al. 1985; Debellé et al. 1986). The NodC synthase is responsible for biosynthesis of the β-1,4-linked *N*-acetyl-D-glucosamine oligomeric backbone. NodB deacetylates glucosamine at the nonreducing end, after which NodA acylates the free amino group of the terminal glucosamine (Spaink 2000; D'Haeze and Holsters 2002). Modifications of the backbone structure are carried out by the enzymatic products of host-specificity nodulation genes (*nodFE*, *nodH*, *nodG*, *nodPQ*, and several others) whose products modify the *N*-acylglucosamine backbone by adding species-specific residues involved in host specificity determination (Perret et al. 2000). Residues such as L-fucosyl, 2-*O*-Me-fucosyl, 4-*O*-Ac-fucosyl, acetyl, or sulfate esters are present at the reducing end of NF, and *N*-methyl, *O*-acetyl, and *O*-carbamoyl residues are found at the nonreducing end (Spaink 2000; D'Haeze and Holsters 2002). Structural modifications of NFs at the reducing or nonreducing ends have been shown to be important determinants of host specificity, e.g., the sulfate group at the reducing end of NF from *S. meliloti* is required for nodulation of *M. sativa*. The mutant strain producing non-sulfated NF gains the ability to form nodules on *Vicia sativa*, a nonhost plant of *S. meliloti* (Roche et al. 1991). Variation in the Nod factor structure does not fully explain the host-range specificity, which is more complex, and some other signals specify the host range of rhizobia (Perret et al. 2000).

The key role in the *Rhizobium*-legume infection is played by the regulatory *nodD* gene encoding a protein belonging to the LysR family of transcriptional regulators (Fisher et al. 1988; Kondorosi et al. 1989; Spaink 2000). NodD, in a complex with a flavonoid, binds conservative sequences upstream of *nod* operons, called *nod* boxes, acting as a transcriptional activator of several *nod*, *nol*, and *noe* gene promoters (Peck et al. 2006). Negative regulation of *nod* genes has also been described. The NolR protein, which binds to *nod* promoters, negatively controls their expression in the presence of luteolin, a specific *nod* gene inducer in *S. meliloti* (Cren et al. 1995). The products of *nodD*, *nodABC* genes producing core NF are also required for the establishment of the three-dimensional structure of the biofilm formed by *S. meliloti*, which is enhanced by the flavonoid luteolin—an inducer of *nod* genes (Fujishige et al. 2008). The core NF facilitates bacterial adhesion to the roots until an optimum concentration of the host-specific NF is produced in the presence of plant flavonoid inducers, and plant developmental processes are initiated. The essential function of NF in biofilm formation on the root surfaces is different from the established role as a morphogen for inducing legume nodule development (Fujishige et al. 2008; Faure et al. 2009).

The substituents of NFs also significantly affect their stability in the rhizosphere and protect against degradation by plant chitinases and other glycosyl hydrolases (Schultze and Kondorosi 1998; Staehelin et al. 2000). Legume plants synthesize specific enzymes that rapidly cleave and inactivate the NFs, whereas they are very stable in the rhizosphere of non-host plants. This suggests that NF degradation can be a prerequisite for a compatible symbiotic interaction (Staehelin et al. 1995). *N*-acetylglucosamine oligomers forming a core structure of NFs are essential for their biological activity, because deacetylation of NFs of *S. meliloti* caused only vestigial biological activity in the root hair deformation assay compared with an intact Nod factor.

Environmental stress factors such as elevated temperature, acidity, high osmolarity, or oxidative stress can also affect rhizobial symbiosis adversely. In *R. tropici* CIAT899, at a high NaCl concentration, synthesis of different NF structures occurred, and NFs were produced at a higher level than under normal conditions (Morón et al. 2005; Estévez et al. 2009). Moreover, Guasch-Vidal et al. (2013) found that in the presence of a high NaCl concentration (300 mM), the *nod* genes of *R. tropici* CIAT899 were induced, and synthesis of several structurally different but biologically active NFs occurred even in the absence of the flavonoid apigenin inducer. Purified structurally different NFs induced pseudonodules on the host plant in the absence of apigenin. Under this condition, part of the synthesized NFs had a changed structure as *N*-methyl substitution at the nonreducing end, but they were still structurally similar to Nod factors produced under apigenin induction. A majority of rhizobial species certainly require NF for initiation of nodulation; however, there are some exceptions (Masson-Boivin et al. 2009; Toth and Stacey 2015). Giraud et al. (2007) reported that photosynthetic *Bradyrhizobium* strains BTAi1 and ORS278 forming root and stem nodules on some *Aeschynomene* plants species can specifically nodulate plants in the absence of common nodulation genes, which are required for NF synthesis. In other experiment, *B. elkanii* mutant unable to produce NF has also been reported to form nodules on a nodulation-defective *nfr1* mutant of the soybean (Okazaki et al. 2013). The nodulation by this mutant was dependent on active effector proteins of the type III secretion system (T3SS), which functions in many but not all leguminous plants. The participation of the T3SS and T4SS secretion systems in the modulation of *nod* gene expression and plant reactions to rhizobial infection has earlier been reported for several *Bradyrhizobium*, *Mesorhizobium*, and *Sinorhizobium* strains (Fauvart and Michiels 2008; Deakin and Broughton 2009). In earlier studies, induction of the effector proteins of type III secretion (T3SS) in response to flavonoid exudates has been described (Krishnan et al. 2003).

The structural modifications of NF influence its binding to plant receptors like kinases or receptor proteins localized at the epidermal cells of root hairs. The extracellular region of these proteins can be composed of lysin motif (LysM)-domains and/or leucine-rich repeats, both of which are involved in microbe detection. Mutations in these genes significantly alter the nodulation capability of the legume host. Recognition of the symbiont triggers the calcium-dependent signaling pathway during initiation of infection (Oldroyd et al. 2011; Toth and Stacey 2015). Shortly after Nod factor recognition by plant receptors, oscillations in the calcium concentrations, termed calcium spiking, in the nucleoplasm and nuclear-associated cytoplasm and alterations in the root hair cytoskeleton are induced (Timmers et al. 1999). The synthesis of early nodulins (ENODs) is induced, and Ca^{2+} calmodulin-dependent protein kinase (CCaMK) is involved in cortical cell responses, which include reinitiation of the meristematic activity of cortical cells to form nodule primordia and establishment of symbiosis. Downstream, the protein nodulation signaling pathway NSP1 and NSP2 (two GRAS family transcriptional regulators) are required (extensively reviewed by Oldroyd and Downie 2008; Madsen et al. 2010; Oldroyd et al. 2011; Suzaki et al. 2015).

In recent research, the transcriptomics approach to study gene expression in root hairs from *M. truncatula* has shed light on the developmental events during rhizobial infection and the underlying hormone responses (Breakspear et al. 2014). This approach has revealed the induction of several cyclins, which activate the cell-division machinery in rhizobial infection. Changes in the cell cycle in plants are governed by hormones, in particular, auxin and cytokinin. One of the first responses of legumes to rhizobial infection is initial downregulation and then increasing synthesis of auxins in the infection site by auxin encoding genes (Tiwari et al. 2003; Breakspear et al. 2014). A main role of auxins is stimulation of cell wall acidification, thereby increasing its extensibility. Cell wall loosening required for root hair growth promotes formation of infection pockets and initiation of growth of infection threads (Esseling et al. 2003; Oldroyd et al. 2011). In *M. truncatula* root infection by *S. meliloti* or treatment with a purified Nod factor, significant induction of auxin encoding genes such as GH3.1 and ARF16a was specific to infected root hairs but not to uninfected ones, showing two possible responses of plants. Then, in the presence of the Nod factor, several auxin-transport and auxin-responsive genes were activated during root hair infection (Breakspear et al. 2014).

The NF is a potent mitogen, and during root infection it induces expression of genes encoding specific types of cyclin genes and many other proteins involved in DNA replication. A purified Nod factor can induce many signaling events in the absence of bacteria (Geurts et al. 2005); however, contrary to infection by rhizobia, NFs do not form infection threads in the root hairs (Timmers et al. 1999) showing two different plant responses to NFs. The NFs action is limited to triggering the entry of root hairs into the cell cycle, and, only in the presence of infecting rhizobia, cells enter into mitosis, and the infection thread is formed (Breakspear et al. 2014).

Besides flavonoids, plant hormones such as gibberellic acid, brassinosteroids, and strigolactone together with auxins promote cell expansion and growth of an infection thread during infection (Breakspear et al. 2014). Gibberellins are mainly involved in regulating plant cell division and elongation, and they influence almost all stages of plant growth, including seed germination, stem and leaf growth, floral induction, and fruit growth (Spaepen et al. 2009). Earlier, these compounds were reported to positively affect nodule organogenesis in several legumes (Soto et al. 2010; Foo et al. 2013; Liu et al. 2015). Ethylene is one of the hormones mediating defense responses. An ethylene-insensitive mutant (sickle) showed an altered expression of several putative defense-related proteins (Penmetsa and Cook 1997). Salicylic acid (SA) and jasmonic acid (JA) also play a role in regulating defense responses, and there is evidence that Nod factors downregulate defense responses mediated by SA (Martinez-Abarca et al. 1998) and that JA biosynthesis is enhanced during the early stages of infection (Kouchi et al. 2004). While Nod factor stimulates hormone synthesis, jasmonic acid (JA) production is repressed during early stages of infection, which is correlated with downregulation of plant defenses necessary for rhizobial colonization (Ballare 2011). During *S. meliloti* infection and Nod factor treatment, degradation of NF by specific chitinase *NFH1* was observed in *M. truncatula*, which interrupted the flavonoid-Nod factor action (Tian et al. 2013; Breakspear et al. 2014). Also, NF receptors *NFP* and *LYK3* were repressed, which indicated the host regulation of Nod factor synthesis and perception. As a result of complex interactions of several signaling factors such as flavonoids, NFs, hormones, and plant proteins, functional nodules infected by rhizobia are formed.

3.5 Flavonoids as Biofertilizers in Agriculture

The main problem in using rhizobia as microbial inoculant in agricultural practice is their lack of competitiveness with indigenous rhizobial strains, which are generally abundant and better adapted to a range of environmental conditions (Andrade et al. 2002; Laguerre et al. 2006; Yates et al. 2011). Traditionally, the rhizobial inoculants have provided better results when the local indigenous rhizobial populations are low or absent due to lack of the specific host plant; hence, the inoculant strain does not face strong competition. This can only be observed in the case of fallow soils, where there are no wild legumes, or in soils where plants from this group, e.g., soybeans in North America, have never been grown. In such areas, live cultures of *B. japonicum* are used successfully (Souleimanov et al. 2002).

In some cases, flavonoid pretreatment of rhizobial inoculants can facilitate the process of nodule infection and influence their symbiotic activity. Specific flavonoids or clover seed exudates (mixtures of flavonoids) stimulated green wet mass and the number of nodules in clover plants inoculated with *R. leguminosarum* bv. *trifolii* strains treated with flavonoids under laboratory conditions (Maj et al. 2010). Pretreatment of *R. leguminosarum* bv. *viciae* pIJ1477 with flavonoid hesperetin, an inducer of *nod* genes, improved the efficiency of the symbiosis with pea and lentil at low temperature (17 °C) in laboratory experiments (Begum et al. 2001). Both legume plants, after the application of the inoculant treated with the flavonoid, showed a statistically significant increase in the number of root nodules and dry biomass. A stimulation effect of flavonoids on legume growth was observed in the case of bean nodulation by *R. leguminosarum* bv. *phaseoli* and *R. tropici* strains pretreated with flavonoids (quercetin, malvidin glucoside) (Hungria and Phillips 1993). Pretreatment of *B. japonicum* with genistein resulted in an increase in total protein and grain yields of soybean (10–40%), compared to conventional inoculants under laboratory conditions (Zhang and Smith 1995, 1996).

The effect of pre-activation of rhizobial inoculants with specific flavonoids on legumes was also studied under field conditions. Some flavonoids have already been used as commercial products in promoting legume yields by stimulating the natural *rhizobium*-legume plant symbiosis (Hungria and Stacey 1997; Mabood et al. 2008). For example, genistein and daidzein, i.e., inducers of *B. japonicum nod* genes, are a component of a biofertilizer known under the tradename SoyaSignal. The use of this product increased the yield of soybean by 10% (Leibovitch et al. 2001). Siczek et al. (2015) observed that the application of flavonoids on pea seeds can improve root-microbial interactions through enhancement of the number of PGPR microbes and fungal populations in the soil as well as their activity in the pea rhizosphere. It is known that flavonoids can act as chemoattractants (Caetano-Anollés et al. 1988) and a carbon source for many microorganisms, including *Pseudomonas*, *Bacillus*, and *Rhizobium*.

3.6   Practical Use of Nod Factors in Legume and Nonlegume Cultivation

Rhizobial Nod factors (alone or together with bacterial cultures) have often been used as biofertilizers to improve the effectiveness of BNF, making the process of recognition of symbiosis partners more effective. Under laboratory experiments, application of Nod factors directly to legume seeds accelerated plant growth at early developmental stages and increased the number of nodules (Prithiviraj et al. 2003; Macchiavelli and Belles-Marino 2004). Generally, Nod factors enhance the plant yield by facilitating seed germination and increasing shoot and root mass and the number of lateral roots of plants (Souleimanov et al. 2002; Oláh et al. 2005; Maj et al. 2009). It was shown that Nod factors isolated from *Rhizobium* sp. NGR234 positively stimulated colonization of lablab bean (*Lablab purpureus*) roots by mycorrhizal fungi (Xie et al. 1997). Moreover, it was noted that Nod factor treatment reduced the incidence of *Microsphaera diffusa* infection resulting in powdery mildew of soybean and hence improved plant health (Duzan et al. 2005). Probably, this is connected with the correlation between Nod factors and the phytoalexin concentration in plant tissues (Dakora 2003). The effect of biofertilizers containing the rhizobial Nod factors is not limited to legume plants. Similar root cell surface receptors, which are responsible for recognition of molecular signals produced by mycorrhizal fungi, were also found in other groups of plants (Maillet et al. 2011). Although NFs synthesized both by mycorrhizal fungi (Myc factors) and by symbiotic rhizobia (Nod factors) use slightly different signal transduction pathways in plant cells, they are structurally similar, and also non-legume plants can respond to the presence of rhizobial NFs (Maillet et al. 2011). It suggests a possibility of a quite universal use of Nod factors, not only in legume cultivation. Rhizobial Nod factors are morphogens that stimulate meristem activity of various plant tissues at nano- and micromolar concentrations. Moreover, the production of NF-containing extracts could be quite simple; therefore, such biofertilizers might be an inexpensive alternative for artificial nitrogen fertilizers.

Even though the structure, biosynthesis, and mechanisms of action of rhizobial Nod factors have been widely studied, yet the reports on the use of NFs in greenhouse or field experiments are scarce. The potential agricultural use of Nod factors was indicated two decades ago in a patent application describing formulations of different NFs (Lerouge et al. 1996; Hungria and Stacey 1997). A general recommendation for preparation of plant-treatment solutions was proposed, but neither a detailed composition of the mixture nor experimental data describing the effect of the formulation on plant growth were presented. A few years later, some experiments showing a seedling emergence-promoting effect of NFs under field conditions were described in another patent application (Smith et al. 2005a, b), and a beneficial effect of rhizobial Nod factors on leguminous (soybean) and non-leguminous (corn, cotton, beet) plants was presented.

Agro-technologies taking advantage of NFs were mainly developed in North America, and they are focused on soybean and corn. Multiple-site trials conducted in North and South America during the last decade revealed that NF treatment of soybean gave a positive response in almost three-fourths of the conducted experiments. The application of NFs often improved plant growth via an increase in nitrogen and sugar concentrations, finally resulting in better seed production. The increase in soybean yield could rarely reach up to 25%; however, the average increase was 3–4%. The beneficial effect of NFs was dependent on the presence of rhizobia in the soil, and, in sites where *B. japonicum* were not detected in the soil, NF treatment resulted only in a slight and insignificant increase in harvest (Smith et al. 2015). Rhizobial NFs were also used in plant cultivation together with other compounds. The growth of different soybean cultivars was improved by preparations containing bradyrhizobial Nod factors and different flavonoids (e.g., genistein, daidzein, formononetin, naringenin, hesperetin, luteolin, or apigenin), which were used for improving plant-microbial signal exchange and for activation of rhizobia in symbiosis (Smith and Osburn 2009). The other modification involved application of chitin or chitosan together with NFs. It was demonstrated that foliar or seed treatment of soybean with such compounds slightly increased the beneficial effect of Nod factors, but the detailed mechanism was not elucidated (Smith and Osburn 2010). Moreover, numerous concepts concerning the combined use of NFs with selected fungicides, insecticides, and their combination were also presented with a moderate effect (Smith et al. 2005a, b).

Enhanced growth of plants and increased yield were the main goals to be attained after Nod factor treatment; however, some subsidiary and advantageous effects were also noted. It was reported that foliar application of NFs alleviated soybean yield losses under water-deficient conditions by extending the root system and providing access to more soil water (Atti et al. 2005).

NFs produced by *Bradyrhizobium* cultures and application thereof in soybean cropping have been best studied, but there are some reports concerning other legumes and bacterial species. Promotion of nodulation and growth of red clover and vetch was observed after treatment of seeds with Nod factors isolated from *R. leguminosarum* bv. *trifolii* (Maj et al. 2009) and *R. leguminosarum* bv. *viciae* strains, respectively (Kidaj et al. 2012). Some greenhouse experiments showed that foliar or seed application of NFs produced by *R. leguminosarum* bv. *viciae* cultures increased nodulation and photosynthetic activity, thus significantly improving the yield of different pea genotypes (Podleśny et al. 2014a, b). A similar but more profitable effect was observed in pot and field trials when specific *R. leguminosarum* bv. *viciae* Nod factors were sprayed on pea leaves together with sulfur, which positively influenced the efficiency of nitrogen fixation as well as several physiological features such as photosynthesis and transpiration intensity of pea plants, and enhanced important agronomic properties such as pod and seed numbers (Podleśna et al. 2015). In other field experiments, the application of a preparation of Nod factors on pea seeds resulted in increased nitrogenase activity in nodules and increased plant yield, but the positive responses were not found in each year of cultivation (Siczek et al. 2014). Rhizobial LCOs were also used in nonleguminous plant cultivations. Foliar application to field-grown tomatoes resulted in a significant increase in the number of fruits and yield (Chen et al. 2006). Interestingly, this result was similar to an effect observed for pea (Podleśny et al. 2014a), in which the increased yield was a consequence of an increase in the pod number, suggesting changes in the flowering capacity of plants after such treatment.

Conclusion

The legume-rhizobia symbiosis has enormous ecological and agronomic importance. The symbioses can successfully be optimized by applying the most competitive and highly nitrogen fixation-efficient rhizobial strains even under soils having low number and poor competitiveness of autochthonous rhizobia. Research conducted so far has shown that the productivity of symbioses can be improved both by competitive rhizobia and by manipulating the bacteria and plant signal compounds. Advances in molecular studies have provided some new and exciting information about the function of molecular signals in the symbiotic interactions of rhizobia with legume plants. The signaling factors, such as plant flavonoids, and bacterial Nod factors initiate symbiosis and determine successful development of effective symbiosis. It was found that specific flavonoids applied to rhizobia or plant seeds can stimulate the symbioses. Also, Nod factors acting as an initial signal and potent morphogen in symbiosis positively affected the yield of several legumes. A significant increase in the productivity of legumes has recently been achieved due to application of Nod factors to seed and leaves. New-formulated biofertilizers containing flavonoids or isolated Nod factors have been developed and patented for use in legume production.

Acknowledgments   This study is included in the framework of following Research Project from Polish National Centre for Research and Development PBS3/A8/28/2015 SEGENMAS.

References

Andrade DS, Murphy PJ, Giller KJ (2002) The diversity of *Phaseolus*-nodulating rhizobial populations is altered by liming of acid soils planted with *Phaseolus vulgaris* L. in Brazil. Appl Environ Microbiol 68:4025–4034

Atti S, Bonnell R, Prasher S, Smith DL (2005) Response of soybean (Glycine max (L.) Merr.) under chronic water deficit to LCO application during flowering and pod filling. Irrig Drain 54:15–30

Ballare CL (2011) Jasmonate-induced defences: a tale of intelligence, collaborators and rascals. Trends Plant Sci 16:249–257

Becker A, Fraysse N, Sharypova L (2005) Recent advances in studies on structure and symbiosis-related function of rhizobial K-antigens and lipopolysaccharides. Mol Plant Microbe Interact 18:899–905

Begum AA, Leibovitch S, Migner P, Zhang F (2001) Specific flavonoids induced nod gene expression and pre-activated nod genes of *Rhizobium leguminosarum* increased pea (*Pisum sativum* L.) and lentil (*Lens culinaris* L.) nodulation in controlled growth chamber environments. J Exp Bot 152:1537–1543

Breakspear A, Liu C, Roy S, Stacey N, Rogers C, Trick M, Morieri G, Mysore KS, Wen J, Oldroyd GE, Downie JA, Murray JD (2014) The root hair "Infectome" of *Medicago truncatula* uncovers changes in cell cycle genes and reveals a requirement for auxin signaling in rhizobial infection. Plant Cell 26:4680–4701

Caetano-Anollés G, Crist-Estes DK, Bauer WD (1988) Chemotaxis of *Rhizobium meliloti* on the plant flavone luteolin requires functional nodulation genes. J Bacteriol 170:3164–3169

Cebolla A, Vinardell JM, Kiss E, Oláh B, Roudier F, Kondorosi A, Kondorosi E (1999) The mitotic inhibitor ccs52 is required for endoreduplication and ploidy-dependent cell enlargement in plants. EMBO J 18:4476–4484

Cesco S, Neumann G, Tomasi N, Pinton R, Weisskopf L (2010) Release of plant-borne flavonoids into the rhizosphere and their role in plant nutrition. Plant Soil 329:1–25

Chen H-C, Feng J, Hou B-H, Li F-Q, Li Q, Hong G-F (2005) Modulating DNA bending affects NodD-mediated transcriptional control in *Rhizobium leguminosarum*. Nucleic Acids Res 33:2540–2548

Chen C, McIver J, Yang Y, Bai Y, Schultz B, McIver A (2006) Foliar application of lipo-chitooligosaccharides (Nod factors) to tomato (*Lycopersicon esculentum*) enhances flowering and fruit production. Can J Plant Sci 87:365–372

Cooper JE (2004) Multiple responses of rhizobia to flavonoids during legume root infection. Adv Bot Res 41:1–62

Cooper JE (2007) Early interactions between legumes and rhizobia: disclosing complexity in a molecular dialogue. J Appl Microbiol 103:1355–1365

Cren M, Kondorosi A, Kondorosi E (1995) NolR controls expression of the *Rhizobium meliloti* nodulation genes involved in the core Nod factor synthesis. Mol Microbiol 15:733–747

Dakora FD (2003) Defining new roles for plant and rhizobial molecules in sole and mixed plant cultures involving symbiotic legumes. New Phytol 158:39–49

D'Haeze W, Holsters M (2002) Nod factor structures, responses, and perception during initiation of nodule development. Glycobiology 12:79R–105R

Deakin WJ, Broughton WJ (2009) Symbiotic use of pathogenic strategies: rhizobial protein secretion systems. Nat Rev Microbiol 7:312–320

Debellé F, Rosenberg C, Vasse J, Maillet F, Martinez E, Denarie J, Truchet G (1986) Assignment of symbiotic developmental phenotypes to common and specific nodulation (*nod*) genetic loci of *Rhizobium meliloti*. J Bacteriol 168:1075–1086

Dixon R, Kahn D (2004) Genetic regulation of biological nitrogen fixation. Nat Rev Microbiol 2:621–631

Duzan HM, Mabood F, Zhou X, Souleimanov A, Smith DL (2005) Nod factor induces soybean resistance to powdery mildew. Plant Physiol Biochem 43:1022–1030

Esseling J, Lhuissier F, Emons A (2003) Nod factor-induced root hair curling: continuous polar growth towards the point of nod factor application. Plant Physiol 132:1982–1988

Estévez J, Soria-Díaz ME, Fernández de Córdoba F, Morón B, Manyani H, Gil A, Thomas-Oathes J, van Brussel AAN, Dardanelli MS, Sousa C, Megías M (2009) Different and new Nod factors produced by *Rhizobium tropici* CIAT899 following Na⁺ stress. FEMS Microbiol Lett 293:220–231

Faure D, Vereecke D, Leveau JH (2009) Molecular communication in the rhizosphere. Plant Soil 321:279–303

Fauvart M, Michiels J (2008) Rhizobial secreted proteins as determinants of host specificity in the rhizobium-legume symbiosis. FEMS Microbiol Lett 285:1–9

Ferguson BJ, Indrasumunar A, Hayashi S, Lin MH, Lin YH, Reid DE, Gresshoff PM (2010) Molecular analysis of legume nodule development and autoregulation. J Integr Plant Biol 52:61–76

Fisher RF, Egelhoff TT, Mulligan JT, Long SR (1988) Specific binding of *Rhizobium meliloti* extracts containing *nodD* to DNA sequences upstream of inducible nodulation genes. Genes Dev 2:282–293

Foo E, Yoneyama K, Hugill C, Quittenden LJ, Reid JB (2013) Strigolactones: Internal and external signals in plant symbioses? Plant Signal Behav 8:e23168

Fournier J, Timmers AC, Sieberer BJ, Jauneau A, Chabaud M, Barker DG (2008) Mechanism of infection thread elongation in root hairs of *Medicago truncatula* and dynamic interplay with associated rhizobial colonization. Plant Physiol 148:1985–1995

Fraysse N, Couderc F, Poinsot V (2003) Surface polysaccharide involvement in establishing the *rhizobium*-legume symbiosis. Eur J Biochem 270:1365–1380

Fujishige NA, Lum MR, De Hoff PL, Whitelegge JP, Faull KF, Hirsch AM (2008) *Rhizobium* common *nod* genes are required for biofilm formation. Mol Microbiol 67:504–515

Gage DJ (2004) Infection and invasion of roots by symbiotic, nitrogen-fixing rhizobia during nodulation of temperate legumes. Microbiol Mol Biol Rev 68:280–300

Geurts R, Fedorova E, Bisseling T (2005) Nod factor signaling genes and their function in the early stages of *Rhizobium* infection. Curr Opin Plant Biol 8:346–352

Gibson KE, Kobayashi H, Walker GC (2008) Molecular determinants of a symbiotic chronic infection. Annu Rev Genet 42:413–441

Giraud E, Moulin L, Vallenet D, Barbe V, Cytryn E, Avarre JC, Jaubert M, Simon D, Cartieaux F, Prin Y, Bena G, Hannibal L, Fardoux J, Kojadinovic M, Vuillet L, Lajus A, Cruveiller S, Rouy Z, Mangenot S, Segurens B, Dossat C, Franck WL, Chang WS, Saunders E, Bruce D, Richardson P, Normand P, Dreyfus B, Pignol D, Stacey G, Emerich D, Vermeglio A, Medigue C, Sadovsky M (2007) Legume symbioses: absence of *nod* genes in photosynthetic bradyrhizobia. Science 316:1307–1312

Graham PH, Vance CP (2003) Legumes: importance and constraints to greater utilization. Plant Physiol 131:872–877

Guasch-Vidal B, Estévez J, Dardanelli MS, Soria-Díaz ME, de Córdoba FF, Balog CI, Manyani H, Gil-Serrano A, Thomas-Oats J, Hensbergen PJ, Deeler AM, Megías M, van Brussel AA (2013) High NaCl concentrations induce the *nod* genes of *Rhizobium tropici* CIAT899 in the absence of flavonoid inducers. Mol Plant Microbe Interact 26:451–460

Haag AF, Arnold MF, Myka KK, Kerscher B, Dall'Angelo S, Zanda M, Mergaert P, Ferguson GP (2013) Molecular insights into bacteroid development during the *Rhizobium*-legume symbiosis. FEMS Microbiol Rev 37:364–383

Harborne JB, Williams CA (2000) Advances in flavonoid research since 1992. Phytochemistry 55:481–504

Hassan S, Mathesius U (2012) The role of flavonoids in root-rhizosphere signaling: opportunities and challenges for improving plant-microbe interactions. J Exp Bot 63:3429–3444

Herridge DF, Peoples MB, Boddey RM (2008) Global inputs of biological nitrogen fixation in agricultural systems. Plant Soil 311:1–18

Hungria M, Phillips DA (1993) Effects of a seed color mutation on rhizobial *nod*-gene-inducing flavonoids and nodulation in common bean. Mol Plant Microbe Interact 6:418–422

Hungria M, Stacey G (1997) Molecular signals exchanged between host plants and rhizobia: basic aspects and potential application in agriculture. Soil Biol Biochem 29:819–830

Jacobs TW, Egelhoff TT, Long SR (1985) Physical and genetic map of a *Rhizobium meliloti* nodulation gene region and nucleotide sequence of *nodC*. J Bacteriol 162:469–476

Jones KM, Kobayashi H, Davies BW, Taga ME, Walker GC (2007) How rhizobial symbionts invade plants: the *Sinorhizobium–Medicago* model. Nat Rev Microbiol 5:619–633

Kawaharada Y, Kelly S, Wibroe Nielsen M, Hjuler CT, Gysel K, Muszynski A, Carlson MB, Thygesen RW, Sandal N, Asmussen MH, Vinther M, Andersen SU, Krusell L, Thirup S, Jensen KJ, Ronson CW, Blaise M, Radutoiu S, Stougaard J (2015) Receptor-mediated exopolysaccharide perception controls bacterial infection. Nature 523:308–312

Kidaj D, Wielbo J, Skorupska A (2012) Nod factors stimulate seed germination and promote growth and nodulation of pea and vetch under competitive conditions. Microbiol Res 167:144–150

Kondorosi E, Gyuris J, Schmidt J, John E, Hofmann DB, Schell J, Kondorosi A (1989) Positive and negative control of *nod* gene expression in *Rhizobium meliloti* is reqiured for optimal nodulation. EMBO J 8:1331–1340

Kouchi H, Shimomura K, Hata S, Hirota A, Wu G-J, Kumagai H, Tajima S, Suganuma N, Suzuki A, Aoki T, Hayashi M, Yokoyama T, Ohyama T, Asamizu E, Kuwata C, Shibata D, Tabata S (2004) Large-scale analysis of gene expression profiles during early stages of root nodule formation in a model legume, *Lotus japonicus*. DNA Res 11:263–274

Krishnan HB, Lorio J, Kim WS, Jiang GQ, Kim KY, DeBoer M, Pueppke SG (2003) Extracellular proteins involved in soybean cultivar-specific nodulation are associated with pilus-like surface appendages and exported by a type III protein secretion system in *Sinorhizobium fredii* USDA257. Mol Plant Microbe Interact 16:617–625

Laguerre G, Courdr L, Nouaim R, Lamy I, Revellin C, Breuil MC, Chaussod R (2006) Response of rhizobial populations to moderate copper stress applied to an agricultural soil. Microb Ecol 52:426–435

Lerouge P, Roche P, Faucher C, Maillet F, Denarie J, Prome J-C, Truchet G (1996) Substance with lipo-oligosachcaride structure capable of acting as plant-specific symbiotic signals, processes for producing them and their applications. US Patent 5,549,718, filed 30 Sept 1994 and issued 27 Aug 1996

Leibovitch S, Migner P, Smith DL (2001) Evaluation of the effect of SoyaSignal technology on soybean yield [*Glycine max*. (L.) Merr.] under field conditions over 6 years in Eastern Canada and the Northern United States. J Agron Crop Sci 187:281–292

Liu C-W, Breakspear A, Roy S, Murray JD (2015) Cytokinin responses counterpoint auxin signaling during rhizobial infection. Plant Signal Behav 10(6):e1019982

Long SR (1996) *Rhizobium* symbiosis: Nod factors in perspective. Plant Cell 8:1885–1898

Mabood F, Jung WJ, Smith DL (2008) Signals in the underground: microbial signaling and plant productivity. In: Nautiyal CS, Dion PE, Chopra VL (eds) Molecular mechanism of plant and microbe coexistence. Springer-Verlag, Berlin, Heidelberg, pp 291–318

Macchiavelli RE, Brelles-Marino G (2004) Nod factor-treated *Medicago truncatula* roots and seeds show an increased number of nodules when inoculated with a limiting population of *Sinorhizobium meliloti*. J Exp Bot 55:2635–2640

Madsen LH, Tirichine L, Jurkiewicz A, Sullivan JT, Heckmann AB, Bek AS, Ronson CW, James EK, Stougaard J (2010) The molecular network governing nodule organogenesis and infection in the model legume *Lotus japonicus*. Nat Commun 1:10

Maillet F, Poinsot V, Andre O, Puech-Pages V, Haouy A, Gueunier M, Cromer L, Giraudet D, Formey D, Niebel A, Martinez EA, Driguez H, Becard G, Denarie J (2011) Fungal lipochitooligosaccharide symbiotic signals in arbuscular mycorrhiza. Nature 469:58–63

Maj D, Wielbo J, Marek-Kozaczuk M, Skorupska A (2009) Pretreatment of clover seeds with Nod factors improves growth and nodulation of *Trifolium pratense*. J Chem Ecol 35:479–487

Maj D, Wielbo J, Marek-Kozaczuk M, Skorupska A (2010) Response to flavonoids as a factor influencing competitiveness and symbiotic activity of *Rhizobium leguminosarum*. Microbiol Res 165:50–60

Maroti G, Kondorosi E (2014) Nitrogen-fixing Rhizobium-legume symbiosis: are polyploidy and host peptide governed symbiont differentiation general principles of endosymbiosis? Front Microbiol 5(326):1–6

Martinez-Abarca F, Herrera-Cervera JA, Bueno P, Sanjuan J, Bisseling T, Olivares J (1998) Involvement of salicylic acid in the establishment of the *Rhizobium meliloti* – alfalfa symbiosis. Mol Plant Microbe Interact 11:153–155

Masson-Boivin C, Giraud E, Perret X, Batut J (2009) Establishing nitrogen-fixing symbiosis with legumes: how many rhizobium recipes? Trends Microbiol 17:458–466

Mergaert P, Uchiumi T, Alunni B, Evanno G, Cheron A, Catrice O, Mausset A.-E., Barloy-Hubler F, Galibert F, Kondorosi A, Kondorosi E (2006) Eukaryotic control on bacterial cell cycle and differentiation in the Rhizobium-legume symbiosis. Proceedings of the National Academy of Sciences 103(13):5230–5235

Morón B, Soria-Díaz ME, Ault J, Verroios G, Noreen S, Rodríguez-Navarro DN, Gil-Serrano A, Thomas-Oathes J, Megías M, Sousa C (2005) Low pH changes the profile of nodulation factors produced by *Rhizobium tropici* CIAT899. Chem Biol 12:1029–1040

Moulin L, Munive A, Dreyfus B, Boivin-Masson C (2001) Nodulation of legumes by members of the beta-subclass of proteobacteria. Nature 411:948–950

Murray JD (2011) Invasion by invitation: rhizobial infection in legumes. Mol Plant Microbe Interact 24:631–639

Newton WE (2007) Physiology, biochemistry and molecular biology of nitrogen fixation. In: Bothe H, Ferguson S, Newton WE (eds) Biology of nitrogen cycle. Elsevier, Amsterdam, pp 109–130

Okazaki S, Kaneko T, Sato S, Saeki K (2013) Hijacking of leguminous nodulation signaling by the rhizobial type III secretion system. Proc Natl Acad Sci U S A 110:17131–17136

Oláh B, Brière C, Bécard G, Dénarié J, Gough C (2005) Nod factors and a diffusible factor from arbuscular mycorrhizal fungi stimulate lateral root formation in *Medicago truncatula* via the DMI1/DMI2 signalling pathway. Plant J 44:195–207

Oldroyd GE, Downie JA (2008) Coordinating nodule morphogenesis with rhizobial infection in legumes. Annu Rev Plant Biol 59:519–546

Oldroyd GE, Murray JD, Poole PS, Downie JA (2011) The rules of engagement in the legume-rhizobial symbiosis. Annu Rev Genet 45:119–144

Oono R, Schmitt I, Sprent JI, Denison RF (2010) Multiple evolutionary origins of legume traits leading to extreme rhizobial differentiation. New Phytol 187:508–520

Peck MC, Fisher RF, Long SR (2006) Diverse flavonoids stimulate NodD1 binding to nod gene promoters in *Sinorhizobium meliloti*. J Bacteriol 188:5417–5427

Peck MC, Fisher RF, Bliss R, Long SR (2013) Isolation and characterization of mutant *Sinorhizobium meliloti* NodD1 proteins with altered responses to luteolin. J Bacteriol 195:3714–3723

Penmetsa RV, Cook DR (1997) A legume ethylene-insensitive mutant hyperinfected by its rhizobial symbiont. Science 275:527–530

Penterman J, Abo RP, De Nisco NJ, Arnold MFF, Longhi R, Zanda M, Walker GC (2014) Host plant peptides elicit a transcriptional response to control the *Sinorhizobium meliloti* cell cycle during symbiosis. Proc Natl Acad Sci U S A 111:3561–3566

Peoples MB, Brockwell J, Herridge DF, Rochester IJ, Alves BJR, Urquiaga S, Boddey RM, Dakora FD, Bhattarai S, Maskey SL, Sampet C, Rerkasem B, Khan DF, Hauggaard-Nielsen H, Jensen ES (2009) The contribution of nitrogen-fixing crop legumes to the productivity of agricultural systems. Symbiosis 48:1–17

Perret X, Staehelin C, Broughton WJ (2000) Molecular basis of symbiotic promiscuity. Microbiol Mol Biol Rev 64:180–201

Peters NK, Frost JW, Long SR (1986) A plant flavone, luteolin, induces expression of *Rhizobium meliloti* nodulation genes. Science 233:917–1008

Podleśny J, Wielbo J, Podleśna A, Kidaj D (2014a) The pleiotropic effects of extract containing rhizobial Nod factors on pea growth and yield. Centr Eur J Biol 9:396–409

Podleśny J, Wielbo J, Podleśna A, Kidaj D (2014b) The responses of two pea genotypes to Nod factors (LCOs) treatment. J Food Agric Environ 12:554–558

Podleśna A, Wielbo J, Podleśny J, Kidaj D (2015) Effect of sulfur and Nod factors (LCOs) on some physiological features and yield of pea (*Pisum sativum* L.) In: De Kok LJ et al (eds) Molecular physiology and ecophysiology of sulfur proceedings of the international plant sulfur workshop. Springer International Publishing, Switzerland, pp 221–226

Prell J, Poole P (2006) Metabolic changes of rhizobia in legume nodules. Trends Microbiol 14:161–168

Prithiviraj B, Zhou X, Souleimanov A, Kahn WM, Smith DL (2003) A host-specific bacteria-to-plant signal molecule (Nod factor) enhances germination and early growth of diverse crop plants. Planta 21:437–445

Promé (1996) Signaling events elicited in plants by defined oligosaccharide structures. Curr Opin Struct Biol 6:671–678

Redmond JW, Batley M, Djordjevic MA, Innes RW, Kuempel PL, Rolfe BG (1986) Flavones induce expression of nodulation genes in *Rhizobium*. Nature 323:632–634

Roche P, Debellé F, Maillet F, Lerouge P, Faucher C, Truchet G, Dénarie J, Promé J-C (1991) Molecular basis of symbiotic host specificity in *Rhizobium meliloti*: *nodH* and *nodPQ* genes encode the sulfation of lipo-oligosaccharide signals. Cell 67:1131–1143

Ruvkun GB, Ausubel FM (1980) Interspecies homology of nitrogenase genes. Proc Natl Acad Sci U S A 77:191–195

Schultze M, Kondorosi A (1998) Regulation of symbiotic root nodule development. Annu Rev Genet 32:33–57

Siczek A, Frąc M, Nawrocka A, Wielbo J, Kidaj D (2015) The response of rhizosphere microbial properties to flavonoids and Nod factors. Acta Agric Scand B 65:125–131

Siczek A, Lipiec J, Wielbo J, Kidaj D, Szarlip P (2014) Symbiotic activity of pea (*Pisum sativum*) after application of nod factors under field conditions. Int J Mol Sci 15:7344–7351

Skorupska A, Białek U, Urbanik-Sypniewska T, Van Lammeren A (1995) Two types of nodules induced on *Trifolium pratense* by mutants of *Rhizobium leguminosarum* bv. *trifolii* deficient in exopolysaccharide production. J Plant Physiol 147:93–100

Skorupska A, Janczarek M, Marczak M, Mazur A, Król J (2006) Rhizobial exopolysaccharides: genetic control and symbiotic functions. Microb Cell Fact 5:1–19

Skorupska A, Wielbo J, Maj D, Marek-Kozaczuk M (2010) Enhancing *rhizobium*–legume symbiosis using signaling factors. In: Khan MS et al (eds) Microbes for legume improvement. Springer-Verlag, Wien, pp 27–54

Smith DL, Bo P, Deng Y, Migner P, Zhang F, Prithiviraj B, Habib A (2005a) Composition for accelerating seed germination and plant growth. US Patent 6,979,664B1, filed 21 July 1999 and issued 27 December 2005

Smith S, Habib A, Kang Y, Legget M, Diaz-Zorita M (2015) LCO applications provide improved response with legumes and nonlegumes. In: De Bruijn FJ (ed) Biological nitrogen fixation. John Wiley & Sons, New York, pp 1077–1085

Smith RS, Osburn RM (2009) Lipo-oligosaccharide and flavonoid combination for enhanced plant growth and yield. Patent WO2009/049747A2, filed 23 September 2008 and issued 23 April 2009

Smith RS, Osburn RM (2010) Lipo-chitooligosaccharide compositions for enhanced plant growth and yield. US Patent 2010/0093537A1, filed 8 January 2008 and issued 15 April 2010

Smith RS, Osburn RM, Kosanke JW (2005b) Methods and compositions providing agronomically beneficial effects in legumes and non-legumes. Patent WO2005062899, filed 23 December 2004 and issued 3 November 2005

Soto MJ, Fernández-Aparicio M, Castellanos-Morales V, García-Garrido JM, Okampo JA, Delgado MJ, Vierheilig H (2010) First indications for the involvement of strigolactones on nodule formation in alfalfa (*Medicago sativa*). Soil Biol Biochem 42:383–385

Souleimanov A, Prithiviraj B, Smith DL (2002) The major Nod factor of *Bradyrhizobium japonicum* promotes early growth of soybean and corn. J Exp Bot 376:1929–1934

Spaepen S, Vanderleyden J, Okon Y (2009) Plant growth promoting actions of rhizobacteria. Adv Bot Res 51:283–320

Spaink HP (2000) Root nodulation and infection factors produced by rhizobial bacteria. Annu Rev Microbiol 54:257–288

Staehelin C, Schultze M, Kondorosi E, Kondorosi A (1995) Lipo-chitooligosaccharide nodulation signals from *Rhizobium meliloti* induce their rapid degradation by the host plant alfalfa. Plant Physiol 108:1607–1614

Staehelin C, Schultze M, Tokuyasu K, Poinsot V, Promé J-C, Kondorosi E, Kondorosi A (2000) N-deacetylation of *Sinorhizobium meliloti* Nod factors increases their stability in the *Medicago sativa* rhizosphere and decreases their biological activity. Mol Plant Microbe Interact 13:72–79

Subramanian S, Stacey G, Yu O (2006) Endogenous isoflavones are essential for the establishment of symbiosis between soybean and *Bradyrhizobium japonicum*. Plant J 48:261–273

Subramanian S, Stacey G, Yu O (2007) Distinct, crucial roles of flavonoids during legume nodulation. Trends Plant Sci 12:282–285

Suzaki T, Yoro E, Kawaguchi M (2015) Leguminous plants: inventors of root nodules to accommodate symbiotic bacteria. Int Rev Cell Mol Biol 316:111–158

Tian Y, Liu W, Cai J, Zhang LY, Wong KB, Feddermann N, Boller T, Xie ZP, Staehelin C (2013) The nodulation factor hydrolase of *Medicago truncatula*: characterization of an enzyme specifically cleaving rhizobial nodulation signals. Plant Physiol 163:1179–1190

Timmers AC, Auriac MC, Truchet G (1999) Refined analysis of early symbiotic steps of the *Rhizobium*–*Medicago* interaction in relationship with microtubular cytoskeleton rearrangements. Development 126:3617–3628

Tiwari SB, Hagen G, Guilfoyle T (2003) The roles of auxin response factor domains in auxin-responsive transcription. Plant Cell 15:533–543

Toth K, Stacey G (2015) Does plant immunity play a critical role during initiation of the legume-rhizobium symbiosis? Front Plant Sci 6:401

Van de Velde W, Zehirov G, Szatmari A, Debreczeny M, Ishihara H, Kevei Z, Farkas A, Mikulass K, Nagy A, Tiricz H, Satiat-Jeunemaître B, Alunni B, Bourge M, Kucho K, Abe M, Kereszt A, Maroti G, Uchiumi T, Kondorosi E, Mergaert P (2010) Plant peptides govern terminal differentiation of bacteria in symbiosis. Science 327:1122–1126

Vasse J, de Billy F, Camut S, Truchet G (1990) Correlation between ultrastructural differentiation of bacteroids and nitrogen fixation in alfalfa nodules. J Bacteriol 172:4295–4306

Wasson AP, Pellerone FI, Mathesius U (2006) Silencing the flavonoid pathway in *Medicago truncatula* inhibits root nodule formation and prevents auxin transport regulation by rhizobia. Plant Cell 18:1617–1629

Weston LA, Mathesius U (2013) Flavonoids: Their structure, biosynthesis and role in the rhizosphere, including allelopathy. J Chem Ecol 39:283–297

Xie Z-P, Müller J, Wiemken A, Broughton WJ, Boller T (1997) Nod factors and triiodobenzoic acid stimulate mycorrhizal colonization and affect carbohydrate partitioning in mycorrhizal roots of *Lablab purpureus*. New Phytol 139:361–366

Yates RJ, Howieson JG, Reeve WG, O'Hara GW (2011) A re-appraisal of the biology and terminology describing rhizobial success in nodule occupancy of legumes in agriculture. Plant Soil 348:255–267

Yeh KC, Peck MC, Long SR (2002) Luteolin and GroESL modulate *in vitro* activity of NodD. J Bacteriol 184:525–530

Zaat SA, Wijffelman CA, Mulders IHM, van Brussel AAN, Lugtenberg BJJ (1988) Root exudates of various host plants of Rhizobium leguminosarum contain different sets of inducers of *Rhizobium* nodulation genes. Plant Physiol 86:1298–1303

Zaat SA, Schripsema J, Wijffelman CA, van Brussel AAN, Lugtenberg BJJ (1989) Analysis of the major inducers of the *Rhizobium nodA* promoter from *Vicia sativa* root exudate and their activity with different *nodD* genes. Plant Mol Biol 13:175–188

Zhang F, Smith DL (1995) Preincubation of *Bradyrhizobium japonicum* with genistein accelerates nodule development of soybean at suboptimal root zone temperatures. Plant Physiol 108:961–968

Zhang F, Smith DL (1996) Inoculation of soybean (*Glycine max* (L.) Merr.) with genistein-preincubated *Bradyrhizobium japonicum* or genistein directly applied into soil increases soybean protein and dry matter yield under short season conditions. Plant Soil 179:233–241

Zhang J, Subramanian S, Stacey G, Yu O (2009) Flavones and flavonols play distinct critical roles during nodulation of *Medicago truncatula* by *Sinorhizobium meliloti*. Plant J 57:171–183

Zuanazzi JAS, Clergeot PH, Quirion JC, Husson HP, Kondorosi A, Ratet P (1998) Production of *Sinorhizobium meliloti nod* gene activator and repressor flavonoids from *Medicago sativa* roots. Mol Plant Microbe Interact 11:784–794

Role of Ethylene and Bacterial ACC-Deaminase in Nodulation of Legumes

4

Azeem Khalid, Zulfiqar Ahmad, Shahid Mahmood, Tariq Mahmood, and Muhammad Imran

Abstract

Rhizobia-legume symbiosis is a complex process involving a number of plant and bacterial genes that lead to the formation and development of root nodules. Plant hormone ethylene plays an important role in nodule development and nodule signaling networks in response to a wide range of biotic and abiotic stresses. Ethylene is known as a negative regulator of nodulation. Inoculation of rhizobia leads to a temporal stimulation of ethylene production that suppresses nodule formation. In contrast, inhibitors of ethylene synthesis or its physiological action promote nodule formation in legumes. 1-Aminocyclopropane-1-carboxylate (ACC)-deaminase is a biological inhibitor of ethylene synthesis. The rhizosphere bacteria containing ACC-deaminase can increase nodulation in legumes by degrading ACC (an immediate precursor of ethylene) and, thus, by lowering ethylene concentration in the plant. Similarly, some rhizobia also have shown ACC-deaminase activity and improvement in nodulation by regulating the concentration of ethylene in plant tissues. In this chapter, the role of ethylene and bacterial ACC-deaminase in nodulation of legumes is reviewed and discussed.

A. Khalid (✉) • Z. Ahmad • S. Mahmood • T. Mahmood
Department of Environmental Sciences, PMAS Arid Agriculture University, Rawalpindi, Pakistan
e-mail: azeem@uaar.edu.pk; azeemuaf@yahoo.com

M. Imran
Department of Soil Science, Muhammad Nawaz Shareef University of Agriculture, Multan, Pakistan

© Springer International Publishing AG 2017
A. Zaidi et al. (eds.), *Microbes for Legume Improvement*,
DOI 10.1007/978-3-319-59174-2_4

4.1   Introduction

The relationship between rhizobia and legume plants is a classic example of mutu-alistic association between the two partners. This symbiotic association involves a series of interactions between a microsymbiont such as *Rhizobium* and its legume host plant, resulting in the formation of nodules in legumes. Nodule organogenesis is accompanied by significant changes in gene expression in plants. Several hun-dreds of genes were found to be strongly and specifically up and downregulated during nodulation process (El Yahyaoui et al. 2004; Benedito et al. 2008). Legume biological nitrogen fixation (BNF) is particularly sensitive and perturbed by envi-ronmental stresses such as drought, salt stress, defoliation, continuous darkness, and cold stress. Adverse environmental conditions affect nodule structure and function-ing and induce drastic metabolic and molecular modifications that ultimately lead to a stress-induced senescence (Dupont et al. 2012).

Plant growth regulators (PGRs)play a very imperative role in the processes of nod-ule formation and development (Nagata and Suzuki 2014; Jones et al. 2015). Hormone signaling is integrated at several levels during growth and nodulation developmental process (Stepanova and Alonso 2009). Ethylene, a gaseous plant hormone, regulates many physiological processes of plants, ranging from germination of seeds to senes-cence of various organs and in many responses to environmental stresses (El-Maarouf-Bouteau et al. 2015; Ullah et al. 2016a; Yoong et al. 2016). It also acts as an autoregulator to control nodule formation and development during symbiosis (Ligero et al. 1991; Lohar et al. 2009). It is effective in evoking physiological responses in plants even when present in extremely low concentrations. Even though ethylene is crucial for many physiological processes, yet it inhibits nodulation in numerous plant species when produced by the plants in excessive amounts (Saleem et al. 2007; Shaharoona et al. 2011). Previous studies reveal that ethylene may play a positive role in nodule senescence, just as they do in the senescence of other plant tissues. The positive role of ethylene is illustrated by the upregulation of ethylene response factors (ERF) and ethylene biosynthetic genes, such as S-adenosyl methionine (SAM) synthetase and ACC oxidase (Dupont et al. 2012). Ethylene is produced in plant tissues from the pre-cursor, ACC, which is converted to ethylene by the enzyme ACC oxidase. Ethylene production in plant tissues is directly related to stress conditions. During growth, plants are commonly exposed to various environmental stresses (Bari and Jones 2009), and resultantly more ethylene is produced. In the process of nodule formation, in addition to abiotic stresses, infection of roots with microsymbiont imposes biotic stress and results in increased ACC in the infected roots and consequently ethylene levels in plant tissues. The high concentration of ACC and ethylene in root tissues serves as negative regulators of nodulation in legume plants (Barnawal et al. 2014). Any factor or stimu-lus which changes ethylene levels of a plant, either by altering its synthesis endoge-nously or in the close vicinity of the roots, can also affect nodulation process. Previously, chemical inhibitors of ethylene synthesis (aminoethoxyvinylglycine, AVG, and cobalt, Co^{2+}) and action (silver, Ag^+) have been used to lower the production of ethylene and promote growth and nodulation of various legumes (Li et al. 2014; Jones et al. 2015). Several studies also show that ethylene production can also be suppressed biologically by converting ACC, partially or completely, into other products instead of ethylene (Baig et al. 2014; Nadeem et al. 2016, 2017). In plants, subjected to stress, high levels of ACC accumulate in the plant tissues and are excreted into the rhizosphere where presumably it is reabsorbed by growing root tips such as occur with many organic acids or converted into ethylene by rhizosphere microflora. At the same time, some plant growth-promoting rhizobacteria (PGPR) are able to lower the plant's ethylene concentration by taking up ACC and destroying the precursor using it as nitrogen (N) source. These bacteria carry gene which are involved in the synthesis of ACC-deaminase. The ACC-deaminase then hydrolyzes ACC into ammonia and α-ketobutyric acid and thus promotes root growth and nodulation of certain plant species by suppressing ethylene biosynthesis (Shaharoona et al. 2006; Shahzad et al. 2013). Furthermore, proliferation of primary or lateral roots through lowering of ethylene as a result of bacterial ACC-deaminase activity provides more infection sites and contact to rhizobia for nodule formation (Shahzad et al. 2010, 2013). This implies that nodulation in legumes could be enhanced through the modulation of ethylene levels in plants as a result of bacterial ACC-deaminase activity. Recently, some studies on the molecular level have shown that nitrogen-fixing genes (nifA) also play a role in regulating the expression of ACC-deaminase in the nodules (Nascimento et al. 2014). The nifA binding sites are present in the upstream region of ACC-deaminase (acdS) gene, suggesting a regulatory role of rhizobial species for ACC-deaminase activity (Nascimento et al. 2012). Here, the role of ethylene as a negative regulator of nodulation and how the bacterial ACC-deaminase activity promotes nodulation in legumes via modulation of ethylene biosynthesis is discussed.

4.2 Ethylene vs. Nodulation

Several factors such as physicochemical soil conditions, host-microsymbiont compatibility, and the presence of known and unknown biomolecules can affect nodulation process, but the role of ethylene in the formation and development of nodules on the roots of legumes is very critical (Arshad and Frankenberger 2002; Ding and Oldroyd 2009). In the last decade, identification of a number of ethylene-insensitive mutants in different legumes has provided genetic evidence for the involvement of ethylene signaling in nodulation (Gresshoff et al. 2009; Prayitno and Mathesius 2010). Ethylene is involved both in nodulation process and in senescence. And hence, any suppression in ethylene production aids in maintaining nodules by delaying senescence (Nukui et al. 2004). Generally, nodulation response to ethylene is variable and depends on the plant species as well as the concentration of ethylene released in root tissues during nodulation.

4.2.1 Ethylene as a Negative Regulator of Nodulation

Several authors have reported that ethylene affects nodulation negatively (Musarrat et al. 2009; Foo et al. 2016; Kawaharada et al. 2017). Ethylene controls the epidermal responses during the nodulation process and thus negatively regulates multiple epidermal responses in order to inhibit rhizobial infection (Nukui et al. 2004; Sugawara et al. 2006). Although not all legumes respond similarly, addition of exogenous ethylene to the most of nodulating plants reduces the frequency of nodule primordium formation (Guinel and Geil 2002; Lohar et al. 2009). In the presence of ethylene, the number of infected root hairs did not change; however, many infection threads were aborted, and the epidermis or outer cortex and nodule primordia did not form (Lee and LaRue 1992b). This leads to reduction in infection process and consequently the number of nodules in legumes. It has been observed that ethylene production significantly increases in roots infected by *Rhizobium* or *Bradyrhizobium* and decreases the number of nodules that form on the infected plants (Gonzalez-Rizzo et al. 2006; Middleton et al. 2007).

As reviewed previously, ethylene has been found to reduce nodule numbers in several legumes, while ethylene synthesis or sensing inhibitors generally increase nodule numbers (Shaharoona et al. 2011; Heckmann et al. 2011). Effects of ethylene on nodulation were demonstrated through the application of inhibitors of ethylene synthesis (e.g., AVG, L-α(2-aminoethoxyvinyl)) and perception (e.g., silver ions), by an increased nodule number on *Medicago sativa* (*M. Sativa*) (Peters and Crist-Esters 1989; Caba et al. 1998), *Pisum. sativum* (*P. Sativum*) (Ferguson et al. 2011; Jones et al. 2015) and *Lotus japonicus* (*L. Janponicus*) (Heckmann et al. 2011; Li et al. 2014). In contrast, several reports have shown that exogenous treatments of soybean with ethylene or ethylene inhibitors did not affect nodule number (Suganuma et al. 1995; Schmidt et al. 1999). The variable impact of ethylene or ethylene inhibitors could be due to differences in culture systems, level of treatment, or plant genotype. For example, Xie et al. (1996) demonstrated that soybean cultivars differed in their natural ethylene responsiveness. Also, Lee and LaRue (1992a) indicated that the formation of nodules on soybean roots was less sensitive to exogenous ethylene than they were on other leguminous plants, perhaps a reflection of the determinate vs. indeterminate nodule ontogeny. Like chemical inhibitors, ACC-deaminase-positive bacterial strains lower ethylene synthesis by degrading its precursor ACC and, concurrently, promote nodulation in leguminous plants (Kong et al. 2015; Nascimento et al. 2016). ACC-deaminase activity in bacteroids may delay nodule senescence by reducing ethylene production and, as a consequence, prolong N fixation, which may represent another strategy to improve the efficiency of rhizobial inoculation (Tittabutr et al. 2015).

In order to understand the molecular basis that how decreased levels of ethylene enhance nodulation, several models considering the relationships between signal transduction, ethylene sensing, and nodule development have been proposed (Sun et al. 2006). As an example, Nukui et al. (2004) in a study transformed *L. japonicus* B-129 with a mutated ethylene receptor gene Cm-ERS1/H70A. A point mutation was introduced into the melon ethylene receptor Cm-ERS1 by abolishing its ethylene-binding ability. The *L. japonicus* transgenic plants exhibited low sensitivity to ethylene and produced substantially higher numbers of infection threads and nodule primordia on their roots than did either wild-type or azygous plants without the transgene. Moreover, the amount of transcripts of NIN, a gene governing formation of infection threads, increased in the inoculated transgenic plants as compared to wild-type plants. These results imply that endogenous ethylene in *L. japonicus* roots inhibits the formation of nodule primordia, as well as other infection processes. Earlier studies have clearly demonstrated that ethylene can serve as a negative regulator of nodulation and reduction in ethylene concentration stimulates formation and development of nodules in legumes (Heckmann et al. 2011; Jones et al. 2015). However, the mechanism of selective ethylene inhibition on nodulation, as compared to root growth, is still unknown, and ethylene may be involved at various stages of nodule development.

4.2.2 Accelerated Ethylene Biosynthesis During Nodulation Process

Generally, inoculated nodule-forming legumes produce more ethylene than non-inoculated/non-nodulated legumes (Ligero et al. 1999). The higher production of ethylene during nodulation is most likely a plant response to the nodulating bacteria (Zaat et al. 1989). Accordingly, several authors have reported that ethylene release was stimulated after inoculation, for instance, in alfalfa (Caba et al. 1998), vicia (van Workum et al. 1995), and soybean (Suganuma et al. 1995). In a similar study, production of ethylene by soybean roots was facilitated by inoculation with *Bradyrhizobium japonicum* (*B. japonicum*), and stimulation was maximum three days after bacterization (Suganuma et al. 1995). The rate of ethylene synthesis thereafter dropped to the extent as observed for uninoculated plants. Caba et al. (1999) compared ethylene evolution activity in roots of soybean cv. Bragg (wild type) vs. the supernodulating mutants "*nts* 382 and *nts* 1007" after inoculation and treatment with ACC or ethephon. They observed that ethylene release was greater in inoculated Bragg than its mutants in the absence of ACC or ethephon. The skl mutant is an ethylene-insensitive legume mutant showing a hypernodulation phenotype when inoculated with its symbiont *Sinorhizobium meliloti* (*S. meliloti*) (Prayitno et al. 2006). The skl mutant was used to study the ethylene-mediated protein changes during nodule development in *Medicago truncatula* (*M. truncatula*). The root proteome of the skl mutant was compared to its wild type in response to the ethylene precursor ACC. Then, the proteome of skl roots were compared to its wild type after *Sinorhizobium* inoculation to identify differentially displayed proteins during nodule development after one and three days of inoculation. Six proteins (pprg-2, Kunitz proteinase inhibitor, and ACC oxidase isoforms) were downregulated in skl roots, while three protein spots were upregulated (trypsin inhibitor, albumin 2, and CPRD49). ACC induced stress-related proteins in wild-type roots. For example, pprg-2, ACC oxidase, proteinase inhibitor, ascorbate peroxidase, and heat-shock proteins were stimulated in response to ACC. However, the expression of stress-related proteins, such as pprg-2, Kunitz proteinase inhibitor, and ACC oxidase, was downregulated in inoculated skl roots. It was hypothesized that during early nodule development, the plant induces ethylene-mediated stress responses to limit nodule numbers. When a mutant defective in ethylene signaling, such as skl, is inoculated with rhizobia, the plant stress response is reduced, resulting in more nodule formation (Prayitno et al. 2006). The rate of ethylene production increases in plants growing under stressed environment. Various types of stresses including temperature, light, nutrition, gravity, and biological stressors enhance ethylene production in plant tissues (Siddikee et al. 2011; Ali et al. 2014). However, the mechanism of ethylene biosynthesis is the same under stress such as under optimum conditions. Under nutrient stress conditions, plant accumulates reactive oxygen species (ROS) and results in oxidative burst inside the plant cells. This leads to activation of mitogen-activated protein kinase cascade in response to the oxidative burst and induces the activation/phosphorylation of ACC synthase: the enzyme involved in conversion of S-adenosyl-l-methionine (SAM) to ACC (Liu and Zhang 2004). The phosphorylated ACC synthase becomes stabilized and subsequently enhances the production of ethylene (Iqbal et al. 2013).

Nitrate (NO_3^-) has also been documented to increase biosynthesis of ethylene by roots and affect nodulation (Okamoto et al. 2009; Reid et al. 2011). Also, a positive correlation between NO_3^- concentrations and the quantity of ethylene released from roots of alfalfa inoculated with *Rhizobium meliloti* (*R. meliloti*) is reported (Ligero et al. 1987). In a follow-up study, Ligero et al. (1999) compared NO_3^- and inoculation-induced ethylene biosynthesis in soybean genotypes Bragg (wild type) and its supernodulating *(nts 382* and *nts* 1007) and non-nodulated *(nod* 49 and *nod* 139) mutants. They found that regardless of NO_3^- treatment, inoculation with *B. japonicum* significantly increased the release of ethylene in roots. The highest production of ethylene was observed between 24 and 48 h after inoculation. They suggested that the response could be due to the infection process and nodule development as the treatment with Ag^+ at the time of inoculation substantially increased nodule numbers of Bragg under both low and high NO_3^- concentrations. The availability and development of legume mutants varying in sensitivity and responsiveness against ethylene or lacking autoregulation may provide excellent tools to explore the role of ethylene in nodulation.

4.2.3   Effect of Exogenously Applied Ethylene on Nodulation

In leguminous plants, ethylene negatively affects nodulation process (Heckmann et al. 2011; Jones et al. 2015), and there are numerous reports on effects of exogenously applied ethylene gas or ethylene-releasing compounds such as ACC and 2-chloroethylphosphonic acid (ethephon/ethrel) on nodule formation and development (Gour et al. 2012; Imin et al. 2013; Li et al. 2014), as presented in Table 4.1. Exogenous ethylene has been found to inhibit nodulation in sweet clover and pea mutants which either were hypernodulating or had ineffective nodules. The externally applied ethylene though did not decrease the number of infections per cm of lateral pea roots, but the infections were almost completely blocked when the infection thread was in the basal epidermal cell or in the outer cortical cells (Lee and LaRue 1992c). Similarly, a significant reduction in the number of mature nodules on roots of mung bean following 100 μM ethephon is reported (Duodu et al.

Table 4.1 Effect of ethylene gas or ethylene releasing compounds on nodulation of legume crops

Plant	Treatment	Responses	References
Lotus japonicus	ACC	Decrease in nodule number	Li et al. (2014)
Trigonella foenumgraecum	Ethephon	Increase in nodule number	Gour et al. (2012)
Medicago truncatula	ACC	The expression of MtCEP1 gene (modulates lateral root and nodule numbers) reduced	Imin et al. (2013)
Lotus japonicus	ACC	Decrease in nodule number	Heckmann et al. (2011)
Discaria trinervis	ACC and ethephon	Decrease in nodule number	Valverde and Wall (2005)
Glycine max	ACC	Decrease in nodule number	Schmidt et al. (1999)
Glycine max L.	Ethylene gas	Decrease in nodule number	Xie et al. (1996)
	ACC	Stunted root growth but no effect on nodulation	Schmidt et al. (1999)
	ACC/ethephon	Decrease in nodule number	Caba et al. (1999)
	ACC	No effect on nodule number	Nukui et al. (2000)
Lotus japonicus L.	ACC	Decrease in nodule number	Nukui et al. (2000)
Macroptilium atropurpureum L.	ACC	Suppress nodulation	Yuhashi et al. (2000)
	ACC	Decrease in nodule number	Nukui et al. (2000)
Medicago sativa L.	ACC	Decrease in nodule number	Nukui et al. (2000)
	ACC	Decrease nodulation	Charon et al. (1999)
	ACC	Control of the position of nodule primordium formation and hyperinfection	Penmetsa et al. (2003)
	ACC	Blockage of Ca spiking in root cells, fewer infection threads, and decrease in nodule numbers	Oldroyd et al. (2001)
Melilotus alba L.	Ethylene gas	Decrease in nodule number	Lee and LaRue (1992c)
Phaseolus vulgaris L.	Ethephon	Decrease in nodule number	Tamimi and Timko (2003)
	2-Chloroethyl phosphonic acid	Decrease in nodule number	Drennan and Norton (1972)
	Ethylene	Decrease in nodule number	Lee and LaRue (1992a)
	Ethylene	Decrease in nodule number	Lee and LaRue (1992c)
	Ethrel	Decrease in nodule number	Drennan and Norton (1972)
	Ethylene gas	Blockage of infection thread elongation in inner cortex and decrease nodule number	Heidstra et al. (1997)

(continued)

Table 4.1 (continued)

Plant	Treatment	Responses	References
Sesbania rostrata L.	Ethephon/ACC	Decrease in nodule number and induction of indeterminate nodule	Fernandez-Lopez et al. (1998)
Vicia sativa L.	Ethephon	Tsr (thick, root and shoot) phenotype	van Spronsen et al. (1995)
Vigna radiata	Ethephon	Decrease in nodule number	Duodu et al. (1999)

1999). Apart from ethylene, its precursor (ACC) also inhibits nodule formation. For instance, ACC decreased the number of nodules formed on soybean wild-type Hobbit 87 roots, but had nonsignificant effect on the ethylene-insensitive mutant *etrl-l* line (Schmidt et al. 1999). However, from this study it was suggested that control of nodule numbers is independent of ethylene signaling, and the effect of ACC on nodule numbers may be attributed to the stunted growth of Hobbit 87 roots. In a recent study, the application of ACC was also found to inhibit the nodulation process in *L. japonicus* (Li et al. 2014). In this experiment, application of 10 μM ACC inhibited nodules up to 70% in wild type and over 80% in rel3 mutant, whereas 100 μM ACC completely abolished nodulation in both types of plants, as also reported by Heckmann et al. (2011). Valverde and Wall (2005) investigated regulatory function of ethylene in nodulation in the actinorhizal symbiosis between *Discaria trinervis* (*D. trinervis*) and Frankia BCU110501. Roots of axenic *D. trinervis* seedlings had abnormal growth and reduced elongation rate in the presence of ethylene-releasing compounds ACC and 2-chloroethylphosphonic acid (CEPA) in growth pouch studies. In contrast, AVG or Ag$^+$ did not modify root growth, indicating that the development of *D. trinervis* roots is sensitive to elevated ethylene levels. Although drastic response to higher ethylene levels did not result in a systemic impairment of root nodule development, changes in the nodulation pattern of the taproots were observed. As a result of root exposure to CEPA, less nodules were developed in older portions of the taproot, while a slight increase in nodulation of the mature regions of the taproot was observed in response to chemical inhibitors of ethylene. These results suggest that ethylene is involved in modulating the susceptibility for nodulation of the basal portion of *D. trinervis* seedling roots. In another study, to determine whether ethylene has any regulatory effect on spontaneous nodulation of *snf* mutants of *Mesorhizobium loti* (*M. loti*), different concentrations of ACC were applied in nodulation plate tests (Tirichine et al. 2006). Five weeks after germination, nodule numbers in *snf* mutants and *M. loti*-inoculated wild type declined with increasing concentrations of the ethylene precursor ACC. Spontaneous nodulation was totally inhibited at 10 μM ACC, while nodulation of wild-type plants was reduced to 50% (Tirichine et al. 2006). Imin et al. (2013) reported that ACC suppresses the expression of MtCEP1 gene that modulates lateral root and nodule numbers in *M. truncatula*. Some studies show an intricate web of molecular mechanism underlying the plant control over nodulation. In this sense, ethylene acts as a major participant in the autoregulation of nodulation process (Guinel and Geil 2002; Ferguson et al. 2010).

4.2.4   Effect of Ethylene on *Nod* Factor(s)

The rhizobia signals that initiate development of the nodule organ are specific lipo-chitin oligosaccharides called *Nod* factors (Arshad and Frakenberger 2002). *Nod* factors induce root hair deformation by inducing tip growth in existing root hairs and also activate cortical cells to resume mitosis resulting in nodule primordia (Truchet et al. 1991). Perception of *Nod* factor in the plant leads to the activation of a number of rhizobial-induced genes (Middleton et al. 2007; Oldroyd and Downie 2008). Several studies have shown that ethylene can inhibit numerous steps of the nodulation process (Ferguson and Mathesius 2014). Conclusively, ethylene may affect various stages of symbiosis, including the initial response to bacterial Nod factors, nodule development, senescence, and abscission (Csukasi et al. 2009; Patrick et al. 2009). Oldroyd et al. (2001) further suggested that ethylene inhibits the calcium spiking process responsible for the perception of bacterial Nod factors in *M. truncatula*. Nonetheless, mechanisms involved in the regulation of nodule development are poorly understood, and to date, very few regulatory genes have been cloned and characterized (Vernie et al. 2008). Charon et al. (1999) examined the effect of alteration of *enod 40*, a nodulation gene associated with the earliest phases of nodule organogenesis, on nodule development in transgenic *M. truncatula*. They observed that *enod 40* actions could be partially imitated by treatment of the infected root with the ethylene inhibitor, AVG. Similarly, Vernie et al. (2008) investigated the role of EFD, a gene that is upregulated during nodulation in *M. truncatula*. EFD is an ethylene response factor required for nodule differentiation and also involved in *Nod* factor signaling. The studies indicated that EFD is a negative regulator of root nodulation and infection by rhizobium. Goormachtig et al. (2004) reported plenty of root hairs in *Sesbanic rostrata* (*S. rostrata*) roots under nonaquatic conditions in contrast to hydroponic roots. Root hair infection was inhibited by ethylene and required more stringent Nod factor features than intercellular invasion. The addition of AVG enhanced the number of nodules. Similar results were obtained with Ag_2SO_4. On the other hand, ethylene has been shown to have no or a negative effect on the root hair invasion process (Guinel and Geil 2002). D'Haeze et al. (2003) reported that ethylene mediates *Nod* factor responses and is required for nodule initiation. It was found that application of purified Nod factors triggered cell division, and both *Nod* factors and ethylene induced cavities and cell death features in the root cortex. Thus, in *S. rostrata*, ethylene acts downstream from the *Nod* factors in pathways that lead to formation of infection pockets and initiation of nodule primordia (D'Haeze et al. 2003).

It has been observed that some strains of *Rhizobium* induce the formation of thick, short roots (Tsr) in common vetch (*Vicia sativa*) just like the exogenous ethylene, and this response is eliminated by AVG (Zaat et al. 1989). Such type of root phenotype as well as root hair induction and root hair formation are induced by a factor(s) produced by the bacterium in response to plant flavonoids. Root growth inhibition and root hair induction but not root hair formation could be mimicked by an ethephon treatment (Zaat et al. 1989). The addition of AVG to bacterized vetch plants suppressed the development of Tsr and restored nodulation. Similarly, van Spronsen et al. (1995) reported the development of Tsr phenotype in *Vicia sativa* ssp. *nigra* plants upon inoculation with *R. leguminosarum* bv. *viciae*. The Tsr phenotype can be mimicked by addition of ethephon and inhibited by AVG, suggesting that the Tsr phenotype is caused by excessive ethylene production. The ethylene-related localized changes were also observed during infection thread formation (Chan et al. 2013). These phenomena inhibit nodulation of the main root by preventing formation of preinfected threads and by reducing formation of root nodule primordia.

4.3   Ethylene Regulation and Its Effects on Nodulation

Any factor or substance which reduces ethylene level in plant tissues may have a positive effect on nodule formation and development. Both chemical and biological molecules have been reported to suppress ethylene levels in plants and their subsequent effect on nodulation, which are discussed in the following section.

4.3.1   Effect of Chemicals on Ethylene Regulation and Nodulation

The premise that ethylene inhibits nodule development in legumes is supported by the observations that use of all the chemicals which suppress the biosynthesis of ethylene and/or inhibit ethylene action within plant enhances nodulation (Prayitno and Mathesius 2010; Ferguson et al. 2011; Jones et al. 2015). Among chemicals, silver [Ag(I)] is a well-established inhibitor of ethylene action, while AVG, aminooxyacetic acid (AOA), and rhizobitoxine (Rtx) are well-known inhibitors of ethylene biosynthesis in plants (Tirichine et al. 2006; Ding and Oldroyd 2009; Jones et al. 2015). For action/response, ethylene binds to particular receptors for downstream signaling for which a copper cofactor is required. Ag(1) replaces copper from the ethylene binding site resulting in the failure of downstream signaling (Kumar et al. 2009), thus inhibiting ethylene action. The response of applying Ag(1) on nodulation in legumes has been described in Table 4.2. The application of Ag(1) restores partially or completely the NO_3^- or ethylene suppressed nodulation. For example, Ag(1) treatment increased nodulation in alfalfa at all NO_3^- concentrations (Caba et al. 1998). Later on, maximum stimulation of nodulation in *Bradyrhizobium* inoculated soybean plants was observed when it was grown in the presence of Ag(1) (Ligero et al. 1999). Recently, application of Ag_2SO_4 (10 μM) increased the nodulation by 40% in both wild-type and mutant pea lines compared to untreated plants (Jones et al. 2015). Aminoethoxyvinylglycine is yet another important ethylene inhibitor which has been found to stimulate nodule formation in *R. meliloti*-inoculated *M. sativa* plants (Peters and Crist-Esters 1989). Stimulation of nodule formation by AVG showed a similar concentration-dependent inhibiting effect on endogenous ethylene biosynthesis, suggesting that the primary action of AVG is in the inhibition of the endogenous ethylene biosynthesis. Jones et al. (2015) observed a huge increase in nodule numbers of wild-type pea *cv.* Sparkle plants; however, no significant improvement in nodule number was observed in *E151* mutant (low and

Table 4.2 Effect of chemical inhibitors of ethylene synthesis or action on nodulation of legume crops

Plant	Treatment[a]	Responses	References
Pisum sativum L.	AVG/Ag$_2$SO$_4$	Increase in nodule number. Both chemical inhibitors had similar effect on nodulation	Jones et al. (2015)
Lotus japonicus	AVG	Nodule number increased at 1 μM AVG while no improvement at 10 μM AVG	Li et al. (2014)
Pisum sativum L.	AVG	Nodule number increased by 36-fold in high-ethylene-producing mutant plants while insignificant increase in nodule number of wild-type plants	Ferguson et al. (2011)
Medicago truncatula L.	AVG	The number of nodules increased three times in wild-type plants, but no change occurred in ethylene-insensitive *sickle* mutant plants	Prayitno and Mathesius (2010)
Lotus japonicus	AVG	Increase in nodule number	Heckmann et al. (2011)
Lotus japonicus	STS/AVG	Nodulation was stimulated in wild-type plants, but no significant effect was observed in its mutant species	Ooki et al. (2005)
Discaria trinervis	AVG/STS	Plants had more nodules than control plants	Valverde and Wall (2005)
Sesbania rostrata	AVG	Number of nodules significantly increased	Goormachtig et al. (2004)
Lotus japonicus L.	AVG/STS	Enhancement of *NIN* gene expression	Nukui et al. (2004)
Phaseolus vulgaris L.	AVG/AOA/Cobalt	Increase in nodule number	Tamimi and Timko (2003)
Pisum sativum L.	AVG/Ag$_2$SO$_4$	Restoration of nodule in *Brz* mutant	Guinel and Geil (2002)
Medicago truncatula L.	AVG	More infection threads and nodule number	Oldroyd et al. (2001)
Pisum sativum L.	AVG/Ag$_2$SO$_4$	Increase in nodule number and changed nodule distribution	Lorteau et al. (2001)
Macroptilium atropurpureum	Rhizobitoxine	Rhizobitoxine-producing strain (*Bradyrhizobium elkanii* USDA94) significantly produced more nodules than its rhizobitoxine-deficient mutant (RTS2)	Yuhashi et al. (2000)
Lotus japonicus L.	AVG/STS	Increase in nodule number	Nukui et al. (2000)
Glycine max L.	AVG/STS	No effect on nodule number	Nukui et al. (2000)

(continued)

Table 4.2 (continued)

Plant	Treatment[a]	Responses	References
Macroptilium atropurpureum L.	AVG/STS	Increase in nodule number	Nukui et al. (2000)
Medicago sativa L.	AVG/STS	Increase in nodule number	Nukui et al. (2000)
Pisum sativum L.	AVG	Acceleration of nodule development	Guinel and Sloetjes (2000)
Pisum sativum L.	AVG/Ag_2SO_4	Control of the position of nodule primordium formation	Guinel and Sloetjes (2000)
Vigna radiata	Rhizobitoxine	A positive role of rhizobitoxine on nodulation was observed in the symbiosis between *Bradyrhizobium elkanii* USDA61 and *Vigna radiata*	Duodu et al. (1999)
Vigna radiata	STS/Cobalt	Increase in nodule number	Duodu et al. (1999)
Glycine max L.	STS	Restoration of nitrate inhibition of nodulation	Ligero et al. (1999)
Sesbania rostrata L.	Ag_2SO_4	Induction of determinate nodule	Fernandez-Lopez et al. (1998)
Medicago sativa L.	STS	Restoration of nitrate inhibition of nodulation	Caba et al. (1998)
Vicia sativa L.	AVG	Restoration of normal root phenotype	van Spronsen et al. (1995)
Medicago sativa L.	AVG	Restoration of nitrate inhibition of nodulation	Ligero et al. (1991)
Pisum sativum L.	AVG/Ag_2SO_4/ $Co(NO_3)_2$	Restoration of nodule in *sym*5 mutant	Fearn and LaRue (1991)

[a]*AOA* aminooxyacetic acid, *AVG* aminoethoxyvinylglycine, *STS* silver thiosulfate

delay nodulator). Ferguson et al. (2011) developed high-ethylene-producing mutants of pea (*na-1*) and investigated the role of ethylene and AVG on nodulation. The AVG did not significantly affect the nodule number in wild-type pea plants, while nodule number increased by 36-fold in mutant plants which shows that AVG is more effective in high-ethylene-producing plants. In *M. truncatula* L., AVG application promoted the nodule number by three times in wild-type plants. In contrast, no change in nodule number was observed in the ethylene-insensitive *sickle* mutant plants (Prayitno and Mathesius 2010). Similar enhancement in nodulation in pea (Fearn and LaRue 1991), *L. japonicus* (Tirichine et al. 2006), common bean (Tamimi and Timko 2003), and other legumes (Kang et al. 2010) when grown in the presence of varying concentration of AVG and AOA is reported.

Rhizobitoxine, an enol-ether amino acid [2-amino-4-(2-amino-3-hydroxypropoxy)-trans-3-butenoic acid], is a structural analog of AVG and inhibits ethylene synthesis (Tittabutr et al. 2008). Rhizobitoxine blocks ethylene synthesis in two ways: firstly, it inhibits the activity of β-cystathionase involved in methionine biosynthesis (Sugawara et al. 2006) and, secondly, it strongly inhibits ACC synthase in the ethylene biosynthesis pathway (Sugawara et al. 2006). It is produced by the legume microsymbiont, like *Bradyrhizobium elkanii* (*B. elkanii*), and because of its inhibitory effects on ethylene (Owens et al. 1972), Rtx has been shown to enhance nodulation in legumes (Sugawara et al. 2006; Vijayan et al. 2013). Yuhashi et al. (2000) reported a significantly less number of nodules on *Macroptilium atropurpureum* by the inoculation of Rtx-deficient mutant *B. elkanii* USDA94 than of Rtx-producing wild-type bacterium. This fact was supported that Rtx-deficient mutant synthesized more ethylene than Rtx-producing wild-type *B. elkanii* USDA94 which inhibited the nodulation process. Similarly, higher nodule number was observed in green gram inoculated with Rtx-producing *B. elkanii* USDA61 than uninoculated plants (Duodu et al. 1999). However, the effect of Rtx on nodulation has been contradictory and legume and rhizobia dependent. For example, nodulation of soybean is generally not sensitive to ethylene (Xie et al. 1996; Schmidt et al. 1999), while nodulation of green gram is sensitive (Duodu et al. 1999). Similarly, some reports have shown that there is not a significant difference in nodule number of legumes inoculated rhizobia and plants inoculated with Rtx-deficient mutants (Xiong and Fuhrmann 1996). Among bradyrhizobia able to produce Rtx, *B. elkanii* accumulates Rtx in cultures and in nodules, while *B. japonicum* does not (Kuykendall et al. 1992). In this regard, nodulation experiments using *B. elkanii* USDA61 and its Rtx -minus mutants revealed that the efficient nodulation occurring in *A. edgeworthii* but not in *A. bracteata* is highly dependent on Rtx production (Parker and Peters 2001). Therefore, such variation in the performance of Rtx seems probably be due to the differences in the abilities of the legume genotypes and rhizobial strains forming symbiosis with their host plant.

4.3.2 Effect of Bacterial ACC-Deaminase on Ethylene Regulation and Nodulation

The bacterial ACC-deaminase enzyme can improve nodulation in legumes by modulating ethylene concentration in the plant (Glick et al. 2007; Khan et al. 2009; Khalid et al. 2009). This enzyme acts as biological inhibitor of ethylene biosynthesis and thus reduces ethylene concentration in plants. The production of ethylene in plant tissues has been found directly related to the amount of ACC synthesized by the plant (Penrose and Glick 2001). Being the precursor of ethylene, ACC is immediately converted to ethylene by the enzyme ACC oxidase. However, the uptake and cleavage of ACC by the bacteria containing ACC-deaminase outside the germinating seeds or growing roots reduce the amount of ACC as well as ethylene, by acting as a sink for ACC. It is well established that higher concentration of ethylene suppresses plant growth, whereas reduction in ethylene levels in plant tissues (as a result of bacterial ACC-deaminase activity) can promote plant growth (Andrea et al. 2007; Shaharoona et al. 2007). Furthermore, plants inoculated with bacteria containing ACC-deaminase have been found resistant to the harmful effects of stress ethylene, usually generated in high amounts under undesirable environments

(Mayak et al. 2004; Bonfante and Anca 2009). Several bacteria including both rhizobia and free-living rhizobacteria have been found to facilitate plant growth and nodulation through ACC-deaminase activity (Kong et al. 2015; Gopalakrishnan et al. 2015; Chaudhary and Sindhu 2015; Nadeem et al. 2016). The potential application of bacteria carrying ACC-deaminase for nodulation improvement has been reviewed comprehensively in Sect. 4.4.

4.3.3 ACC-Deaminase-Encoding Gene (*acd*S) and Its Role in Nodulation

Genes encoding ACC-deaminase have been reported in many bacterial species (Duan et al. 2009; Farajzadeh et al. 2010). The studies on *acd*S gene encoding ACC-deaminase activity show that such gene improves symbiotic efficiency and increases nodulation in legumes. For example, *S. meliloti* containing ACC-deaminase gene (*acd*S) derived from *R. leguminosarum* showed increased ability to nodulate alfalfa (Ma et al. 2004). Tittabutr et al. (2008) investigated the role of ACC-deaminase in nodulation and growth of *Leucaena leucocephala*. The *acd*S genes encoding ACC-deaminase were cloned from *Rhizobium* sp. strain TAL1145 and *Sinorhizobium* sp. BL3 BL3 in multicopy plasmids and transferred to TAL1145. The BL3-*acd*S gene greatly enhanced ACC-deaminase activity in TAL1145 compared to the native *acd*S gene. The resulting transconjugants of TAL1145 containing the native and BL3-acdS genes formed greater (in number) and bigger nodules on *L. leucocephala* than by TAL1145 besides yielding higher root biomass. Similarly, Ma et al. (2003) isolated ACC-deaminase gene to examine its regulatory role in nodulation. Mutants with bacterial ACC-deaminase gene (*acd*S) and a leucine-responsive regulatory protein-like gene (*lrp*L) were constructed to assess their abilities to nodulate pea cv. Sparkle. Both mutants were then unable to synthesize ACC-deaminase. A decrease in nodulation efficiency was observed in response to inoculation with both mutants compared to parental strain. The study demonstrated that the presence of ACC-deaminase activity in bacteria enhanced the nodulation of pea L. cv. Sparkle, by modulating ethylene levels in the plant roots during the early stages of nodule development. Nukui et al. (2006) reported the regulation of the *acd*S gene encoding ACC-deaminase in bacteria during symbiosis in *L. japonicus*. A glucuronidase (GUS) gene was introduced into *acd*S to show GUS under control of the *acd*S promoter. Another mutant was generated with mutation in a *nif*A gene (a nitrogen-fixing regulatory gene). Two homologous *nif*A genes, mll5857 and mll5837 (designated as *nif*A1 and *nif*A2, respectively), were observed in the symbiosis island of *M. loti*. The *nif*A2 disruption resulted in considerably reduced expression of *acd*S, *nif*H, and *nif*A1 in bacteroid cells, while *nif*A1 disruption slightly promoted expression of the *acd*S transcripts and suppressed *nif*H. The study illustrated that the *acd*S gene and other symbiotic genes were positively regulated by the NifA2 protein, but not by the NifA1 protein, in *M. loti*. Furthermore, it was suggested that *M. loti acd*S participates in the establishment and/or maintenance of mature nodules by interfering with the production of ethylene. In comparison, Uchiumi et al. (2004) found that inactivation of the *acd*S gene in *M. loti* reduced number of nodules on *L. japonicus*, compared to the number of nodules formed by the wild-type strain.

4.3.4   Effect of Nodulation on ACC-Deaminase Expression

Some molecular level studies reveal that NifA protein can regulate expression of ACC-deaminase inside nodules (Nukui et al. 2006; Nascimento et al. 2014). The expression of ACC-deaminase without acdR genes shows that the presence of the gene acdR is not necessary for acdS transcription in symbiotic association of rhizobia and legume. The NifA binding sites are present in the immediate upstream region of acdS gene, suggesting that this regulatory mechanism is widespread in rhizobium species (Nascimento et al. 2012). In *M. loti*, the DNA sequence of the upstream region of acdS and nifH showed nifA1 and nifA2 (N_2-fixation regulators) and r54 RNA polymerase sigma recognition sites. The N_2-fixing regulator nifA2 encodes NifA2 protein that interacts with r54 RNA polymerase sigma recognition factor and initiates transcription of the gene acdS (Nukui et al. 2006). It has been assumed that expression of gene acdS within N_2-fixing nodules involves diminishing the effect of senesce induced by ethylene in the nodules, increasing the endurance of nodules (Gontia-Mishra et al. 2014). The NifA binding site (59-TGT-N9–11-ACA-39) is also very similar to the cyclic AMP receptor protein (CRP) binding site (59-TGTGA-N6-TCACA-39). The presence of CRP and FNR (fumarate-nitrate reductase) binding sites in upstream region could also be used for acdS gene expression and regulation in *Proteobacteria* (Grichko and Glick, 2000; Prigent-Combaret et al. 2008; Nascimento et al. 2014). Nascimento et al. (2012) found that the nitrogenase activity in *Mesorhizobium* LMS-1 was more prominent in the nodules occupied by a transconjugant (which increases ACC-deaminase activity) than in nodules occupied by wild-type species, after inoculation. While, Slater et al. (2009) reported another mechanism of acdS transcription in *Mesorhizobium* strains through plasmid gene integration into the ancestral chromosome. In *Proteobacteria* and in other rhizobial strains, acdS genes are often located on plasmids (Young et al. 2006; Kuhn et al. 2008; Kaneko et al. 2010). In *R. leguminosarum* bv. *viciae* 3841, the acdS gene is located on pRL10 plasmid near N_2-fixation genes cluster (Young et al. 2006). Recently, Kong et al. (2015) reported that *S. meliloti* (strain CCNWSX0020) encompasses a functional acdS gene in its symbiotic plasmid and contains a moderate level of ACC-deaminase activity. This arrangement has also been observed in *S. meliloti* BL225C on pSINMEB01 plasmid (Lucas et al. 2011), whereas in *B. japonicum* USDA110 and *R. leguminosarum* bv. *viciae* 128C53 K, the acdS genes are regulated by LRP-like protein and r70 promoter (Ma et al. 2003). In *Pseudomonas putida* UW4, the DNA sequence of the upstream region of acdS gene contains a CRP binding site, a FNR regulatory protein binding site (known as anaerobic transcription regulator), and a promoter sequence which controls ACC-deaminase regulatory gene (Gontia-Mishra et al. 2014).

4.4   Role of Bacterial ACC-Deaminase in *Rhizobium*-Legume Symbiosis

4.4.1   Effectiveness of Rhizobial Strains Containing ACC-Deaminase for Nodulation

Rhizobial strains with ACC-deaminase activity have been found very effective in nodulating the host legumes (Duan et al. 2009; Musarrat et al. 2009; Nascimento et al. 2016); however, the extent of ACC-deaminase activity in different strains of rhizobia varies greatly. Ma et al. (2003) found that the ACC-deaminase-producing *R. leguminosarum* bv. *viciae* 128C53K enhanced the nodulation in pea *P. sativum* L. cv. Sparkle, by modulating the ethylene levels in plant roots during the early stages of nodule development. Singh and Patel (2016) reported that ACC-deaminase-producing *R. meliloti* increased nodule number in fenugreek and its performance was better than 20 kg urea ha^{-1}. Kong et al. (2015) observed that inoculation of *S. meliloti* strain containing ACC-deaminase resulted in reduction of ethylene production in roots and increased nodulation. Indeed, ACC-deaminase-producing rhizobial cells reduce ethylene concentrations in the infection threads and increase the persistence of infection threads by suppressing the defense signals in the plant cells (Ma et al. 2004). Consequently, greater numbers of nodules are formed on roots of the inoculated plants. However, relatively low ACC-deaminase activities have been observed in rhizobia than free-living rhizobacteria.

4.4.2   Effectiveness of Free-Living Rhizobacteria Containing ACC-Deaminase for Nodulation

Inoculation with PGPR other than rhizobia has been shown to increase nodulation in legumes either by changing root architecture to facilitate root infection with rhizobia or by suppressing ethylene biosynthesis in legume roots. Several authors have reported that co-inoculation with rhizobacteria containing ACC-deaminase promotes nodulation of legumes by lowering ethylene concentrations (Shaharoona et al. 2011; Zafar-ul-Hye et al. 2013; Subramanian et al. 2015). As an example, Shaharoona et al. (2006) while evaluating the effectiveness of PGPR possessing ACC-deaminase activity on nodulation on mung bean demonstrated that the co-inoculation of PGPR with *Bradyrhizobium* enhanced the nodulation to an extent of 48% compared with only *Bradyrhizobium* inoculated legume. It was, therefore, concluded from this study that improvement in nodulation was most likely due to lowering of ethylene as a result of ACC-deaminase activity of the PGPR. Similarly, of the total nine *Pseudomonas* strains containing ACC-deaminase, three isolates (PGPR1, PGPR2, and PGPR4) resulted in a significantly higher pod yield, N and P contents of peanut in a pot trial experiment (Dey et al. 2004). Under field conditions, these PGPR significantly enhanced nodule dry weight (up to 24%) over control in 3 years' trials. Other biological traits like root length, pod numbers, and nodule numbers were also enhanced. Three rhizobacterial strains with ACC-deaminase were evaluated for improving nodulation in chickpea, both under pot and field conditions (Shahzad et al. 2008). Inoculation with ACC-deaminase-producing bacteria resulted in a highly significant increase (87%) in number of nodules per plant compared to control. Recently, Subramanian et al. (2015) found that co-inoculation of *B. japonicum* with *B. megaterium* and *M. oryzae* enhanced nodule numbers in pot-grown soybean compared to the single rhizobial inoculation. Co-inoculation also increased nodule activity measured in terms of nodule leghemoglobin content, nodulated root ARA, and the total content of the plant N. Similarly, Ullah et al. (2016b) reported the efficacy of *M. ciceri* inoculation in improving nodulation in chickpea, with and without ACC-deaminase-producing rhizobacteria. The co-inoculation increased the nodule number by 83% and 32% compared to the control and rhizobial treatment, respectively. Prakamhang et al. (2015) reported that co-inoculation was more effective for nodulation in soybean compared with single inoculation. Moreover, the amounts of poly-ß-hydroxybutyrate (PHBs) remained in mature nodules of co-inoculation treatment, while nodules senescence observed in single inoculation. Induction of soybean root could increase nodulation signaling and then trigger the accumulation of trehalose and transport of carbon that represents an increase in PHB accumulation, thereby enhancing nodulation and N_2 fixation in soybean.

Conclusion Symbiotic association between rhizobia and host legume plant is severally affected by various environmental stresses. Under any biotic or abiotic stress, the endogenous ethylene levels are increased, which negatively affects nodulation process. However, reduction in ethylene levels through the use of certain inhibitors restores nodulation. Also, ethylene concentration can be reduced by ACC-deaminase-positive symbiotic and free-living PGPR. Therefore, isolation and screening of rhizobia and free-living PGPR endowed with high ACC-deaminase activity could be a promising strategy to improve the efficiency of rhizobia-legume symbiosis. Furthermore, higher nodulation efficiency can be achieved by reducing the negative effects caused by various environmental stresses on the nodulation process. Alternatively, the insertion of genes for ACC-deaminase in rhizobia or co-inoculation of rhizobium with PGPR containing ACC-deaminase would enhance their symbiotic interactions with host legumes. Recently, it has been reported that ACC-deaminase activity is also regulated during nodulation by *nif*A protein of rhizobia; however, intensive work is required to understand such regulatory mechanism(s).

Acknowledgment The authors greatly acknowledge the contributions of Professor Muhammad Arshad (Late) in the field of plant hormones and bacterial ACC-deaminase biotechnology. Professor Arshad, a leading scientist and highly respected academician world over, laid the foundation of this chapter when he and his team of eminent teachers contributed to the first edition of the book *Microbes for Legume Improvement*, published in 2010. As editor of this book, I (MSK) pay a great tribute to Prof. Arshad for his timely and outstanding contribution to the first edition of this book.

References

Ali S, Charles TC, Glick BR (2014) Amelioration of high salinity stress damage by plant growth-promoting bacterial endophytes that contain ACC deaminase. Plant Physiol Biochem 80:160–167

Andrea JF, Vesely S, Nero V, Rodriguez H, McCormack K, Shah S, Dixon DG, Glick BR (2007) Tolerance of transgenic canola plants (*Brassica napus*) amended with plant growth-promoting bacteria to flooding stress at a metal-contaminated field site. Environ Poll 147:540–545

Arshad M, Frakenberger WT Jr (2002) Ethylene: agricultural sources and applications. Kluwer/Academic Publishers, New York

Baig KS, Arshad M, Khalid A, Hussain S, Abbas MN, Imran M (2014) Improving growth and yield of maize through bioinoculants carrying auxin production and phosphate solubilizing activity. Soil Environ 33:159–168

Bari R, Jones JDG (2009) Role of plant hormones in plant defense responses. Plant Mol Biol 69:473–488

Barnawal D, Bharti N, Maji D, Chanotiya CS, Kalra A (2014) ACC deaminase-containing Arthrobacter protophormiae induces NaCl stress tolerance through reduced ACC oxidase activity and ethylene production resulting in improved nodulation and mycorrhization in *Pisum sativum*. J Plant Physiol 171(11):884–894

Benedito VA, Torres-Jerez I, Murray JD, Andriankaja A, Allen S, Kakar K, Wandrey M, Verdier J, Zuber H, Ott T, Moreau S, Niebel A, Frickey T, Weiller G, He J, Dai X, Zhao PX, Tang Y, Udvardi MK (2008) A gene expression atlas of the model legume *Medicago truncatula*. Plant J 55:504–513

Bonfante P, Anca A (2009) Plants mycorrhizal fungi, and bacteria: a network of interactions. Annu Rev Microbiol 63:363–383

Caba JM, Recalde L, Ligero F (1998) Nitrate-induced ethylene biosynthesis and the control of nodulation in alfalfa. Plant Cell Environ 21:87–93

Caba JM, Poveda JL, Gresshoff PM, Ligero F (1999) Differential sensitivity of nodulation to ethylene in soybean cv. Bragg and a super-nodulating mutant. New Phytol 142:233–242

Chan PK, Biswas B, Gresshoff PM (2013) Classical ethylene insensitive mutants of the Arabidopsis EIN2 orthologue lack the expected 'hypernodulation' response in *lotus japonicas*. J Integr Plant Biol 55:395–408

Charon C, Sousa C, Crespi M, Kondorosi A (1999) Alteration of *enod40* expression modifies *Medicago truncatula* root nodule development induced by *Sinorhizobium meliloti*. Plant J 11:1953–1965

Chaudhary D, Sindhu SS (2015) Inducing salinity tolerance in chickpea (*Cicer arietinum* L.) by inoculation of 1-aminocyclopropane-1-carboxylic acid deaminase-containing *Mesorhizobium* strains. Afr J Microbiol Res 9:117–124

Csukasi F, Merchante D, Valpuesta V (2009) Modification of plant hormone levels and signaling as a tool in plant biotechnology. Biotechnol J 4:1293–1304

D'Haeze W, Rycke RD, Mathis R, Goormachtig S, Pagnotta S, Verplancke C, Capoen W, Holsters M (2003) Reactive oxygen species and ethylene play a positive role in lateral root base nodulation of a semi aquatic legume. PNAS 100:11789–11794

Dey R, Pal KK, Bhatt DM, Chauhan SM (2004) Growth and yield enhancement of peanut (*Arachis hypogaea* L.) by application of plant growth promoting rhizobacteria. Microbiol Res 159:371–394

Ding Y, Oldroyd GED (2009) Positioning the nodule, the hormone dictum. Plant Signal Behav 4:89–93

Drennan DSH, Norton C (1972) The effect of ethrel on nodulation in *Pisum sativum* L. Plant and Soil 36:53–57

Duan J, Müller K, Charles T, Vesely S, Glick BR (2009) 1Aminocyclopropane-1-carboxylate (ACC) deaminase genes in rhizobia from Southern Saskatchewan. Microb Ecol 57:423–436

Duodu S, Bhuvaneswari TV, Stokkermans TJ, Peters NK (1999) A positive role for rhizobitoxine in *Rhizobium*-legume symbiosis. Mol Plant Microbe Interact 12:1082–1089

Dupont L, Alloing G, Pierre O, El Msehli S, Hopkins J, Hérouart D, Frendo P (2012) The legume root nodule: from symbiotic nitrogen fixation to senescence. In: Nagata T (ed) Senescence. InTech, pp 137–168

El Yahyaoui F, Kuster F, Ben Amor H, Hohnjec B, Puhler N, Becker A, Gouzy A, Vernie J, Gough T, Niebel C, Godiard A, Gamas PL (2004) Expression profiling in *Medicago truncatula* identifies more than 750 genes differentially expressed during nodulation, including many potential regulators of the symbiotic program. Plant Physiol 136:3159–3176

El-Maarouf-Bouteau H, Sajjad Y, Bazin J, Langlade N, Cristescu SM, Balzergue S, Baudouin E, Bailly C (2015) Reactive oxygen species, abscisic acid and ethylene interact to regulate sunflower seed germination. Plant Cell Environ 38:364–374

Farajzadeh D, Aliasgharzad N, Bashir NS, Yakhchali B (2010) Cloning and characterization of a plasmid encoded ACC deaminase from an indigenous *Pseudomonas fluorescens* FY32. Curr Microbiol 61:37–43

Fearn JC, LaRue TA (1991) Ethylene inhibitors restore nodulation of sim-5 mutants of *Pisum sativum* L. cv. Sparkle. Plant Physiol 96:239–246

Ferguson BJ, Indrasumunar A, Hayashi S, Lin M, Lin Y, Reid DE, Gresshoff PM (2010) Molecular analysis of legume nodule development and autoregulation. J Integr Plant Biol 52:61–76

Ferguson BJ, Foo E, Ross JJ, Reid JB (2011) Relationship between gibberellin, ethylene and nodulation in *Pisum sativum*. New Phytol 189:829–842

Ferguson BJ, Mathesius U (2014) Phytohormone regulation of legume-rhizobia interactions. J Chem Ecol 40:770–790

Fernandez-Lopez M, Goormachtig S, Gao M, D'Haeze W, Van Montagu M, Holsters M (1998) Ethylene-mediated phenotypic plasticity in root nodule development on Sesbania rostrata. Proc Natl Acad Sci 95:12724–12728

Foo E, McAdam EL, Weller JL, Reid JB (2016) Interactions between ethylene, gibberellins, and brassinosteroids in the development of rhizobial and mycorrhizal symbioses of pea. J Exp Bot 67:2413–2424

Glick BR, Todorovic B, Czarny J, Cheng Z, Duan J, McConkey B (2007) Promotion of plant growth by bacterial ACC deaminase. Crit Rev Plant Sci 26:227–242

Gontia-Mishra I, Sasidharan S, Tiwari S (2014) Recent developments in use of 1-aminocyclopropane-1carboxylate (ACC) deaminase for conferring tolerance to biotic and abiotic stress. Biotechnol Lett 36:889–898

Gonzalez-Rizzo S, Crespi M, Frugler F (2006) The *Medicago truncatula* CRE1 cytokinin receptor regulates lateral root development and early symbiotic interaction with *Sinorhizobium meliloti*. Plant Cell 18:2680–2693

Goormachtig S, Capoen W, James EK, Holsters M (2004) Switch from intracellular to intercellular invasion during water stress-tolerant legume nodulation. PNAS 101:6303–6308

Gopalakrishnan S, Sathya A, Vijayabhharathi R, Varshney RK, Gowda CL, Krishnamurthy L (2015) Plant growth promoting rhizobia: challenges and opportunities. Biotech 5:355–377

Gour K, Patel BS, Mehta RS (2012) Yield and nodulation of fenugreek (*Trigonella foenumgraecum*) as influenced by growth regulators and vermi-wash. Indian J Agr Res 46:91–93

Gresshoff PM, Lohar D, Chan PK, Biswas B, Jiang Q, Reid D, Ferguson B, Stacey G (2009) Genetic analysis of ethylene regulation of legume nodulation. Plant Signal Behav 4:818–823

Grichko VP, Glick BR (2000) Identification of DNA sequences that regulate the expression of the *Enterobacter cloacae* UW4 1-aminocyclopropane-1-carboxylic acid deaminase gene. Can J Microbiol 46:1159–1165

Guinel FC, Sloetjes LL (2000) Ethylene is involved in the nodulation phenotype of Pisum sativum R50 (sym16), a pleiotropicmutant that nodulates poorly and has pale green leaves. J Exp Bot 51:885–894

Guinel FC, Geil RD (2002) A model for the development of the rhizobial and arbuscular mycorrhizal symbioses in legumes and its use to understand the roles of ethylene in the establishment of these two symbioses. Can J Bot 80(7):695–720

Heckmann AB, Sandal N, Bek AS, Madsen LH, Jurkiewicz A, Nielsen MW, Stougaard J (2011) Cytokinin induction of root nodule primordia in *Lotus japonicus* is regulated by a mechanism operating in the root cortex. Mol Plant Microbe Interact 24:1385–1395

Heidstra RW, Yang WC, Yalcin Y, Peck S, Emons AM, van Kammen A, Bisseling T (1997) Ethylene provides positional information on cortical cell division but is not involved in Nod factor-induced root hair tip growth in Rhizobium-legume interaction. Development 124:1781–1787

Imin N, Mohd-Radzman NA, Ogilvie HA, Djordjevic MA (2013) The peptide-encoding CEP1 gene modulates lateral root and nodule numbers in *Medicago truncatula*. J Exp Bot 64:5395–5409

Iqbal N, Trivellini A, Masood A, Ferrante A, Khan NA (2013) Current understanding on ethylene signaling in plants: the influence of nutrient availability. Plant Physiol Biochem 73:128–138

Jones JM, Clairmont L, Macdonald ES, Weiner CA, Emery RN, Guinel FC (2015) E151 (sym15), a pleiotropic mutant of pea (*Pisum sativum* L.), displays low nodule number, enhanced mycor-rhizae, delayed lateral root emergence, and high root cytokinin levels. J Exp Bot 66:4047–4059

Kaneko T, Minamisawa K, Isawa T, Nakatsukasa H, Mitsui H, Kawaharada Y, Nakamura Y, Watanabe A, Kawashima K, Ono A, Shimizu Y (2010) Complete genomic structure of the cultivated rice endophyte *Azospirillum* sp. B510. DNA Res 17:37–50

Kang BG, Kim WT, Yun HS, Chang SC (2010) Use of plant growth-promoting rhizobacteria to control stress responses of plant roots. Plant Biotechnol Rep 4:179–183

Kawaharada Y, James EK, Kelly S, Sandal N, Stougaard J (2017) The ethylene responsive factor required for nodulation 1 (ERN1) transcription factor is required for infection-thread formation in Lotus japonicus. Mol Plant Microbe Interact 30:194–204

Khalid A, Arshad M, Shaharoona B, Mahmood T (2009) Plant growth promoting rhizobacteria and sustainable agriculture. In: Khan MS, Zaidi A, Musarat J (eds) Microbial strategies for crop improvement. Springer-Verleg, Berlin, pp 133–160

Khan MS, Zaidi A, Musarrat J (2009) Microbial strategies for crop improvement. Springer-Verleg, Berlin

Kong Z, Glick BR, Duan J et al (2015) Effects of 1-aminocyclopropane-1-carboxylate (ACC) deaminase-over producing *Sinorhizobium meliloti* on plant growth and copper tolerance of *Medicago lupulina*. Plant Soil 391:383–398

Kuhn S, Stiens M, Puhler A, Schluter A (2008) Prevalence of pSmeSM11a-like plasmids in indigenous *Sinorhizobium meliloti* strains isolated in the course of a field release experiment with genetically modified *S. meliloti* strains. FEMS Microbiol Ecol 63:118–131

Kumar V, Giridhar P, Ravishankar GA (2009) $AgNO_3$-a potential regulator of ethylene activity and plant growth modulator. Electron J Biotechnol 12:1–15

Kuykendall LD, Saxena B, Devine TE, Udell SE (1992) Genetic diversity in *Bradyrhizobium japonicum* Jordan 1982 and a proposal for *Bradyrhizobium elkanii* sp. nov. Can J Microbiol 38:501–505

Lee KH, LaRue TA (1992a) Inhibition of nodulation of pea by ethylene. Plant Physiol 99:108

Lee KH, LaRue TA (1992b) Exogenous ethylene inhibits nodulation of *Pisum sativum* L. cv Sparkle. Plant Physiol 100:11759–11763

Lee KH, LaRue TA (1992c) Ethylene as a possible mediator of light and nitrate induced inhibition of nodulation of *Pisum sativum* L. cv. *Sparkle*. Plant Physiol 100:1334–1338

Li X, Lei M, Yan Z, Wang Q, Chen A, Sun J, Luo D, Wang Y (2014) The REL3-mediated TAS3 ta-siRNA pathway integrates auxin and ethylene signaling to regulate nodulation in *Lotus japonicus*. New Phytol 201:531–544

Ligero F, Lluch C, Olivares J (1987) Evolution of ethylene from roots and nodulation rate of alfalfa (*Medicago sativa* L.) plants inoculated with *Rhizobium meliloti* as affected by the presence of nitrate. J Plant Physiol 129:461–467

Ligero F, Caba JM, Lluch C, Olivares J (1991) Nitrate inhibition of nodulation can be overcome by the ethylene inhibitor aminoethoxyvinylglycine. Plant Physiol 97:1221–1225

Ligero F, Poveda JL, Gresshoff PM, Caba JM (1999) Nitrate inoculation in enhanced ethylene biosynthesis in soybean roots as a possible mediator of nodulation control. J Plant Physiol 154:482–488

Liu Y, Zhang S (2004) Phosphorylation of 1-aminocyclopropane-1-carboxylic acid synthase by MPK6, a stress-responsive mitogen-activated protein kinase, induces ethylene biosynthesis in Arabidopsis. Plant Cell 16:3386–3399

Lohar D, Stiller J, Kam J, Stacey G, Gresshoff PM (2009) Ethylene insensitivity conferred by a mutated Arabidopsis ethylene receptor gene alters nodulation in transgenic *Lotus japonicus*. Ann Bot 104:277–285

Lorteau MA, Ferguson BJ, Guinel FC (2001) Effects of cytokinin on ethylene production and nodulation in pea (Pisum sativum) cv. Sparkle. Physiol Planta 112:421–428

Lucas S, Han J, Lapidus A et al (2011) Complete sequence of plasmid 1 of Sinorhizobium meliloti BL225C. Submitted (09MAY-2011) to the EMBL/GenBank/DDBJ databases

Ma W, Guinel FC, Glick BR (2003) *Rhizobium leguminosarum* biovar *viciae* 1-aminocyclopropane-1-carboxylate deaminase promotes nodulation of pea plants. Appl Environ Microbiol 69:4396–4402

Ma W, Charles TC, Glick BR (2004) Expression of an exogenous 1-aminocyclopropane-1-carboxylate deaminase gene in *Sinorhizobium meliloti* increases its ability to nodulate alfalfa. Appl Environ Microbiol 70:5891–5897

Mayak S, Tirosh T, Glick BR (2004) Plant growth promoting bacteria that confer resistance in tomato to salt stress. Plant Physiol Biochem 42:565–572

Middleton PH, Jakab J, Penmetsa RV, Starker CG, Doll J, Kalo P, Prabhu R, Marsh JF, Mitra RM, Kereszt A, Dudas B, Bosch KV, Long SR, Cook DR, Kiss GB, Oldroyda GED (2007) An ERF transcription factor in *Medicago truncatula* that is essential for Nod factor signal transduction. Plant Cell 19:1221–1234

Musarrat J, Al Khedhairy AA, Al-Arifi S, Khan MS (2009) Role of 1-aminocyclopropane-1-carboxylate deaminase in *Rhizobium*-legume symbiosis. In: Khan MS, Zaidi A, Musarat J (eds) Microbial strategies for crop improvement. Springer-Verleg, Berlin, pp 63–83

Nadeem SM, Ahmad M, Naveed M, Imran M, Zahir ZA, Crowley DE (2016) Relationship between in vitro characterization and comparative efficacy of plant growth-promoting rhizobacteria for improving cucumber salt tolerance. Arch Microbiol 198:379–387

Nadeem SM, Imran M, Naveed M, Khan MY, Ahmad M, Zahir ZA, Crowley DE (2017) Synergistic use of biochar, compost and plant growth-promoting rhizobacteria for enhancing cucumber growth under water deficit conditions. J Sci Food Agric. doi:10.1002/jsfa.8393

Nagata M, Suzuki A (2014) Effects of phytohormones on nodulation and nitrogen fixation in leguminous plants. In: Ohyama T (ed) Agricultural and biological sciences: advances in biology and ecology of nitrogen fixation. InTech, pp 111–128

Nascimento FX, Brigido C, Glick BR, Oliveira S (2012) ACC deaminase genes are conserved among Mesorhizobium species able to nodulate the same host plant. FEMS Microbiol Lett 336:26–37

Nascimento FX, Rossi MJ, Soares CRFS, McConkey BJ, Glick BR (2014) New insights into 1-aminocyclopropane 1carboxylate (ACC) deaminase phylogeny, evolution and ecological significance. PLoS One 9:99168

Nascimento FX, Brígido C, Glick BR, Rossi MJ (2016) The role of rhizobial ACC deaminase in the nodulation process of leguminous plants. Int J Agron 2016:1369472 9p

Nukui N, Ezura H, Yohsshi K, Yasuta T, Minamisawa K (2000) Effects of ethylene precursor and inhibitors for ethylene biosynthesis and perception on nodulation in Lotus japonicus and Macroptilium atropurpureum. Plant Cell Physiol 41:893–897

Nukui N, Ezura H, Minamisawa K (2004) Transgenic *Lotus japonicus* with an ethylene receptor gene Cm-ERS1/H70A enhances formation of infection threads and nodule primordia. Plant Cell Physiol 45:427–435

Nukui N, Minamisawa K, Ayabe SI, Aoki T (2006) Expression of the 1-aminocyclopropane-1-carboxylate deaminase gene requires symbiotic nitrogen fixing regulator gene *nifA2* in *Mesorhizobium loti* MAFF303099. Appl Environ Microbiol 72:4964–4969

Okamoto S, Ohnishi E, Sato S, Takahashi H, Nakazono M, Tabata S, Kawaguchi M (2009) Nod factor/nitrate-induced CLE genes that drive HAR1-mediated systemic regulation of nodulation. Plant Cell Physiol 50:67–77

Oldroyd GE, Downie JA (2008) Coordinating nodule morphogenesis with rhizobial infection in legumes. Annu Rev Plant Biol 59:519–546

Oldroyd GED, Engstrom EM, Long SR (2001) Ethylene inhibits the Nod factor signal transduction pathway of *Medicago truncatula*. Plant Cell 13:1835–1849

Ooki Y, Banba M, Yano K, Maruya J, Sato S, Tabata S, Hata S (2005) Characterization of the Lotus japonicus symbiotic mutant lot1 that shows a reduced nodule number and distorted trichomes. Plant Physiol 137:1261–1271

Owens LD, Thompson JF, Fennessy PV (1972) Dihydrorhizobitoxine, a new ether amino acid from *Rhizobium japonicum*. J Chem Soc Chem Commun 1972:715

Parker MA, Peters NK (2001) Rhizobitoxine production and symbiotic compatibility of *Bradyrhizobium* from Asian and North American lineages of *Amphicarpaea*. Can J Microbiol 47:1–6

Patrick A, Gusti A, Cheminant S, Alioua M, Dhondt S, Coppens F, Beemster GTS, Genschik P (2009) Gibberellin signaling controls cell proliferation rate in Arabidopsis. Curr Biol 19:1188–1193

Penmetsa RV, Frugoli JA, Smith LS, Long SR (2003) Dual genetic pathways controlling nodule number in Medicago truncatula. Plant Physiol 131:998–1008

Penrose DM, Glick BR (2001) Levels of ACC and related compounds in exudate and extracts of canola seeds treated with ACC-deaminase containing plant growth promoting bacteria. Can J Microbiol 47:368–372

Peters NK, Crist-Esters DK (1989) Nodule formation is stimulated by the ethylene inhibitor aminoethoxyvinylglycine. Plant Physiol 91:690–693

Prakamhang J, Tittabutr P, Boonkerd N, Teamtisong K, Uchiumi T, Abe M, Teaumroong N (2015) Proposed some interactions at molecular level of PGPR coinoculated with *Bradyrhizobium diazoefficiens* USDA110 and *B. japonicum* THA6 on soybean symbiosis and its potential of field application. Appl Soil Ecol 85:38–49

Prayitno J, Mathesius U (2010) Differential regulation of the nodulation zone by silver ions, L-α-(2-amino-ethoxyvinyl)-glycine, and the skl mutation in *Medicago truncatula*. HAYATI J Biosci 17:15–20

Prayitno J, Rolfe BG, Mathesius U (2006) The ethylene-insensitive sickle mutant of *Medicago truncatula* shows altered auxin transport regulation during nodulation. Plant Physiol 142:168–180

Prigent-Combaret C, Blaha D, Pothier JF, Vial L, Poirier M-A et al (2008) Physical organization and phylogenetic analysis of acdR as leucine-responsive regulator of the 1-aminocyclopropane-1-carboxylate deaminase gene acdS in phytobeneficial *Azospirillum lipoferum* 4B and other *Proteobacteria*. FEMS Microbiol Ecol 65:202–219

Reid DE, Ferguson BJ, Gresshoff PM (2011) Inoculation-and nitrate-induced CLE peptides of soybean control NARK-dependent nodule formation. Mol Plant Microb Interact 24:606–618

Saleem M, Arshad M, Hussain S, Bhatti A (2007) Perspective of plant growth promoting rhizobacteria (PGPR) containing ACC-deaminase in stress agriculture. J Ind Microbiol Biotechnol 34:635–648

Schmidt JS, Harper JE, Hoffman TK, Bent AF (1999) Regulation of soybean nodulation independent of ethylene signalling. Plant Physiol 119:951–959

Shaharoona B, Arshad M, Zahir ZA (2006) Effect of plant growth promoting rhizobacteria containing ACC-deaminase on maize (*Zea mays* L.) growth under axenic conditions and on nodulation in mung bean (*Vigna radiata* L.) Lett Appl Microbiol 42:155–159

Shaharoona B, Arshad M, Khalid A (2007) Differential response of etiolated pea seedling to 1-aminocyclopropane-1-carboxylate and/or l-methionine utilizing rhizobacteria. J Micrbiol 45:15–20

Shaharoona B, Imran M, Arshad M, Khalid A (2011) Manipulation of ethylene synthesis in roots through bacterial ACC deaminase for improving nodulation in legumes. Crit Rev Plant Sci 30(3):279–291

Shahzad MS, Khalid A, Arshad M, Khalid M, Mehboob I (2008) Integrated use of plant growth promoting bacteria and P-enriched compost for improving growth, yield and nodulation of chickpea. Pak J Bot 40:1735–1144

Shahzad SM, Khalid A, Arshad M, Tahir J, Mahmood T (2010) Improving nodulation, growth and yield of *Cicer arietinum* L. through bacterial ACC-deaminase induced changes in root architecture. Eur J Soil Biol 46:342–347

Shahzad SM, Arif MS, Riaz M, Iqbal Z, Ashraf M (2013) PGPR with varied ACC-deaminase activity induced different growth and yield response in maize (*Zea mays* L.) under fertilized conditions. Eur J Soil Biol 57:27–34

Siddikee MA, Glick BR, Chauhan PS, Yim WJ, Sa T (2011) Enhancement of growth and salt tolerance of red pepper seedlings (*Capsicum annuum* L.) by regulating stress ethylene synthesis with halotolerant bacteria containing 1-aminocyclopropane-1-carboxylic acid deaminase activity. Plant Physiol Biochem 49:427–434

Singh NK, Patel DB (2016) Performance of fenugreek bioinoculated with *Rhizobium meliloti* strains under semi-arid condition. J Environ Biol 37:31

Slater SC, Goldman BS, Goodner B, Setubal JC, Farrand SK, Nester EW, Burr TJ, Banta L, Dickerman AW, Paulsen I, Otten L (2009) Genome sequences of three *Agrobacterium biovars* help elucidate the evolution of multichromosome genomes in bacteria. J Bacteriol 191:2501–2511

Stepanova AN, Alonso JM (2009) Ethylene signaling and response: where different regulatory modules meet. Curr Opin Plant Biol 12:548–555

Subramanian P, Kim K, Krishnamoorthy R, Sundaram S, Sa T (2015) Endophytic bacteria improve nodule function and plant nitrogen in soybean on co-inoculation with *Bradyrhizobium japonicum* MN110. Plant Growth Regul 76:327–332

Suganuma N, Yamauchi H, Yamamoto K (1995) Enhanced production of ethylene by soybean roots after inoculation with *Bradyrhizobium japonicum*. Plant Sci 111:163–168

Sugawara M, Okazaki S, Nukui N, Ezura H, Mitsui H, Minamisawa K (2006) Rhizobitoxine modulates plant microbe interactions by ethylene inhibition. Biotechnol Adv 24:382–388

Sun J, Cardoza V, Mitchell DM, Bright L, Oldroyd G, Harris JM (2006) Crosstalk between jasmonic acid, ethylene and Nod factor signaling allows integration of diverse inputs for regulation of nodulation. Plant J 46:961–970

Tamimi SM, Timko MP (2003) Effects of ethylene and inhibitors of ethylene synthesis and action on nodulation in common bean (*Phaseolus vulgaris* L.) Plant Soil 257:125–131

Tirichine L, Sandal N, Madsen LH, Radutoiu S, Albrektsen AS, Sato S, Asamizu E, Tabata S, Stougaard J (2006) A gain-of-function mutation in a cytokinin receptor triggers spontaneous root nodule organogenesis. Sci Mag 315:104–107

Tittabutr P, Awaya JD, Li QX, Borthakur D (2008) The cloned 1-aminocyclopropane-1-carboxylate (ACC) deaminase gene from *Sinorhizobium* sp. strain BL3 in *Rhizobium* sp. strain TAL1145 promotes nodulation and growth of *Leucaena leucocephala*. Syst Appl Microbiol 31:141–150

Tittabutr P, Sripakdi S, Boonkerd N, Tanthanuch W, Minamisawa K, Teaumroong N (2015) Possible role of 1-aminocyclopropane-1-carboxylate (ACC) deaminase activity of *Sinorhizobium* sp. BL3 on symbiosis with mung bean and determinate nodule senescence. Microbes Environ 30:310

Truchet G, Roche P, Lerouge P, Vasse J, Camut S, De Billy F, Promé JC, Dénarié J (1991) Sulphated lipo-oligosaccharide signals of *Rhizobium meliloti* elicit root nodule organogenesis in alfalfa. Nature 351:670–673

Uchiumi T, Oowada T, Itakura M, Mitsui H, Nukui N, Dawadi P, Kaneko T, Tabata S, Yokoyama T, Tejima T, Saeki K, Oomori H, Hayashi M, Maekawa T, Sriprang R, Murooka Y, Tajima S, Simomura K, Nomura M, Suzuki A, Shimoda S, Sioya K, Abe M, Minamisawa K (2004) Expression islands clustered on symbiosis island of *Mesorhizobium loti* genome. J Bacteriol 186:2439–2448

Ullah S, Raza MS, Imran M, Azeem M, Awais M, Bilal MS, Arshad M (2016a) Plant growth promoting rhizobacteria amended with *mesorhizobium ciceri* inoculation effect on nodulation and growth of chickpea (*Cicer arietinum* L.) Am Res Thoughts 3:3408–3420

Ullah U, Ashraf M, Shehzad SM, Siddiqui AR, Piracha MA, Suleman M (2016b) Growth behavior of tomato (*Solanum lycopersicum*) under drought stress in the presence of silicon and plant growth promoting rhizobacteria. Soil Environ 35:65–75

Valverde C, Wall LG (2005) Ethylene modulates the susceptibility of the root for nodulation in actinorhizal *Discaria trinervis*. Physiol Planta 124:121–131

Van Spronsen PC, Van Brussel AA, Kijne JW (1995) Nod factors produced by Rhizobium *legumi-nosarum biovar viciae* induce ethylene-related changes in root cortical cells of *Vicia sativa* ssp. *nigra*. Eur J Cell Biol 68:463–469

van Workum WAT, Van Brussel AAN, Tak T, Wijffelman CA, Kijne WJ (1995) Ethylene prevents nodulation of *Vicia sativa ssp. nigra* by exopolysaccharides deficient mutants of *Rhizobium leguminosarum* bv viciae. Mol Plant Microbe Interact 8:278–285

Vernie T, Moreau S, de Billy F, Plet J, Combier J-P, Rogers C, Vernie GO (2008) Factor involved in the control of nodule number and differentiation in *Medicago truncatula*. Plant Cell 20:2696–2713

Vijayan R, Palaniappan P, Tongmin SA, Padmanaban E, Natesan M (2013) Rhizobitoxine enhances nodulation by inhibiting ethylene synthesis of *Bradyrhizobium elkanii* from Lespedeza species: validation by homology modeling and molecular docking study. World. J Pharm Pharm Sci 2:4079–4094

Xie ZP, Staehelin C, Wiemken A, Bolle T (1996) Ethylene responsiveness of soybean cultivars characterized by leaf senescence, chitinase induction and nodulation. J Plant Physiol 149:690–694

Xiong K, Fuhrmann JJ (1996) Comparison of rhizobitoxine-induced inhibition of β-cystathionase from different bradyrhizobia and soybean genotypes. Plant Soil 186:53–61

Yoong FY, O'Brien LK, Truco MJ, Huo H, Sideman R, Hayes R, Michelmore RW, Bradford KJ (2016) Genetic variation for thermotolerance in lettuce seed germination is associated with temperature-sensitive regulation of ethylene response factor1 (ERF1). Plant Physiol 170:472–488

Young JPW, Crossman LC, Johnston AWB et al (2006) The genome of *Rhizobium leguminosarum* has recognizable core and accessory components. Genome Biol 7:R34

Yuhashi KI, Ichikawa N, Ezuura H, Akao S, Minakawa Y, Nukui N, Yasuta T, Minamisawa K (2000) Rhizobitoxine production by *Bradyrhizobium elkanii* enhances nodulation and competitiveness on *Macroptilium atropurpureum*. Appl Environ Microbiol 66:2658–2663

Zaat SA, Van Brussel AA, Tak T, Lugtenberg BJ, Kijne JW (1989) The ethylene inhibitor aminoethoxyvinylglycine restores normal nodulation by *Rhizobium leguminosarum* biovar Viciaeon *Vicia sativa* ssp. *nigra* by suppressing the thick and short roots phenotype. Planta 177:141–150

Zafar-ul-Hye M, Ahmad M, Shahzad SM (2013) Synergistic effect of rhizobia and plant growth promoting rhizobacteria on the growth and nodulation of lentil seedlings under axenic conditions. Soil Environ 32:79–86

Rhizobial Exopolysaccharides: A Novel Biopolymer for Legume-Rhizobia Symbiosis and Environmental Monitoring

5

Rabindranath Bhattacharyya, Sandip Das, Raktim Bhattacharya, Madhurima Chatterjee, and Abhijit Dey

Abstract

Extracellular polymeric substances produced by microorganisms are a complex mixture of biopolymers mainly consisting of polysaccharides along with fewer amounts of proteins, nucleic acids, uronic acids, lipids, and humic substances. Biopolymers secreted by microorganisms are considered as a potential alternative over conventional chemical polymers because of their easy biodegradability, nontoxicity, and renewable nature. Exopolysaccharides (EPSs) released by rhizobia play a pivotal role in both establishment of effective symbiosis with leguminous plants and adaptation to environmental stresses. Moreover, low-molecular-weight fraction of this polysaccharide acts as a signal molecule in the symbiotic dialogue. Besides these, EPSs extracted from different microbes have been recognized as a sustainable flocculant for their application in different types of wastewater treatment. EPS has also been considered as a good bioemulsifier for different hydrocarbons. Microbial EPSs have also been found useful in removal of pollutants from contaminated sites. In this chapter, the role of rhizobial EPS in developing effective legume-rhizobia symbiosis is discussed. Also, the role of EPS secreted by root-nodulating bacteria in remediation of heavy metals and hydrocarbon degradation has been highlighted.

R. Bhattacharyya (✉) • S. Das • R. Bhattacharya • M. Chatterjee • A. Dey
Microbiology Laboratory, Department of Life Sciences, Presidency University, 86/1 College Street, Kolkata 700073, West Bengal, India
e-mail: rabindranathbpc@yahoo.co.in

© Springer International Publishing AG 2017
A. Zaidi et al. (eds.), *Microbes for Legume Improvement*,
DOI 10.1007/978-3-319-59174-2_5

5.1   Introduction

Legumes are one of the most important sources of protein in human dietary system because in many developing countries, animal proteins are expensive and are not readily accepted as food by some layers of society. In addition, they are source of oils, fibers, and raw materials for many products. The cultivation of legumes is therefore incredibly valuable. And so, the cultivation of nitrogen-rich leguminous crops in distinctly variable farming practices across the globe plays an important role in enhancing the fertility of soils especially in the newly reclaimed lands or soils which are nutrient poor/deficient (Zahra et al. 1990). Another challenging aspect before the scientists is the rapidly increasing environmental contamination which is primarily due to massive population burst, industrialization, and urbanization. Release of toxic pollutants, for example, heavy metals and hydrocarbons from these industries into the environments, has threatened the sustainability of the environment very seriously. Due to the deposition of such pollutants in agricultural field across the world, the fertility of soils, the production of crops, and indirectly the human health via food chain/web have been affected adversely (Chaterjee et al. 1995; Das et al. 2004), largely because heavy metals are nondestructive and persist in soil (Khan et al. 2009). So, one of the major challenges in environmental biotechnology is the bioremediation of heavy metal pollution from contaminated sites and restoring soil quality for future crop production. At present for absorption of heavy metals, various chemically synthesized flocculants (viz., polysaccharide derivatives or polyethylene amine) are widely used in industrial fields for wastewater treatment and drinking water purification (Shih et al. 2001). These synthetic flocculants are being used widely because of cost-effectiveness. However, they have some major drawback. For example, they are not biodegradable, and even few of their degraded monomers such as acrylamide are neurotoxic and carcinogenic (Yokoi et al. 1995). Alternatively, EPSs of different microbiological sources, including *Rhizobium*, applied in drinking and wastewater treatment are considered as good bioflocculant, because they are biodegradable and harmless and do not cause any secondary pollution (He et al. 2004). During the last decade, biosurfactants have been investigated as potential alternative for synthetic surfactants and are expected to have many potential industrial and environmental applications related to emulsification, foaming, detergency, wetting, dispersion, and solubilization of hydrophobic compounds (Banal et al. 2000; Luma et al. 2013). Presently, EPSs from different microbial sources are gaining importance as bioemulsifiers and increase the solubilities of hydrocarbons and the efficiency of hydrocarbon degradation (Han et al. 2014; Huang et al. 2012) apart from their bioflocculating (Han et al. 2014) and bioadsorption properties of heavy metals from wastewater and natural water (Shuhong et al. 2014).

Modern agriculture is however facing numerous challenges. One of the major challenges is the loss of soil fertility. Apart from soil fertility, fluctuating climatic factors and pest attack are also major concern in this regard. Sustainability and environmental safety of agricultural production rely on the use of eco-friendly approaches like biofertilizers, biopesticides, and crop residue return. In this context, rhizobia play an important role in agriculture and crop production as they induce nitrogen-fixing nodules on the roots of leguminous plants. Rhizobia promote the growth and yield of legumes by secreting siderophores (Datta and Chakrabartty 2014), phosphate solubilization (Kranthi Kumar and Raghu Ram 2014), phytohormones (Ferguson and Mathesius 2014), and exopolysaccharides (EPSs) (Janczarek et al. 2015). Of these biomolecules, EPS produced by rhizobia plays an indispensable role in elongation of infection threads, a special tubular structures through which rhizobia colonize root nodules (Cheng and Walker 1998; Jaszek et al. 2014), and consequently forms an effective legume-rhizobia symbiosis (Skorupska et al. 2006). This chapter presents the features of rhizobial EPS with an aim to establish their role in the development of effective legume-rhizobia symbiosis. The role of rhizobial EPS in some process of environmental monitoring especially in heavy metal bioremediation and emulsification of hydrocarbons has also been discussed.

5.2   Microbial Exopolysaccharides

Exopolysaccharides are organic macromolecules that are formed by polymerization of similar or identical building blocks, which may be arranged as repeated units within the polymer (Sutherland 2001). The bacterial EPS was first detected in *Leuconostoc mesenteroides*. The bacterial EPSs are found in (a) capsule (or capsular polysaccharides) and (b) slime (slime polysaccharides). In capsular polysaccharides, the polymer is closely associated with the cell surface, whereas slime polysaccharides are loosely associated with cell surface. EPS-producing bacteria are present in a variety of ecological niches. Therefore, the physiological role of these exopolysaccharides is also diverse. However, bacterial EPSs have a wide range of applications, viz., food products, pharmaceuticals, and bioemulsifiers (Xie et al. 2013), bioflocculants (Sathiyanaryanan et al. 2013), chemical products (Shah et al. 2008), biosorption of heavy metals (Mohamad et al. 2012), and antibiofilm agents (Rendueles et al. 2013). The rhizobial EPSs on the contrary have a significant role in the development of effective legume-rhizobia symbiosis. Polysaccharides can be extracted from biomass resources like algae and higher-ordered plants or recovered from the fermentation broth of bacteria (Gram-positive bacteria) or fungal cultures (Öner 2013; Han et al. 2016). Regarding bacteria it might be Gram positive (Yuksekdag and Aslim 2008; Ismail and Nampoothiri 2010) or Gram negative (Vu et al. 2009; Freitas et al. 2011; Janczarek et al. 2015) including some extreme marine bacteria (Poli et al. 2010; Liu et al. 2013). However, for sustainable and economical production of bioactive polysaccharides at industrial scale, microbial sources are preferred over plants and algae as they are very fast in respect of production under fully controlled fermentation conditions. So, due to increased demand of microbial biopolymer for various industrial and biotechnological applications, scientists are very much interested for searching new organisms for newer exopolysaccharides.

5.2.1 Composition of Exopolysaccharides

Polysaccharides are a major fraction of the EPS matrix. Most EPSs are long molecules, linear or branched, with a molecular mass of 0.5×10^6 to 2×10^6 Da. Both α- and β-*Proteobacteria* are able to produce EPS that can be classified as homo- and heteropolysaccharides. Homopolysaccharides are generally neutral glucans, whereas heteropolysaccharides are mostly polyanionic compounds, due to the presence of uronic acid. Several exopolysaccharides are homopolysaccharides, including the sucrose-derived glucans and fructans produced by the streptococci in oral biofilms and cellulose formed by *Rhizobium* spp., *Gluconacetobacter xylinus*, *Agrobacterium tumefaciens*, and different species from the *Pseudomonadaceae* and *Enterobacteriaceae* families (Wingender et al. 2001; Zogaj et al. 2001). A mixture of neutral and charged sugar residues are the major components of most heteropolysaccharides. They can contain organic or inorganic substituents that greatly affect their physicochemical properties. Many known exopolysaccharides, including xanthan, alginate, and colonic acid, are polyanionic owing to the occurrence of uronic acids. Polycationic exopolysaccharides also exist, for instance, intercellular adhesin, which is composed of β-1,6-linked *N*-acetylglucosamine with partly deacetylated residues. This adhesin was discovered in important nosocomial pathogens such as *Staphylococcus aureus* and *S. epidermidis*. It has since been detected in a range of other bacteria (Götz 2002; Jefferson 2009).

Exopolysaccharides can be strain specific; for instance, various *Streptococcus thermophilus* strains produce heteropolysaccharides of different monomer compositions and ratios and different molecular masses (Vaningelgem et al. 2004). *Pseudomonas aeruginosa*, one of the best studied models, produces at least three distinct exopolysaccharides, alginate, Pel, and Psl, which contribute to biofilm development and architecture (Ryder et al. 2007). The rhizobial EPSs are mostly species- or strain-specific heteropolysaccharides and are formed from repeat units of hexose residues such as glucose, galactose, rhamnose, mannose, galacturonic and glucuronic acids with pyruvate, acetyl, and succinyl and hydroxybutanoic substitutions (Lepek and D'antuono 2005). The rhizobia EPSs are extremely diverse, varying in the type of sugars and their linkage in the single subunit, repeat unit size, and polymerization degree, as well as noncarbohydrate decoration (Laus et al. 2005; Fraysse et al. 2003; Skorupska et al. 2006). Examples of rhizobial EPSs are (1) succinoglycan or EPS I, best known rhizobial EPS, and (2) galactoglucan or EPS II. Both EPSs are produced by several *Sinorhizobium meliloti* strains.

Strains of *R. leguminosarum*, despite having different biovars (*trifolii, viciae,* and *phaseoli*) and nodulating different host plants, have conserved EPS composed of glucose, glucuronic acid, and galactose at a ratio of 5:2:1 (Robertsen et al. 1981; O'Neill et al. 1991) (Fig. 5.1a). However, some strains secreted EPS with different sugar contents and chain length. In *R. leguminosarum* bv. *trifolii* 4S, EPS subunit is composed of seven sugars, and the galactose molecule is absent in this chain (Amemura et al. 1983) (Fig. 5.1a). In *R. leguminosarum* bv. viciae 248, the EPS subunit has an additional glucuronic acid (Cremers et al. 1991) (Fig. 5.1b). EPS of *R. tropici* CIAT899T composed of subunits consisting of glucose and galactose sugars at a ratio of 6:2 (Gil-Serrano et al. 1990) (Fig. 5.1c).

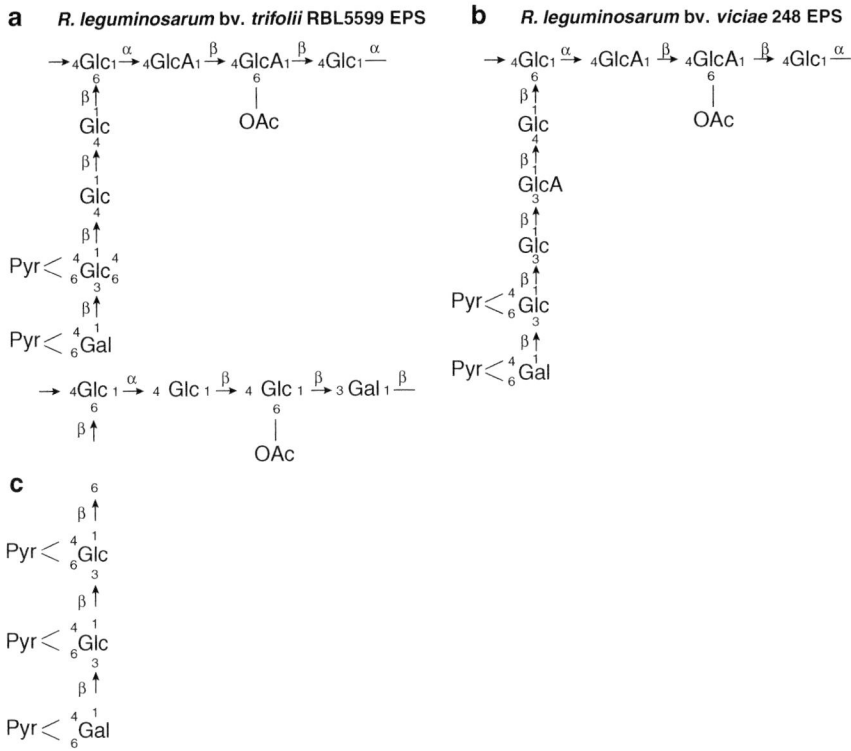

Fig. 5.1 Primary structure of EPS produced by *R. leguminosarum* bv. *trifolii* (**a**), *R. leguminosarum* bv. *viciae* (**b**), and *R. tropici* (**c**); units/subunits used are *Glc* glucose, *Gal* galactose, *GlcA* glucuronic acid, *GalA* galacturonic acid (Janczarek 2011)

5.2.2 Extraction and Purification of EPS

The identification of EPS components depends on the isolation method used. However, efficient EPS isolation is challenging, particularly for EPS from environmental biofilms, which can contain a variable range of components and that each component will require different extraction methods. Furthermore, it is extremely difficult to quantitatively isolate EPS from a given biofilm, because some of the EPS fraction remains bound to the bacteria. Also, the isolation procedure damages cells, causing intracellular material to leak into the matrix. Centrifugation, filtration, heating, blending, sonication, and treatment with complexing agents and with ion-exchange resins are some of methods which have been/are being used for EPS isolation. According to our knowledge and lab work experience, solvent extraction method with slide modification is the cost-effective technique for rhizobial EPS extraction (Bhattacharyya and Das 2015). For the extraction of rhizobial EPS through the solvent extraction method, sample of 24 h culture broth was centrifuged (11,200 rpm × 20 min), and supernatant was collected and mixed with three times volume of chilled MB grade ethanol (Himedia) and kept overnight at 6 °C prior to centrifugation. The resulting precipitate was collected via centrifugation at

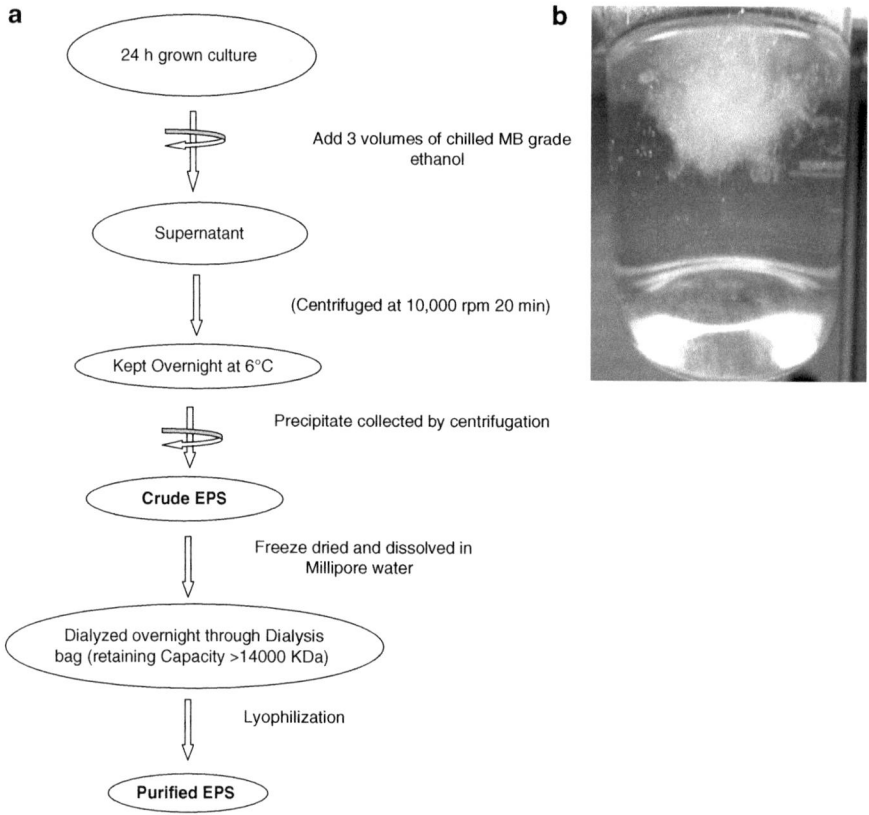

Fig. 5.2 (a) Flowchart depicting isolation of EPS by solvent extraction method. (b) Isolation of EPS

10,000 rpm × 10 min, redissolved in Millipore water, and dialyzed through a cellulose membrane (Sigma-Aldrich, retaining MW > 12,400 Da) against deionized distilled water for 24 h to discard low-molecular-weight materials. The dialyzed material was lyophilized to obtain purified EPS (Fig. 5.2a and b).

5.3   Role of Rhizobial EPS in the Development of Effective Legume-Rhizobia Symbiosis

The development of legume-rhizobia symbiosis is an effective biological process in the sustainable agriculture system. Due to consistently increasing human populations, the current agricultural systems are under tremendous pressure basically for these reasons: (1) modernization and urbanization, (2) nonjudicious use of chemical pesticides and fertilizers, and (3) the human food demand is on rise. Therefore, to overcome these problems, well-directed and concerted efforts are needed in order to use the full potential of agroecosystems especially. Nitrogen, an important plant nutrient, is one of the key factors which plays an important role in crop improvement. The chemical fertilizers which are the source of nitrogen are important in this regard. However, Martensson (1992) reported that most of the cultivated legumes are exposed to agrichemicals and chemical fertilizers which not only contain essential nutrients but also contaminants like heavy metals. Mussarat and Haseeb (2000) reported that agrichemicals may protect rhizobial recognition sites on root surface of legumes. As a result, the biological nitrogen fixation and consequently the yield of leguminous crop will be reduced in a substantial amount due to poor nodulation. So, the development of efficient legume-rhizobia symbiosis is an essential need for the legume crop improvement. The success of legume-rhizobia symbiosis development depends on several factors, for example, the efficacy of rhizobial strains to fix atmospheric nitrogen, the survivability of the microsymbiont in the field, etc.

For the establishment of the effective legume-rhizobia symbiosis, polysaccharides secreted by rhizobia play a pivotal role. Exopolysaccharides (EPSs) play a key role in both adaptation to environmental conditions and establishment of effective symbiosis with leguminous plants (Downie 2010). Rhizobial EPS may also be involved in invasion and nodule development, bacterial release from infection threads, bacterial development, suppression of plant defense response, and plant antimicrobial compounds (Skorupska et al. 2006). Djordjevic et al. (1987) reported that mutant strains deficient in EPS production were unable to promote the formation of efficient nodules on the host plants. However, the restoration of the development of functional nodule with the mutant can be done by adding purified EPS from the parental strain. Again, a mutant of *Sinorhizobium meliloti* Rm 2011 unable to produce EPS I only induced the formation of pseudonodules that did not contain infection thread or bacteroids in *Medicago sativa* (Niehaus et al. 1993). Cheng and Walker (1998) also reported that an intact structure of the EPS I of *S. meliloti* Rm 1021 is required for the initial infection thread formation and elongation, suggesting that EPS functions as a signaling molecule that recognizes complex receptors present in plants. EPSs also play an important role in suppression of plant defense response. EPS mutant strains (*exo⁻*) of *Bradyrhizobium japonicum* stimulated the accumulation of phytoalexins in the early stages of interaction with *Glycine max*. In contrast accumulation of phytoalexin is tenfold less when inoculated with wild-type strain (Parniske et al. 1994). During lateral root base nodulation, the mutant microsymbiont *Azorhizobium caulinodans* which are deficient in EPS production were unable to penetrate the tissue of host (*Sesbania rostrata*) due to loss of protection by EPS upon exposure to H_2O_2, produced by host as a defense mechanism (D'Haeze et al. 2004). Considering all these information, it is clear that rhizobial polysaccharide is an important factor for the establishment of effective legume-rhizobia symbiosis.

5.4 Role of Rhizobial EPS in Wastewater Treatment

Apart from their role in development of effective legume *Rhizobium* symbiosis, the rhizobial EPS is currently utilized for the treatment of wastewater as bioflocculant (Fig. 5.3). The interaction between microbial anionic polymers and heavy metals

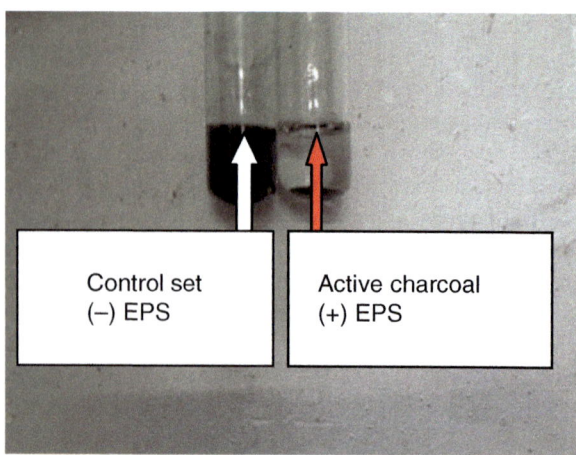

Fig. 5.3 Bioflocuulation activity of activated charcoal treat with and without EPS

Table 5.1 Flocculating activity of microbial EPSs of different origin

EPS producers	Activity tested on	Bioflocculating activity (%)	References
Paenibacillus polymyxa	High-ash Indian coals	60	Liang and Wang (2015)
Klebsiella sp. PB12	Activated charcoal	80	Mandal et al. (2013)
Acinetobacter junii BB1A		90	Yadav et al. (2012)
Rhizobium tropici		90	Das et al. (Unpublished)

plays important ecological and practical uses. It can be useful for removing toxic metals from solutions. Presently, different types (inorganic, organic synthetic, and natural occurring and microbial) of flocculating agents are used in industrial processes such as wastewater treatment, downstream processing, and food and fermentation processes (Zhang et al. 2007). However naturally occurring flocculants and microbial flocculants have several advantages over inorganic and organic synthetic flocculants. For example, inorganic flocculant, aluminum, can induce Alzheimer's disease. The organic synthetic flocculants—polyacrylamide derivatives—are carcinogenic, neurotoxic, as well as nonbiodegradable in nature (Yokoi et al. 1995). Presently different species of *Rhizobium* are being utilized for the wastewater treatment using the exopolysaccharides secreted by them.

Bioflocculation of activated charcoal treated with EPS of *R. tropici* as viewed in Fig. 5.3. Foster et al. (2000) reported that the ability of *Rhizobium etli* M4 and its EPS can bind a variety of metals and can be used as a potential strength for heavy metal bioremediation. A comparison of the flocculating activity of microbial EPS of different origins is shown in Table 5.1

Though application of bacterial EPS spans area such as industry (textile, dairy, cosmetics, etc.), health (medicine, pharmaceuticals), and environment (remediation, flocculation, etc.), its application in the flocculation process will be a significant milestone in health promotion and eco-friendly uses especially in municipal and wastewater treatment process. As mentioned earlier, that is, the disadvantage of synthetic flocculants, the application of microbial flocculants meant for safe alternative. This quest has been the driving force behind the investigation.

5.5   Emulsification and Hydrocarbon Degradation

Petroleum-degrading bacteria were first investigated by Hino et al. in 1997, and they found that the main reason behind this property is the production of an exopolysaccharide (EPS). By using this EPS, these bacteria use petroleum as their energy source (Ta-Chen et al. 2008). Degradation of hydrocarbon will be very much useful in case of oil spill in ocean, where the spillage most often leads to catastrophe. In case of ocean, the most of the dissolved organic matter exists as biopolymer (approximately 10–25% of total oceanic dissolved organic matter) (Verdugo 1994; Verdugo et al. 2004). And most of this dissolved organic matter is produced by different marine microorganisms (Decho 1990; Santschi et al. 1998). It has been found that the EPS produced by the marine bacteria is rich in uronic acid than their nonmarine counterparts (Ford et al. 1991). And this uronic acid confers the ability to the EPS to bind with different hydrocarbons (Janecka et al. 2002; Gutierrez et al. 2008, 2009). The amphiphilic characteristic of the EPS is due to the presence of different polymers, especially different amino acids and peptides in the bacterial EPS (Decho, 1990; Wolfaardt et al. 1999). Though the exact nature of this marine bacterial EPS and its uses in case of a major oil spill (Gutierrez et al. 2013) is still unknown, it has been hypothesized that during a major spill, the number of EPS-producing bacteria is enriched in the region of spill which leads to the formation of oil aggregates. This leads to deployment of some other indigenous oil-degrading bacterial communities (Passow et al. 2012; Ziervogel et al. 2012), though the identity of the microbial species involved in this process is not elucidated (Gutierrez et al. 2013). Rhizobial EPSs have, however, shown better emulsifying activity, 86.66% with olive oil, 83.33% with kerosene, 73.33% in n-hexane, and 76.66% in toluene (Fig. 5.4a, than commercial surfactant like Tween 80 and Tween 20 (Fig. 5.4b and c, respectively).

The exopolysaccharide secreted by the microbes can act as a bioemulsifying agent; they can reduce the surface tension and interfacial tension of bacteria and increase the cell surface hydrophobicity of bacteria, thereby resulting in the enhancement of dispersal, emulsification, and degradation of hydrocarbon contents in the contaminated site (Jhang and Miller 1992; Yakimov et al. 1998). Huang et al. (2012) reported that out 40 root-nodulating bacteria, the better diesel-emulsifying abilities are recorded from three rhizobial strains.

Fig. 5.4 (**a**) Emulsifying activity of Rhizobial EPS, (**b**) Tween 80, and (**c**) Tween 20 with different oils

Conclusion

Heterogeneously distributed microbial communities involving agronomically important nitrogen-fixing genera rhizobia have been shown to possess the ability to synthesize and excrete incredibly higher concentrations of EPS. These complex active biomolecules differ widely in structure and physiological function. It has now a well-established fact that rhizobial EPS plays some critical role in the development of effective legume-rhizobia symbiosis. The mutant rhizobial strains which are deficient in EPS production in contrast fail to develop functional nodule and concomitantly a poor legume production. Apart from their role in nitrogen fixation, rhizobial EPS plays important roles in abatement of contamination of soil/water, a threatening environmental problem. Rhizobial EPS is considered safe because it is obtained from a nonpathogenic rhizobial strain. On the other hand, rhizobial EPS can also play a major role as bioemulsifier and increase the solubility of hydrocarbons and have the capacity of emulsifying of different types of oils (viz., diesel, kerosene, toluene, etc.). So, rhizobia can be considered as an unexplored source of microbial EPS and a highly promising tool for different industrial applications and for solving emerging environmental problems.

References

Amemura A et al (1983) Structural studies on the extracellular acidic polysaccharide from *Rhizobium trifolii* 4S. Carbohydr Res 115:165–174

Banal IM, Makkar RS, Camcotra SS (2000) Potential commercial applications of microbial surfactants. Appl Microbiol Biotechnol 53:495–508

Bhattacharyya R, Das S (2015) Exopolysaccharide production by a *Rhizobium* sp. from root nodules of *Phaseolus mungo* (L.) in heavy metal stress condition. Indian Biol 47:53–59

Chaterjee A, Das D, Mandal BK (1995) Arsenic in groundwater in six districts of West Bengal, India. The biggest arsenic calamity in the world. Part 1. Arsenic species in drinking water and urine of affected people. Analyst 120:643–650

Cheng HP, Walker GC (1998) Succinoglycan is required for initiation and elongation of infection threads during nodulation of alfalfa by *Rhizobium meliloti*. J Bacteriol 180:5183–5191

Cremers HCC, Stevens K, Lugtenberg BJ, Wijffelman CA, Batley M, Redmond JW, Zevenhuizen LP (1991) Unusual structure of the exopolysaccharide of *Rhizobium leguminosarum* bv. viciae strain 248. Carbohydr Res 218:185–200

D'Haeze W, Glushka J, DeRycke R, Holsters M, Carlson RW (2004) Structural characterization of extracellular polysaccharides of *Azorhizobium caulinodans* and importance for nodule initiation on *Sesbania rostrata*. Mol Microbiol 52:485–500

Das HK, Mitra AK, Sengupta PK, Hossain A, Islam F, Rabbani GH (2004) Arsenic concentration in rice, vegetables, and fish in Bangladesh: a preliminary study. Environ Int 30:383–387

Datta B, Chakrabartty PK (2014) Siderophore biosynthesis genes of *Rhizobium* sp. isolated from *Cicer arietinum* L. 3 Biotech 4:391–401

Decho AW (1990) Microbial exopolymer secretions in ocean environments: their role(s) in food webs and marine processes. In: Barnes M (ed) Oceanography marine biology annual review. Aberdeen University Press, Aberdeen, pp 73–153

Djordjevic SP, Chen H, Batley M, Redmond JW, Rolfe BG (1987) Nitrogen fixation ability of exopolysaccharide synthesis mutants of Rhizobium sp. strain NGR234 and Rhizobium trifolii is restored by the addition of homologous exopolysaccharides. J Bacteriol 169:53–60

Downie JA (2010) The roles of extracellular proteins, polysaccharides and signals in the interactions of rhizobia with legume roots. FEMS Microbiol Rev 34(2):150–170

Ferguson BJ, Mathesius U (2014) Phytohormone regulation of legume-rhizobia interactions. J Chem Ecol 40:770–790

Ford T, Sacco E, Black J, Kelley T, Goodacre R (1991) Characterization of exopolymers of aquatic bacteria by pyrolysis-mass spectrometry. Appl Environ Microbiol 57:1595–1601

Foster LJR, Moy YP, Rogers PL (2000) Metal binding capabilities of Rhizobium etli and its extracellular polymeric substances. Biotechnol Lett 22:1757–1760

Fraysse N, François C, Verena P (2003) Surface polysaccharide involvement in establishing the *Rhizobium*–legume symbiosis. Eur J Biochem 270:1365–1380

Freitas F, Alves VD, Reis MAM (2011) Advances in bacterial exopolysaccharides: from production to biotechnological applications. Trends Biotechnol 29(8):388–398

Gil-Serrano A et al (1990) Structure of the extracellular polysaccharide secreted by Rhizobium leguminosarum var. phaseoli CIAT 899. Carbohydr Res 204:103–107

Götz F (2002) Staphylococcus and biofilms. Mol Microbiol 43(6):1367–1378

Gutierrez T, Shimmield T, Haidon C, Black K, Green DH (2008) Emulsifying and metal ion binding activity of a glycoprotein exopolymer produced by *Pseudoalteromonas* sp. strain TG12. Appl Environ Microbiol 74:4867–4876

Gutierrez T, Morris G, Green DH (2009) Yield and physicochemical properties of EPS from *Halomonas* sp. strain TG39 identifies a role for protein and anionic residues (sulfate and phosphate) in emulsification of n-hexadecane. Biotechnol Bioeng 103:207–216

Gutierrez T, Berry D, Yang T, Mishamandani S, McKay L, Teske A, Aitken MD (2013) Role of bacterial exopolysaccharides (EPS) in the fate of the oil released during the Deepwater Horizon oil spill. PLoS One 8:e67717

Han M, Du C, Xu Z, Qian H, Zhang W (2016) Rheological properties of phosphorylated exopolysaccharide produced by *Sporidiobolus pararoseus* JD-2. Int J Biol Macromol 88:603–613

Han P, Sun Y, Wu X, Yuan Y, Dai Y, Jia S (2014) Emulsifying, flocculating, and physiochemical properties of exopolysaccharide produced by cyanobacterium *Nostoc flagelliforme*. Appl Biochem Biotechnol 172:36–49

He N, Li Y, Chen J (2004) Production of a novel polygalacturonic acid bioflocculant REA-11 by *Corynebacterium glutamicum*. Bioresour Technol 94:99–105

Huang KH, Chen BY, Shen FT, Young CC (2012) Optimization of exopolysaccharide production and diesel oil emulsifying properties in root nodulating bacteria. World J Microbiol Biotechnol 28:1367–1373

Ismail B, Nampoothiri KM (2010) Production, purification and structural characterization of an exopolysaccharide produced by a probiotic *Lactobacillus plantarum* MTCC 9510. Arch Microbiol 192:1049–1057

Janczarek M (2011) Environmental signals and regulatory pathways that influence exopolysaccharide production in rhizobia. Int J Mol Sci 12:7898–7933

Janczarek M, Rachwal K, Cieśla J, Ginalska G, Bieganowski A (2015) Production of exopolysaccharide by *Rhizobium leguminosarum* bv. Trifolii and its role in bacterial attachment and surface properties. Plant and Soil 388:211–227

Janecka J, Jenkins MB, Brackett NS, Lion LW, Ghiorse WC (2002) Characterization of a *Sinorhizobium* isolate and its extracellular polymer implicated in pollutant transport in soil. Appl Environ Microbiol 68:423–426

Jaszek M, Janczarek M, Kuczynski K, Piersiak T, Grywnowicz K (2014) The response of the Rhizobium leguminosarum bv. trifolii wild type and exopolysachharide deficient mutants to oxidative stress. Plant and Soil 376:75–94

Jefferson KK (2009) Bacterial polysaccharides. In: Ullrich M (ed) Current innovations and future trends. Caister Academic, Norfolk, pp 175–186

Khan MS, Zaidi A, Wani PA, Oves M (2009) Role of plant growth promoting rhizobacteria in the remediation of metal contaminated soils. Environ Chem Lett 7:1–19

Kranthi Kumar G, Raghu Ram M (2014) Phosphate solubilizing Rhizobia isolated from *Vigna trilobata*. Am J Microbiol Res 2(3):105–109

Laus MC, Van Brussel AA, Kijne JW (2005) Role of cellulose fibrils and exopolysaccharides of *Rhizobium leguminosarum* in attachment to and infection of *Vicia sativa* root hairs. Mol Plant Microbe Interact 18:533–538

Lepek VC, D'Antuono AL (2005) Bacterial surface polysaccharides and their role in the rhizobia-legume association. Lotus Newslett 35:93–105

Liang TW, Wang SL (2015) Recent advances in exopolysachharides from Paenibacillus spp.: production, isolation, structure, and bioactivities. Mar Drugs 13:1847–1863

Liu S, Chen X, He H, Zhang X, Xie B, Yu Y, Chen B, Zhou B, Zhang Y (2013) Structure and ecological roles of a novel exopolysaccharide from the Arctic sea ice bacterium *Pseudoalteromonas* sp. strain SM20310. Appl Environ Microbiol 79:224–230

Luma JH, Rufino RD, Sarubbo LA, Campos-Takaki GM (2013) Characterization, surface properties and biological activity of a biosurfactant produced from industrial waste by *Candida sphaerica* UCP0995 for application in the petroleum industry. Colloids Surf B Biointerfaces 102:202–209

Mandal AK, Yadav KK, Sen IK, Kumar A, Chakraborti S, Ilam SS, Chakraborty R (2013) Partial characterization and flocculating behavior of an exopolysaccharide produced in nutrient-poor medium by a facultative oligotroph Klebsiella sp. PB12. J Biosci Bioeng 115:76–81

Martensson AM (1992) Effects of agrochemicals and heavy metals on fast-growing rhizobia and their symbiosis with small-seeded legumes. Soil Biol Biochem 24:435–445

Mohamad OA, Hao X, Xie P, Hatab S, Lin Y, Wei G (2012) Biosorption of copper (II) from aqueous solution using non-living *Mesorhizobium amorphae* strain CCNWGS0123. Microbes Environ 27(3):234–241

Mussarat J, Haseeb A (2000) Agrochemicals as antagonists of lectin mediated Rhizobium legume symbiosis: Paradigms and prospects. Curr Sci 78:793–797

Niehaus K, Kapp D, Puhler A (1993) Plant defense and delayed infection of alfalfa pseudonodules induced by an exopolysaccharide (EPS) deficient *Rhizobium meliloti* mutant. Planta 190:415–425

O'Neill MA, Darvill AG, Albersheim P (1991) The degree of esterification and points of substitution by O-acetyl and O-(3-hydroxybutanoyl) groups in the acidic extracellular polysaccharides secreted by *Rhizobium leguminosarum* biovars *viciae*, *trifolii*, and *phaseoli* are not related to host range. J Biol Chem 266:9549–9555

Öner ET (2013) Microbial production of extracellular polysaccharides from biomass. In: Fang Z (cd) Pretreatment techniques for biofuels and biorefineries, green energy and technology. Springer, Berlin, pp 35–56

Parniske M, Schimdt PE, Kosch K, Muller P (1994) Plant defense response of host plants with determinate nodules induced by EPS defective exo B mutants of *Bradyrhizobium japonicum*. Mol Plant Microbe Interact 7:631–638

Passow U, Ziervogel K, Asper V, Diercks A (2012) Marine snow formation in the aftermath of the Deepwater Horizon oil spill in the Gulf of Mexico. Environ Res Lett 7(3):035301

Poli A, Anzelmo G, Nicolaus B (2010) Bacterial exopolysaccharides from extreme marine habitats: production, characterization and biological activities. Mar Drugs 8:1779–1802

Rendueles O, Kaplan JB, Ghingo JM (2013) Antibiofilm polysaccharides. Environ Microbiol 15:334–346

Robertsen BK et al (1981) Host-symbiont interactions: V. The structure of acidic extracellular polysaccharides secreted by *Rhizobium leguminosarum* and *Rhizobium trifolii*. Plant Physiol 67:389–400

Ryder C, Byrd M, Wozniak DJ (2007) Role of exopolysaccharides in *Pseudomonas aeruginosa* biofilm development. Curr Opin Microbiol 10:644–648

Santschi PH, Guo L, Means JC, Ravichandran M (1998) Natural organic matter binding of trace metal and trace organic contaminants in estuaries. In: Bianchi TS, Pennock JR, Twilley R (eds) Biogeochemistry of Gulf of Mexico estuaries. Wiley, New York, pp 347–380

Sathiyanarayanan G, Kiran GS, Selvin J (2013) Synthesis of silver nanoparticles by polysaccharides bioflocculant produced from marine *Bacillus subtilis* MSBN17. Colloids Surf B Biointerfaces 102:13–20

Shah AA, Hasan F, Hameed A, Ahmed S (2008) Biological degradation of plastics: a comprehensive review. Biotechnol Adv 26:246–265

Shih IL, Van YT, Yeh LC, Lin HG, Chang YN (2001) Production of a biopolymer flocculant from *Bacillus licheniformis* and its flocculation properties. Bioresour Technol 78:267–272

Shuhong Z, Meiping Z, Hong Y, Han W, Shan X, Yan L, Jihui W (2014) Biosorption of Cu^{2+}, Pb^{2+}, and Cr^{2+} by a novel exopolysaccharide from *Arthrobacter* ps-5. Carbohydr Polym 101:50–56

Skorupska A, Janczarek M, Marezak M, Mazur A, Krol J (2006) Rhizobial exopolysaccharides: genetic control and symbiotic functions. Microb Cell Fact 5:7

Sutherland IW (2001) Microbial polysaccharides from gram negative bacteria. Int Dairy J 11:663–674

Ta-Chen L, Chang JS, Young CC (2008) Exopolysaccharides produced by Gordonia alkanivorans enhance bacterial degradation activity for diesel. Biotechnol Lett 30:1201–1206

Vaningelgem F et al (2004) Biodiversity of exopolysaccharides produced by *Streptococcus thermophilus* strains is reflected in their production and their molecular and functional characteristics. Appl Environ Microbiol 70:900–912

Verdugo P (1994) Polymer gel phase transition in condensation decondensation of secretory products. Adv Polym Sci 110:145–156

Verdugo P, Alldredge AL, Azam F, Kirchman DL, Passow U (2004) The oceanic gel phase: a bridge in the DOM-POM continuum. Mar Chem 92:67–85

Vu B, Chen M, Crawford RJ, Ivanova EP (2009) Bacterial extracellular polysaccharides involved in biofilm formation. Molecules 14:2535–2554

Wingender J, Strathmann M, Rode A, Leis A, Flemming HC (2001) Isolation and biochemical characterization of extracellular polymeric substances from *Pseudomonas aeruginosa*. Methods Enzymol 336:302–314

Wolfaardt GM, Lawrence JR, Korber DR (1999) Function of EPS. In: Wingender J, Neu TR, Flemming H-C (eds) Microbial extracellular polymeric substances: characterization, structure and function. Springer, New York, pp 171–200

Xie P, Hao X, Mohamad OA, Liang J, Wei C (2013) Comparative study of chromium biosorption by *Mesorhizobium amorphae* strain CCNWGS0123 in single and binary mixtures. Appl Biochem Biotechnol 169:570–587

Yadav KK, Mandal AK, Sen IK, Chakraborti S, Islam S, Chakraborty R (2012) Flocculating property of extracellular polymeric substances produced by a biofilm forming bacterium Acinetobacter junii BB1A. Appl Biochem Biotechnol 168:1621–1634

Yakimov MM, Golyshin PN, Lang S, Moore ER, Abraham WR, Lunsdorf H, Timmis KN (1998) Alcanivorax borkumensis gen. nov., sp. nov., a new, hydrogen-degrading and surfactant-producing marine bacterium. Int J Syst Evol Microbiol 48:339–348

Yokoi H, Natsuda O, Hirose J, Hayashi S, Takasaka Y (1995) Characteristics of a biopolymer flocculant produced by *Bacillus subtilis*. J Ferment Bioeng 79:378–380

Yuksekdag ZN, Aslim B (2008) Influence of different carbon sources on exopolysaccharide production by *Lactobacillus delbrueckii* Subsp. *bulgaricus* (B3, G12) and *Streptococcus thermophilus* (W22). Braz Arch Biol Technol 51:581–585

Zahra MK, Fayey M, Hassan ME, Ghalal NM (1990) Symbiosis between *Bradyrhizobium japoni-cum* and soybean (*Glycine max*) in different soils. Egypt J Microbiol 25:181–196

Zhang ZQ, Bo L, Xia SQ, Wang XJ, Yang AM (2007) Production and application of a novel bio-flocculant by multiple-microorganism consortia using brewery wastewater as carbon source. J Environ Sci 19:667–673

Ziervogel K, McKay L, Rhodes B, Osburn CL, Dickson-Brown J (2012) Microbial activities and dissolved organic matter dynamics in oil contaminated surface seawater from the Deepwater Horizon oil spill site. PLoS One 7:e34816

Zogaj X et al (2001) The multicellular morphotypes of *Salmonella typhimurium* and *Escherichia coli* produce cellulose as the second component of the extracellular matrix. Mol Microbiol 39(6):1452–1463

Rhizobial Diversity for Tropical Pulses and Forage and Tree Legumes in Brazil

6

Mario Andrade Lira Junior, Giselle Gomes Monteiro Fracetto, Ademir Sérgio Ferreira Araújo, Felipe José Cury Fracetto, Rafaela Simão Abrahão Nóbrega, Krisle Silva, and Adeneide Candido Galdino

Abstract

The current information on Brazilian rhizobial diversity, concentrating especially on the microbial symbionts of tropical pulses, forage legumes, and legume trees, some of which are native to Brazil or of which Brazil is a major producer, is highlighted. These legume species are nodulated by a large number of currently known rhizobial genera, including both alpha and beta rhizobia, with widely varying nitrogen-fixing efficiencies. The rhizobial diversity is strongly affected by soil and climatic factors, as well as genetic variation among pulses. The greater diversity among rhizobia may allow the selection of more effective nitrogen-fixing strains which could be used as inexpensive inoculants to substitute/to reduce the use of nitrogen fertilizers.

M.A. Lira Junior (✉) • G.G.M. Fracetto • F.J.C. Fracetto • A.C. Galdino
Departamento de Agronomia, Universidade Federal Rural de Pernambuco, Dom Manoel de Medeiros, s/n, Dois Irmãos, 52171-900 Recife, PE, Brazil
e-mail: mario.alirajr@ufrpe.br; mariolirajunior@gmail.com; giselle.fracetto@ufrpe.br; curyfelipe@hotmail.com; adeneide.candido@gmail.com

A.S.F. Araújo
Centro de Ciências Agrárias, Universidade Federal do Piauí, Campus da Soccopo, 64000-000 Teresina, PI, Brazil
e-mail: asfaruaj@yahoo.com.br

R.S.A. Nóbrega
Centro de Ciências Agrárias, Ambientais e Biológicas, Universidade Federal do Recôncavo da Bahia, Rui Barbosa, 710, Centro, 44380000 Cruz das Almas, BA, Brazil
e-mail: rafaela.nobrega@gmail.com

K. Silva
Empresa Brasileira de Pesquisa Agropecuária, Embrapa Florestas Estrada da Ribeira, Km 111 - Bairro Guaraituba Caixa Postal: 319, 83411000 Colombo, PR, Brazil
e-mail: krisle.silva@embrapa.br

© Springer International Publishing AG 2017
A. Zaidi et al. (eds.), *Microbes for Legume Improvement*,
DOI 10.1007/978-3-319-59174-2_6

135

6.1 Introduction

Biological nitrogen fixation (BNF) through the legume-rhizobia symbiosis is the most important nitrogen (N) source for most agroecosystems (Herder et al. 2010). Currently, these bacteria are classified as *Allorhizobium, Aminobacter, Azorhizobium, Bradyrhizobium, Devosia, Ensifer (Sinorhizobium), Mesorhizobium, Methylobacterium, Microvirga, Ochrobacterium, Phyllobacterium, Rhizobium,* and *Shinella* among the *α-Proteobacteria* and *Burkholderia, Cupriavidus,* and *Herbaspirillum* among the *β-Proteobacteria* (Vinuesa 2015). At least one *Pseudomonas* sp. is also among the *γ-Proteobacteria* (Shiraishi et al. 2010). This genus-level diversity is repeated on lower taxonomic levels, since a single soil may harbor several species and strains of a single species (Guimarães et al. 2015), while strains from a single species may be found at faraway points (Martins et al. 2015). Phenotypic characteristics are generally used for initial characterization and screening of rhizobia (Oliveira et al. 2011; Rufini et al. 2014), but molecular characterization has mostly replaced the more traditional phenotypic characteristics due to sensitivity of techniques and precision in results. Among molecular tools, fingerprinting techniques using conserved and repetitive DNA oligonucleotides sequences such as BOX (genomic box elements), ERIC (enterobacterial repetitive intergenic consensus), and REP (repetitive extragenic palindromic) are frequently used in rhizobial diversity research (Guimarães et al. 2012; Bianco et al. 2013). For example, a polyphasic approach based on phenotypic characterization and BOX, ERIC, and REP was used to evaluate *Mimosa caesalpiniifolia* rhizobial diversity between different regions of the Northeast Brazil (Martins et al. 2015). At the same time, the 16S rRNA housekeeping gene is no longer considered to be sufficiently discriminatory between closely related rhizobial species (Menna et al. 2006; Delamuta et al. 2012) or for intraspecific analysis. This conclusion led to the increased use of the 16S-23S rRNA intergenic transcribed spacer (ITS) (Tesfaye and Holl 1998; van Berkum and Fuhrmann 2000), but the ribosomal gene proximity may lead to erroneous phylogenetic conclusions if horizontal gene transference occurs (van Berkum et al. 2003) as it is relatively common among rhizobial species.

Since no single gene, even among the housekeeping ones, can reliably avoid horizontal gene transfer, nowadays the multilocus sequence analysis (MLSA) of several housekeeping genes is being increasingly used for rhizobial phylogenetic and taxonomic identification (Zilli et al. 2014). This technique is based on the sequencing and linking of several housekeeping genes dispersed over at least 100 kb of the genome and thus should be largely immune to horizontal gene transfer effects (Martens et al. 2008; Ribeiro et al. 2015a). For example, MLSA has found high *Bradyrhizobium* strain diversity from several legume species and land use systems which was not described earlier (Guimarães et al., 2015). Even when just soybean rhizobial diversity was evaluated from soils of different ecological regions of Brazil, new species were found from both the Northeast (tropical) and Southeast (subtropical to tropical) regions, some of which were highly efficient for BNF (Ribeiro et al. 2015a) when 16S rRNA and five housekeeping genes were sequenced. This is a major breakthrough, since it might lead to more efficient inoculants for this culture, which under Brazilian conditions does not receive any nitrogen fertilizer.

Another toolkit frequently used in rhizobial diversity and phylogeny research is functional gene analysis/sequencing, most frequently *nif*H which is highly conserved among diazotrophs and codes for the Fe-protein of the nitrogenase complex (Coelho et al. 2009). As an example, several unknown species as well as strains from *Bradyrhizobium* and *Sinorhizobium* were found (Roesch et al. 2007) even though the authors used maize colms for DNA extraction while at the same time finding that soil clay content affects diazotroph diversity, while *nifH* pyrosequencing from a gradient of agricultural soils found that both diversity and dynamics of diazotroph communities are affected by soil chemical characteristics (Collavino et al. 2014). Since there is a huge scope for these studies, this review will concentrate its efforts on some legume groups which are somewhat less studied abroad than in Brazil. This differential concentration might be due to any of several equally important reasons, ranging from the lower importance of the crop to the endemic or indigenous nature of the species. As such, we decided to cover *Phaseolus vulgaris* (common or French beans), *P. lunatus* (lima beans), *Vigna unguiculata* (cowpeas), and tropical legume trees and forage species while, at the same time, not including soybean, peas, or any of the temperate climate forage legumes.

6.2   *Phaseolus vulgaris*

Common or French beans (*Phaseolus vulgaris* L.) figure among the main protein sources in Latin America (Torres et al. 2009) and may fix nitrogen with a wide range of rhizobial species (Table 6.1). There are several reports that indicate some promising results from field inoculation in several bean-growing regions in Brazil (Raposeiras et al. 2006; Lombardi et al., 2009; Torres et al. 2009). *R. tropici*, in particular, is usually described as highly efficient, genetically stable, and tolerant to environmental stresses and is commonly found in Brazilian soils (Hungria et al. 2000; Mostasso et al. 2002), while *R. etli* is usually dominant and has been frequently found in Brazil (Mostasso et al. 2002; Soares et al. 2006a; Giongo et al. 2007), including when plants were grown under environmental stresses such as high temperature, aluminum stress, and low pH, together with *R. leguminosarum* strains (Soares et al. 2006a; Grange et al. 2007; Stocco et al. 2008). The prevalence of *R. etli* may be linked to the different centers of origin of *P. vulgaris*, since strains from the Northeast region of Brazil were genetically closer to a Mexican strain than those from the South region, based on 16S rRNA (Grange et al. 2007). This link between bean cultivar and rhizobial diversity has also been found for strains from the Mesoamerican and Andean centers of origin, with higher diversity for the first than the second center (Oliveira et al. 2011). Another study in South Brazil found that 32.5% of the strains were *R. leguminosarum* (Stocco et al. 2008) and this species was also found in several other studies in Brazil (Giongo et al. 2007; Pinto et al. 2007) and Columbia (Eardly et al. 1995). Besides the species-level diversity found in Brazil, high strain-level diversity is also found in Brazilian soils. For example,

Table 6.1 Rhizobial species currently known to nodulate *Phaseolus vulgaris* and their known geographical occurrences

Species	Origin	References
α-Proteobacteria		
Rhizobiales		
Rhizobiaceae		
Rhizobium		
R. leguminosarum/ biovares	Europe, South, Central and North America, Asia	Andrade et al. (2002); Soares et al. (2006a); Giongo et al. (2007); Grange et al. (2007); Pinto et al. (2007)
R. paranaense	South America	Dall'agnol et al. (2014)
R. etli	Europe, South and North America, Asia	Hungria et al. (2003); Aguilar et al. (2004); Grange et al. 2007; Stocco et al. (2008)
R. ecuadorense	South America	Ribeiro et al. (2015b)
R. tropici	South and North America, Europe, Asia	Mostasso et al. (2002); Lombardi et al. (2009); Torres et al. (2009)
R. giardinii	South America, North Africa, Asia	Mhamdi et al. (2002); Torres et al. (2009)
R. gallicum	North Africa, Europe	Rodriguez-Navarro et al. (2000); Mhamdi et al. (2002)
R. galegae	South America, Europe	Laguerre et al. (2001); Melloni et al. (2006)
R. phaseoli	Europe	Atzorn et al. (1988)
R. lusitanum	Europe	Valverde et al. (2006)
R. freirei	South America	Dall'agnol et al. (2013)
R. mongolense	South America, Africa do Norte	Andrade et al. (2002); Mhamdi et al. (2002)
R. meliloti	Europe	Bromfield and Barran (1990)
Sinorhizobium		
S. americanum	North Africa	Mnasri et al. (2012)
Bradyrhizobiaceae		
Bradyrhizobium		
B. japonicum	South America	Michiels et al. (1998)
B. elkanii	Europe	Laguerre et al. (2001)
Xanthobacteriaceae		
Azorhizobium		
A. caulinodans	South America	Melloni et al. (2006)
Phyllobacteriaceae		
Mesorhizobium		
M. loti	Europe	Laguerre et al. (2001)
M. tianshanense	Asia	Chen et al. (1995)
β-pProteobacteria		
Burkholderiales		
Burkholderiaceae		

(continued)

Table 6.1   (continued)

Species	Origin	References
Burkholderia		
B. tuberum	North Africa	Elliott et al. (2007)
B. phymatum	North Africa	Elliott et al. (2007)
B. caribensis	America do Norte	Estrada-De Los Santos et al. (2012)
B. cepacia	South America	Peix et al. (2001)
Paraburkholderia		
P. nodosa	South America	Dall'agnol et al. (2016)

62 strains from Amazon agricultural soils analyzed by BOX-PCR resulted in 50 genotypes with 70% similarity and 21 genotypes with 30% similarity (Guimarães et al. 2012). Although 16S rRNA sequencing indicated higher prevalence of *Bradyrhizobium*, species from *Rhizobium*, *Burkholderia*, and *Achromobacter* (Guimarães et al. 2012) were also identified.

6.3   *Phaseolus lunatus*

Lima bean (*Phaseolus lunatus*) is an important species of plant for humans in tropical regions. This legume originated in Peru; and archeological evidence supports the hypothesis that lima beans were domesticated in Mesoamerica and South America (Salgado et al. 1995). The lima bean seed is considered a main crop and an important source of protein for people of South America, Africa, and Mexico. The rustic quality of lima bean and its capacity to resist to long, dry periods are important characteristics for the semiarid region of Northeast Brazil (Azevedo et al. 2003). The rhizobia associated with this crop have scarcely been studied. In the old host-based classification scheme, symbionts of *P. lunatus* were included in the same group as rhizobia associated with slow-growing cowpea (*Vigna unguiculata*). This group was a diverse assemblage of strains that were later included in the genus *Bradyrhizobium*. The rhizobial isolates for *P. lunatus* were obtained from areas where this legume is not native, and research has only focused on analysis of symbiotic characteristics, such as infectiveness and effectiveness (Ormeño-Orrillo et al. 2006). Antunes et al. (2011) evaluated the symbiotic effectiveness of 17 rhizobial isolates of lima bean in Northeast Brazil, and they compared the isolates with two reference *Rhizobium* strains CIAT 899 and NGR 234. They found eight isolates with higher N accumulation and N_2-fixation efficiency compared with the reference strains CIAT 899 and NGR 234. The morphological and biochemical characteristics of these isolates revealed that six isolates belonged to genera *Bradyrhizobium* and two isolates to *Rhizobium*.

Interestingly, lima beans are nodulated by both fast- and slow-growing rhizobia (Santos et al. 2011), and some studies found *Bradyrhizobium* and *Rhizobium* as symbionts of this legume (Thies et al. 1991; Santos et al. 2011). A very few studies on the genetic diversity of rhizobia from lima bean have been conducted using collection from geographic locations including Mexico and Brazil (Ormeño et al. 2007; Lopez-Lopez et al. 2013). Ormeño-Orrillo et al. (2006) evaluated the molecular diversity of rhizobial isolates associated with lima bean in Peru. They found divergent bradyrhizobial lineages (*Bradyrhizobium yuanmingense* and *Bradyrhizobium* sp.) according to PCR-RFLP of the *rpoB* gene and sequence analysis of the 16S rDNA and *dnaK, nifH*, and *nodB* genes. Lopez-Lopez et al. (2013) described the nodule bacteria from native lima beans from Mexico. The bacterial diversity of isolates from nitrogen-fixing nodules of *P. lunatus*, using ERIC-PCR and PCR-RFLP of *rpoB* genes and sequencing of *recA, nodZ*, and *nifH* genes, shows that nodule bacteria correspond to *Bradyrhizobium*. According to the authors, this is the first report of nodule bacteria from *P. lunatus* in its Mesoamerican site of origin and domestication, and it may confirm that *Bradyrhizobium* is the main nodulating group of lima bean. However, the fast-growing rhizobia that nodulate *Phaseolus* commonly belong to the genus *Rhizobium* and *Sinorhizobium*, and Ormeño et al. (2007) found a strain of *S. meliloti* isolated from lima bean in Peru, and they suggested that this rhizobial species may also nodulate this legume. A study about the genetic diversity of native rhizobia that nodulate lima bean from Brazil shows a broad spectrum of rhizobial groups associated with lima bean (Santos et al. 2011). In this study, rhizobia isolates were obtained and placed into groups based on the differences in their morphological, physiological, and genetic characteristics. The restriction patterns obtained with endonucleases MboI, HaeIII, and NheI showed sufficient variability to discriminate isolates identified as species from the genera *Bradyrhizobium, Mesorhizobium*, and *Rhizobium*. Araujo et al. (2015) further sequenced the 16S rDNA of the above isolates and found that species that nodulate lima bean belonged to the genus *Bradyrhizobium, Sinorhizobium*, and *Rhizobium*. These results confirm that lima bean may be nodulated by diverse rhizobia species.

6.4   *Vigna unguiculata*

Cowpea (*Vigna unguiculata* (L.) Walp.) is an African pulse, traditionally grown in the tropical regions of Africa, America, and Asia and is used mainly as a protein source (Carvalho et al. 2012). It has high genetic variability, and some genotypes are tolerant to water deficit (Nascimento et al. 2011) and pests (Torres et al. 2016), as well as they adapt well to other environmental stresses, such as low soil fertility (Ferreira et al. 2013). It fixes nitrogen in association with several genera of rhizobia (Table 6.2) including *Rhizobium* (Jaramillo et al. 2013), *Mesorhizobium* (Moreira 2008), *Microvirga* (Marinho et al. 2014; Radl et al. 2014), *Achromobacter* (Guimarães et al. 2012), *Burkholderia* (Moreira 2008), *Brevibacillus* (Costa et al. 2013), *Sinorhizobium* (Moreira 2008), *Acinetobacter* (Marra et al. 2012), *Azorhizobium* (Moreira 2008), *Ralstonia* (Sarr et al. 2009), and *Allorhizobium* (Moreira 2008), leading to its frequent use as a bait crop to trap the largest diversity of soil rhizobia in diversity studies. *Bradyrhizobium* species have been found to frequently nodulate cowpea in Africa, America, and Asia, with large strain

Table 6.2 Bacterial genera forming symbiosis with *Vigna unguiculata* in Brazil and other regions of the world

Genera	Location	Source
Bradyrhizobium	India	Appunu et al. (2009)
Bradyrhizobium	Europe and Africa	Bejarano et al. (2014)
Bradyrhizobium, Rhizobium, Bacillus, Paenibacillus	Brazil (Cerrado)	Costa et al. (2011)
Bradyrhizobium, Rhizobium, Burkholderia; Achromobacter	Brazil (Amazon)	Guimarães et al. (2012)
Rhizobium, Ochrobacterium, Paenibacillus, Bosea, Bacillus, Enterobacter, and *Stenotrophomonas*	Brazil (Amazon)	Jaramillo et al. (2013)
Bacillus, Firmicutes, Acinetobacter, Rhizobium, Microbacterium, and *Paenibacillus*	Brazil	Marra et al. (2012)
Bradyrhizobium	Botswana Ghana, Africa do Sul	Pule-Meulenberg et al. (2010)
Microvirga	Brazil (Semiarid)	Radl et al. (2014)
Bradyrhizobium	Brazil (Cerrado)	Rufini et al. (2014)
Bradyrhizobium and *Ralstonia*	Japan	Sarr et al. (2009)
Bradyrhizobium	Japan	Sarr et al. (2011)
Bradyrhizobium, Klebsiella, Rhizobium, and *Enterobacter*	Brazil (Amazon)	Silva et al. (2012)
Bradyrhizobium	Africa	Steenkamp et al. (2008)

diversity, even where the crop is not traditionally grown. For example, in Japan, cowpea-nodulating *Bradyrhizobium* diversity is geographically diverse, indicating some effect of temperature, vegetation, and soil type and pH on the prevalence of this bacterium (Sarr et al. 2011). In yet another study, a total of 1010 rhizobial strains were recovered from Western Amazon (Nóbrega 2006), of which 148 were obtained from an agroforestry system and which were dominated by *Bradyrhizobium* strains (Jaramillo et al. 2013), whereas another set of 119 was isolated from areas with annual crops and included species of *Bradyrhizobium, Rhizobium, Burkholderia,* and *Achromobacter* (Guimarães et al. 2012). At the same time, nonsymbiotic endophytic bacteria have been isolated from superficially disinfested nodules. For example, Meyer et al. (2015) found a large diversity of these nonsymbiotic bacteria from unconventional legumes, based on partial 16S rRNA sequencing, including members of *Alphaproteobacteria, Betaproteobacteria, Gammaproteobacteria, Actinobacteria, Firmibacteria, Flavobacteria,* and *Sphingobacteria,* with close to 18% being *Bacillus* and 16% being *Pseudomonas*. Some of these unconventional endophytic rhizobacteria such as *Pseudomonas* (Li et al. 2008), *Paenibacillus* (Marra et al. 2012), *Bacillus* (Marra et al. 2012; Jaramillo et al. 2013), *Enterobacter* (Costa et al. 2013), and *Pontibacter* (Dastager et al. 2011) were also isolated from cowpeas. The identification of such unconventional endophytic rhizobacteria opens a new area where these bacteria can be tested for their nitrogen-fixing efficiency using legumes as a host plant. And hence, they can be applied under field environment as microbial inoculants for enhancing legume production.

Under Brazilian law, inoculant producers can use strains suggested only by the Agricultural Ministry, and in accordance to this law, only four *Bradyrhizobium* strains including INPA 3-11B and UFLA 3-84 of *Bradyrhizobium* (Soares et al. 2006b; Moreira 2008) and BR 3267 and BR 3262 belonging to *Bradyrhizobium yuanmingense* and *B. pachyrhizi*, respectively (Simões-Araújo et al. 2016a, b), have been approved for cowpea inoculation (Brasil 2011). These inoculants have resulted into sufficiently higher yields due to sufficient N availability to crops cultivated in several field experiments in different ecosystems ranging from the semiarid to the Amazon, from 2°N to 23°S (Zilli et al. 2009; Chagas Junior et al. 2010; Almeida et al. 2010; Costa et al. 2011; Ferreira et al. 2013; Marinho et al. 2014).

6.5   Legume Trees

There are 147 genera and 1190 species of leguminous tree (Garcia and Fernandes 2015) in Brazil including 65 Faboideae (Papilonoideae), 54 Caesalpinioideae, and 28 Mimosoideae subfamilies. Many surveys conducted on nodulation of legume trees have reported numerous highly efficient rhizobial strains (Franco and Faria 1997; Menna et al. 2009). From these studies, several rhizobial strains were officially recommended as inoculants for 43 native and exotic species of leguminous trees (Brasil 2011). However, the diversity among such rhizobial species is still insufficiently understood. The first study on rhizobial diversity of legume trees in Brazil was conducted by Moreira and coworkers (Moreira et al. 1992, 1993), and a total of 800 rhizobial strains were isolated from the Amazonian and Atlantic forests. After phenotypic characterization, 171 strains were selected for total protein analysis using polyacrylamide gel electrophoresis (PAGE) (Moreira et al. 1993), followed by partial 16S rRNA gene sequence analysis. Using these techniques, *Azorhizobium*, *Bradyrhizobium*, *Rhizobium*, *Sinorhizobium*, and *Mesorhizobium* genera were identified (Moreira et al. 1998). One of these species was later described as a new species, *Azorhizobium doebereinerae* (Moreira et al. 2006). Strains already recommended for commercial inoculant production for several legume tree species (*Acacia*; *Albizia*; *Clitoria*; *Dalbergia*; *Enterolobium*; *Falcataria*; *Gliricidia*; *Prosopis, Leucaena*; *Mimosa*; *Ormosia*; *Piptadenia*; *Sesbania*; and *Tipuana*) were identified as *Bradyrhizobium*, *Rhizobium*, *Sinorhizobium*, *Burkholderia*, and *Azorhizobium* (Menna et al. 2006, 2009) including probable new rhizobial species.

Brazil is a major diversification center for *Mimosa* (Simon and Proença 2000), with 323 species (Garcia and Fernandes 2015), and its rhizobial diversity is well studied (Bontemps et al. 2010; Reis Junior et al. 2010; Bournaud et al. 2013). *Mimosa* genus is mostly nodulated by *Betaproteobacteria* including the genera *Burkholderia* and *Cupriavidus* (Gyaneshwar et al. 2011). However, in Brazil *Mimosa* have a particular association with *Burkholderia*, but *Cupriavidus* was not found (Bontemps et al. 2010; Reis Junior et al. 2010). An evaluation of just 143 bacteria from root nodules of 47 native species of *Mimosa*, evaluating 16S rRNA and *recA* gene sequences, found that 98% of isolates were *Burkholderia* grouped in seven clades (Bontemps et al. 2010). The isolates of three of these clades present sequence distant from those of all type strains of the species, indicating a large diversity of *Burkholderia* in Brazil. Later, four (three obtained from *M. cordistipula* and one from *M. misera*) and three isolates (one obtained from *M. candollei*, one from *M. tenuiflora*, and one *M. pudica*) were described as *B. symbiotica* (Sheu et al. 2012) and *B. diazotrophica* (Sheu et al. 2013), respectively. In a study conducted on *M. caesalpiniifolia* Benth., four geographically distant woodlots were sampled, in regions ranging from tropical semiarid to subhumid, sea level to about 600 m elevation, achieving 47 isolates of *Burkholderia*, some of which presented low similarity in 16S rRNA gene with the type strains, indicating the possibility of new species (Martins et al. 2015). The predominance and high diversity of *Burkholderia* isolates were also observed with *M. scabrella* plants in a subtropical humid forest known as Araucaria forest (Lammel et al. 2013). Actually, at least five *Burkholderia* species, namely, *B. mimosarum* (Chen et al. 2006), *B. nodosa* (Chen et al. 2007), *B. sabiae* (Chen et al. 2008), *B. symbiotica* (Sheu et al. 2012), and *B. diazotrophica* (Sheu et al. 2013), able to nodulate *Mimosa* plants were described from Brazil. Genera of the tribe Mimoseae, such as *Piptadenia*, *Parapiptadenia*, *Pseudopiptadenia*, *Pityrocarpa*, *Anadenanthera*, and *Microlobius*, all of which phylogenetically close to Mimosa, were also found in symbiosis with strains of *B. sabiae*, *B. phymatum*, *B. caribensis*, *B. diazotrophica*, *B. nodosa*, *B. phenoliruptrix*, and possible new species large diversity of *Burkholderia* (Bournaud et al. 2013). Phylogenetic analyses of neutral and symbiotic markers showed that symbiotic genes in *Burkholderia* from the tribe Mimoseae have evolved mainly through vertical transfer but also by horizontal transfer in two species (Bournaud et al. 2013). *Inga* (Mimosoideae) is another genus of leguminous tree adapted to acid and low fertility soils that establish symbiosis with rhizobia, of which 131 species are present in Brazil (Garcia and Fernandes 2015). However, very little is known about the diversity of rhizobia associated with this genus in Brazil. While there are two *Bradyrhizobium* spp. which are officially recommended for *Inga marginata* Willd inoculation (Franco and Faria 1997; Menna et al. 2009), 17 strains were obtained from root nodules of *Inga laurina* (Sw.) Willd in 2008, naturally growing in the savannah of Roraima state, in the Amazon region (Silva ct al. 2014). Six representative strains were subjected to detailed polyphasic taxonomic studies and were named as *Bradyrhizobium ingae*. Recently, 178 nitrogen-fixing bacteria were isolated from root nodules of *Centrolobium paraense* Tul (Faboideae), a neotropical legume tree from the northern Brazilian Amazon (Baraúna et al. 2014). The most common rhizobia belonged to genus *Bradyrhizobium*, but *Rhizobium* and *Burkholderia* were also found among the isolates. This result was confirmed by the *rpoB* gene sequencing (Baraúna et al. 2014). This new species was later named as *Bradyrhizobium neotropicale* (Zilli et al. 2014). Interestingly, these strains presented a discordance in the 16S rRNA phylogeny compared with the ITS phylogeny, which was also confirmed by MLSA. While the 16S rRNA gene sequence analysis placed the strains in the subgroup I (*B. elkanii*), the ITS and concatenated MLSA trees placed the strains in the subgroup II (*B. japonicum*). This result indicates a high diversity of *Bradyrhizobium* strains associated with *C. paraense*. Brazilian rhizobial diversity was also evaluated for *Chamaecrista*, *Dimorphandra*, and *Tachigali* (Moreira et al. 1998; Fonseca et al. 2012). *Chamaecrista ensiformis* (Vell.) H.S.Irwin and Barneby has a *Mesorhizobium* strain recommended as inoculant (Moreira et al. 1998). *Tachigali paniculata* Aubl. is associated to *Bradyrhizobium* (Moreira et al. 1998); *Dimorphandra parviflora* Spruce ex Benth., *D. exalata* Schott, and *D. wilsonii* Rizzini are associated to *Bradyrhizobium* (Menna et al. 2009; Fonseca et al. 2012); and *D. mollis* Benth. is associated to *Rhizobium* and *Sinorhizobium* (Moreira et al. 1998).

6.6 Forage Legumes

Although there is a large legume diversity in Brazil, with over 2800 species identified (Garcia and Fernandes 2015), majority of them are not consistently used as forage. However, species of *Stylosanthes*, *Arachis*, *Centrosema*, *Macroptilium*, *Desmodium*, *Desmanthus*, *Leucaena*, and *Calopogonium* are some of the most commonly used forage legumes (Pereira 2001; Valle 2002). Unfortunately, relatively little work is done on rhizobial diversity of these legumes. However, one strain is already recommended for inoculant production (Menna et al. 2006) for 32 different forage species. This paucity of research leads to an unclear picture of the rhizobial diversity. For example, fast-growing strains were isolated from *Arachis*, *Stylosanthes*, and *Aeschynomene* plants grown under tropical humid conditions in the Northeast Brazil (Santos et al. 2007; Guimarães et al. 2012). Species currently recommended for inoculant production for *Stylosanthes* are identified as *Bradyrhizobium* (Menna et al. 2006). Similar results were found for rhizobial isolates from *Calopogonium mucunoides*, in this case from a tropical subhumid area also in the Northeast Brazil. Most of the 1575 isolated strains were fast growers (Calheiros et al. 2013, 2015), while the currently recommended strain is a *B. japonicum* (Menna et al. 2006), although *B. stylosanthis* has been described as a new species which is also used for inoculation of this legume. *Macroptilium atropurpureum* is widely recognized as a very promiscuous legume and is also found in Brazil. For example, a study conducted in the Amazon region revealed species from at least six genera (*Rhizobium*, *Bradyrhizobium*, *Azorhizobium*, *Burkholderia*, *Mesorhizobium*, and *Sinorhizobium*), based on just 88 strains (Lima et al. 2009), while soil from seasonally dry mountain tops also had *Burkholderia* and *Paenibacillus* strains. Most of the *Burkholderia*, however, did not nodulate, and those which nodulated were ineffective (Araújo 2014) as reported earlier by Moulin et al. (2001).

Conclusion

A large number of new rhizobial species are still being discovered for lesser studied legume species, particularly those found in tropical regions. However, there is greater need to identify and evaluate some novel rhizobia for their potential growth-promoting activities so that they could be used as inoculants to reduce the application of chemical nitrogen fertilizer in agronomic production. Also, efforts should be directed to assess the impact of environmental variables such as temperature and other stress factors on the survival of rhizobia besides taking into account the genetic variations among legumes before recommending the rhizobial strains for use in farming practices especially in legume cultivation.

References

Aguilar OM, Riva O, Peltzer E (2004) Analysis of *Rhizobium etli* and of its symbiosis with wild *Phaseolus vulgaris* supports coevolution in centers of host diversification. Proc Natl Acad Sci U S A 101:13548–13553

Almeida ALG, Alcântara RMCM, Nóbrega RSA, Nóbrega JC, Leite LF, Silva JÁ (2010) Produtividade do feijão-caupi cv BR 17 Gurguéia inoculado com bactérias diazotróficas simbióticas no Piauí. Rev Bras Cienc Agrar 5:364–369

Andrade DS, Murphy PJ, Giller KE (2002) The diversity of *Phaseolus*-nodulating rhizobial populations is altered by liming of acid soils planted with *Phaseolus vulgaris* L. in Brazil. Appl Environ Microbiol 68:4025–4034

Antunes JEL, Gomes RLF, Lopes ACA, Araújo ASF, Lyra MCCP, Figueiredo MVB (2011) Eficiência simbiótica de isolados de rizóbio noduladores de feijão-fava (*Phaseolus lunatus* L.) Rev Bras Ciênc Solo 35:751–757

Appunu C, N'Zoue A, Moulin L, Depret G, Laguerre G (2009) *Vigna mungo*, *V. radiata* and *V. unguiculata* plants sampled in different agronomical-ecological-climatic regions of India are nodulated by *Bradyrhizobium yuanmingense*. Syst Appl Microbiol 32:460–470

Araújo KS (2014) Eficiência simbiótica e identificação de estirpes de *Burkholderia* oriundas de campos rupestres. Dissertation (Ms in Agricultural Microbiology). Microbiologia Agrícola, UFLA, Lavras, p 126

Araujo ASF, Lopes ACA, Gomes RLF, Beserra Junior JEA, Antunes JEL, Lyra MDCCP, Figueiredo MVB (2015) Diversity of native rhizobia-nodulating *Phaseolus lunatus* in Brazil. Legume Res 38:653–657

Atzorn R, Crozier A, Wheeler CT, Sandberg G (1988) Production of gibberellins and indole-3-acetic-acid by *Rhizobium phaseoli* in relation to nodulation of *Phaseolus vulgaris* roots. Planta 175:532–538

Azevedo JN, Franco JD, Araújo ROC (2003) Composição química de sete variedades de feijão-fava. Norte, E.M. Teresina: Embrapa Meio Norte

Baraúna AC, Silva KD, Pereira GMD, Kaminski PE, Perin L, Zilli JE (2014) Diversity and nitrogen fixation efficiency of rhizobia isolated from nodules of *Centrolobium paraense*. Pesq Agropec Bras 49:296–305

Bejarano A, Ramirez-Bahena MH, Velazquez E, Peix A (2014) *Vigna unguiculata* is nodulated in Spain by endosymbionts of Genisteae legumes and by a new symbiovar (vignae) of the genus *Bradyrhizobium*. Syst Appl Microbiol 37:533–540

van Berkum P, Fuhrmann JJ (2000) Evolutionary relationships among the soybean bradyrhizobia reconstructed from 16s rRNA gene and internally transcribed spacer region sequence divergence. Int J Syst Evol Microbiol 50:2165–2172

van Berkum P, Terefework Z, Paulin L, Suomalainen S, Lindstrom K, Eardly BD (2003) Discordant phylogenies within the *rrn* loci of Rhizobia. J Bacteriol 185:2988–2998

Bianco L, Angelini J, Fabra A, Malpassi R (2013) Diversity and symbiotic effectiveness of indigenous rhizobia-nodulating *Adesmia bicolor* in soils of Central Argentina. Curr Microbiol 66:174–184

Bontemps C, Elliott GN, Simon MF, Reis Júnior FB, Gross E, Lawton RC, Elias Neto N, Loureiro MF, Faria SM, Sprent JI, James EK, Young JPW (2010) *Burkholderia* species are ancient symbionts of legumes. Mol Ecol 19:44–52

Bournaud C, Faria SM, Santos JMF, Tisseyre P, Silva M, Chaintreuil C, Gross E, James EK, Prin Y, Moulin L (2013) *Burkholderia* species are the most common and preferred nodulating symbionts of the Piptadenia group (tribe Mimoseae). PLoS One 8:E63478–E63478

Brasil, S.D.D.A.-M.D.A.P.E.A. Instrução Normativa N°13, de 24 de março de 2011. Diário Oficial Da União – Seção 1. Brasília: Imprensa Nacional. 58, 25 de março de 2011: 3-7 P. 2011

Bromfield ESP, Barran LR (1990) Promiscuous nodulation of *Phaseolus vulgaris*, *Macroptilium atropurpureum*, and *Leucaena leucocephala* by indigenous *Rhizobium meliloti*. Can J Microbiol 36:369–372

Calheiros AS, Lira Junior MA, Soares DM, Figueiredo MVB (2013) Symbiotic capability of calopo rhizobia from an agrisoil with different crops in Pernambuco. Rev Bras Ciênc Solo 37:869–876

Calheiros AS, Lira Júnior MA, Santos MVF, Lyra MDCP (2015) Symbiotic effectiveness and competitiveness of calopo rhizobial isolates in an Argissolo vermelho-amarelo under three vegetation covers in the dry forest zone of Pernambuco. Rev Bras Ciênc Solo 39:367–376

Carvalho AFU, Sousa NM, Farias DF, Rocha-Bezerra LCB, Silva RMP, Viana MP, Gouveia ST, Saker-Sampaio S, Sousa MB, Lima GPG, Morais SM, Barros CC, Freire FR (2012) Nutritional ranking of 30 Brazilian genotypes of cowpeas including determination of antioxidant capacity and vitamins. J Food Compos Anal 26:81–88

Chagas Junior AF, Rahmeier W, Fidelis RR, Santos GR, Chagas LFB (2010) Eficiência agronômica de estirpes de rizóbio inoculadas em feijão-caupi no Cerrado, Gurupi-TO. Rev Cienc Agron 41:709–714

Chen WX, Wang E, Wang SY, Li YB, Chen XQ, Li YB (1995) Characteristics of *Rhizobium tianshanense* sp-nov, a moderately and slowly growing root-nodule bacterium isolated from an arid saline environment in Xinjiang, Peoples Republic of China. Int J Syst Bacteriol 45:153–159

Chen WM, James EK, Coenye T, Chou JH, Barrios E, Faria SM, Elliott GN, Sheu SY, Sprent JI, Vandamme P (2006) *Burkholderia mimosarum* isolated from root nodules of *Mimosa* spp. from Taiwan and South America. Int J Syst Evol Microbiol 56:1847–1851

Chen WM, Faria SM, James EK, Elliott GN, Lin KY, Chou JH, Sheu SY, Cnockaert M, Sprent JI, Vandamme P (2007) *Burkholderia nodosa* sp. nov., isolated from root nodules of the woody Brazilian legumes *Mimosa bimucronata* and *Mimosa scabrella*. Int J Syst Evol Microbiol 57:1055–1059

Chen WM, Faria SM, Chou JH, James EK, Elliott GN, Sprent JI, Bontemps C, Young JPW, Vandamme P (2008) *Burkholderia sabiae* sp. nov., isolated from root nodules of *Mimosa caesalpiniifolia*. Int J Syst Evol Microbiol 58:2174–2179

Coelho MRR, Marriel IE, Jenkins SN, Lanyon CV, Seldin L, O'Donnell AG (2009) Molecular detection and quantification of *nifH* gene sequences in the rhizosphere of sorghum (*Sorghum bicolor*) sown with two levels of nitrogen fertilizer. Appl Soil Ecol 42:48–53

Collavino MM, Tripp HJ, Frank IE, Vidoz ML, Calderoli PA, Donato M, Zehr JP, Aguilar OM (2014) *nifH* pyrosequencing reveals the potential for location-specific soil chemistry to influence N_2-fixing community dynamics. Environ Microbiol 16:3211–3223

Costa EM, Nóbrega RSA, Martins LV, Amaral FHC, Moreira FMS (2011) Nodulação e produtividade de *Vigna unguiculata* (L.) walp. por cepas de rizóbio em Bom Jesus, PI. Rev Cienc Agron 42:1–7

Costa EM, Nóbrega RSA, Carvalho F, Trochmann A, Ferreira LVM, Moreira FMS (2013) Promoção do crescimento vegetal e diversidade genética de bactérias isoladas de nódulos de feijão-caupi. Pesq Agropec Bras 48:1275–1284

Dall'agnol RF, Ribeiro RA, Ormeno-Orrillo E, Rogel MA, Delamuta JRM, Andrade DS, Martinez-Romero E, Hungria M (2013) *Rhizobium freirei* sp. nov., a symbiont of *Phaseolus vulgaris* that is very effective at fixing nitrogen. Int J Syst Evol Microbiol 63:4167–4173

Dall'agnol RF, Ribeiro RA, Delamuta JRM, Ormeno-Orrillo E, Rogel MA, Andrade DS, Martinez-Romero E, Hungria M (2014) *Rhizobium paranaense* sp. nov., an effective N_2-fixing symbiont of common bean (*Phaseolus vulgaris* L.) with broad geographical distribution in Brazil. Int J Syst Evol Microbiol 64:3222–3229

Dall'agnol RF, Plotegher F, Souza RC, Mendes IC, dos Reis Junior FB, Béna G, Moulin L, Hungria M (2016) *Paraburkholderia nodosa* is the main N^2-fixing species trapped by promiscuous common bean (*Phaseolus vulgaris* L.) in the Brazilian "Cerradão". FEMS Microbiol Ecol 92. doi:10.1093/femsec/fiw108

Dastager SG, Deepa CK, Pandey A (2011) Plant growth promoting potential of *Pontibacter niistensis* in cowpea (*Vigna Unguiculata* (L.) Walp.) Appl Soil Ecol 49:250–255

Delamuta JRM, Ribeiro RA, Menna P, Bangel EV, Hungria M (2012) Multilocus sequence analysis (MLSA) of *Bradyrhizobium* strains: revealing high diversity of tropical diazotrophic symbiotic bacteria. Braz J Microbiol 43:698–710

Eardly BD, Wang FS, Whittam TS, Selander RK (1995) Species limits in *Rhizobium* populations that nodulate the common bean (*Phaseolus vulgaris*). Appl Environ Microbiol 61: 507–512

Elliott GN, Chen WM, Chou JH, Wang HC, Sheu SY, Perin L, Reis VM, Moulin L, Simon MF, Bontemps C, Sutherland JM, Bessi R, Faria SM, Trinick MJ, Prescott AR, Sprent JI, James EK (2007) *Burkholderia phymatum* is a highly effective nitrogen-fixing symbiont of *Mimosa* spp. and fixes nitrogen ex planta. New Phytol 173:168–180

Estrada-de Los Santos P, Martinez-Aguilar L, Lopez-Lara IM, Caballero-Mellado J (2012) *Cupriavidus alkaliphilus* sp. nov., a new species associated with agricultural plants that grow in alkaline soils. Syst Appl Microbiol 35:310–314

Ferreira LDVM, Nóbrega RSA, Nóbrega JCA, Aguiar FL, Moreira FMS, Pacheco LP (2013) Biological nitrogen fixation in production of *Vigna unguiculata* (L.) walp, family farming in Piauí, Brazil. J Agric Sci 5:153–160

Fonseca MB, Peix A, Faria SM, Mateos PF, Rivera LP, Simões-Araújo JL, Franca MGC, Isaias RMD, Cruz C, Velazquez E, Scotti MR, Sprent JI, James EK (2012) Nodulation in *Dimorphandra wilsonii* Rizz. (Caesalpinioideae), a threatened species native to the Brazilian Cerrado. PLoS One 7:e4952

Franco AA, Faria SM (1997) The contribution of N_2-fixing tree legumes to land reclamation and sustainability in the tropics. Soil Biol Biochem 29:5–6

Garcia FCP, Fernandes JM (2015) Inga. Lista de espécies da flora do Brasil, Rio de Janeiro. http://floradobrasil.jbrj.gov.br/2012/FB022803. Cited 20 Apr 2016

Giongo A, Passaglia LMP, Freire JRJ, Sa ELS (2007) Genetic diversity and symbiotic efficiency of population of rhizobia of *Phaseolus vulgaris* L. in Brazil. Biol Fertil Soils 43:593–598

Grange L, Hungria M, Graham PH, Martinez-Romero E (2007) New insights into the origins and evolution of rhizobia that nodulate common bean (*Phaseolus vulgaris*) in Brazil. Soil Biol Biochem 39:867–876

Guimarães AA, Jaramillo PMD, Nóbrega RSA, Florentino LA, Silva KB, Moreira FMS (2012) Genetic and symbiotic diversity of nitrogen-fixing bacteria isolated from agricultural soils in the Western Amazon by using cowpea as the trap plant. Appl Environ Microbiol 78: 6726–6733

Guimarães AA, Florentino LA, Almeida KA, Lebbe L, Silva KB, Willems A, Moreira FMDS (2015) High diversity of *Bradyrhizobium* strains isolated from several legume species and land uses in Brazilian tropical ecosystems. Syst Appl Microbiol. doi:10.1016/j.syapm.2015.06.006

Gyaneshwar P, Hirsch AM, Moulin L, Chen WM, Elliott GN, Bontemps C, Estrada-de Los Santos P, Gross E, Reis FB, Sprent JI, Young JPW, James EK (2011) Legume-nodulating Betaproteobacteria: diversity, host range, and future prospects. Mol Plant Microbe Interact 24:1276–1288

Herder GD, Van Isterdael G, Beeckman T, De Smet I (2010) The roots of a new green revolution. Trends Plant Sci 15:600–607

Hungria M, Andrade DS, Chueire LMO, Probanza A, Guttierrez-Mañero FJ, Megias M (2000) Isolation and characterization of new efficient and competitive bean (*Phaseolus vulgaris* L.) rhizobia from Brazil. Soil Biol Biochem 32:1515–1528

Hungria M, Campo RJ, Mendes IC (2003) Benefits of inoculation of the common bean (*Phaseolus vulgaris*) crop with efficient and competitive *Rhizobium tropici* strains. Biol Fertil Soils 39:88–93

Jaramillo PMD, Guimarães AA, Florentino LA, Silva KB, Nóbrega RSA, Moreira FMS (2013) Symbiotic nitrogen-fixing bacterial populations trapped from soils under agroforestry systems in the Western Amazon. Sci Agric 70:397–404

Laguerre G, Nour SM, Macheret V, Sanjuan J, Drouin P, Amarger N (2001) Classification of rhizobia based on *nodc* and *nifh* gene analysis reveals a close phylogenetic relationship among *Phaseolus vulgaris* symbionts. Microbiology 147:981–993

Lammel DR, Cruz LM, Carrer H, Cardoso EJBN (2013) Diversity and symbiotic effectiveness of beta-rhizobia isolated from sub-tropical legumes of a Brazilian Araucaria Forest. World J Microbiol Biotechnol 29:2335–2342

Li JH, Wang ET, Chen WF, Chen WX (2008) Genetic diversity and potential for promotion of plant growth detected in nodule endophytic bacteria of soybean grown in Heilongjiang province of China. Soil Biol Biochem 40:238–246

Lima AS, Nóbrega RSA, Barberi A, Silva K, Ferreira DF, Moreira FMS (2009) Nitrogen-fixing bacteria communities occurring in soils under different uses in the Western Amazon region as indicated by nodulation of siratro (*Macroptilium atropurpureum*). Plant Soil 319:127–145

Lombardi MLCD, Moreira M, Ambrosio LA, Cardoso EJBN (2009) Occurrence and host specificity of indigenous rhizobia from soils of São Paulo State, Brazil. Sci Agric 66:543–548

Lopez-Lopez A, Negrete-Yankelevich S, Rogel MA, Ormeno-Orrillo E, Martinez J, Martinez-Romero E (2013) Native bradyrhizobia from Los Tuxtlas in Mexico are symbionts of *Phaseolus lunatus* (Lima bean). Syst Appl Microbiol 36:33–38

Marinho RDCN, Nóbrega RSA, Zilli JE, Xavier GR, Santos CAF, Aidar SDT, Martins LMV, Fernandes Júnior PI (2014) Field performance of new cowpea cultivars inoculated with efficient nitrogen-fixing rhizobial strains in the Brazilian Semiarid. Pesqui Agropecu Bras 49: 395–402

Marra LM, Soares CRFS, Oliveira SM, Ferreira PAA, Soares BL, Carvalho RF, Lima JM, Moreira FMS (2012) Biological nitrogen fixation and phosphate solubilization by bacteria isolated from tropical soils. Plant Soil 357:289–307

Martens M, Dawyndt P, Coopman R, Gillis M, de Vos P, Willems A (2008) Advantages of multilocus sequence analysis for taxonomic studies: a case study using 10 housekeeping genes in the genus *Ensifer* (including former *Sinorhizobium*). Int J Syst Evol Microbiol 58:200–214

Martins PGS, Lira Junior MA, Fracetto GGM, Silva MLRB, Vincentin RP, Lyra MCCP (2015) *Mimosa caesalpiniifolia* rhizobial isolates from different origins of the Brazilian Northeast. Arch Microbiol 197:459–469

Melloni R, Moreira FMS, Nóbrega RSA, Siqueira JO (2006) Eficiência e diversidade fenotípica de bactérias diazotróficas que nodulam caupi [*Vigna unguiculata* (L.) walp] e feijoeiro (*Phaseolus vulgaris* L.) em solos de mineração de bauxita em reabilitação. Rev Bras Ciênc Solo 30: 235–246

Menna P, Hungria M, Barcellos FG, Bangel EV, Hess PN, Martínez-Romero E (2006) Molecular phylogeny based on the 16s rRNA gene of elite rhizobial strains used in Brazilian commercial inoculants. Syst Appl Microbiol 29:315–332

Menna P, Barcellos FG, Hungria M (2009) Phylogeny and taxonomy of a diverse collection of *Bradyrhizobium* strains based on multilocus sequence analysis of the 16s rRNA gene, ITS region and *glnII, recA, atpD* and *dnaK* genes. Int J Syst Evol Microbiol 59:2934–2950

Meyer SE, Beuf K, Vekeman B, Willems AA (2015) Large diversity of non-rhizobial endophytes found in legume root nodules in Flanders (Belgium). Soil Biol Biochem 83:1–11

Mhamdi R, Laguerre G, Aouani ME, Mars M, Amarger N (2002) Different species and symbiotic genotypes of field rhizobia can nodulate *Phaseolus vulgaris* in Tunisian soils. FEMS Microbiol Ecol 41:77–84

Michiels J, Dombrecht B, Vermeiren N, Xi C, Luyten E, Vanderleyden J (1998) *Phaseolus vulgaris* is a non-selective host for nodulation. FEMS Microbiol Ecol 26:193–205

Mnasri B, Saidi S, Chihaoui SA, Mhamdi R (2012) *Sinorhizobium americanum* symbiovar mediterranense is a predominant symbiont that nodulates and fixes nitrogen with common bean (*Phaseolus vulgaris* L.) in a Northern Tunisian field. Syst Appl Microbiol 35:263–269

Moreira FMS (2008) Bactérias fixadoras de nitrogênio em leguminosas. In: Moreira FMS, Siqueira JO et al (eds) Biodiversidade do solo em ecossistemas brasileiros. UFLA, Lavras, pp 621–680

Moreira FMS, Silva MF, Faria SM, Silva MF, Faria SG (1992) Occurrence of nodulation in legume species in the Amazon region of Brazil. New Phytol 121:563–570

Moreira FMS, Gillis M, Pot B, Kersters K, Franco AA (1993) Characterization of rhizobia isolated from different divergence groups of tropical leguminosae by comparative polyacrylamide gel electrophoresis of their total proteins. Syst Appl Microbiol 16:135–146

Moreira FMS, Haukka K, Young JPW (1998) Biodiversity of rhizobia isolated from a wide range of forest legumes in Brazil. Mol Ecol 7:889–895

Moreira FMS, Cruz L, Faria SM, Marsh T, Martínez-Romero E, Pedrosa FO, Pitard RM, Young JPW (2006) *Azorhizobium doebereinerae* sp. nov. microsymbiont of *Sesbania virgata* (Caz.) Pers. Syst Appl Microbiol 29:197–206

Mostasso L, Mostasso FL, Dias BG, Vargas MAT, Hungria M (2002) Selection of bean (*Phaseolus vulgaris* L.) rhizobial strains for the Brazilian Cerrados. Field Crop Res 73:121–132

Moulin L, Munive A, Dreyfus B, Boivin-Masson C (2001) Nodulation of legumes by members of the beta-subclass of Proteobacteria. Nature 411:948–950

Nascimento SP, Bastos EA, Araújo ECE, Freire Filho FR, Silva EM (2011) Tolerância ao déficit hídrico em genótipos de feijão-caupi. Rev Bras Eng Agric Ambient 15:853–860

Nóbrega RSA (2006). Efeito de sistemas de uso da terra na amazônia sobre atributos do solo, ocorrência, eficiência e diversidade de bactérias que nodulam caupi (*Vigna unguiculata* (L.) Walp), p 188. Ph.D. thesis (Dr.). Programa De Pós-Graduação Em Ciência Do Solo, Universidade Federal De Lavras, Lavras

Oliveira JP, Galli-Terasawa LV, Enke CG, Cordeiro VK, Armstrong LCT, Hungria M (2011) Genetic diversity of rhizobia in a Brazilian oxisol nodulating Mesoamerican and Andean genotypes of common bean (*Phaseolus vulgaris* L.) World J Microbiol Biotechnol 27:643–650

Ormeño E, Torres R, Mayo J, Rivas R, Peix A, Velázquez E, Zúñiga D (2007) *Phaseolus lunatus* is nodulated by a phosphate solubilizing strain of *Sinorhizobium meliloti* in a peruvian soil. In: Velázquez EE, Rodríguez-Barrueco C (eds) First international meeting on microbial phosphate solubilization. Springer, Dordrecht, Netherlands, pp 143–147

Ormeño-Orrillo E, Vinuesa P, Zuñiga-Davila DZ, Martínez-Romero E (2006) Molecular diversity of native bradyrhizobia isolated from lima bean (*Phaseolus lunatus* L.) in Peru. Syst Appl Microbiol 29:253–262

Peix A, Mateos PF, Rodriguez-Barrueco C, Mart¡Nez-Molina E, Velazquez E (2001) Growth promotion of common bean (*Phaseolus vulgaris* L.) by a strain of *Burkholderia cepacia* under growth chamber conditions. Soil Biol Biochem 33:1927–1935

Pereira JM (2001) Produção e persistência de leguminosas em pastagens tropicais. Simpósio de Forragicultura e Pastagens, Universidade Federal de Lavras, pp 111–142

Pinto FGS, Hungria M, Mercante FM (2007) Polyphasic characterization of Brazilian *Rhizobium tropici* strains effective in fixing N_2 with common bean (*Phaseolus vulgaris* L.) Soil Biol Biochem 39:1851–1864

Pule-Meulenberg F, Belane A, Krasova-Wade T, Dakora F (2010) Symbiotic functioning and bradyrhizobial biodiversity of cowpea (*Vigna unguiculata* L. Walp.) in Africa. BMC Microbiol 10:89

Radl V, Simões-Araujo JL, Leite J, Passos SR, Martins LMV, Xavier GR, Rumjanek NG, Baldani JI, Zilli JE (2014) *Microvirga vignae* sp. nov., a root nodule symbiotic bacterium isolated from cowpea grown in Semi-Arid Brazil. Int J Syst Evol Microbiol 64:725–730

Raposeiras R, Marriel IE, Muzzi MRS, Paiva E, Pereira IA, Carvalhais LC, Passos RVM, Pinto PP, Sá NMH (2006) *Rhizobium* strains competitiveness on bean nodulation in Cerrado soils. Pesq Agropec Bras 41:439–447

Reis Junior FB, Simon MF, Gross E, Boddey RM, Elliott GN, Neto NE, Loureiro MF, Queiroz LP, Scotti MR, Chen W, Norén A, Rubio MC, Faria SM, Bontemps C, Goi SR, Young JPW, Sprent JI, James EK (2010) Nodulation and nitrogen fixation by *Mimosa* spp. in the Cerrado and Caatinga biomes of Brazil. New Phytol 186:934–946

Ribeiro PRA, Santos JV, Costa EM, Lebbe L, Assis ES, Louzada MO, Guimarães AA, Willems A, Moreira FMS (2015a) Symbiotic efficiency and genetic diversity of soybean bradyrhizobia in Brazilian soils. Agric Ecosyst Environ 212:85–93

Ribeiro RA, Martins TB, Ormeno-Orrillo E, Delamuta JRM, Rogel MA, Martinez-Romero E, Hungria M (2015b) *Rhizobium ecuadorense* sp. nov., an indigenous N$_2$-fixing symbiont of the ecuadorian common bean (*Phaseolus vulgaris* L.) genetic pool. Int J Syst Evol Microbiol 65:3162–3169

Rodriguez-Navarro DN, Buendia AM, Camacho M, Lucas MM, Santamaria C (2000) Characterization of *Rhizobium* spp. bean isolates from South-West Spain. Soil Biol Biochem 32:1601–1613

Roesch LFW, Passaglia LMP, Bento FM, Triplett EW, Camargo FAO (2007) Diversidade de bactérias diazotróficas endofíticas associadas a plantas de milho. Rev Bras Ciênc Solo 31:1367–1380

Rufini M, Silva MAP, Ferreira PAA, Cassetari AS, Soares BL, Andrade MJB, Moreira FMS (2014) Symbiotic efficiency and identification of rhizobia that nodulate cowpea in a Rhodic Eutrudox. Biol Fertil Soils 50:115–122

Salgado AG, Gepts P, Debouck DG (1995) Evidence for two gene pools of the Lima bean, *Phaseolus lunatus* L., in the Americas. Genet Resour Crop Evol 42:15–28

Santos CERS, Stamford NP, Neves MCP, Rumjanek NG, Borges WL, Bezerra RV, Freitas ADS (2007) Diversidade de rizóbios capazes de nodular leguminosas tropicais. Rev Bras Cienc Agrar 2:249–256

Santos JO, Antunes JEL, Araujo ASF, Lyra MCCP, Gomes RLF, Lopes ACA, Figueiredo MVB (2011) Genetic diversity among native isolates of rhizobia from *Phaseolus lunatus*. Ann Microbiol 61:437–444

Sarr PS, Yamakawa T, Fujimoto S, Saeki Y, Thao HTB, Myint AK (2009) Phylogenetic diversity and symbiotic effectiveness of root-nodulating bacteria associated with cowpea in the South-West area of Japan. Microbes Environ 24:105–112

Sarr PS, Yamakawa T, Saeki Y, Guisse A (2011) Phylogenetic diversity of indigenous cowpea bradyrhizobia from soils in Japan based on sequence analysis of the 16S-23S rRNA internal transcribed spacer (ITS) region. Syst Appl Microbiol 34:285–292

Sheu SY, Chou JH, Bontemps C, Elliott GN, Gross E, James EK, Sprent JI, Young JPW, Chen WM (2012) *Burkholderia symbiotica* sp. nov., isolated from root nodules of *Mimosa* spp. native to north-east Brazil. Int J Syst Evol Microbiol 62:2272–2278

Sheu SY, Chou JH, Bontemps C, Elliott GN, Gross E, Reis Junior FB, Melkonian R, Moulin L, James EK, Sprent JI, Young JPW, Chen WM (2013) *Burkholderia diazotrophica* sp. nov., isolated from root nodules of *Mimosa* spp. Int J Syst Evol Microbiol 63:435–441

Shiraishi A, Matsushita N, Hougetsu T (2010) Nodulation in black locust by the Gammaproteobacteria *Pseudomonas* sp. and the Betaproteobacteria *Burkholderia* sp. Syst Appl Microbiol 33:269–274

Silva F, Simões-Araújo J, Silva Júnior J, Xavier G, Rumjanek N (2012) Genetic diversity of rhizobia isolates from Amazon soils using cowpea (*Vigna unguiculata*) as trap plant. Braz J Microbiol 43:682–691

Silva K, Meyer SE, Rouws LF, Farias EN, Santos MA, O'Hara G, Ardley JK, Willems A, Pitard RM, Zilli JE (2014) *Bradyrhizobium ingae* sp. nov., isolated from effective nodules of *Inga laurina* grown in Cerrado soil. Int J Syst Evol Microbiol 64:3395–3401

Simões-Araújo JL, Leite J, Marie Rouws LF, Passos SR, Xavier GR, Rumjanek NG, Zilli JE (2016a) Draft genome sequence of *Bradyrhizobium* sp. strain BR 3262, an effective microsymbiont recommended for cowpea inoculation in Brazil. Braz J Microbiol 47:783–784

Simões-Araújo JL, Leite J, Passos SR, Xavier GR, Rumjanek NG, Zilli JE (2016b) Draft genome sequence of *Bradyrhizobium* sp. strain BR 3267, an elite strain recommended for cowpea inoculation in Brazil. Braz J Microbiol 47:781–782

Simon MF, Proença C (2000) Phytogeographic patterns of *Mimosa* (Mimosoideae, Leguminosae) in the Cerrado biome of Brazil: an indicator genus of high-altitude centers of endemism? Biol Conserv 96:279–296

Soares ALL, Ferreira PAA, Pereira J, Vale HMM, Lima AS, Andrade MJB, Moreira FMS (2006a) Eficiência agronômica de rizóbios selecionados e diversidade de populações nativas nodulíferas em Perdões (MG) I – caupi. Rev Bras Ciênc Solo 30:803–811

Soares ALL, Pereira JPAR, Ferreira PAV, Vale HMM, Lima AS, Andrade MJB, Moreira FMS (2006b) Eficiência agronômica de rizóbios selecionados e diversidade de populações nativas noduliferas em Perdões (MG). I – Caupi. Rev Bras Ciênc Solo 30:795–802

Steenkamp ET, Stepkowski T, Przymusiak A, Botha WJ, Law IJ (2008) Cowpea and peanut in southern Africa are nodulated by diverse *Bradyrhizobium* strains harboring nodulation genes that belong to the large pantropical clade common in Africa. Mol Phylogenet Evol 48:1131–1144

Stocco P, Santos JCP, Vargas VP, Hungria M (2008) Assessment of biodiversity in rhizobia symbionts of common bean (*Phaseolus vulgaris* L.) in Santa Catarina, Brazil. Rev Bras Ciênc Solo 32:1107–1120

Tesfaye M, Holl FB (1998) Group-specific differentiation of *Rhizobium* from clover species by PCR amplification of 23s rDNA sequences. Can J Microbiol 44:1102–1105

Thies JE, Singleton PW, Bohlool BB (1991) Influence of the size of indigenous rhizobial populations on establishment and symbiotic performance of introduced rhizobia on field-grown legumes. Appl Environ Microbiol 57:19–28

Torres AR, Cursino L, Muro-Abad JI, Gomes EA, Araújo EF, Hungria M, Cassini STA (2009) Genetic diversity of indigenous common bean (*Phaseolus vulgaris* L.) rhizobia from the state of Minas Gerais, Brazil. Braz J Microbiol 40:852–856

Torres EB, Nóbrega RS, Fernandes-Júnior PI, Silva LB, Carvalho GDS, Marinho RDCN, Pavan BE (2016) The damage of *Callosobruchus maculatus* on cowpea grains is dependent of the plant genotype. J Sci Food Agr 96:4276–4280

Valle CB (2002) Recursos genéticos de forrageiras para áreas tropicais. In: Conferência virtual global sobre produção orgânica de bovinos de corte, EMBRAPA-Pantanal, Corumbá – MS. http://www.cpap.embrapa.br/agencia/congressovirtual/pdf/portugues/03pt09.pdf. Cited 22 Apr 2016

Valverde A, Igual JM, Peix A, Cervantes E, Velazquez E (2006) *Rhizobium lusitanum* sp. nov. a bacterium that nodulates *Phaseolus vulgaris*. Int J Syst Evol Microbiol 56:2631–2637

Vinuesa P (2015). Rhizobial taxonomy up-to-date [Online]. Mexico: Universidad Nacional Autonoma de Mexico. http://edznaCcgUnamMx/Rhizobial-Taxonomy/. Cited 22 Apr 2016

Zilli JE, Marson LC, Marson BF, Rumjanek NG, Xavier GR (2009) Contribuição de estirpes de rizóbio para o desenvolvimento e produtividade de grãos de feijão-caupi em Roraima. Acta Amazon 39:749–757

Zilli JE, Barauna AC, Silva K, Meyer SE, Farias ENC, Kaminski PE, Costa IB, Ardley JK, Willems A, Camacho NN, Dourado FD, O'hara G (2014) *Bradyrhizobium neotropicale* sp. nov., isolated from effective nodules of *Centrolobium paraense*. Int J Syst Evol Microbiol 64:3950–3957

Potential of Rhizobia as Plant Growth-Promoting Rhizobacteria

7

Luciano Kayser Vargas, Camila Gazolla Volpiano,
Bruno Brito Lisboa, Adriana Giongo, Anelise Beneduzi,
and Luciane Maria Pereira Passaglia

Abstract

Nitrogen-fixing plant growth-promoting rhizobacteria collectively known as rhizobia have been extensively investigated due to their exceptional quality to establish functional symbiosis with legumes. As a result of this incredible interaction, they supply nitrogen to plants, which is one of the major nutrient elements. Rhizobia are capable of colonizing the rhizosphere of nonhost plants (nonlegumes) thus living within plant tissues as endophytes. Due to these properties and their ability to secrete phytohormones and siderophores, and solubilize insoluble phosphate, besides eliciting plant defense reactions against phytopathogens, rhizobia have been placed along the organisms with high potential to act as efficient plant growth-promoting rhizobacteria (PGPR). Here, the mechanisms adopted by rhizobia to facilitate plant growth and yields are highlighted. In addition, the application of rhizobia as PGPR in farming practices is underlined. The information available on rhizobial application and the number of rhizobia stored in different culture collection centers around the world may provide an important microbiological resource to reduce the use of expensive synthetic fertilizers and pesticides in agricultural practices.

L.K. Vargas (✉) • B.B. Lisboa
Fundação Estadual de Pesquisa Agropecuária, Laboratory of Agricultural Chemistry,
Porto Alegre, Brazil
e-mail: luciano@fepagro.rs.gov.br

C.G. Volpiano • A. Beneduzi • L.M.P. Passaglia
Department of Genetics, Institute of Biosciences, Universidade Federal do Rio Grande do Sul, Porto Alegre, Brazil

A. Giongo
Institute of Petroleum and Natural Resources, Pontifical Catholic University of Rio Grande do Sul, Porto Alegre, RS, Brazil

© Springer International Publishing AG 2017
A. Zaidi et al. (eds.), *Microbes for Legume Improvement*,
DOI 10.1007/978-3-319-59174-2_7

153

7.1 Introduction

Plant growth-promoting rhizobacteria (PGPR) involving symbiotic and free-living PGPR have been applied in farming practices to enhance growth and yields of legumes (Ahmad et al. 2013; Khaitov et al. 2016) and nonlegume crops (Antoun et al. 1998; García-Fraile et al. 2012; Ziaf et al. 2016). Of the versatile PGPR group, bacteria that form root nodules on leguminous plants and transform atmospheric nitrogen (N_2) into usable N (ammonia) are collectively known as rhizobia (Lindström and Martinez-Romero 2005). Currently, rhizobia consist of more than 98 species distributed over 13 genera (Weir 2016) which is expected to increase even further with the discovery of geographically new dispersed host plants (Willems 2006) coupled with advent of modern molecular-based tools of rhizobial identification. These reports include some betaproteobacteria such as *Burkholderia* and *Cupriavidus* (Barrett and Parker 2006; Chen et al. 2007), while rhizobia included in alphaproteobacteria group has the genera *Rhizobium, Bradyrhizobium, Sinorhizobium, Mesorhizobium,* and *Azorhizobium* (Sahgal and Johri 2003). Nodule-forming rhizobia are widely used as microbial inoculant to enhance legume production in different production systems and are considered the best option of microbial technology in agricultural practices. Only in Brazil, inoculation of soybean (*Glycine max*) fields with rhizobia has been reported to supply up to 300 kg/ha of N resulting in saving of N fertilizers estimated to be around US\$ 3 billion (Hungria and Campo 2005). Apart from supplying exclusively N to legumes and companion crops, rhizobia may also facilitate plant growth by other means, acting as PGPR: a concept introduced in the late 1970s (Kloepper 1978; Kloepper et al. 1980). Initially, nitrogen-fixing rhizobia was not included in the PGPR group, but later on Gray and Smith in 2005 included rhizobia in intracellular category of PGPR (iPGPR): bacteria living in specialized nodular structures (Gray and Smith 2005). However, some rhizobia have also been grouped in extracellular PGPR (ePGPR) category, which forms associations with nonlegumes, and there are also rhizobia-legume association without nodule formation (endophytic but not symbiotic). Broadly, PGPR can be defined as any root-colonizing bacteria capable of exerting beneficial effects on plant development by direct or indirect mechanisms or in many cases the combination of both.

Numerous soil microbiota belonging to different genera have now been identified as efficient PGPR. Of these, the most widely exploited genera are *Pseudomonas* (Adesemoye and Ugoji 2009; Cattelan et al. 1999), *Bacillus* (Probanza et al. 2001; Recep et al. 2009), and *Azospirillum* (Sivasakthivelan and Saranraj 2013). In addition to these exclusively extracellular PGPR, some rhizobial strains across different genera and species have also received greater attention due to their positive growth promontory effects both on host legume plants (Ahmad et al. 2013; Imen et al. 2015; Khaitov et al. 2016) and on nonlegumes (García-Fraile et al. 2012). Rhizobial strains are reported to possess many distinct plant growth-promoting traits (Bhattacharjee et al. 2012; Datta and Chakrabartty 2014; Ghosh et al. 2015) and exert diverse positive effects over many important crops (Flores-Félix et al. 2013; Zaidi et al. 2015). Due to the enormous information available on rhizobia, the large populations of rhizobial strains stocked in culture collections around the world, the nonpathogenic nature of rhizobia and great genetic variability within species and even within strains, etc., rhizobia are considered as one of the most exciting and promising groups of PGPR for application against a range of crops cultivated in different farming practices.

7.2 Colonization and Establishment of PGPR

Depending on their ability to colonize and ability to influence plant growth, microbial communities in general have been grouped as (1) beneficial: positive impact on plants, (2) deleterious: adverse effect on plants, and (3) neutral: exert no effect. The first category includes the PGPR which involve rhizobia. In order to facilitate plant growth, PGPR must be able to colonize and survive in the rhizosphere, the roots in competitive conditions (Kloepper 2003), and, in some cases even, the phyllosphere (Nandi et al. 1982). Some PGPR also colonize, penetrate, and establish within plant tissues and act as endophyte PGPR (Vessey 2003). As a consequence, the intense interaction occurring between PGPR and plants becomes practically more important largely due to rhizobacteria which, on one hand, play a role as plant growth promoter, and, on the other hand, plants may exert a selective control over bacterial diversity and composition in varying habitats. During this interaction, plant roots release organic compounds into their surrounding environment (a process called rhizodeposition), which triggers an increase in microbial density/quantity (Foster 1998). In terms of microbial diversity/quality, the rhizosphere is one of the richest ecological zones (Prashar et al. 2014). The root colonization can be viewed as one of the first steps during which PGPR express their plant growth promontory activity (Kloepper and Beauchamp 1992). Among PGPR, rhizobia have been found as competent rhizospheric bacteria which could survive well in soil for longer duration, even in the absence of their host legumes (Batista et al. 2007). As an example, population of bradyrhizobia with greater genetic diversity and capable of nodulating soybean survived in a field kept in fallow for more than 30 years even without reinoculation and in the absence of host plants. The ability of rhizobia to survive and multiply in nonlegume rhizosphere has been explored as an alternative way to increase the desirable population of efficient bradyrhizobia in soil, by inoculating winter cereals seeds, prior to soybean sowing (Domit et al. 1990). Nevertheless, more than just increasing its population in soil, the inoculation of nonlegumes may result in intense colonization of roots by rhizobia (Schloter et al. 1997). As an example, Chabot et al. (1996) bioprimed maize (*Zea mays*) and lettuce (*Lactuca sativa*) plants using two strains of *Rhizobium phaseoli*. The results revealed a strong colonization of root surface of both crops, and 4.1 CFU/g (fresh weight) of rhizobial populations was recorded on maize root surface 4 weeks after seeding, while it was 3.7 CFU/g on lettuce roots 5 weeks after seeding. The endophytic colonization of nonlegumes by rhizobia was also reported by Sabry et al. (1997), who studied the interaction between *Azorhizobium caulinodans* and wheat (*Triticum aestivum*) and observed the invasion of azorhizobia between cells of the cortex. Other authors (Gutierrez-Zamora and Martınez-Romero 2001; Yanni et al. 1997) report similar endophytic colonization of nonlegumes by rhizobia. More recently, employing transgenic tobacco (*Nicotiana tabacum*), rape *(Brassica napus* L. var. *napus*), and tomato (*Solanum lycopersicum*) with pea (*Pisum sativum*) lectin gene, Vershinina et al. (2012) demonstrated that it is possible to increase the effectiveness of specific *R. leguminosarum* colonization of nonlegume plants roots. The more expressive and pronounced result was observed for transformed roots of rape plants which had significantly higher (1369.7 $\times$ 10^6 CFU/g fresh weight) population of *R. leguminosarum* compared to those recorded for control plants (36.4 $\times$ 10^6 CFU/g). In a similar experiment, the number of *R. leguminosarum* colonized onto the roots of transformed tobacco and rape plants was 14- and 37-folds, respectively, higher relative to the control plants (Vershinina et al. 2011). Mechanistically, rhizobial penetration into nonlegumes tissues occurs primarily via cracks in epidermal cells of the roots and in fissure sites where lateral roots have emerged (Dazzo and Yanni 2006; Prayitno et al. 1999). Nonetheless, the rhizobial endophytic establishment is a dynamic process which begins with colonization at lateral root emergence, crack entry into the root interior through separated epidermal cells, followed by endophytic ascending migration up to the stem base, leaf sheath, and leaves where they grow rapidly to high local population densities (Chi et al. 2005; Dazzo and Yanni 2006). In summary, rhizobial population is build up within plant tissues, while rhizobia act as PGPR, promoting the growth of plants employing various mechanisms.

7.3   Mechanisms of Plant Growth Promotion by Rhizobia

Free-living and symbiotic nitrogen-fixing PGPR can influence growth and yield of plants through direct or indirect mechanisms. Through direct mechanism, PGPR benefits plant by providing compounds that are synthesized by the bacterium or facilitating the uptake of certain nutrients from the environment (Glick 1995). The most usual and common direct mechanisms of plant growth promotion include fixation of atmospheric nitrogen (Machado et al. 2013), nutrient absorption and mobilization (Reimann et al. 2008; Yu et al. 2012), secretion of plant growth-promoting substances (Ghosh et al. 2015; Sahasrabudhe 2011), and the solubilization of inorganic insoluble phosphates and mineralization of organic phosphates (Abd-Alla 1994a; Kumar and Raghu Ram 2014; Prasad et al. 2014). Nonsymbiotic N_2 fixation is also one of the most attractive PGPR characteristics by which some rhizobia fix N in association even with nonlegume plants. For example, nitrogen-fixing activity, measured by acetylene reduction assay, was detected in wild rice (*Oryza breviligulata*) plants inoculated with a photosynthetic bradyrhizobia (Chaintreuil et al. 2000). However, in most of the cases, nitrogen fixation in nonlegumes by rhizobia is rare and negligible, if not inexistent. The indirect mechanisms in contrast involve the release of bioactive molecules by PGPR which minimize or cease the harmful effects of phytopathogens (Datta and Chakrabartty 2014; Gandhi et al. 2009). Conclusively, functionally variable PGPR under all environmental conditions and in different production systems may facilitate growth and development of plants employing either one or combination of these mechanisms.

7.3.1 Production of Plant Growth Regulators

Plant growth regulators are organic molecules analogous to plant hormones, which, at low concentrations, cause a physiological response and influence plant development. Because of chemical structures and their effects, such compounds have been grouped into six different categories: (1) auxins; (2) cytokinins; (3) gibberellins; (4) ethylene; (5) a group called inhibitors, which includes abscisic acid, phenolics, and alkaloids (Ferguson and Lessenger 2006; Mishra et al. 2006); and (6) brassinosteroids (Bajguz and Tretyn 2003; Rao et al. 2002). Perhaps, majority of PGPR synthesize and secrete these compounds, which, however, vary in concentration from organisms to organisms.

7.3.1.1 Auxins

The production of auxins is considered a common PGPR feature, and more than 80% of the soil bacteria are able to produce auxins, especially the indoleacetic acid (IAA), indolebutyric acid, or other similar compounds derived from tryptophan metabolism (Loper and Schroth 1986; Solano et al. 2008). Auxins are plant growth hormone that stimulates cell division and elongation, and its production by PGPR including rhizobia (Antoun et al. 1998; Bhagat et al. 2014; Hafeez et al. 2004; Schlindwein et al. 2008) is one of the most widely assayed phytohormones and, perhaps, the most effective biomolecules involved in plant growth promotion. Vargas et al. (2009) in a study found a considerably lower frequency of auxin producers (23%) while evaluating a population of clover-nodulating *R. trifolii*. However, they noticed a very distinct behavior between strains isolated from arrow leaf clover (*Trifolium vesiculosum*) and those isolated from white clover (*T. repens*) nodules. In the first group, IAA production was much more frequent accounting for more than 90% of the isolates. On the contrary, IAA production was considerably less frequent (only 15%) in rhizobia isolated from white clover nodules.

Auxins produced by rhizobia have been reported to influence nodulation process. Hence, IAA-producing rhizobia have been found to nodulate more intensely than IAA-negative mutants (Boiero et al. 2007). IAA produced by rhizobia has been found to modify root morphogenesis increasing the size and weight, branching number, and the surface area of roots in contact with soil leading eventually in the development of a more expansive root architecture of nonlegume plants (Dazzo and Yanni 2006). Inoculation with auxin-producing bacteria may also result in the formation of adventitious roots (Solano et al. 2008). Such changes in root system increase its ability to absorb more nutrient from soil and, therefore, improves plant growth (Probanza et al. 1996). Similarly, Biswas et al. (2000) observed that the inoculation of rice with *R. trifolii* increased dry matter and grain production, besides an increment in N, P, K, and Fe content in plant tissue. All these effects were credited to IAA accumulation in rhizosphere by rhizobial inoculation, resulting in physiological changes in root system with consequent improvement in nutrient uptake.

Like any other plant hormones, IAA exhibit positive effect even at a very low concentration while overproduction/higher concentration of IAA is reported to have either unresponsive effect or show adverse impact on plant growth (Biswas et al. 2000; Schlindwein et al. 2008). For instance, Schlindwein et al. (2008), while analyzing the production of IAA by rhizobial strains, found that the *R. trifolii* strain produced 171.1 µg/ml IAA in media enriched with tryptophan. Lettuce seeds treated with strain TV-13 did not germinate normally, and those, which germinated, were without radical protrusion and with precocious opening of the cotyledons (Fig. 7.1). The amount of IAA produced by a given bacterium strain, however, varies with the composition of the growth medium, and in most of the cases it is tryptophan dependent. The deleterious effect of inoculation with TV-13 ceased when the isolate was grown in yeast mannitol (YM) medium without tryptophan supplementation (Schlindwein et al. 2008). In such condition, IAA production by TV-13 was not detected, and germination rates equaled to the non-inoculated control. Similarly, Bhattacharjee et al. (2012) observed IAA production by *R. trifolii* increased from 2 to 25 µg/ml when 100 µg/ml of tryptophan was added to the media.

7.3.1.2 Cytokinins and Gibberellins

Cytokinins also affect cell division and cell enlargement, besides affecting seed dormancy, flowering, fruiting, and plant senescence (Ferguson and Lessenger 2006). However, cytokinins production by PGPR is often inadequately reported largely due to the lack of methods for cytokinins measurement. Despite such limitations, there are reports that at least some strains of *Rhizobium* produce cytokinins in culture, though they have not been fully quantified and characterized (Sturtevant and Taller 1989; Wang et al. 1982) Also, despite data on the secretion of cytokinins by rhizobia and for the involvement of cytokinins in legume nodulation (Frugier et al. 2008), neither the nature of cytokinin production by rhizobia nor the role of these

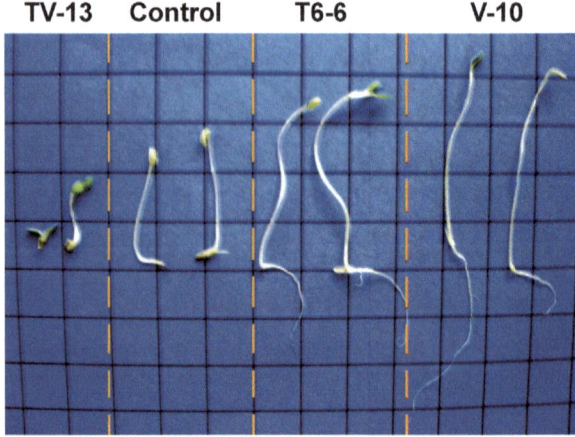

Fig. 7.1 Seeds of lettuce inoculated with *R. trifolii* IAA overproducing strain TV-13 and *Bradyrhizobium* sp. T6-4 and V-10, producers of small concentrations of IAA (adapted from Schlindwein et al. 2008)

cytokinins in the rhizobia-legume symbiosis has adequately been described. Gibberellins are other plant growth regulators that enhance seed germination (Miransari and Smith 2009), stimulate extensive growth of plants, and delay aging (Ferguson and Lessenger 2006). The production of gibberellins in high concentrations is considered very rare, but production of lower concentration of gibberellins has been reported for nodule-forming *Rhizobium* (Solano et al. 2008).

Several reports have demonstrated that free-living rhizobial bacteria have the capacity to produce some amount of gibberellin-like substances. However, it is not known whether bacteria contribute significantly to the amount of gibberellins within the nodule or it is just imported from some remote host plant tissue (Dobert et al. 1992). In spite of this, role of gibberellin in a *Rhizobium*-legume symbiosis that may have important implications to endophytic colonization of nonlegumes by rhizobia is described. For example, *A. caulinodans* infects the semiaquatic legume *Sesbania rostrata* via the intercellular crack entry, a process mediated by gibberellins. Considering that crack entry is the main process of endophytic colonization of non-legumes by rhizobia, the production of gibberellins by the bacterium may facilitate this process (Lievens et al. 2005). There is no known function for gibberellins in bacteria; thus probably PGPR have the ability to produce gibberellin only as a mean of manipulating the plants (Hayashi et al. 2014).

7.3.1.3 Abscisic Acid and Ethylene

Abscisic acid (ABA) acts adversely to gibberellins (Miransari and Smith 2009; Yang et al. 2009), inhibits growth and germination, and promotes seed dormancy (Ferguson and Lessenger 2006). In addition, ABA plays an important function in mediating plant tolerance to abiotic stresses. When plants are exposed to drought stress, they change their plant hormone balance, increasing ABA content in the leaves, accompanied by a reduction in endogenous cytokinin levels, which in turn elicits stomata closure (Yang et al. 2009). Some rhizobial strains, such as *B. japonicum* USDA110, are reported to produce ABA (Boiero et al. 2007) and contribute in some way to plant tolerance to drought stress. However, increases of ABA concentration also have been shown to negatively regulate rhizobium nodulation. In *Trifolium repens* and *Lotus japonicas*, Suzuki et al. (2004) observed that exogenous application of ABA inhibited root nodule formation after inoculation with *R. trifolii*. Working with a *Lotus japonicus* mutant that has lower sensitivity to ABA, Tominaga et al. (2010) reported an enhanced nodule formation in *Mesorhizobium loti* inoculated plants.

Similar to ABA, ethylene is also a stress hormone produced under both biotic and abiotic stresses and negatively regulates the growth of plants. Ethylene also affects ripening and senescence in plants (Ferguson and Lessenger 2006). Boiero et al. (2007) found that the strains of *B. japonicum* E109, USDA110, and SEMIA5080, the most commonly used for inoculation of soybean and nonlegumes in the USA, were able to produce ethylene in yeast extract mannitol medium amended with methionine. On other hand, some bacteria are able to decrease the levels of ethylene in plant root tissue mediated by the bacterial enzyme ACC deaminase, which competes with plant ACC oxidase. According to Glick et al. (1998), the bacterial enzyme acts in rhizosphere and degrades ACC exuded by plant roots to ammonia and α-ketobutyrate, resulting in lowering the level of ACC outside of the plant, forming a gradient from the interior of the plant to its exterior. In order to maintain the equilibrium between internal and external ACC levels, the plant must exude increasing amounts of ACC. Consequently, the level of ACC within the plant is reduced, and, hence, inhibitory action of ethylene is decreased. Thus, plants influenced by ACC deaminase-positive PGPR are supposed to have longer roots and possibly shoots as well (Glick et al. 1997). Additionally, PGPR with ACC deaminase activity have a competitive advantage about other bacteria, as these organisms use ACC as a source of nitrogen (Glick 2005). The reduction of ethylene levels in plant tissues derived from the ACC deaminase activity can cause significant morphological changes in root tissue, such as changes in root hair length and increases in root mass, accompanied with the consequent improvement in nutrient uptake. The morphological changes are greater when ACC deaminase action is combined with production of auxins by PGPR. However, IAA stimulates the activity of ACC synthase. It has been observed that some rhizobia may reduce plant ethylene levels by means of ACC deaminase activity and results in enhanced nodulation in host legumes (Ma et al. 2003) or modifications in root system of nonlegumes. However, most of rhizobial strains are low ACC deaminase-expressing organisms; thus they are not able to lower the overall level of ethylene in the plant but, rather, prevent a localized rise in ethylene levels caused by infection and nodulation (Glick 2005). For instance, strains of *R. leguminosarum* bv. *viciae* and *M. loti* increased the number of lateral roots in *Arabidopsis thaliana* because of this plant growth-promoting mechanism (Contesto et al. 2008).

7.3.1.4 Rhizobitoxine and Brassinosteroids

Rhizobitoxine [2-amino-4-(2-amino-3-hydropropoxy)-*trans*-but-3-enoic acid] is an ethylene synthesis inhibitor produced by bacteria including rhizobia (LA Favre and Eaglesham 1986), for example, *B. elkanni* (Minamisawa 1990; Yuhashi et al. 2000), and acts in a way similar to ACC deaminase: it strongly inhibits ACC synthase and decreases ethylene levels. Since ethylene is known to inhibit or downregulate nodule development, rhizobitoxine plays a positive role in nodule development by inhibiting ethylene biosynthesis (Duodu et al. 1999). However, unlike ACC deaminase, rhizobitoxine is not expected to promote plant growth. Its most usual effect is deleterious, since it causes chlorosis in plant leaves (Duodu et al. 1999; Okazaki et al. 2007). The only presumed role of rhizobitoxine as a mechanism of plant growth promotion has been the protection of soybean roots from *Macrophomina phaseolina* infection, once rhizobitoxine was shown to be antifungal (Chakraborty and Purkayastha 1984). Brassinosteroids are steroidal substances involved in plant growth and resistance against abiotic stresses, which have been considered as a new group of hormones. Brassinosteroids influence various developmental processes such as seed germination, rhizogenesis, flowering, senescence, abscission, and maturation (Rao et al. 2002). Vardhini and Ram Rao (1999) have shown that application of brassinosteroids increased nodulation and nitrogen fixation in groundnut (*Arachis hypogaea*) grown in natural soil (without application of *Rhizobium* inoculum).

Working with pea mutants possessing root systems deficient in gibberellins or brassinosteroids, Ferguson et al. (2005) observed that *R. leguminosarum* bv. *viciae* inoculated plants exhibited a reduction in nodule organogenesis. The transcription of genes involved in brassinosteroids biosynthesis is also positively influenced by inoculation (Breakspear et al. 2014).

7.3.2 Solubilization of Phosphates

Phosphorus (P) is one of the most important nutrients required by plants whose deficiency restricts the crop production severely. Phosphorus can be found in organic and inorganic forms which are not available for uptake by plants. Less than 5% of the total soil P content is available to plants (Dobbelaere et al. 2003). Therefore, depending on the P deficiency of soil, chemically synthesized phosphatic fertilizers can be applied to fulfill P requirement of crops. However, after application, a major portion of the soluble phosphatic fertilizers is quickly immobilized due to complex formation with iron and aluminum oxides in acid soils and with calcium in calcareous soils. Hence, it becomes unavailable for uptake by plants (Chacon et al. 2006; Khan et al. 2009). Phosphate-solubilizing bacteria (PSB) including rhizobia, for example, *Mesorhizobium*, R. *leguminosarum*, *Bradyrhizobium*, and *Sinorhizobium meliloti* (Kumar and Raghu Ram 2014; Machado et al. 2013; Peix et al. 2001), are able to solubilize insoluble P. The solubilization of mineral P generally occurs via the production of organic acids (Zaidi et al. 2009), which acidify the surrounding soil. Due to this reason, solubilization of mineral P is thought to be more efficient in basic soils than in naturally acid soils (Khan et al. 2010; Solano et al. 2008). However, acidification does not seem to be the only mechanism used by P-solubilizing bacteria for mineral P. For solubilization of organic P, the presence of phosphatases (Abd-Alla 1994b) and phytases (López-López et al. 2010) was also reported for rhizobial strains. Microbial phytases are interesting because plants generally have a limited capacity to obtain P directly from inositol hexaphosphate (phytate) (Richardson et al. 2001) which is a major component of organic P in soil. According to Rodríguez and Fraga (1999), the genus *Rhizobium* is one of the major P solubilizers, along with bacteria belonging to the genera *Pseudomonas* and *Bacillus*, as also reported by Vargas et al. (2009).

Among the PGPR traits of 252 isolates of *R. trifolii* evaluated by Vargas et al. (2009), solubilization of P was the most usual characteristics. This trait was identified in 42% of all the isolates and in 100% of the isolates from one of the sampling sites (Porto Alegre, Brazil). Like *Rhizobium* species, other rhizobia also possess this PGPR trait. For example, Alikhani et al. (2007), while working with 446 bacteria belonging to the genera *Bradyrhizobium*, *Mesorhizobium*, *Sinorhizobium*, and *Rhizobium*, evaluated the solubilization of inorganic and organic P under in vitro conditions. They observed that 44% of the isolates solubilized tricalcium phosphate, while 76% solubilized phytate. However, the rhizobial isolates differed in their P-solubilizing ability. Of these, *R. leguminosarum* bv. *viciae* was most prominent P solubilizer which was followed by *M. ciceri*, *M. mediterraneum*, *S. meliloti*, and *R. phaseoli*. However, none of the 70 strains of *Bradyrhizobium* tested were able to solubilize inorganic P, confirming the observation of Antoun et al. (1998) who also found only one P-solubilizer strain out of the 18 tested *B. japonicum* strains. The genus *Bradyrhizobium* is characterized by the production of alkali in growth media, a possible reason to explain poor P solubilization by this organism. On the other hand, when the authors analyzed the mineralization of organic P, a process mediated by phosphatases, *Bradyrhizobium* sp. strains were the most effective ones, even though *B. japonicum* were the less efficient strains. It was concluded from this study that many rhizobia isolated from Iranian soils are able to mobilize P from both inorganic and organic sources, and, hence, the probable beneficial effects of such bacteria need to be tested with crops before they are recommended for use in field environments.

Phosphate-solubilizing bacteria are commonly reported to promote plant growth. For example, in a field experiment, Messele and Pant (2012) assessed the effects of inoculation of *Sinorhizobium ciceri* and PSB on the performance of chickpea using three levels of NP fertilizer and four levels of inoculants. The result revealed that inoculation of *S. ciceri* alone increased dry matter yield by 157% and nodule number by 118% over control, whereas the addition of 18/20 kg NP/ha as urea and diammonium phosphate resulted in 150% increase in dry matter accumulation in plants and 144% increase in nodule number per plant relative to non-inoculated control. There was also a marked increase in nodule dry weight which increased substantially by 200% due to inoculation with *S. ciceri* and 18/20 kg NP/ha indicating the importance of P in nodule tissue development. *Pseudomonas* sp. in the presence of 18/20 kg NP/ha also increased nodule dry weight, nodule number, nodule volume, and seed yield by 240%, 189%, 152%, and 143%, respectively, compared to control, highlighting the potential of this bacterium in solubilizing P. In contrast, the consortium of *S. cicero* and *Pseudomonas* sp. when applied with 18/20 kg NP/ha markedly enhanced nodule number per plant by 209%, nodule dry weight by 220%, nodule volume by 221%, and dry matter by 172% over non-inoculated control at mid-flowering stage of chickpea. Moreover, the consortium inoculation increased nodule number, nodule dry weight, nodule volume, and dry matter by 272%, 220%, 242%, and 181%, respectively, relative to control plants at mid-flowering stage. In a similar field experiment, the effect of *Rhizobium* and PSB inoculants on symbiotic traits, nodule leghemoglobin, and yield of five genotypes of chickpea was found variable (Tagore et al. 2013). Among microbial inoculants, *Rhizobium* together with PSB was found most effective and substantially enhanced nodule number (27.66 nodules/plant), nodule fresh weight (144.90 mg/plant), nodule dry weight (74.30 mg/plant), shoot dry weight (11.76 g/plant), and leghemoglobin content (2.29 mg/g of fresh nodule) and also showed its positive effect in enhancing all the yield attributing parameters, grain and straw yields.

In a follow-up study, two field experiments were conducted by Diep et al. (2016) to evaluate the impacts of rhizobia and PSB on soybean cultivated on ferralsols. The first experiment consisted of five treatments: (1) control (without fertilizer and without inoculant), (2) 400 kg NPK/ha, (3) rhizobial inoculant + 20 kg N/ha applied at 10 days after sowing (DAS), (4) PSB inoculant + 20 kg N/ha at 10 DAS, and (5) rhizobial and PSB inoculant + 20 kg kg N/ha at 10 DAS. The second experiment consisted of four treatments: (1) control (without fertilizer and without inoculant), (2) 100 kg/ha of thermophosphate (15% P_2O_5) + 25 kg NPK/ha applied at 20 and 40 at DAS, (3) rhizobial and PSB inoculant + 200 kg biofertilizer + 20 kg kg N/ha at 10 DAS and rhizobial inoculant, and (4) PSB inoculant + 400 kg biofertilizer + 20 kg kg N/ha at 10 DAS. Importantly, the biofertilizer formulation consisted of organic matter (35%), thermophosphate (5%), dolomite (0.5% P_2O_5, 50% $CaCO_3$, 10% $MgCl_2$) (45%), ground black rice-hull ash (15%), and PSB inoculant. The results revealed that application of rhizobial inoculant and/ or PSB inoculant significantly increased yield component and grain yield than control but did not differ from 400 kg/ha NPK in the first experiment. The biofertilizer application, in general, produced higher soybean grain yield and oil, and protein in seed than control, and was equivalent to the treatment of 100 kg/ha of thermophosphate (15% P_2O_5) + 25 kg NPK/ha in the second experiment. It was concluded from this study that biofertilizers can be considered as a replacement for part of chemical fertilizers used in soybean cultivated in ferralsols.

7.3.3 Biological Control of Plant Pathogens

PGPR adopt different mechanisms to control plant pathogens. The most common mechanisms by which PGPR control phytopathogens include (1) production of siderophores (Lukkani and Reddy 2014; Susilowati and Syekhfani 2014), (2) release of antibiotic substances (Chen et al. 2012) and cyanogenic compounds, (3) secretion of hydrolytic enzymes (Saravanakumar et al. 2007), and (4) the induction of systemic resistance in plant (Sangeetha et al. 2010). The mechanisms/ substances involved in diseases suppression may act independently or simultaneously.

Among PGPR, rhizobia are considered the most effective and promising biocontrol agent, and there are many reports where rhizobia have been used to manage plant diseases. For example, rhizobia belonging to genera *Rhizobium*, *Bradyrhizobium*, and *Sinorhizobium* displayed variable growth inhibitory activities against fungal isolates, but *R. phaseoli* 6-3 among rhizobia significantly reduced the dried weights of mycelium of all the fungal isolate (Chao 1990). In a similar study, some rhizobial strains dissolved the fungal mycelium at the initial stage and used the lyzed contents as nutrients for their growth and development (Hossain and Mårtensson 2008). Similarly, of the 42 rhizobial stains used against *Rhizoctonia solani*, 24 bacterial isolates could effectively inhibited the growth of fungi in vitro, while two rhizobial isolates did reduce the disease more than 80% under glasshouse conditions (Hemissi et al. 2011). In addition, ten rhizobial strains solubilized insoluble P, and 13 strains produced volatile compounds. The production of bacterial organic volatile compounds (VOCs) has been shown to be involved in biocontrol traits of PGPR (Bhagat et al. 2014). The synthesis and release of VOCs, measured through GC-MS, have been reported from *Pseudomonas* sp. strains and from some unknown bacterial stains isolated from canola and soybean plants. These inhibitory compounds were identified as benzothiazole, cyclohexanol, n-decanal, dimethyl tri-sulfide, 2-ethyl 1-hexanol, and nonanal (Fernando et al. 2005) which were able to completely inhibit mycelial growth or sclerotia formation of the soilborne phyto-pathogenic fungi *Sclerotinia sclerotiorum*.

7.3.3.1 Siderophores

One of the extensively investigated and well-established mechanisms of pathogens suppression by rhizobia and PGPR in general is the production of siderophores (Panhwar et al. 2014): a low molecular weight (400–1000 Da) iron-chelating com-pound (Gray and Smith 2005; Solano et al. 2008). The production of siderophores by PGPR such as *Azotobacter* (Muthuselvan and Balagurunathan 2013; Prasad et al. 2014) and rhizobia, for example, *Rhizobium* BICC 651 (Datta and Chakrabartty 2014) and *Mesorhizobium* spp. (Bhagat et al. 2014), can promote plant growth either by directly improving iron acquisition by plants or by inhibiting growth of pathogens in rhizosphere by limiting the availability of iron to the pathogens (Solano et al. 2008). Another possible positive effect of siderophore production is the com-plexation of toxic aluminum. In a study, Roy and Chakrabartty (2000) evaluated the production of siderophores by a *Rhizobium* sp. influenced by the concentration of Al_3^+. Besides increasing iron availability, rhizobial siderophores also reduced Al_3^+ toxicity to the bacterium, once the metal was complexed by the organic molecule. Similarly, Rogers et al. (2001) proved the effectiveness of the hydroxamate sidero-phore vicibactin produced by *R. leguminosarum* bv. *viciae* in alleviating aluminum toxicity. The complex siderophore aluminum may eventually be taken up into the bacterial cytoplasm. However, it is unlikely to become toxic intracellularly because aluminum cannot be released from the complex by reduction, and the complex therefore simply accumulates as a nontoxic species or even, if it is released, Al_3^+ will precipitate as $Al(OH)_3$ at the slightly alkaline cytoplasmic pH (O'hara et al. 1989; Rogers et al. 2001).

There are many reports of effective suppression of plant pathogens by siderophore-producing rhizobia (Ahemad and Khan 2010; Chandra et al. 2007). Later on, Deshwal et al. (2003) evaluated ten strains of peanut (*Arachis hypogaea*) nodulating *Bradyrhizobium* and found three strains (AHR-2, AHR-5, and AHR-6) able to produce siderophores, besides synthesizing IAA and solubilizing P in vitro. These strains also efficiently fixed N and showed antagonistic action against *M. phaseolina*, the causal agent of charcoal rot of peanut. Consequently, the inocula-tion with selected rhizobia may not only provide N to the host legume plants but also promote plant sanity by controlling pathogens. All the three bradyrhizobial strains inhibited radial growth of the fungus in vitro and declined its population in rhizosphere soil of bradyrhizobia inoculated peanut. However, the relation between the production of siderophores and the suppression of plant pathogens is not clear. For example, Vargas et al. (2009) tested ten isolates of *R. trifolii* for antagonism against a phytopathogenic fungus *Verticillium* sp. All rhizobial isolates showed some level of antagonism against the test fungus (Fig. 7.2). The greatest level of inhibition was achieved by the isolates CXS-12, AGR-3, ELD-15, VAC-12, and DPE-12. Two isolates (CXS-12 and AGR-3) with greatest antagonistic activity that

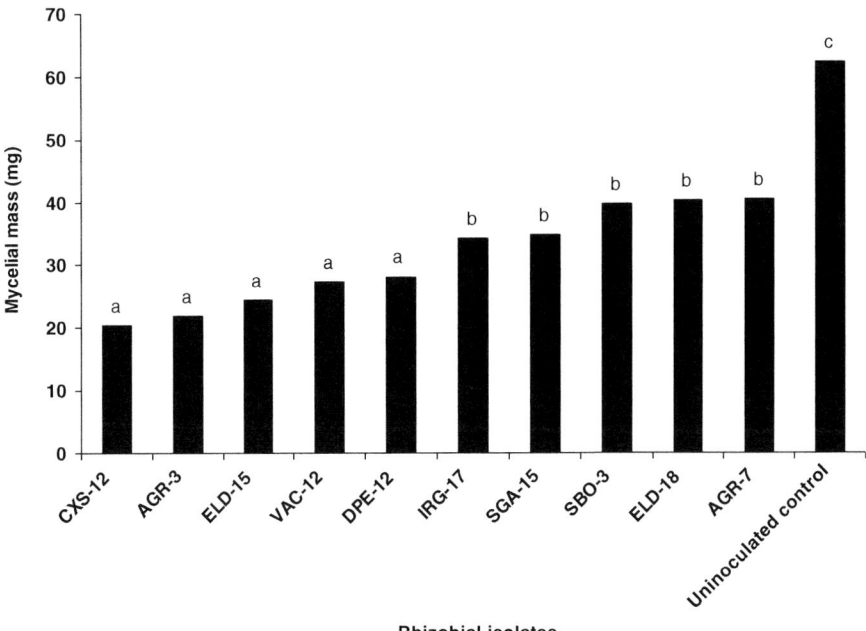

Fig. 7.2 Inhibition of *Verticillium* sp. mycelial mass production by rhizobial isolates. Means followed by the *same letter* did not differ significantly at $P \leq 0.05$ (Scott-Knot test) (adapted from Vargas et al. 2009)

decreased mycelial growth (about 65%) were also siderophore producers. However, two other siderophore-producing isolates (IRG-17 and SBO-3) displayed less pronounced antagonistic effect compared to other non-siderophore producers. The variation in antagonisms could probably be due to the differences in the type of siderophores produced by each isolate. Accordingly, Matthijs et al. (2007) observed that *Pseudomonas fluorescens* ATCC 17400 produced two siderophores, pyoverdine and thioquinolobactin, with thioquinolobactin showing a much more intense antifungal activity than pyoverdine.

A very similar result was obtained by Omar and Abd-Alla (1998) while evaluating the antagonism of 20 isolates of *Bradyrhizobium* and *Rhizobium* against the phytopathogenic fungi *Fusarium solani*, *M. phaseolina*, and *R. solani*. All the isolates possessed antagonistic activity against fungi in both iron-deficient and iron-rich media. In such a case, ability to produce siderophores seems to act more as a competitive advantage, which allows the PGPR to colonize the rhizosphere more efficiently, than as a direct suppressive mechanism against pathogen by iron deprivation.

7.3.3.2 Antibiotics, Cyanogenic Compounds, and Hydrolytic Enzymes

When iron is not limiting, other mechanisms are more important in plant pathogen suppression than siderophore production. The secretion of metabolites such as antibiotics (Keel et al. 1992; Robleto et al. 1998) and hydrocyanic acid (Ahemad and Khan 2009; Prasad et al. 2014; Ruangsanka 2014; Sahasrabudhe 2011;) and the production of lytic enzymes (Nabti et al. 2014) such as β-1,3-glucanase, proteases (Compant et al. 2005), and chitinases (Kacem et al. 2009) are important mechanisms of biological control of plant pathogenic microorganisms adopted by PGPR. Many rhizobia are able to produce antibiotics, especially bacteriocins. For example, while evaluating 32 rhizobial isolates from nodules of horse gram plants (*Macrotyloma uniflorum*), Prabhavati and Anthony (2012) reported that all the isolates produced bacteriocins against the remaining isolates. Bacteriocins are antibiotic-like proteinaceous toxins produced by bacteria, which differ from traditional antibiotics in their relatively narrow killing spectrum. Historically, bacteriocins have been believed to inhibit the growth of similar or closely related bacterial strains, conferring competitive advantage to bacteriocin-producer strains (Hafeez et al. 2005; Yanni et al. 2001). However, the activity spectrum of a bacteriocin depends on the environmental conditions and may not be restricted to species taxonomically close. For example, Warda et al. (2014) observed that bacteriocin-like substances produced by rhizobial strains are also able to inhibit *Clostridium* sp., *Vibrio* sp., and *Enterobacter* sp. Robleto et al. (1998) described the effects of trifolitoxin, a narrow-spectrum peptide antibiotic produced by a *R. etli*, on the microbial composition on the rhizosphere of common bean (*Phaseolus vulgaris*). They observed a significant reduction in the genetic diversity of alphaproteobacteria, with little visible effect on most microbes. Though bacteriocins are a narrow-spectrum antimicrobial compound, yet it is an effective metabolite that inhibits bacterial plant pathogens (Cladera-Olivera et al. 2006). *Rhizobium* sp. strains ORN 24 and ORN 83, isolated from Algerian soil, were found to produce bacteriocins with antimicrobial activities against *Pseudomonas savastanoi*, the agent responsible for olive knot disease.

7.3.3.3 Triggering of Induced Systemic Resistance (ISR)

Rhizobia are also able to elicit reactions of plant defense against pathogens, as demonstrated by Elbadry et al. (2006). The authors verified the occurrence of induced systemic resistance (ISR) against bean yellow mosaic potyvirus (BYMV) in faba bean (*Vicia faba*) inoculated with *P. fluorescens* FB11 and *R. leguminosarum* bv. *viceae* FBG05. Plants inoculated showed a pronounced and significant reduction in percent disease incidence and a significant reduction in virus concentration. Since the PGPR inoculants and the pathogen remained spatially separated, it could be concluded that the tested *Pseudomonas* or *Rhizobium* strains induced systemic resistance in faba bean against BYMV. The activation of ISR by PGPR can be optimized when more than one microorganism are used as elicitors as reported by Dutta et al. (2008) who evaluated the occurrence of the process in pigeon pea (*Cajanus cajan*). The authors exposed, separately, part of plant root systems to the pathogenic fungus *Fusarium udum* and part to PGPR *B. cereus* or *P. aeruginosa* and after that evaluated the interaction of these PGPR with *Rhizobium* sp. It was evidenced by an enhancement of resistance in treated plants, mainly when PGPR strains were associated to *Rhizobium*. Plants with mixture of PGPR and *Rhizobium* survived longer and showed higher level of defense-related enzymes than individual organism and non-bacterized control.

Conclusion

Rhizobia among many PGPR are the most extensively and practically investigated organisms due largely to their ability to form effective symbiosis with leguminous plants. Apart from their beneficial impact on pulses, rhizobia in recent times have also been found extremely attractive and useful in enhancing the quantity and quality of nonlegume crops. Since rhizobia have been found to possess virtually all plant growth-promoting traits similar to those expressed by exclusively nonsymbiotic PGPR, rhizobia, being a nonpathogenic organism, can safely be used in agricultural practices for increasing the production of both legumes and nonlegume crops while reducing the dependence on chemicals used in intensive agricultural practices. However, despite considerable research and success achieved so far, further more research is required to reveal some unidentified characteristics of rhizobia that could be practically valuable in achieving maximum benefits of such a naturally versatile organism.

References

Abd-Alla MH (1994a) Use of organic phosphorus by *Rhizobium leguminosarum* biovar. viceae phosphatases. Biol Fertil Soils 18:216–218

Abd-Alla MH (1994b) Phosphatases and the utilization of organic phosphorus by *Rhizobium leguminosarum* biovar *viceae*. Lett Appl Microbiol 18:294–296

Adesemoye AO, Ugoji EO (2009) Evaluating *Pseudomonas aeruginosa* as plant growth-promoting rhizobacteria in West Africa. Arch Phytopathol Plant Prot 42:188–200

Ahemad M, Khan MS (2009) Effect of insecticide-tolerant and plant growth-promoting *Mesorhizobium* on the performance of chickpea grown in insecticide stressed alluvial soils. J Crop Sci Biotech 12:213–222

Ahemad M, Khan MS (2010) Comparative toxicity of selected insecticides to pea plants and growth promotion. Crop Protect. doi:10.1016/j.cropro.2010.01.005

Ahmad E, Khan MS, Zaidi A (2013) ACC deaminase producing *Pseudomonas putida* strain PSE3 and *Rhizobium leguminosarum* strain RP2 in synergism improves growth, nodulation and yield of pea grown in alluvial soils. Symbiosis 61:93–104

Alikhani H, Saleh-Rastin N, Antoun H (2007) Phosphate solubilization activity of rhizobia native to Iranian soils. In: Velázquez E, Rodríguez-Barrueco C (eds) First international meeting on microbial phosphate solubilization. Springer, Dordrecht, pp 35–41

Antoun H, Beauchamp CJ, Goussard N, Chabot R, Lalande R (1998) Potential of *Rhizobium* and *Bradyrhizobium* species as plant growth promoting rhizobacteria on non-legumes: effect on radishes (*Raphanus sativus* L.) Plant Soil 204:57–67

Bajguz A, Tretyn A (2003) The chemical characteristic and distribution of brassinosteroids in plants. Phytochemistry 62:1027–1046

Barrett CF, Parker MA (2006) Coexistence of *Burkholderia*, *Cupriavidus*, and *Rhizobium* sp. nodule bacteria on two *Mimosa* spp. in Costa Rica. Appl Environ Microbiol 72:1198–1206

Batista JSS, Hungria M, Barcellos FG, Ferreira MC, Mendes IC (2007) Variability in *Bradyrhizobium japonicum* and *B. elkanii* seven years after introduction of both the exotic microsymbiont and the soybean host in a Cerrados soil. Microb Ecol 53:270–284

Bhagat D, Sharma P, Sirari A, Kumawat KC (2014) Screening of *Mesorhizobium* spp. for control of *Fusarium* wilt in chickpea *in vitro* conditions. Int J Curr Microbiol Appl Sci 3:923–930

Bhattacharjee RB, Jourand P, Chaintreuil C, Dreyfus B, Singh A, Mukhopadhyay SN (2012) Indole acetic acid and ACC deaminase-producing *Rhizobium leguminosarum* bv. *trifolii* SN10 promote rice growth, and in the process undergo colonization and chemotaxis. Bio Fert Soils 48:173–182

Biswas J, Ladha J, Dazzo F (2000) Rhizobia inoculation improves nutrient uptake and growth of lowland rice. Soil Sci Soc Am J 64:1644–1650

Boiero L, Perrig D, Masciarelli O, Penna C, Cassán F, Luna V (2007) Phytohormone production by three strains of *Bradyrhizobium japonicum* and possible physiological and technological implications. Appl Microbiol Biotechnol 74:874–880

Breakspear A, Liu C, Roy S, Stacey N, Rogers C, Trick M, Morieri G, Mysore KS, Wen J, Oldroyd GE (2014) The root hair "infectome" of *Medicago truncatula* uncovers changes in cell cycle genes and reveals a requirement for auxin signaling in rhizobial infection. Plant Cell 26:4680–4701

Cattelan A, Hartel P, Fuhrmann J (1999) Screening for plant growth–promoting rhizobacteria to promote early soybean growth. Soil Sci Soc Am J 63:1670–1680

Chabot R, Antoun H, Kloepper JW, Beauchamp CJ (1996) Root colonization of maize and lettuce by bioluminescent *Rhizobium leguminosarum* biovar *phaseoli*. Appl Environ Microbiol 62:2767–2772

Chacon N, Silver WL, Dubinsky EA, Cusack DF (2006) Iron reduction and soil phosphorus solubilization in humid tropical forests soils: the roles of labile carbon pools and an electron shuttle compound. Biogeochemistry 78:67–84

Chaintreuil C, Giraud E, Prin Y, Lorquin J, Bâ A, Gillis M, de Lajudie P, Dreyfus B (2000) Photosynthetic bradyrhizobia are natural endophytes of the African wild rice *Oryza breviligulata*. Appl Environ Microbiol 66:5437–5447

Chakraborty U, Purkayastha R (1984) Role of rhizobitoxine in protecting soybean roots from *Macrophomina phaseolina* infection. Can J Microbiol 30:285–289

Chandra S, Choure K, Dubey RC, Maheshwari DK (2007) Rhizosphere competent *Mesorhizobiumloti* MP6 induces root hair curling, inhibits *Sclerotinia sclerotiorum* and enhances growth of Indian mustard (*Brassica campestris*). Braz J Microbiol 38:124–130

Chao WL (1990) Antagonistic activity of *Rhizobium* spp. against beneficial and plant pathogenic fungi. Lett Appl Microbiol 10:213–215

Chen W-M, De Faria SM, James EK, Elliott GN, Lin K-Y, Chou J-H, Sheu S-Y, Cnockaert M, Sprent JI, Vandamme P (2007) *Burkholderia nodosa* sp. nov., isolated from root nodules of the woody Brazilian legumes *Mimosa bimucronata* and *Mimosa scabrella*. Int J Syst Evol Microbiol 57:1055–1059

Chen N, Jin M, Qu HM, Chen ZQ, Chen ZL, Qiu ZG, Wang XW, Li JW (2012) Isolation and characterization of *Bacillus* sp. producing broad-spectrum antibiotics against human and plant pathogenic fungi. J Microbiol Biotechnol 22:256–563

Chi F, Shen S-H, Cheng H-P, Jing Y-X, Yanni YG, Dazzo FB (2005) Ascending migration of endophytic rhizobia, from roots to leaves, inside rice plants and assessment of benefits to rice growth physiology. Appl Environ Microbiol 71:7271–7278

Cladera-Olivera F, Caron GR, Motta AS, Souto AA, Brandelli A (2006) Bacteriocin-like substance inhibits potato soft rot caused by *Erwinia carotovora*. Can J Microbiol 52:533–539

Compant S, Duffy B, Nowak J, Clément C, Barka EA (2005) Use of plant growth-promoting bacteria for biocontrol of plant diseases: principles, mechanisms of action, and future prospects. Appl Environ Microbiol 71:4951–4959

Contesto C, Desbrosses G, Lefoulon C, Béna G, Borel F, Galland M, Gamet L, Varoquaux F, Touraine B (2008) Effects of rhizobacterial ACC deaminase activity on *Arabidopsis* indicate that ethylene mediates local root responses to plant growth-promoting rhizobacteria. Plant Sci 175:178–189

Datta B, Chakrabartty PK (2014) Siderophore biosynthesis genes of *Rhizobium* sp. isolated from *Cicer arietinum* L. 3 Biotech 4:391–401

Dazzo FB, Yanni YG (2006) The natural *Rhizobium*-cereal crop association as an example of plant-bacterial interaction. In: Uphoff N, Ball AS, Fernandes E, Herren H, Husson O, Laing M, Palm C, Pretty J, Sanchez P, Sanginga N, Thies J (eds) Biological approaches to sustainable soil systems. CRC Press, Boca Raton, pp 109–127

Deshwal V, Dubey R, Maheshwari D (2003) Isolation of plant growth-promoting strains of *Bradyrhizobium* (*Arachis*) sp. with biocontrol potential against *Macrophomina phaseolina* causing charcoal rot of peanut. Curr Sci 84:443–448

Diep CN, So DB, Trung NB, Lam PVH (2016) Effects of rhizobia and phosphate-solubilizing bacteria on soybean (*Glycine max* L. Merr.) cultivated on ferralsols of Daklak Province, Vietnam. Int J Pharm Pharm Sci 5:318–333

Dobbelaere S, Vanderleyden J, Okon Y (2003) Plant growth-promoting effects of diazotrophs in the rhizosphere. Crit Rev Plant Sci 22:107–149

Dobert RC, Rood SB, Blevins DG (1992) Gibberellins and the legume-*Rhizobium* symbiosis: I. Endogenous gibberellins of Lima Bean (*Phaseolus lunatus* L.) stems and nodules. Plant Physiol 98:221–224

Domit L, Costa J, Vidor C, Pereira J (1990) Inoculation of cereal seeds with *Bradyrhizobium japonicum* and its effect on soyabeans grown in succession. R Bras Ci Solo 14:313–319

Duodu S, Bhuvaneswari T, Stokkermans TJ, Peters NK (1999) A positive role for rhizobitoxine in *Rhizobium*-legume symbiosis. Mol Plant Microbe Interact 12:1082–1089

Dutta S, Mishra A, Kumar BD (2008) Induction of systemic resistance against fusarial wilt in pigeon pea through interaction of plant growth promoting rhizobacteria and rhizobia. Soil Biol Biochem 40:452–461

Elbadry M, Taha R, Eldougdoug K, Gamal-Eldin H (2006) Induction of systemic resistance in faba bean (*Vicia faba* L.) to bean yellow mosaic potyvirus (BYMV) via seed bacterization with plant growth promoting rhizobacteria. J Plant Dis Protect 113:247–251

Ferguson L, Lessenger JE (2006) Plant growth regulators. In: Lessenger JE (ed) Agricultural medicine. Springer, New York, pp 156–166

Ferguson BJ, Ross JJ, Reid JB (2005) Nodulation phenotypes of gibberellin and brassinosteroid mutants of pea. Plant Physiol 138:2396–2405

Fernando WD, Ramarathnam R, Krishnamoorthy AS, Savchuk SC (2005) Identification and use of potential bacterial organic antifungal volatiles in biocontrol. Soil Biol Biochem 37:955–964

Flores-Félix JD, Menéndez E, Rivera LP, Marcos-García M, Martínez-Hidalgo P, Mateos PF, Martínez-Molina E, Velázquez ME, García-Fraile P, Rivas R (2013) Use of *Rhizobium leguminosarum* as a potential biofertilizer for *Lactuca sativa* and *Daucus carota* crops. J Plant Nutr Soil Sci 176:876–882

Foster RC (1998) Microenvironments of soil microorganisms. Bio Fert Soils 6:189–203

Frugier F, Kosuta S, Murray JD, Crespi M, Szczyglowski K (2008) Cytokinin: secret agent of symbiosis. Trends Plant Sci 13:115–120

Gandhi PM, Narayanan K, Naik P, Sakthivel N (2009) Characterization of *Chryseobacterium aquaticum* strain PUPC1 producing a novel antifungal protease from rice rhizosphere soil. J Microbiol Biotechnol 19:99–107

García-Fraile P, Carro L, Robledo M, Ramírez-Bahena MH, Flores-Félix JD, Fernández MT, Mateos PF, Rivas R, Igual JM, Martínez-Molina E, Peix A, Velázquez E (2012) *Rhizobium* promotes non-legumes growth and quality in several production steps: towards a biofertilization of edible raw vegetables healthy for humans. PLoS One 7:38122

Ghosh PK, Kumar De T, Maiti TK (2015) Production and metabolism of indole acetic acid in root nodules and symbiont (*Rhizobium undicola*) isolated from root nodule of aquatic medicinal legume *Neptunia oleracea* Lour. J Bot 2015. Article ID 575067

Glick BR (1995) The enhancement of plant growth by free-living bacteria. Can J Microbiol 41:109–117

Glick BR (2005) Modulation of plant ethylene levels by the bacterial enzyme ACC deaminase. FEMS Microbiol Lett 251:1–7

Glick BR, Liu C, Ghosh S, Dumbroff EB (1997) Early development of canola seedlings in the presence of the plant growth-promoting rhizobacterium *Pseudomonas putida* GR12-2. Soil Biol Biochem 29:1233–1239

Glick BR, Penrose DM, Li J (1998) A model for the lowering of plant ethylene concentrations by plant growth-promoting bacteria. J Theor Biol 190:63–68

Gray E, Smith D (2005) Intracellular and extracellular PGPR: commonalities and distinctions in the plant–bacterium signaling processes. Soil Biol Biochem 37:395–412

Gutierrez-Zamora M, Martınez-Romero E (2001) Natural endophytic association between *Rhizobium etli* and maize (*Zea mays* L.) J Biotechnol 91:117–126

Hafeez F, Safdar M, Chaudhry A, Malik K (2004) Rhizobial inoculation improves seedling emergence, nutrient uptake and growth of cotton. Anim Prod Sci 44:617–622

Hafeez FY, Naeem FI, Naeem R, Zaidi AH, Malik KA (2005) Symbiotic effectiveness and bacteriocin production by *Rhizobium leguminosarum* bv. *viciae* isolated from agriculture soils in Faisalabad. Environ Exp Bot 54:142–147

Hayashi S, Gresshoff PM, Ferguson BJ (2014) Mechanistic action of gibberellins in legume nodulation. J Integr Plant Biol 56:971–978

Hemissi I, Mabrouk Y, Abdi N, Bouraoui M, Saidi M, Sifi B (2011) Effects of some Rhizobium strains on chickpea growth and biological control of Rhizoctonia solani. Afr J Microbiol Res 5:4080–4090

Hossain MS, Mårtensson A (2008) Potential use of *Rhizobium* spp. to improve fitness of non-nitrogen-fixing plants. Acta Agric Scand 58:352–358

Hungria M, Campo R (2005) Fixação biológica do nitrogênio em sistemas agrícolas. In: Congresso brasileiro de ciência do solo. SBCS, UFPE Embrapa Solos Pernambuco Rio de Janeiro

Imen H, Neila A, Adnane B, Manel B, Mabrouk Y, Saidi M, Bouaziz S (2015) Inoculation with phosphate solubilizing *Mesorhizobium* strains improves the performance of chickpea (*Cicer aritenium* L.) under phosphorus deficiency. J Plant Nutr 38:1656–1671

Kacem M, Kazouz F, Merabet C, Rezki M, de Lajudie P, Bekki A (2009) Antimicrobial activity of *Rhizobium* sp. strains against *Pseudomonas savastanoi*, the agent responsible for the olive knot disease in Algeria. Grasas Aceites 60:139–146

Keel C, Schnider U, Maurhofer M, Voisard C, Laville J, Burger U, Wirthner P, Haas D, Defago G (1992) Suppression of root diseases by *Pseudomonas fluorescens* CHAO: importance of bacterial secondary metabolite, 2,4-diacetylphoroglucinol. Mol Plant Microbe Interact 5:4–13

Khaitov B, Kurbonov A, Abdiev A, Adilov M (2016) Effect of chickpea in association with *Rhizobium* to crop productivity and soil fertility. Eurasian J Soil Sci 5:105–112

Khan M, Zaidi A, Wani P (2009) Role of phosphate solubilizing microorganisms in sustainable agriculture – a review. In: Lichtfouse E, Navarrete M, Debaeke P, Véronique S, Alberola C (eds) Sustainable agriculture. Springer, Netherlands, pp 551–570. doi:10.1007/978-90-481-2666-8_34

Khan MS, Zaidi A, Ahemad M, Oves M, Wani PA (2010) Plant growth promotion by phosphate solubilizing fungi–current perspective. Arch Agron Soil Sci 56:73–98

Kloepper JW (1978) Schroth MN Plant growth-promoting rhizobacteria on radishes. In: Proceedings of the fourth international conference on plant pathogenic bacteria, pp 879–882

Kloepper JA (2003) Review of mechanisms for plant growth promotion by PGPR. In: Sixth international PGPR workshop, pp 5–10

Kloepper JW, Beauchamp CJ (1992) A review of issues related to measuring colonization of plant roots by bacteria. Can J Microbiol 38:1219–1232

Kloepper JW, Leong J, Teintze M, Schroth MN (1980) Enhanced plant growth by siderophores produced by plant growth-promoting rhizobacteria. Nature 286:885–886

Kumar G, Raghu Ram M (2014) Phosphate solubilizing rhizobia isolated from *Vigna trilobata*. Am J Microbiol Res 2:105–109

LA Favre JS, Eaglesham ARJ (1986) Rhizobitoxine: a phytotoxin of unknown function which is commonly produced by bradyrhizobia. Plant Soil 92:443–452

Lievens S, Goormachtig S, Den Herder J, Capoen W, Mathis R, Hedden P, Holsters M (2005) Gibberellins are involved in nodulation of *Sesbania rostrata*. Plant Physiol 139:1366–1379

Lindström K, Martinez-Romero M (2005) International Committee on Systematics of Prokaryotes; Subcommittee on the taxonomy of *Agrobacterium* and *Rhizobium* Minutes of the meeting, 26 July 2004, Toulouse, France. Int J Syst Evol Microbiol 55:1383–1383

Loper J, Schroth M (1986) Influence of bacteria sources of indol-3-acetic acid on root elongation of sugar beet. Phytopathol 76:386–389

López-López A, Rogel MA, Ormeno-Orrillo E, Martínez-Romero J, Martínez-Romero E (2010) *Phaseolus vulgaris* seed-borne endophytic community with novel bacterial species such as *Rhizobium endophyticum* sp. nov. Syst Appl Microbiol 33:322–327

Lukkani NJ, Reddy ECS (2014) Evaluation of plant growth promoting attributes and biocontrol potential of native *fluorescent pseudomonas* spp. against *Aspergillus niger* causing collar rot of ground nut. Int J Plant Anim Environ Sci 4:267–262

Ma W, Guinel FC, Glick BR (2003) *Rhizobium leguminosarum* biovar *viciae* 1-aminocyclopropane-1-carboxylate deaminase promotes nodulation of pea plants. Appl Environ Microbiol 69:4396–4402

Machado RG, Sá ELS, Bruxel M, Giongo A, Santos NS, Nunes AS (2013) Indoleacetic acid producing rhizobia promote growth of Tanzania grass (*Panicum maximum*) and Pensacola grass (*Paspalum saurae*). Int J Agric Biol 15:827–834

Matthijs S, Tehrani KA, Laus G, Jackson RW, Cooper RM, Cornelis P (2007) Thioquinolobactin, a *Pseudomonas* siderophore with antifungal and anti-*Pythium* activity. Environ Microbiol 9:425–434

Messele B, Pant LM (2012) Effects of inoculation of *Sinorhizobium ciceri* and phosphate solubilizing bacteria on nodulation, yield and nitrogen and phosphorus uptake of chickpea (*Cicer arietinum* L.) in Shoa Robit Area. J Biofertil Biopestici 3:129

Minamisawa K (1990) Division of rhizobitoxine-producing and hydrogen-uptake positive strains of *Bradyrhizobium japonicum* by *nifDKE* sequence divergence. Plant Cell Physiol 31:81–89

Miransari M, Smith D (2009) Rhizobial lipo-chitooligosaccharides and gibberellins enhance barley (*Hordeum vulgare* L.) seed germination. Biotechnology 8:270–275

Mishra RP, Singh RK, Jaiswal HK, Kumar V, Maurya S (2006) *Rhizobium*-mediated induction of phenolics and plant growth promotion in rice (*Oryza sativa* L.) Curr Microbiol 52:383–389

Muthuselvan I, Balagurunathan R (2013) Siderophore production from *Azotobacter* sp. and its application as biocontrol agent. Int J Curr Res Rev 5:23–35

Nabti E, Bensidhoum L, Tabli N, Dahel D, Weiss A, Rothballer M, Schmid M, Hartman A (2014) Growth inhibition of barley and biocontrol effect on plant pathogenic fungi by a *Cellulosimicrobium* isolated from salt affected rhizosphere soil in northwestern Algeria. Eur J Soil Biol 61:20–26

Nandi A, Sengupta B, Sen S (1982) Utility of *Rhizobium* in the phyllosphere of crop plants in nitrogen-free sand culture. J Agric Sci 98:167–171

O'hara GW, Goss TJ, Dilworth MJ, Glenn AR (1989) Maintenance of intracellular pH and acid tolerance in *Rhizobium meliloti*. Appl Environ Microbiol 55:1870–1876

Okazaki S, Sugawara M, Yuhashi K-I, Minamisawa K (2007) Rhizobitoxine-induced chlorosis occurs in coincidence with methionine deficiency in soybeans. Ann Bot 100:55–59

Omar S, Abd-Alla M (1998) Biocontrol of fungal root rot diseases of crop plants by the use of *Rhizobia* and *Bradyrhizobia*. Folia Microbiol 43:431–437

Panhwar QA, Naher UA, Jusop S, Othman R, Latif MA, Ismail MR (2014) Biochemical and molecular characterization of potential phosphate solubilizing bacteria in acid sulphate soils and their beneficial effects on rice growth. PLoS One 9:e97241

Peix A, Rivas-Boyero A, Mateos P, Rodriguez-Barrueco C, Martınez-Molina E, Velazquez E (2001) Growth promotion of chickpea and barley by a phosphate solubilizing strain of *Mesorhizobium mediterraneum* under growth chamber conditions. Soil Biol Biochem 33:103–110

Prabhavati E, Anthony J (2012) Bacteriocin production by rhizobia isolated from root nodules of Horse gram. Bangladesh J Med Sci 11:28–32

Prasad JS, Reddy RS, Reddy PN, Rajashekar AU (2014) Isolation, screening and characterization of *Azotobacter* from rhizospheric soils for different plant growth promotion (PGP) & antagonistic activities and compatibility with agrochemicals: an *in vitro* study. Ecol Environ Conserv 20:959–966

Prashar P, Kapoor N, Sachdeva S (2014) Rhizosphere: its structure, bacterial diversity and significance. Rev Environ Sci Biol 13:63–77

Prayitno J, Stefaniak J, McIver J, Weinman J, Dazzo F, Ladha J, Barraquio W, Yanni Y, Rolfe B (1999) Interactions of rice seedlings with bacteria isolated from rice roots. Funct Plant Biol 26:521–535

Probanza A, Lucas J, Acero N, Mañero FG (1996) The influence of native rhizobacteria on European alder (*Alnus glutinosa* (L.) Gaertn.) growth. Plant Soil 182:59–66

Probanza A, Mateos J, García JL, Ramos B, De Felipe M, Mañero FG (2001) Effects of inoculation with PGPR *Bacillus* and *Pisolithus tinctorius* on *Pinus pinea* L. growth, bacterial rhizosphere colonization, and mycorrhizal infection. Microb Ecol 41:140–148

Rao SSR, Vardhini BV, Sujatha E, Anuradha S (2002) Brassinosteroids – a new class of phytohormones. Curr Sci 82:1239–1245

Recep K, Fikrettin S, Erkol D, Cafer E (2009) Biological control of the potato dry rot caused by *Fusarium* species using PGPR strains. Biol Control 50:194–198

Reimann S, Hauschild R, Hildebrandt U, Sikora RA (2008) Interrelationships between *Rhizobium etli* G12 and *Glomus intraradices* and multitrophic effects in the biological control of the root-knot nematode *Meloidogyne incognita* on tomato. J Plant Dis Protect 115:108–113

Richardson A, Hadobas P, Simpson R (2001) Phytate as a source of phosphorus for the growth of transgenic *Trifolium subterraneum*. In: Horst WJ, Schenk MK, Bürkert A, Claassen N, Flessa H, Frommer WB, Goldbach H, Olfs HW, Römheld V, Sattelmacher B, Schmidhalter U, Schubert S, Wirén NV, Wittenmayer L (eds) Plant nutrition. Springer, Netherlands, pp 560–561

Robleto EA, Borneman J, Triplett EW (1998) Effects of bacterial antibiotic production on rhizosphere microbial communities from a culture-independent perspective. Appl Environ Microbiol 64:5020–5022

Rodríguez H, Fraga R (1999) Phosphate solubilizing bacteria and their role in plant growth promotion. Biotechnol Adv 17:319–339

Rogers NJ, Carson KC, Glenn AR, Dilworth MJ, Hughes MN, Poole RK (2001) Alleviation of aluminum toxicity to *Rhizobium leguminosarum* bv. *viciae* by the hydroxamate siderophore vicibactin. Biometals 14:59–66

Roy N, Chakrabartty PK (2000) Effect of aluminum on the production of siderophore by *Rhizobium* sp. (*Cicer arietinum*). Curr Microbiol 41:5–10

Ruangsanka S (2014) Identification of phosphate-solubilizing fungi from the asparagus rhizosphere as antagonists of the root and crown rot pathogen *Fusarium oxysporum*. Science Asia 40:16–20

Sabry SR, Saleh SA, Batchelor CA, Jones J, Jotham J, Webster G, Kothari SL, Davey MR, Cocking EC (1997) Endophytic establishment of *Azorhizobium caulinodans* in wheat. Proc R Soc Lon B 264:341–346

Sahasrabudhe MM (2011) Screening of rhizobia for indole acetic acid production. Ann Biol Res 2:460–468

Sahgal M, Johri B (2003) The changing face of rhizobial systematics. Curr Sci 84:43–48

Sangeetha G, Thangavelu R, Usha Rani S, Muthukumar A, Udayakumar R (2010) Induction of systemic resistance by mixtures of antagonist bacteria for the management of crown rot complex on banana. Acta Physiol Plant 32:1177–1187

Saravanakumar D, Kumar CV, Kumar N, Samiyappan R (2007) PGPR-induced defense responses in the tea plant against blister blight disease. Crop Protect 26:556–565

Schlindwein G, Vargas LK, Lisboa BB, Azambuja AC, Granada CE, Gabiatti NC, Prates F, Stumpf R (2008) Influence of rhizobial inoculation on seedling vigor and germination of lettuce. Cienc Rural 38:658–664

Schloter M, Wiehe W, Assmus B, Steindl H, Becke H, Höflich G, Hartmann A (1997) Root colonization of different plants by plant-growth-promoting *Rhizobium leguminosarum* bv. *trifolii* R39 studied with monospecific polyclonal antisera. Appl Environ Microbiol 63:2038–2046

Sivasakthivelan P, Saranraj P (2013) *Azospirillum* and its formulations: a review. Int J Microbiol Res 4:275–287

Solano BR, Maicas JB, FJG M (2008) Physiological and molecular mechanisms of plant growth promoting rhizobacteria (PGPR). In: Ahmad I, Pichtel J, Hayat S (eds) Plant-bacteria interactions: strategies and techniques to promote plant growth. Wiley, Weinheim, Germany, pp 41–52

Sturtevant DB, Taller BJ (1989) Cytokinin Production by Bradyrhizobium japonicum. Plant Physiol 89:1247–1252

Susilowati LE, Syekhfani S (2014) Characterization of phosphate solubilizing bacteria isolated from Pb contaminated soils and their potential for dissolving tricalcium phosphate. J Degrad Mining Lands Manag 1:57–62

Suzuki A, Akune M, Kogiso M, Imagama Y, Osuki K-i, Uchiumi T, Higashi S, Han S-Y, Yoshida S, Asami T (2004) Control of nodule number by the phytohormone abscisic acid in the roots of two leguminous species. Plant Cell Physiol 45:914–922

Tagore GS, Namdeo SL, Sharma SK, Kumar N (2013) Effect of *Rhizobium* and phosphate solubilizing bacterial inoculants on symbiotic traits, nodule leghemoglobin, and yield of chickpea genotypes. Int J Agron 2013. Article ID 581627

Tominaga A, Nagata M, Futsuki K, Abe H, Uchiumi T, Abe M, Kucho K-i, Hashiguchi M, Akashi R, Hirsch A (2010) Effect of abscisic acid on symbiotic nitrogen fixation activity in the root nodules of *Lotus japonicus*. Plant Signal Behav 5:440–443

Vardhini BV, Ram Rao SS (1999) Effect of brassionosteriods on nodulation and nitrogenase activity in groundnut (*Arachis hypogaea* L.) Plant Growth Regul 28:165–167

Vargas LK, Lisboa BB, Schlindwein G, Granada CE, Giongo A, Beneduzi A, Passaglia LMP (2009) Occurrence of plant growth-promoting traits in clover-nodulating rhizobia strains isolated from different soils in Rio Grande do Sul state. R Bras Ci Solo 33:1227–1235

Vershinina Z, Baimiev AK, Blagova D, Knyazev A, Baimiev AK, Chemeris A (2011) Bioengineering of symbiotic systems: Creation of new associative symbiosis with the use of lectins on the example of tobacco and oil seed rape. Appl Biochem Microbiol 47:304–310

Vershinina ZR, Baymiev AK, Blagova DK, Chubukova OV, Baymiev AK, Chemeris AV (2012) Artificial colonization of non-symbiotic plants roots with the use of lectins. Symbiosis 56:25–33

Vessey JK (2003) Plant growth promoting rhizobacteria as biofertilizers. Plant Soil 255:571–586

Wang TL, Wood EA, Brewin NJ (1982) Growth regulators, and nodulation in peas. The cytokinin content of a wild type and a Ti plasmid containing strain of *R. leguminosarum*. Planta 155:350–355

Warda A, Zoubida B-h, Faiza BZ, Yamina A, Bekki A (2014) Selection and characterization of inhibitor agents (bacteriocin like) produced by rhizobial strains associated to *Medicago* in western Algeria. Int J Agric Crop Sci 7:393

Weir B (2016) The current taxonomy of rhizobia. New Zealand rhizobia website. http://www.rhizobia.co.nz/taxonomy/rhizobia.html. Accessed 22 Jan 2016

Willems A (2006) The taxonomy of rhizobia: an overview. Plant Soil 287:3–14

Yang J, Kloepper JW, Ryu C-M (2009) Rhizosphere bacteria help plants tolerate abiotic stress. Trends Plant Sci 14:1–4

Yanni YG, Rizk R, Corich V, Squartini A, Ninke K, Philip-Hollingsworth S, Orgambide G, De Bruijn F, Stoltzfus J, Buckley D (1997) Natural endophytic association between *Rhizobium leguminosarum* bv. *trifolii* and rice roots and assessment of its potential to promote rice growth. Plant Soil 194:99–114

Yanni YG, Rizk RY, El-Fattah FKA, Squartini A, Corich V, Giacomini A, de Bruijn F, Rademaker J, Maya-Flores J, Ostrom P (2001) The beneficial plant growth-promoting association of *Rhizobium leguminosarum* bv. *trifolii* with rice roots. Funct Plant Biol 28:845–870

Yu X, Liu X, Zhu T, Liu G, Mao C (2012) Co-inoculation with phosphate-solubilzing and nitrogen-fixing bacteria on solubilization of rock phosphate and their effect on growth promotion and nutrient uptake by walnut. Euro J Soil Biol 50:112–117

Yuhashi K, Ichikawa N, Ezura H, Akao S, Minakawa Y, Nukui N, Yasuta T, Minamisawa K (2000) Rhizobitoxine production by *Bradyrhizobium elkanii* enhances nodulation and competitiveness on *Macroptilium atropurpureum*. Appl Environ Microbiol 66:PMC110596

Zaidi A, Khan MS, Ahemad M, Oves M, Wani P (2009) Recent advances in plant growth promotion by phosphate-solubilizing microbes. In: Khan MS, Zaidi A, Musarrat J (eds) Microbial strategies for crop improvement. Springer, Berlin, Heidelberg, pp 23–50

Zaidi A, Ahmad E, Khan MS, Saif S, Rizvi A (2015) Role of plant growth promoting rhizobacteria in sustainable production of vegetables: current perspective. Sci Hort 193:231–239

Ziaf K, Latif U, Amjad M, Shabir MZ, Asghar W, Ahmed S, Ahmad I, Jahangir MM, Anwar W (2016) Combined use of microbial and synthetic amendments can improve radish (*Raphanus sativus*) yield. J Environ Agric Sci 6:10–15

Role of Phosphate-Solubilizing Bacteria in Legume Improvement

8

Almas Zaidi, Mohammad Saghir Khan, Asfa Rizvi,
Saima Saif, Bilal Ahmad, and Mohd. Shahid

Abstract

Heterogeneously distributed microbial communities belonging to different genera enhance the growth and development of many crop plants including legumes. Among the various microbial populations inhabiting different habitats, the heterotrophic organisms endowed with natural phosphate-solubilizing activity and quite often called as phosphate-solubilizing microorganisms (PSM) supply one of the major plant nutrients, phosphorus, to plants and facilitate the growth of legumes. Phosphate-solubilizing microorganisms convert the complex or locked insoluble phosphorus to soluble phosphates by various mechanisms such as acidification, chelation, exchange reactions and polymeric substances formation and make it available to plants. Apart from supplying P to legumes, PSM also promote the growth of legumes by other mechanisms. Therefore, the widespread use of PSM in legume production helps both to reduce the spiralling cost of phosphatic fertilizers and to make soil free from chemical hazards. Considering these, the application of PSM endowed with multiple growth-promoting activities holds greater promise for increasing the productivity of legumes. Symbiotic/ associative nitrogen-fixing bacteria are yet another important group of beneficial microbiota which is known to supply exclusively nitrogen to legumes, but they can also promote legume growth by other direct or indirect mechanisms. The co-inoculation of functionally different microflora such as N_2-fixers, phosphate solubilizers and mycorrhizal fungi has, however, been found more effective than single inoculation of either organism for legume plants under nutrient-deficient soils. Basic and advance aspect of phosphate solubilization, mechanism of plant growth promotion and impact of single or synergistic association of phosphate solubilizers with other beneficial microflora on legumes growing in different

A. Zaidi (✉) • M.S. Khan • A. Rizvi • S. Saif • B. Ahmad • M. Shahid
Department of Agricultural Microbiology, Faculty of Agricultural Sciences, Aligarh Muslim University, Aligarh 202002, Uttar Pradesh, India
e-mail: alma29@rediffmail.com

© Springer International Publishing AG 2017
A. Zaidi et al. (eds.), *Microbes for Legume Improvement*,
DOI 10.1007/978-3-319-59174-2_8

regions are reviewed and discussed. The literatures surveyed in this chapter are likely to help better understand the functional role of PSM in sustainable production of legumes while reducing dependence on use of phosphatic fertilizers in legume production systems.

8.1   Introduction

Due to ever-increasing human populations worldwide, there is tremendous pressure on agriculture to produce more and more foods so that the hunger among masses can be reduced. In order to solve such a daunting problem and to optimize food production, agrarian communities have long been using chemical fertilizers in their farming practices. Indeed, the use of chemical fertilizers in modern high-input agricultural practices has revolutionized the whole agricultural system with concurrent enhancement in food production. The overuse and abuse of chemical fertilizers together with its high cost and environmental problems have, however, forced farmers to avoid its use in many economically disadvantaged countries. Considering such obvious threats resulting from the excessive and uncontrolled application of synthetic fertilizers, scientists are desperate to find a suitable and viable alternative which could minimize, if not completely eliminate, the dependence on fertilizers. In this context, use of inexpensive biological resources especially microbes is being considered as a most feasible strategy to counteract the rapid decline in soil fertility and environment quality which may result from overuse of fertilizers. Among soil microbiota, plant growth-promoting rhizobacteria (PGPR) endowed with multiple plant growth-promoting activities (Khan et al. 2009a, b) have been found as an effective biological tool for enhancing the production of many crops including legumes (Ahmad et al. 2013). Among PGPR, there are certain groups of microbes which specifically supply phosphorus (P) to plants and are often called as phosphate-solubilizing microorganisms (PSM) (Mohammadi 2012; Khan et al. 2007). Phosphorus, next in biological importance to nitrogen (N), is an essential plant nutrient and is needed for various physiological functions of plants (Khan et al. 2009a; Shenoy and Kalagudi 2005). On the contrary, the deficiency of P (the second most important plant nutrient after N) severely restricts plant growth. So, to circumvent the P deficiency, phosphatic fertilizers are frequently applied in agricultural practices to attain maximum yields (Del Campillo et al. 1999; Shenoy and Kalagudi 2005). The externally applied phosphate fertilizer, however, is rapidly fixed as insoluble P and become unavailable to plants (Goldstein 1986; Takahashi and Anwar 2007). For example, the P deficiency during legumes cultivation has been found to have a negative impact on nodule formation, development and function (Robson et al. 1981).

Supplying P to legumes through PSM has, therefore, been found an eco-friendly and viable alternative (Khan et al. 2007, 2010; Zaidi et al. 2009). Apart from free-living PGPR acting as P solubilizers such as *Pseudomonas* (Karpagam and Nagalakshmi 2014), *Bacillus* (Karpagam and Nagalakshmi 2014), *Burkholderia*

(Ghosh et al. 2016; Walpola et al. 2012), *Alcaligenes* (Nandinin et al. 2014), etc., nitrogen-fixing PGPR such as asymbiotic nitrogen fixers, *Azotobacter* (Nosrati et al. 2014) and *Azospirillum* (Tahir et al. 2013) and symbiotic rhizobia (Kumar and Ram 2014; Karpagam and Nagalakshmi 2014; Halder and Chakrabarthy 1993), have also been reported to solubilize insoluble P (Abd-Alla 1994; Antoun et al. 1998; Alikhani et al. 2006; Rivas et al. 2006; Sridevi and Mallaiah 2009) to available P and make it accessible to plants. The use of nitrogen-fixing PGPR as P-solubilizing organisms has many advantages: (1) besides supplying P to plants, they can also supply N to legumes (rhizobia) and nonlegumes (e.g. *Azotobacter*), (2) can provide important plant growth regulators to legumes (Kumar et al. 2014; Patil 2011) and (3) can act synergistically with other free-living PGPR and/arbuscular mycorrhizal fungi (Wani et al. 2007a; Zaidi et al. 2003), and (4) since symbiotic rhizobia remain well protected inside the nodules, they face little competition from other indigenous rhizosphere microflora. Due to these multiple growth-enhancing properties, it is reported that when applied with free-living PGPR, PSB inoculation could reduce P fertilizer application by 50% without any reduction in crop yields (Jilani et al. 2007; Yazdani et al. 2009). And hence, the use of PSB as microbial fertilizer could act as a supplement to chemical P fertilizer for sustainable production of legumes. However, the performance of PSB depends on many factors including soil fertility, plant genotypes, composition of root exudates, etc. Despite all these factors, increase in legume-*Rhizobium* symbiosis and yields such as chickpea, pea (Abid et al. 2016), green gram (Zaidi et al. 2006), etc., due to single or co-inoculation of PSB with other PGPR have been reported.

8.2   Phosphate-Solubilization Mechanism: A Brief Account

Among major plant nutrients, phosphorus in soils is present in two forms, organic and inorganic, and together these account for about 0.05% of soil content. Of these, only 0.1% of the total P is, however, available for uptake by plants (Zou et al. 1992). Organic P accounts for about 20–80% of total soil P (Richardson 1994) and is found associated with decayed plant, animal and microbial tissues and humus. On the contrary, the inorganic forms include tricalcium phosphate $(Ca_3PO_4)_2$, iron phosphate (Fe_3PO_4), aluminium phosphate (Al_3PO_4), etc. Phosphorus in such a complex forms is not available for uptake by plants. However, both organic and inorganic forms of P may be converted to soluble and available P by P-solubilizing bacteria inhabiting different soil ecosystems (Khan et al. 2007, 2010; Song et al. 2008). The inorganic P can be solubilized by microbial populations (Khan et al. 2007) including bacteria (Khan et al. 2010; Buch et al. 2008; Song et al. 2008; Illmer and Schinner 1995; Cunningham and Kuiack 1992). Even though microbes adopt various strategies to solubilize insoluble P (Fig. 8.1) in soils, the secretion of low molecular mass organic acids by many bacteria including free-living phosphate solubilizers, for example, *Bacillus* (Narveer et al. 2014; Maheswar and Sathiyavani 2012), *Pseudomonas* (Oteino et al. 2015) and nitrogen-fixing symbiotic rhizobia (Alikhani et al. 2006) or asymbiotic nitrogen fixers, for example, *Azotobacter*

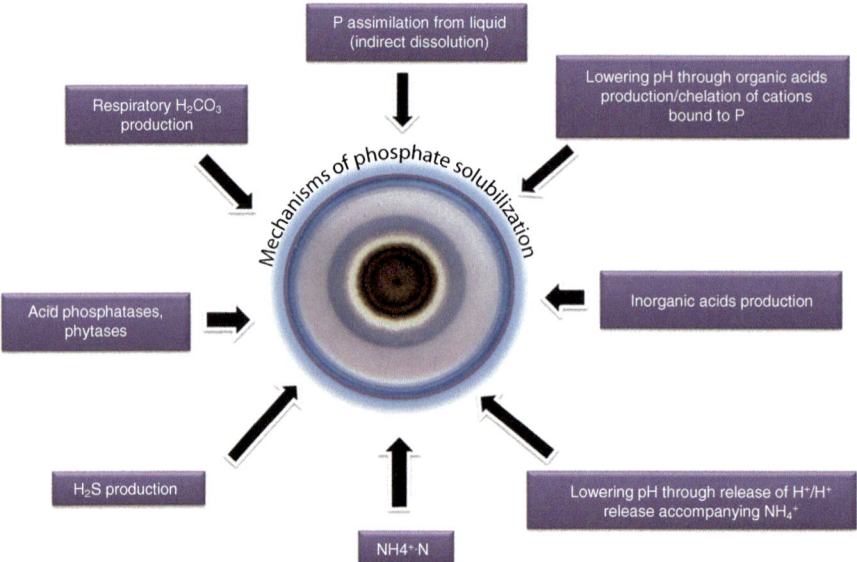

Fig. 8.1 Mechanisms of P solubilization by phosphate-solubilizing microbes (adapted from Khan et al. (2009a, b))

(Nosrati et al. 2014), has been considered as an important mechanism of P solubilization (Chen et al. 2015; Khan et al. 2007). The organic acids secreted both by free-living phosphate solubilizers and nitrogen-fixing phosphate solubilizers (Table 8.1) lower the pH to bring P into solution (Pradhan and Shukla 2005; Maliha et al. 2004). Consequently, the acidification of cells and their environment causes the discharge of P ions from the P mineral by H^+ substitution for Ca^{2+} (Goldstein 1995). The composition of organic acids produced by phosphate-solubilizing bacteria, however, determines the extent of solubilization. It is also reported that insoluble P could be changed into soluble P devoid of organic secretion by microbes (Chen et al. 2006; Ilmer and Schinner 1992, 1995; Asea et al. 1988). Apart from the organic acids, inorganic acids, (Reyes et al. 2001; Richardson 2001) for example, hydrochloric acid (Kim et al. 1997), nitric acid and sulfuric acids (Dugan and Lundgren 1965), excreted by chemoautotrophs and the H^+ pump, for example, in *Penicillium rugulosum*, have also been reported to solubilize the insoluble P (Reyes et al. 1999). Phosphorus in labile organic forms can be enzymatically degraded by PSB, for example, *Pseudomonas*, *Enterobacter* and *Pantoea* (Jorquera et al. 2009), as inorganic P, or it can be incorporated into soil organic matter (Mckenzie and Roberts 1990). The process of mineralization or immobilization carried out by soil microorganisms is greatly affected by moisture and temperature of soil. The process of mineralization and immobilization occurs very rapidly in warm and well-drained soils (Busman et al. 2002).

Table 8.1 Organic acids affecting P solubilization secreted by free-living and symbiotic/associative phosphate-solubilizing bacteria

PS bacteria	Organic acid produced	Reference
Free-living phosphate solubilizers		
Pseudomonas sp.	Gluconic acid	Chen et al. (2015), Oteino et al. 2015)
Enterobacter hormaechei subsp. *steigerwaltii* strain NM23-1	Gluconic acid, succinic acid, acetic acid, glutamic acid, oxalic acid, propionic acid, malic acid, fumaric acid, alpha-ketoglutaric acid	Mardad et al. (2013)
Bacillus and *Enterobacter*	Acetic acid, citric acid and gluconic acid	Tahir et al. (2013)
Pseudomonas trivialis (BIHB 769), *P. poae* (BIHB 808), *Pseudomonas* spp. (BIHB 751)	Gluconic acid, 2α-ketogluconic acid, lactic acid, succinic acid, fumaric acid, malic acid, citric acid, oxalic acid	Vyas and Gulati (2009)
Enterobacter Hy-401, *Arthrobacter* Hy 505, *Azotobacter* Hy -510, *Enterobacter* Hy-402	Oxalic acid, gluconic acid, Lactic acid, Citric acid, succinic acid, malic acid, fumaric acid, tartaric acid	Yi et al. (2008)
Rhodococcus erythropolis (CC-BC11), *Bacillus megaterium* (CC-BC10), BC03), *A. ureafaciens* (CC-BC02), *Serratia marcescens* (CC-BC14)	Gluconic acid, lactic acid, citric acid, propionic acid, succinic acid	Chen et al. (2006)
Delftia (CC-BC21), *Chryseobacterium* (CC-BC05), *Phyllobacterium myrsinacearum* (CC-BC19), *Gordonia* (CC-BC07)	Succinic acid, citric acid, gluconic acid	Chen et al. (2006)
Enterobacter intermedium	2α-Ketogluconic acid	Hwangbo et al. (2003)
B. amyloliquefaciens, *B. atrophaeus*, *B. licheniformis*, *V. proteolyticus*, *P. macerans*	Acetic acid, isobutyric acid, isovaleric acid, lactic acid, succinic acid, propionic acid, valeric acid, fumaric acid, isocaproic acid	Vazquez et al. (2000)
Symbiotic/associative N₂-fixing phosphate solubilizers		
Azospirillum brasilense, *Azospirillum zeae*	Acetic, citric, lactic, malic, and succinic acids	Ayyaz et al. (2016)
Rhizobium RASH6	Succinic and gluconic acids, citric acid	Singh et al. (2014)
Azotobacter spp.	Oxalic acid, malic acid, succinic acid, formic acid, acetic acid, citric acid, lactic acid	Liang et al. (2013)
Azospirillum	Acetic acid, citric acid and gluconic acid	Tahir et al. (2013)
Sinorhizobium meliloti 1021 strain RD64	Malic, succinic and fumaric acids, 2-hydroxyglutaric acid	Bianco and Defez (2010)
Acinetobacter sp.	Gluconic acid, formic acid, oxalic acid	Gulati et al. (2010)

8.3   How Free-Living and Symbiotic/Associative PSB Facilitates Legume Growth?

Phosphate-solubilizing microbes involving both free-living and symbiotic/associative organisms when applied to agronomic practices supply essentially P to plants including legumes (Kumar et al. 2014; Ahmad et al. 2013; Mishra et al. 2009; Zaidi and Khan 2006). Apart from providing P to plants, both free-living and symbiotic/associative phosphate solubilizers also provide other important biomolecules that ultimately enhance the growth of plants (Fig. 8.2). For instance, such beneficial PSB stimulate the efficiency of BNF, enhance the availability of other trace elements (such as iron, zinc), supply important plant growth-promoting substances (Ayyaz et al. 2016; Gusain et al. 2015; Fallo et al. 2015; Kumar et al. 2014) including siderophores (Singh et al. 2014; Datta and Chakrabartty 2014; Parani and Saha 2012) and antibiotics (Lipping et al. 2008; Dilantha et al. 2006) and consequently protect plants from soilborne pathogens (biocontrol) (Parikh and Jha 2012; Saini 2012; El-Mehalawy 2009; Khan et al. 2002). Phosphate-solubilizing bacteria such as Gram-negative

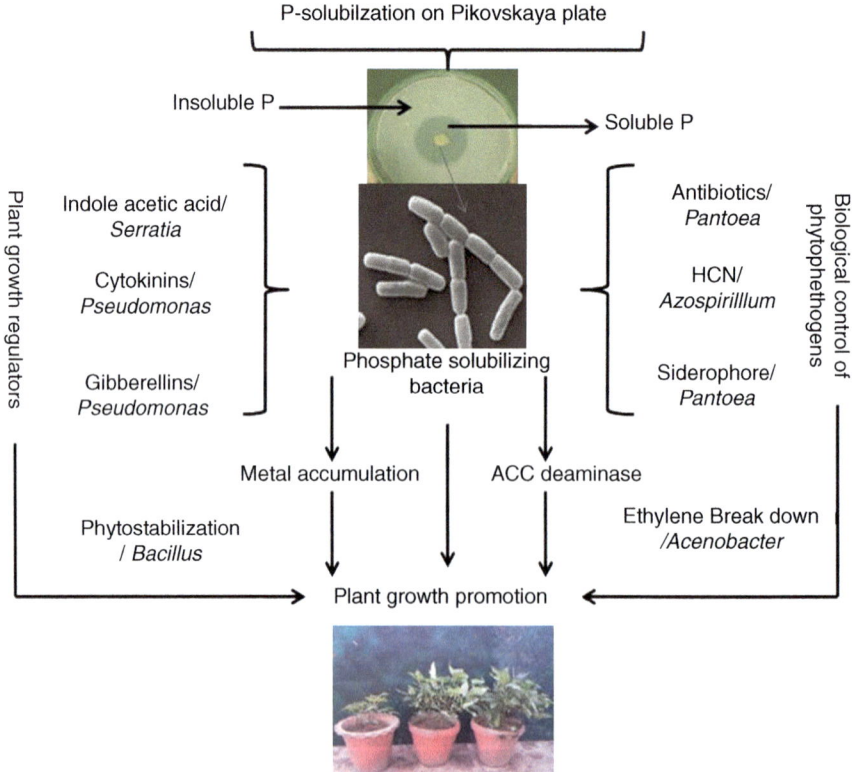

Fig. 8.2 Functional diversity among phosphate-solubilizing bacteria

genera *Azotobacter, Pseudomonas* and *Chromobacterium* also secrete cyanide, a secondary metabolite which is ecologically important (Ponmurugan 2012; Wani et al. 2007a; Siddiqui et al. 2006) and gives a selective advantage to the producing strains (Rudrappa et al. 2008). Also, the secretion of 1-aminocyclopropane-1-carboxylate (ACC) deaminase by free-living phosphate solubilizers, for example, *Bacillus subtilis* (Khan et al. 2016) and symbiotic/associative phosphate solubilizers (Othman and Tamimi 2016), has been found to reduce the level of plant stress hormone ethylene and consequently to enhance the plant growth (Ahmad et al. 2013). Various plant growth regulators released by free-living and symbiotic/associate PSB are briefly summarized in Table 8.2.

Table 8.2 Plant growth-promoting substances released by free-living and symbiotic/associative phosphate-solubilizing bacteria

Phosphate-solubilizing bacteria	Plant growth-promoting substances	References
Free-living P solubilizers		
Bacillus subtilis	IAA, ACC deaminase	Khan et al. (2016)
Pseudomonas koreensis, Arthrobacter nitroguajacolicus strain YB4 and *Klebsiella oxytoca*	IAA, siderophores	Gusain et al.(2015)
Escherichia coli DACG2, *Pseudomonas fluorescens* strain DACG3, *Burkholderia* sp. DACG1	IAA	Dasgupta et al. (2015)
Bacillus sp.	IAA	Fallo et al. (2015)
Burkholderia thailandensis, Sphingomonas pituitosa and *Burkholderia seminalis*	IAA	Panhwar et al. (2014)
Azospirillum, Bacillus and *Enterobacter*	IAA	Tahir et al. (2013)
Pseudomonas fluorescens, P. putida	IAA, siderophore, ACC deaminase	Zabihi et al. (2011)
Pseudomonas sp.	IAA, siderophore, ammonia	Stajković et al. (2011)
Serratia nematodiphila, Pontibacter niistensis	IAA, siderophore, HCN, ACC deaminase	Dastager et al. (2011a, b)
Pantoea agglomerans	IAA	Mishra et al. (2011)
Arthrobacter, Bacillus	IAA, antifungal activity, HCN, NH₃	Banerjee et al. (2010)
Pantoea	IAA, siderophore, antifungal activity	Taurian et al. (2010)
Enterobacter aerogenes, E. cloacae, E. asburiae	IAA, siderophore, HCN	Deepa et al. (2010)
Pseudomonas sp. SRI2, *Psychrobacter* sp. SRS8 and *Bacillus* sp. SN9	IAA, siderophore, ACC deaminase	
Pseudomonas sp.	ACC deaminase, IAA, siderophore	Poonguzhali et al. (2008)

(continued)

Table 8.2 (continued)

Phosphate-solubilizing bacteria	Plant growth-promoting substances	References
Serratia marcescens	IAA, siderophore, HCN	Selvakumar et al. (2008)
Acinetobacter sp., *Pseudomonas* sp.	ACC deaminase, IAA, antifungal activity, N_2 fixation	Indiragandhi et al. (2008)
Burkholderia	ACC deaminase, IAA, siderophore	Jiang et al. (2008)
Fluorescent pseudomonas	IAA, siderophores, HCN, antifungal activity	Shweta et al. (2008)
Symbiotic/associative N_2-fixing phosphate solubilizers		
Rhizobium strains nodulating faba bean	IAA, ACC deaminase	Othman and Tamimi (2016)
Azospirillum brasilense, Azospirillum zeae	IAA	Ayyaz et al. (2016)
Azotobacter	IAA, EPS, HCN production, ammonia	Ahmad et al. (2016)
Bradyrhizobium sp. (vigna)	IAA, EPS, HCN production, ammonia, ACC deaminase	Ahmad et al. (2016)
Rhizobium RASH6	IAA, siderophore, ammonia and HCN production	Singh et al. (2014)
Azotobacter	siderophores, ammonia, hydrogen cyanide, IAA	
Sinorhizobium sp. strain MRR101-KC428651, *Rhizobium* sp. strain 103–JX576499, *Sinorhizobium kostiense* strain MRR104- KC428653	Antifungal activity	Kumar et al. (2014)
Azospirillum spp.	IAA	Oedjijono et al. (2014)
Rhizobium spp.	IAA	Garg and Sharma (2013)
Rhizobium leguminosarum (groundnut), *Rhizobium loti* (chickpea), *Rhizobium meliloti* (lucerne), *Rhizobium meliloti* (fenugreek)	IAA	Madhuri (2011)

8.4 Phosphate-Solubilizing Bacteria and Legume Improvement

Phosphate-solubilizing bacteria especially belonging to free-living PGPR group have been reported to enhance the growth and yield of many agronomically important crops in different production systems (Walpola and Yoon 2012; Qureshi et al. 2012a, b). However, the possible action of phosphate-solubilizing bacteria belonging to symbiotic/associative group in legume improvement has recently been realized (Collavino et al. 2010; Shaharoona et al. 2008; Vikram et al. 2007; Rivas et al. 2006; Peix et al. 2001). Considering the importance of both free-living and symbiotic/associative PSB in agricultural practices, an attempt is made in the following section to highlight the role of inexpensive free-living and symbiotic/associative PSB used either alone or in combination in the improvement of legumes grown in different agroecosystems.

8.4.1 Effect of Free-Living PSB and Nitrogen Fixers on Legumes

Pulses are the second most important group of crops after cereals. Developing countries contribute about 74% to the global pulses production, and the remaining comes from developed countries. However, the production of pulses, in general, is low in most of the pulse-producing countries which is mainly due to imbalanced application of nutrients and use of traditional varieties. In order to optimize pulse production, various strategies including the use of expensive fertilizers are applied. The excessive use of fertilizers is, however, neither economical nor environmentally safe (Lemanski and Scheu 2014). And hence, to avoid such situations, use of microbial inoculants, a cost-effective, eco-friendly and renewable sources of plant nutrients, is recommended and practiced (Khan et al. 2007). In this regard the symbiotic/associative nitrogen fixers and free-living phosphate-solubilizing bacteria (PSB) have shown advantage in enhancing legume productivity (Diep et al. 2016; Tagore et al. 2013; Li et al. 2013). The nitrogen fixers especially the symbiotic rhizobia and PSB assume a great importance on account of their vital role in N_2 fixation and simultaneous P solubilization besides providing some important plant hormones and other growth regulators to developing legumes (Ahmad et al. 2016; Peix et al. 2001). When used alone or in combinations, such beneficial microbiota have been found to reduce the use of chemical fertilizer and the pollution of underground water, renovate the ecological environment of soil and increase the yield and quality of legumes (Ahmad et al. 2013; Chen et al. 2006).

In a study, Rosas et al. (2006) observed that the P-solubilizing strains of symbiotic rhizobia *Sinorhizobium meliloti* used for alfalfa and *Bradyrhizobium japonicum* used for soybean and two strains of free-living PSB, *Pseudomonas putida* (*SP21* and *SP22*), used to inoculate both alfalfa and soybean, had a significant stimulatory effect on these crops. Similarly, Valverde et al. (2006), in a study conducted in two regions of Spain, Castilla León and Andalucía, used an efficient symbiotic bacterium *Mesorhizobium ciceri* strain able to nodulate chickpea (termed as C-2/2) and a powerful *in vitro* phosphate-solubilizing bacterium *Pseudomonas jessenii* strain (termed as PS06) in order to find a suitable biofertilizer for staple grain-legumes (chickpea, ecotype ILC-482) grown under greenhouse and field experiments. Under greenhouse conditions, plants inoculated with *Mesorhizobium ciceri* C-2/2 alone had the highest shoot dry weight. The plants inoculated with *P. jessenii* PS06 yielded a shoot dry weight 14% greater than the uninoculated control plant, but it was not correlated with shoot P contents. The mixture of C-2/2 with PS06, however, resulted in a decrease in shoot dry weight relative to single inoculation of strain C-2/2. Under field conditions, the single and composite inoculation of *M. ciceri* C-2/2 showed an enhanced symbiosis and produced higher nodule mass, nodule number and shoot N content compared to other treatments. Furthermore, the combined application of *M. ciceri* C-2/2 and *P. jessenii* PS06 resulted in highest seed yield which was calculated as 52% increase over uninoculated control. Interestingly, sole application of *P. jessenii* PS06 did not show any significant effect on plant growth. From this finding it was concluded that the free-living P solubilizer such as *P. jessenii* PS06 used in this study can act synergistically with symbiotic rhizobium *M. ciceri* C-2/2 in promoting chickpea growth. Qureshi et al. (2012a, b) in a follow-up study conducted a pot experiment in Faisalabad, Pakistan, to assess the co-inoculation effect of N_2-fixing (*Rhizobium*) and P-solubilizing (*Bacillus* sp.) in the presence of L-TRP on mash bean. Even though the co-inoculation gave the maximum pod and straw yield, the effect was more pronounced with L-TRP. Moreover, the mixture of *Rhizobium* and *Bacillus* sp. increased the length and mass of roots and number and dry matter accumulation in nodule relative to control with L-TRP. The coculture of *Rhizobium* and *Bacillus* sp. in the presence of L-TRP produced 30.87 pod and 32.73 g/pot straw yield which was followed by 30.47 and 31.10 g/pot with rhizobial inoculation, respectively. Co-inoculation produced higher root mass (33.5 g), root length (36 cm), nodule number (34) and nodule mass (0.131 g) which were further enhanced with L-TRP (40.5 g, 49 cm, 48 and 0.145 g, respectively). The nutrient concentration in plants, grain yields and nodulation were found maximum in co-inoculation treatment compared to sole bacterial inoculations. Also, co-inoculation with L-TRP showed higher soil N and available P at harvest as compared to control. This study thus demonstrated that co-inoculation of symbiotic *Rhizobium* and free-living P solubilizer *Bacillus* species was superior than their individual inoculation which was more obvious in the presence of L-TRP.

Inoculation of *Rhizobium* species generally improves the root architecture of legumes by their ability to colonize and fix atmospheric nitrogen. However, when *Rhizobium* and PSB strains are used together, they further improve the nutrient pool and its availability to plant. Considering this beneficial aspect, Abid et al. (2016) evaluated the response of peas to *Rhizobium* and PSB used both individually and together in the presence of 20 and 40 mg P/kg applied in vitro and under pot culture experiment. As expected, the co-inoculation of *Rhizobium* + PSB at 40 mg P/kg soil significantly enhanced the length of shoot and root; number of flowers, pod and nodules produced on each plant; root and shoot dry weight; 100 grain weight; and number of grains per pod up to 37%, 25%, 60%, 220%, 25%, 125%, 34%, 19% and 20%, respectively, over 20 mg P/kg. Also, the N and P concentration of straw and grain and soil N and P contents at harvest were significantly increased by the mixed application of N_2-fixing and P-solubilizing bacteria relative to sole inoculation in the presence of 20 mg/kg P. Results showed that two unrelated bacterial cultures when used together in the presence of 20 and 40 mg kg P can improve the symbiosis, biological properties, grain yield and nutrient uptake by plants. Nandinin et al. (2014) assayed the interactive effect of a phosphate-solubilizing bacterium *Alcaligenes faecalis* on growth and yield of soybean grown under glasshouse conditions by inoculating the bacterium either alone or in combination with *Bradyrhizobium japonicum* and *Bacillus megaterium* or both. In general, the inoculated plants were significantly taller and had more number of leaves, higher numbers of pods, plant dry weight and grain yield compared to uninoculated control. Of the various treatments, the triple inoculation was found superior compared to single and dual inoculations, and N and P content were also higher in plant tissue compared to other treatments suggesting the occurrence of synergism among interacting organisms in soybean rhizosphere.

Nitrogen and phosphorus are the two major elements which play an important role in legume-*Rhizobium* symbiosis. The inoculation with bacterial mixtures consisting of both N_2 fixers and P solubilizers is, therefore, likely to provide a more balanced nutrition for the plants. And hence, the improvement in root uptake of N and P has been found as the major mechanism of interaction between plants and bacteria (Belimov et al. 1995). Considering this, Ahmad et al. (2016) in a recent study observed that the dry matter accumulation in whole plants, symbiotic attributes, nutrient uptake and grain yields of green gram plants were significantly enhanced following co-inoculation of phosphate-solubilizing asymbiotic N_2 fixer *Azotobacter chroococcum* and ACC deaminase-positive symbiotic *Bradyrhizobium* sp. (vigna) under field conditions. The biopriming of *Azotobacter* and *Bradyrhizobium* together maximizes the seed yield by twofolds and resulted in the highest grain protein. A 75% and 52% increase in P concentration in root and shoots, respectively, was observed for *A. chroococcum*, while P uptake was highest (0.52 mg/g) in shoots following combined inoculation of *A. chroococcum* with *Bradyrhizobium* at harvest. The highest N concentration in roots and shoot at harvest was observed with coculture of *A. chroococcum* and *Bradyrhizobium* sp. (vigna). Moreover, the effects of mixed cultures of *A. chroococcum* and *Bradyrhizobium* were relatively greater than the sum of the individual inoculation effects, suggesting synergisms beyond simple additive effects. In addition, the maximum accumulation of N and P in inoculated plants suggested that both asymbiotic and symbiotic bacteria can live favourably in a microhabitat. The overall increase in the performance of green gram following co-inoculation of *Azotobacter* and *Bradyrhizobium* was suggested due to the availability of soluble P and secretion of indole acetic acid, ammonia, cyanogenic compounds and exopolysaccharides by *Azotobacter* and *Bradyrhizobium*. The results suggest that two unrelated bacteria belonging to symbiotic and asymbiotic PS group and capable of facilitating green gram production under field conditions and expressing multiple plant growth regulating potential can be used to develop mixed bioinoculants for upgrading green gram production while saving the use of fertilizers. Similarly, Diep et al. (2016) conducted two field experiments during 2015 at Dak Lak Province to study the effects of rhizobia and PSB on soybean (cv. Cujut) cultivated on ferralsols. The first experiment carried out at Centre for Agricultural Research, Western Highland Agricultural Institute (at Buon Ma Thuot City), included five treatments: (1) control (no fertilizer, no inoculant), (2) 400 kg/ha NPK 15-15-15, (3) rhizobial inoculant [with liquid cover seeds] + 20 kg N/ha applied at 10 days after sowing [DAS], (4) PSB inoculant [with liquid cover seeds] + 20 kg N/ha at 10 DAS and (5) rhizobial and PSB inoculant [with liquid cover seeds] + 20 kg N/ha at 10 DAS. The second experiment carried out at Buonho town included four treatments: (1) control (no fertilizer, no inoculant), (2) 100 kg/ha thermophosphate

(15% P_2O_5) + 25 kg/ha NPK 16-16-8 applied at 20 and 40 at DAS, (3) rhizobial and PSB inoculant [with liquid cover seeds] + 200 kg biofertilizer + 20 kg N/ha at 10 DAS and (4) rhizobial and PSB inoculant [with liquid cover seeds] + 400 kg biofertilizer + 20 kg N/ha at 10 DAS. Application of rhizobial inoculant and/or PSB inoculant produced significantly higher grain yield than control but did not differ from 400 kg/ha NPK 15-15-15 in the experiment 1. In the second experiment, biofertilizer also produced higher yield component, grain yield and oil and protein in seed than control but was equivalent with treatment of 100 kg/ha thermophosphate (15% P_2O_5) + 25 kg/ha NPK 16-16-8. It was concluded from this experiment that biofertilizers can be considered as a replacement for part of chemical fertilizers in soybean cultivation on ferralsols. In a similar experiment, Fernández et al. (2007) reported that the soybean plants bacterized with *Burkholderia* sp. (PER2F) had the highest aerial height and an appropriate N/P ratio, but inoculation with *Enterobacter* sp. and *Bradyrhizobium* sp. did not increase P uptake by plants. They suggested from this finding that PSB inoculation does not necessarily improve P nutrition in soybean. Also, no relationship between soil P and P content in the shoot of soybean raised in greenhouse was observed. Moreover, P-solubilizing fluorescent pseudomonads (PS1 and PS2) isolated from the rhizosphere of groundnut when used as inoculants for groundnut enhanced germination up to 15% and 30% with subsequent increase in grain yield by 66% and 77%, respectively. Conversely, when *M. phaseolina* was tested alone, a 57% decline in yield was noticed. This study thus revealed the potential of the two pseudomonads which acted not only as biocontrol material against *M. phaseolina* but also as an efficient growth enhancer for groundnut (Shweta et al. 2008). In a follow-up study, Rathi et al. (2008) evaluated the P-solubilizing bacterial isolates (4GRP, 25MRP, 27MRP, 28MRP, 33MRP and 34MRP) recovered from the rhizosphere of green gram and mustard on green gram and mustard in a pot experiment. Under pot house conditions, highest increment in dry matter accumulation in both crops was observed with 25MRP with URP followed by 33MRP with URP. In mustard, maximum P uptake was recorded for 25MRP with URP (284%) followed by 4GRP with URP (143%) at 60 DAS. In green gram, highest P uptake was detected in 25MRP with URP (224%) followed by 33MRP with URP (182%) at 60 days after sowing. In a similar study, Gulati et al. (2009) reported a significant increase in the growth of pea, chickpea, maize and barley under both controlled conditions and field trials following P-solubilizing *Acinetobacter rhizosphaerae*. The strain apart from solubilizing inorganic and organic P secreted auxin, ACC deaminase, ammonia and siderophore. The rifampicin mutant of this strain effectively colonized the pea rhizosphere without adversely affecting the resident microbial populations.

Sinorhizobium ciceri is yet another classical example of rhizobial species with specific ability to form symbiosis with chickpea plants which is also influenced by PSB (Messele and Pant 2012). In order to assess the possible action of PSB (*Pseudomonas* sp.) on *Sinorhizobium*-chickpea symbiosis, a field experiment was conducted in Shoa Robit area, Ethiopia, using three levels of NP fertilizer and four levels of inoculants. The inoculation of *Sinorhizobium ciceri* alone increased dry matter yield by 157% and nodule number by 118% over control while the addition of 18/20 kg NP/ha as urea and DCB resulted in 150% increase of dry matter yield and 144% increase in nodule number per plant over uninoculated control. There was also a marked increase in nodule dry weight (200%), as a result of *Sinorhizobium ciceri* + 18/20 kg NP/ha as urea and DCB, indicating the importance of P in nodule tissue development. Likewise, inoculation of phosphate-solubilizing *Pseudomonas* sp. + 18/20 kg NP/ha enhanced symbiotic attributes, for example, nodule dry weight, nodule number, nodule volume and seed yield by 240%, 189%, 152% and 143%, respectively, over control suggesting the efficiency of bacteria in solubilizing insoluble P in DCB. Among all treatments, the composite application of N_2-fixing *Sinorhizobium ciceri* and free-living phosphate-solubilizing *Pseudomonas* sp. along with 18/20 kg NP/ha as urea and DCB was found superior and increased nodule number per plant by 209%, nodule dry weight by 220%, nodule volume by 221% and dry matter by 172% over uninoculated control at midflowering stage of chickpea. Similarly, mixed application of *Sinorhizobium ciceri and Pseudomonas* sp. in the presence of 18/20 kg N (urea) P (DAP)/ha increased nodule number, nodule dry weight, nodule volume and dry matter by 271.59%, 220%, 241.97% 181.40%, respectively, over uninoculated control at midflowering stage.

Phosphate-solubilizing strain of *Mesorhizobium mediterraneum* (PECA21) substantially increased the growth and phosphorus concentration in chickpea plants when grown in soil treated with or without TCP in a growth chamber experiment (Peix et al. 2001). The strain PECA21 could mobilize P efficiently and increased the P concentration by 100%. Moreover, the dry biomass matter, N, K, Ca and Mg content, was markedly enhanced in inoculated plants, grown in soil treated with insoluble P. These results, therefore, suggested that the inoculation of soil with rhizobia should not be considered only for its N_2-fixing potential but also for its ability to solubilize P. Likewise, inoculation of green gram seeds with PS bacteria resulted in maximum nodule numbers, nodule dry weight, shoot and root dry mass, P content and P uptake compared to RP and single superphosphate (SSP) control. However, plant growth-promoting activity of microbial cultures differed substantially considerably (Vikram and Hamzehzarghani 2008). Similarly, Gull et al. (2004) reported that chickpea growth, shoot P and N content, nodulation efficiency and nitrogenase activity were enhanced substantially in the presence of P-solubilizing bacterial strains isolated from rhizosphere, roots and nodules of chickpea. Phosphate-solubilizing strains, CPS-2, CPS-3 and Ca-18, demonstrated the greatest positive effect on shoot length, shoot dry weight and nodulation of chickpea plants. In a similar experiment conducted under field soils, the effect of mixed culture of *Rhizobium* and phosphate-solubilizing bacteria on symbiotic traits, nodule leghaemoglobin and yield of five elite genotypes of chickpea was variable (Tagore et al. 2013). Of the different chickpea genotypes, IG-593 performed better in respect of symbiotic parameters including nodule number, nodule fresh weight, dry biomass of nodules and shoots and yield. Leghaemoglobin content (2.55 mg/g of fresh nodule) was also higher in genotype IG-593. Among microbial inoculants, the mixture of *Rhizobium* and PSB showed maximum increase in nodule numbers (27.66 nodules/plant), fresh weight of nodules (145 mg/plant), nodule dry weight (74.3 mg/plant), shoot dry weight (11.76 g/plant) and leghaemoglobin content (2.29 mg/g of fresh nodule) and also showed its positive effect in enhancing all the yield attributing parameters, grain and straw yields. A similar increase in shoot dry weight and N and P contents in bean (*Phaseolus vulgaris* L.) plants following co-inoculation of *Rhizobium phaseoli* and phosphate-solubilizing *Pseudomonas* sp. or *Bacillus* sp. is reported (Stajković et al. 2011). Among the two P solubilizers, *Pseudomonas* sp. was found better and promoted bean growth and particularly P uptake more efficiently than *Bacillus* sp. The resulting improvement in common bean growth was attributed to the secretion of IAA, ammonia and siderophore production by *Pseudomonas* sp., while *Bacillus* sp. showed only ammonia production.

8.4.2 Synergism Between Free-Living Phosphate-Solubilizing Bacteria and AM-Fungi: Importance in Legume Production

Legume crops are important for human and animal consumption due to high nutritional value as sources of protein, phosphorus, carbohydrate, minerals and different vitamins. The scarcity of food containing high levels of protein, micronutrients and various vitamins sources is an increasing problem affecting human populations in developing countries. In order to overcome the challenge of food scarcity and malnutrition across the world, some new technologies are needed to be devised. Regarding improving the quality and yield of economically important legumes, farmers apply large quantities of chemical fertilizers, which after deposition in soils disrupt the soil fertility and concurrently the legume production. The application of biofertilizers has, however, been suggested as the best solution because it is environmentally friendly and inexpensive (Mia and Shamsuddin 2010). So, like other crops, legume production can also be increased by applying bio-preparation consisting of phosphate-solubilizing bacteria, N_2-fixing *Rhizobium* and arbuscular mycorrhizal (AM) fungi, which together supply enough nutrients especially N and P content to plants (Silveira and Cardoso 2004; Hegde et al. 1999; Vessey 2003).

The microbial plant mutualistic symbionts, mycorrhizal fungi, form a functional symbiosis with the roots of most plant species. Mycorrhizal symbioses can be found in almost all ecosystems worldwide where they enhance plant health and soil quality by improving the P and N plant uptake from soil and also assist plant host in uptake of the minor elements such as Zn, Cu and Fe. Additionally, mycorrhizal interactions improve plant health through increased protection against biotic and abiotic stresses and soil structure through aggregate formation (Garg and Chandel 2010; Lingua et al. 2008; Turnau et al. 2006; Barea et al. 2005a, b; Jeffries et al. 2003). Thus, when mycorrhizal fungi are used together with P-solubilizing bacteria, increase in overall performance of legumes is obvious (Zaidi et al. 2003). For example, Mehdi et al. (2006) in a study evaluated the responses of lentil (*Lens culinaris* cv. 'Ziba') to co-inoculation with AM fungi and some indigenous rhizobial strains varying in P-solubilizing ability in a calcareous soil with high pH and low amounts of soluble P and N. The results revealed that the impact of *Glomus mosseae* and *G. intraradices*, rhizobial strains *(R. leguminosarum* bv. viciae) and a mixed rhizobial inoculant with an effective P-solubilizing activity (*M. ciceri)* and phosphatic fertilizers such as superphosphate and phosphate rock were remarkable for all the measured characteristics, like the dry matter of shoots, plus seeds, their P and N contents and percent of root colonized by AM fungus. The rhizobial strain with P-solubilizing ability showed a more beneficial effect on plant growth and nutrient uptake than the strain without this ability, although both strains had similar effectiveness for N_2 fixation in symbiosis with lentil. Synergistic relationships were observed between AM fungi and some rhizobial strains that related to the compatible pairing of these two microsymbionts. The P uptake efficiency was increased when P fertilizers were applied along with AM fungi and/or P-solubilizing rhizobial strains. Likewise, Zaidi and Khan (2006) and Zaidi et al. (2004) while evaluating the single or combined effects of N_2-fixing [(*Bradyrhizobium* sp. (vigna)], P-solubilizing bacteria (*Bacillus subtilis/P. striata*), P-solubilizing fungus (*Aspergillus awamori and Penicillium variable*) and AM fungus (*G. fasciculatum*) on the biological and chemical properties of green gram plants grown in P-deficient soils observed that the triple inoculation of AM fungus, *Bradyrhizobium* sp. (vigna) and *B. subtilis/P. striata* significantly increased dry matter yield, chlorophyll content in foliage and N and P uptake of plants which in turn resulted in substantial increase in seed yield (24%) relative to the uninoculated plants. Moreover, the symbiotic properties (nodule occupancy) of inoculated plants as determined by indirect enzyme linked immunosorbent assay (ELISA) increased by 77% (*Bradyrhizobium* with *A. awamori*) and 96% (*Bradyrhizobium* used with *G. fasciculatum* and *B. subtilis*) at flowering which decreased considerably at the pod-fill stage. However, a negative effect occurred on all the considered parameters when *P. variable* was added to the combination of *Bradyrhizobium* sp. (vigna) and *G. fasciculatum*. In addition, the available P status of the soil improved by the addition of *P. striata* with *Bradyrhizobium* sp. (vigna) and AM fungus. The N content of the soil, however, did not show appreciable changes after the inoculation. The population of PSM in some treatments, percentage root infection and spore density of the AM fungus in the soil increased between 35 and 50 days of plant growth. The present findings showed that rhizospheric microorganisms can interact positively and promote plant growth synergistically leading to improved grain yield and quality. Furthermore, Toro et al. (2008) in a trial assessed the interactive effects of multiple microbial inoculations and rock phosphate (RP) on N and P acquisition by alfalfa plants using ^{15}N and ^{32}P isotopes. The microbial inocula included a wild-type (WT) *R. meliloti* strain; its genetically modified (GM) derivative, which had an enhanced competitiveness; the arbuscular mycorrhizal (AM) fungus *Glomus mosseae* (Nicol. and Gerd) Gerd and Trappe; and a P-solubilizing bacterium (*Enterobacter* sp.). The inoculated organisms established well inside root tissues and/or in the alfalfa rhizosphere. Of these, GM *Rhizobium* strain did not interfere with AM colonization. Even though the inoculated P-solubilizing bacterium established in the alfalfa rhizosphere, the level of establishment was lower where the natural population of P-solubilizing bacterium was stimulated by AM inoculation and RP application. The stimulation of these indigenous bacteria was also greater in the rhizosphere of alfalfa nodulated by the GM *Rhizobium*. Improvements in N and P accumulation in alfalfa corroborate beneficial effects of the improved GM *Rhizobium* on AM performance, in RP-amended plants. Inoculation with *Enterobacter*, however, did not improve the AM effect on N or P accumulation in the RP-added soil, but it did in the non-RP-amended controls. In addition, ^{15}N:^{14}N ratio in plant shoots indicated enhanced N_2 fixation rates in *Rhizobium*-inoculated AM plants, compared to those obtained by the same *Rhizobium* strain in non-mycorrhizal plants. Regardless of the *Rhizobium* strain and of whether or not RP was added, AM-inoculated plants showed a lower specific activity (^{32}P:^{31}P) than did their comparable non-mycorrhizal controls, suggesting that the plant was using otherwise unavailable P sources. The P-solubilizing, AM-associated, microbiota could in fact release P ions, either from the added RP or from the indigenous 'less-available' P. Additionally, the proportion of plant P derived either from the labelled soil P (labile P pool) or from RP was similar for AM inoculated and non-mycorrhizal controls (without *Enterobacter* inoculation) for each *Rhizobium* strain, but the total P uptake, regardless of the P source, was far higher in AM plants which could probably be due to P activity of *Enterobacter*.

Recently, Mirdhe and Lakshman (2014) conducted a greenhouse pot experiment to evaluate the effect of AM fungi (*Funneliformis mosseae*) along with the dual inoculation of *Funneliformis mosseae* with *Rhizobium* and PSB and a triple inoculation of *Funneliformis mosseae*, *Rhizobium* and PSB on *Vigna unguiculata* (L) Verdc. Results revealed that triple inoculation of *Funneliformis mosseae* + *Rhizobium* + PSB showed an increase in height, dry weight of root and shoot, spore number, per cent root colonization, number of nodules and P and N uptake by *Vigna* plants when compared with dual inoculation. Also, the combined inoculation of bacteria and AM fungi synergistically enhanced the measured parameters.

Conclusion

Nitrogen and phosphorus are the two essential nutrients for plant growth and development. The extensive use of chemical fertilizers to provide these nutrients in agriculture is currently under debate due to environmental concern, and questions are raised regarding the consumers' health. Recent advancements in the field of biofertilizers offer an opportunity to environmental friendly sustainable agricultural practices to reduce dependence on chemical fertilizers and thereby decrease adverse environmental effects. Phosphate-solubilizing bacteria in association with symbiotic/associative nitrogen fixers and arbuscular mycorrhizal fungi have been found to increase legume growth by various mechanisms. The use of such inexpensive and naturally abundant microbes both in isolation and association could be of great practical value in sustainable and low-input legume production systems. Moreover, the development of effective microbial inoculants for raising the productivity of legumes remains a major scientific challenge. And hence, functional properties of interacting microbes together with the development of suitable microbial pairing still require further experimental confirmation in order to achieve optimum benefits of such natural resources. Future research should therefore strive hard towards an improved understanding of the functional mechanisms behind such microbial interactions, so that compatible organisms could be identified and applied as effective inoculants within sustainable legume production systems.

References

Abd-Alla MH (1994) Use of organic phosphorus by *Rhizobium leguminosarum* bv. viciae phosphatases. Biol Fertil Soils 8:216–218

Abid K, Sultan T, Kiani MZ et al (2016) Effect of *Rhizobium* and phosphate solubilizing bacteria at different levels of phosphorus applied on nodulation, growth and yield of peas (*Pisum sativum*). Int J Biosci 8:112–121

Ahmad E, Khan MS, Zaidi A (2013) ACC deaminase producing *Pseudomonas putida* strain PSE3 and *Rhizobium leguminosarum* strain RP2 in synergism improves growth, nodulation and yield of pea grown in alluvial soils. Symbiosis 61:93–104

Ahmad E, Zaidi A, Khan MS (2016) Effects of plant growth promoting rhizobacteria on the performance of greengram under field conditions. Jordan J Biol Sci 9:79–88

Alikhani HA, Rastin NS, Antoun H (2006) Phosphate solubilization activity of rhizobia native to Iranian soils. Plant Soil 287:35–41

Antoun HA, Beauchamp CJ, Goussard N et al (1998) Potential of *Rhizobium* and *Bradyrhizobium* species as plant growth promoting rhizobacteria on nonlegumes: effect on radishes (*Raphanus sativus* L.) Plant Soil 204:57–67

Asea PEA, Kucey RMN, Stewart JWB (1988) Inorganic phosphate solubilization by two *Penicillium* species in solution culture and soil. Soil Biol Biochem 20:459–464

Ayyaz K, Zaheer A, Rasul G et al (2016) Isolation and identification by 16S rRNA sequence analysis of plant growth-promoting azospirilla from the rhizosphere of wheat. Braz J Microbiol 47(3):542–550

Banerjee S, Palit R, Sengupta C et al (2010) Stress induced phosphate solubilization by *Arthrobacter* sp. and *Bacillus* sp. isolated from tomato rhizosphere. Aust J Crop Sci 4:378–383

Barea JM, Pozo MJ, Azcón R et al (2005a) Microbial co-operation in the rhizosphere. J Exp Bot 56:1761–1778

Barea JM, Azcón R, Azcón-Aguilar C (2005b) Interactions between mycorrhizal fungi and bacteria to improve plant nutrient cycling and soil structure. In: Buscot F, Varma S (eds) Microorganisms in soils: roles in genesis and functions. Springer-Verlag, Heidelberg, Germany, pp 195–212

Belimov AA, Kojemiakor AP, Chuvarliyeva CV (1995) Interactions between barley and mixed cultures of nitrogen fixing and phosphate solubilizing bacteria. Plant Soil 173:29–37

Bianco C, Defez R (2010) Improvement of phosphate solubilization and *Medicago* plant yield by an indole-3-acetic acid-overproducing strain of *Sinorhizobium meliloti*. Appl Environ Microbiol 76:4626–4632

Buch A, Archana G, Kumar GN (2008) Metabolic channelling of glucose towards gluconate in phosphate-solubilizing *Pseudomonas aeruginosa* P4 under phosphorus deficiency. Res Microbiol 159:635–642

Busman L, Lamb J, Randall G et al (2002) The nature of phosphorus in soils. University of Minnesota Extension Service, MN

Chen YP, Rekha PD, Arun AB et al (2006) Phosphate solubilizing bacteria from subtropical soil and their tricalcium phosphate solubilizing abilities. Appl Soil Ecol 34:33–41

Chen W, Yang F, Zhang L et al (2015) Organic acid secretion and phosphate solubilizing efficiency of *Pseudomonas* sp. PSB12: effects of phosphorus forms and carbon sources. Geomicrobiol J. doi:10.1080/01490451.2015.1123329

Collavino MM, Sansberro PA, Mroginski LA et al (2010) Comparison of *in vitro* solubilization activity of diverse phosphate-solubilizing bacteria native to acid soil and their ability to promote *Phaseolus vulgaris* growth. Biol Fertil Soils 46:727–738

Cunningham JE, Kuiack C (1992) Production of citric and oxalic acids and solubilization of calcium phosphate by *Penicillium bilaii*. Appl Environ Microbiol 58(5):1451–1458

Dasgupta D, Sengupta C, Paul G (2015) Screening and identification of best three phosphate solubilizing and IAA producing PGPR inhabiting the rhizosphere of *Sesbania bispinosa*. Int J Innov Res Sci Eng Technol 4:3968–3979

Dastager SG, Deepa CK, Pandey A (2011a) Potential plant growth promoting activity of *Serratia nematodiphila* NII- 0928 on black pepper (*Piper nigrum* L.) World J Microbiol Biotechnol 27:259–265

Dastager SG, Deepa CK, Pandey A (2011b) Plant growth promoting potential of *Pontibacter niistensis* in cowpea (*Vigna unguiculata* (L.) Walp.) Appl Soil Ecol 49:250–255

Datta B, Chakrabartty PK (2014) Siderophore biosynthesis genes of *Rhizobium* sp. isolated from *Cicer arietinum* L. 3 Biotech 4:391–401

Deepa CK, Dastager SG, Pandey A (2010) Isolation and characterization of plant growth promoting bacteria from non-rhizospheric soil and their effect on cowpea (*Vigna unguiculata* (L.) Walp.) seedling growth. World J Microbiol Biotechnol 26:1233–1240

Del Campillo MC, Van der Zee SEATM, Torrent J (1999) Modelling long-term phosphorus leaching and changes in phosphorus fertility in excessively fertilized acid sandy soils. Eur J Soil Sci 50:391–399

Diep CN, So DB, Trung NB et al (2016) Effects of rhizobia and phosphate-solubilizing bacteria on soybean (*Glycine max* L. Merr.) cultivated on ferralsols of daklak province, Vietnam. World J Pharm Pharm Sci 5:318–333

Dilantha F, Nakkeeran S, Yilan Z (2006) Biosynthesis of antibiotics by PGPR and its relation in biocontrol of plant diseases. In: PGPR: biocontrol and biofertilization. Springer, Netherlands, pp 67–109

Dugan P, Lundgren DG (1965) Energy supply for the chemoautotroph *Ferrobacillus ferrooxidans*. J Bacteriol 89:825–834

El-Mehalawy AA (2009) Management of plant diseases using phosphate-solubilizing microbes. In: Khan MS, Zaidi A (eds) Phosphate solubilizing microbes for crop improvement. Nova Science Publishers, USA, pp 265–279

Fallo G, Mubarik NR, Triadiati (2015) Potency of auxin producing and phosphate solubilizing bacteria from dryland in rice paddy field. Res J Microbiol 10:246–259

Fernández LA, Zalba P, Gómez MA et al (2007) Phosphate-solubilization activity of bacterial strains in soil and their effect on soybean growth under greenhouse conditions. Biol Fertil Soils 43:805–809

Garg N, Chandel S (2010) Arbuscular mycorrhizal networks: process and functions. Agron Sustain Dev 30:581–599

Garg A, Sharma M (2013) Evaluation of phosphate solubilizing activity and indole acetic acid production of rhizobia. Int J Sci Res 2:314–316

Ghosh R, Barman S, Mukherjee R (2016) Role of phosphate solubilizing *Burkholderia* spp. for successful colonization and growth promotion of *Lycopodium cernuum* L. (Lycopodiaceae) in lateritic belt of Birbhum district of West Bengal, India. Microbiol Res 183:80–91

Goldstein AH (1986) Bacterial solubilization of mineral phosphates: historical perspectives and future prospects. Am J Altern Agric 1:57–65

Goldstein AH (1995) Recent progress in understanding the molecular genetics and biochemistry of calcium phosphate solubilization by Gram negative bacteria. Biol Agric Hortic 12:185–193

Gulati A, Vyas P, Rahi P et al (2009) Plant growth-promoting and rhizosphere-competent *Acinetobacter rhizosphaerae* strain BIHB 723 from the cold deserts of the Himalayas. Curr Microbiol 58:371–377

Gulati A, Sharma N, Vyas P et al (2010) Organic acid production and plant growth promotion as a function of phosphate solubilization by *Acinetobacter rhizosphaerae* strain BIHB 723 isolated from the cold deserts of the trans-Himalayas. Arch Microbiol 192:975–983

Gull M, Hafeez FY, Saleem M et al (2004) Phosphorus uptake and growth promotion of chickpea by co-inoculation of mineral phosphate solubilising bacteria and a mixed rhizobial culture. Aust J Exp Agric 44:623–628

Gusain YS, Kamal R, Mehta CM et al (2015) Phosphate solubilizing and indole-3-acetic acid producing bacteria from the soil of Garhwal Himalaya aimed to improve the growth of rice. J Environ Biol 36:301–307

Halder AK, Chakrabarthy PK (1993) Solubilization of inorganic phosphate by *Rhizobium*. Folia Microbiol 38:325–330

Hegde DM, Dwivedi BS, Sudhakara SN (1999) Biofertilizers for cereal production in India. Indian J Agr Sci 69:73–83

Hwangbo H, Park RD, Kim YW et al (2003) 2-Ketogluconic acid production and phosphate solubilization by *Enterobacter intermedium*. Curr Microbiol 47:87–92

Ilmer P, Schinner F (1992) Solubilization of inorganic phosphates by microorganisms isolated from forest soil. Soil Biol Biochem 24:389–395

Ilmer P, Schinner F (1995) Solubilization of inorganic calcium phosphates-solubilization mechanisms. Soil Biol Biochem 27:257–263

Indiragandhi P, Anandham R, Madhaiyan M et al (2008) Characterization of plant growth-promoting traits of bacteria isolated from larval guts of diamondback moth *Plutella xylostella* (Lepidoptera: Plutellidae). Curr Microbiol 56:327–333

Jeffries P, Gianinazzi S, Perotto S et al (2003) The contribution of arbuscular mycorrhizal fungi in sustainable maintenance of plant health and soil fertility. Biol Fertil Soil 37:1–16

Jiang C, Sheng X, Qian M et al (2008) Isolation and characterization of a heavy metal-resistant *Burkholderia* sp. from heavy metal-contaminated paddy field soil and its potential in promoting plant growth and heavy metal accumulation in metal-polluted soil. Chemosphere 72:157–164

Jilani G, Akram AR, Ali M et al (2007) Enhancing crop growth, nutrients availability, economics and beneficial rhizosphere microflora through organic and biofertilizers. Ann Microbiol 57:177–183

Jorquera MA, Hernández MT, Rengel Z et al (2009) Isolation of culturable phosphobacteria with both phytate-mineralization and phosphate-solubilization activity from the rhizosphere of plants grown in a volcanic soil. Biol Fertil Soils 44:1025–1034

Karpagam T, Nagalakshmi PK (2014) Isolation and characterization of phosphate solubilizing microbes from agricultural soil. Int J Curr Microbiol Appl Sci 3:601–614

Khan MS, Zaidi A, Aamil M (2002) Biocontrol of fungal pathogens by the use of plant growth promoting rhizobacteria and nitrogen fixing microorganisms. Ind J Bot Soc 81:255–263

Khan MS, Zaidi A, Wani PA (2007) Role of phosphate-solubilizing microorganisms in sustainable agriculture—a review. Agron Sustain Dev 27:29–43

Khan AA, Jilani G, Akhtar MS et al (2009a) Phosphorus solubilizing bacteria: occurrence, mechanisms and their role in crop production. J Agric Biol Sci 1:48–58

Khan MS, Zaidi A, Wani PA et al (2009b) Functional diversity among plant growth-promoting rhizobacteria. In: Khan MS, Zaidi A, Musarrat J (eds) Microbial strategies for crop improvement. Springer, Berlin, Heidelberg, pp 105–132

Khan MS, Zaidi A, Ahemad M et al (2010) Plant growth promotion by phosphate solubilizing fungi—current perspective. Arch Agron Soil Sci 56:73–98

Khan AL, Halo BA, Elyassi A et al (2016) Indole acetic acid and ACC deaminase from endophytic bacteria improves the growth of *Solanum lycopersicum* Electron J Biotechnol 21:58–64

Kim KY, Jordan D, Krishnan HB (1997) *Rahnella aquatilis*, bacterium isolated from soybean rhizosphere, can solubilize hydroxyapatite. FEMS Microbiol Lett 153:273–277

Kumar GK, Ram MR (2014) Phosphate solubilizing rhizobia isolated from *Vigna trilobata*. Am J Microbiol Res 2:105–109

Kumar A, Kumar K, Kumar P et al (2014) Production of indole acetic acid by *Azotobacter* strains associated with mungbean. Plant Arch 14:41–42

Lemanski K, Scheu S (2014) Incorporation of ^{13}C labelled glucose into soil microorganisms of grassland: effects of fertilizer addition and plant functional group composition. Soil Biol Biochem 69:38–45

Li JF, Zhang SQ, Huo PH et al (2013) Effect of phosphate solubilizing rhizobium and nitrogen fixing bacteria on growth of alfalfa seedlings under P and N deficient conditions. Pak J Bot 45:1557–1562

Liang Z, Yuhong Y, Qian L et al (2013) Mobilization of inorganic phosphorus from soils by five azotobacters. Acta Ecol Sin 33:2157–2164

Lingua G, Franchin I, Todeschini V et al (2008) Arbuscular mycorrhizal fungi differentially affect the response to high zinc concentrations of two registered poplar clones. Environ Pollut 153:137–147

Lipping Y, Jiatao X, Daohong J et al (2008) Antifungal substances produced by *Penicillium oxalicum* strain PY-1-potential antibiotics against plant pathogenic fungi. World J Microbiol Biotechnol 24:909–915

Madhuri MS (2011) Screening of rhizobia for indole acetic acid production. Ann Biol Res 2:460–468

Maheswar NU, Sathiyavani G (2012) Solubilization of phosphate by *Bacillus* sp. from groundnut rhizosphere (*Arachis hypogaea* L.) J Chem Pharm Res 4:4007–4011

Maliha R, Samina K, Najma A et al (2004) Organic acid production and phosphate solubilization by phosphate solubilizing microorganisms under in vitro conditions. Pak J Biol Sci 7:187–196

Mardad I, Serrano A, Soukri A (2013) Solubilization of inorganic phosphate and production of organic acids by bacteria isolated from a Moroccan mineral phosphate deposit. Afr J Microbiol Res 7:626–635

McKenzie RH, Roberts TL (1990) Soil and fertilizers phosphorus update. In: Proceedings of the Alberta soil science workshop proceedings, Edmonton, Alberta, 20–22 Feb 1990, pp 84–104

Mehdi Z, Nahid SR, Alikhani HA et al (2006) Response of lentil to co-inoculation with phosphate solubilizing rhizobial strains and arbuscular mycorrhizal fungi. J Plant Nutr 29:1509–1522

Messele B, Pant LM (2012) Effects of inoculation of *Sinorhizobium ciceri* and phosphate solubilizing bacteria on nodulation, yield and nitrogen and phosphorus uptake of chickpea (*Cicer arietinum* L.) in Shoa Robit Area. J Biofertil Biopestici 3:129

Mia M, Shamsuddin ZH (2010) *Rhizobium* as a crop enhancer biofertilizers for increased cereal production. Afr J Biotechnol 9:6001–6009

Mirdhe RM, Lakshman HC (2014) Synergistic interaction between arbuscular mycorrhizal fungi, *Rhizobium* and phosphate solubilizing bacteria on *Vigna unguiculata* l verdc. Int J Bioassays 3:2096–2099

Mishra PK, Mishra S, Selvakumar G et al (2009) Coinoculation of *Bacillus thuringeinsis*-KR1 with *Rhizobium leguminosarum* enhances plant growth and nodulation of pea (*Pisum sativum* L.) and lentil (*Lens culinaris* L.) World J Microbiol Biotechnol 25:753–761

Mishra A, Chauhan PS, Chaudhry V et al (2011) Rhizosphere competent *Pantoea agglomerans* enhances maize (*Zea mays*) and chickpea (*Cicer arietinum* L.) growth, without altering the rhizosphere functional diversity. Antonie van Leeuwenhoek 100:405–413

Mohammadi K (2012) Phosphorus solubilizing bacteria: occurrence, mechanisms and their role in crop production. Resources Environ 2:80–85

Nandinin K, Preethi U, Earanna N (2014) Molecular identification of phosphate solubilizing bacterium (*Alcaligenes faecalis*) and its interaction effect with *Bradyrhizobium japonicum* on growth and yield of soybean (*Glycine max* L.) Afr J Biotechnol 13:3450–3454

Narveer, Vyas A, Kumar H et al (2014) *In vitro* phosphate solubilization by *Bacillus* sp. NPSBS 3.2.2 obtained from the cotton plant rhizosphere. Biosci Biotechnol Res Asia 11:401–406

Nosrati R, Owlia P, Saderi H et al (2014) Phosphate solubilization characteristics of efficient nitrogen fixing soil *Azotobacter* strains. Iran J Microbiol 6:285–295

Oedjijono ES, Soetarto S, Moeljopawiro S et al (2014) Promising plant growth promoting rhizobacteria of *Azospirillum* spp. isolated from iron sand soils, Purworejo coast, central Java, Indonesia. Adv Appl Sci Res 5:302–308

Oteino N, Lally RD, Kiwanuka S et al (2015) Plant growth promotion induced by phosphate solubilizing endophytic *Pseudomonas* isolates. Front Microbiol 6:745

Othman H, Tamimi SM (2016) Characterization of rhizobia nodulating faba bean plants isolated from soils of Jordan for plant growth promoting activities and N_2 fixation potential. Int J Adv Res Biol Sci 3:20–27

Panhwar QA, Naher UA, Jusop S et al (2014) Biochemical and molecular characterization of potential phosphate-solubilizing bacteria in acid sulfate soils and their beneficial effects on rice growth. PLoS One 9:e97241. doi:10.1371/journal.pone.0097241

Parani K, Saha BK (2012) Prospects of using phosphate solubilizing *Pseudomonas* as bio fertilizer. Eur J Biol Sci 4:40–44

Parikh K, Jha A (2012) Biocontrol features in an indigenous bacterial strain isolated from agricultural soil of Gujarat, India. J Soil Sci Plant Nutr 12:245–252

Patil V (2011) Production of indole acetic acid by *Azotobacter* sp. Rec Res Sci Technol 3:14–16

Peix A, Rivas-Boyero AA, Mateos PF et al (2001) Growth promotion of chickpea and barley by a phosphate solubilizing strain of *Mesorhizobium mediterraneum* under growth chamber conditions. Soil Biol Biochem 33:103–110

Ponmurugan K (2012) Biological activities of plant growth promoting *Azotobacter* sp. isolated from vegetable crops rhizosphere soils. J Pure Appl Microbiol 6:1–10

Poonguzhali S, Madhaiyan M, Sa T (2008) Isolation and identification of phosphate solubilizing bacteria from chinese cabbage and their effect on growth and phosphorus utilization of plants. J Microbiol Biotechnol 18:773–777

Pradhan N, Shukla LB (2005) Solubilization of inorganic phosphates by fungi isolated from agriculture soil. Afr J Biotechnol 5:850–854

Qureshi MA, Ahmad ZA, Akhtar N et al (2012a) Role of phosphate solubilizing bacteria (PSB) in enhancing p availability and promoting cotton growth. J Anim Plant Sci 22:204–210

Qureshi MA, Iqbal A, Akhtar N et al (2012b) Co-inoculation of phosphate solubilizing bacteria and rhizobia in the presence of L-tryptophan for the promotion of mash bean (*Vigna mungo* L.) Soil Environ 31:47–54

Rathi M, Malik DK, Bhatia P et al (2008) Effect of phosphate solubilizing bacterial strains on plant growth of green gram (*Vigna radiata*) and mustard (*Brassica compestris*). J Pure Appl Microbiol 3:125–133

Reyes I, Bernier L, Simard R et al (1999) Effect of nitrogen source on solubilization of different inorganic phosphates by bacterial strain of *Penicillium rugulosum* and two UV induced mutants. FEMS Microbiol Ecol 28:281–290

Reyes I, Baziramakenga R, Bernier L et al (2001) Solubilization of phosphate rocks and minerals by a wild-type strain and two UV induced mutants of *Penicillium rugulosum*. Soil Biol Biochem 33:1741–1747

Richardson AE (1994) Soil microorganisms and phosphorous availability. In: Pankhurst CE, Doube BM, Gupta VVSR (eds) Soil biota: management in sustainable farming systems. CSIRO, Victoria, Australia, pp 50–62

Richardson AE (2001) Prospects for using soil microorganisms to improve the acquisition of phosphorus by plants. Aust J Plant Physiol 28:897–906

Rivas R, Peix A, Mateos PF et al (2006) Biodiversity of populations of phosphate solubilizing rhizobia that nodulates chickpea in different Spanish soils. Plant Soil 287:23–33

Robson AD, O'Hara GW, Abbott LK (1981) Involvement of phosphorus in nitrogen fixation by subterranean clover (*Trifolium subterraneum* L.) Aust J Plant Physiol 8:427–436

Rosas SB, Andrés JA, Rovera M et al (2006) Phosphate-solubilizing *Pseudomonas putida* can influence the rhizobia–legume symbiosis. Soil Biol Biochem 38:3502–3505

Rudrappa T, Splaine RE, Biedrzycki ML et al (2008) Cyanogenic pseudomonads influence multitrophic interactions in the rhizosphere. PLoS One 3:e2073. doi:10.1371/journal.pone. 0002073

Saini P (2012) Preliminary screening for plant disease suppression by plant growth promoting rhizobacteria. Sci Res Rep 2:246–250

Selvakumar G, Mohan M, Kundu S et al (2008) Cold tolerance and plant growth promotion potential of *Serratia marcescens* strain SRM (MTCC 8708) isolated from flowers of summer squash (*Cucurbita pepo*). Lett Appl Microbiol 46:171–175

Shaharoona B, Naveed M, Arshad M et al (2008) Fertilizer-dependent efficiency of *Pseudomonads* for improving growth, yield, and nutrient use efficiency of wheat (*Triticum aestivum* L.) Appl Microbiol Biotechnol 79:147–155

Shenoy VV, Kalagudi GM (2005) Enhancing plant phosphorus use efficiency for sustainable cropping. Biotechnol Adv 23:501–513

Shweta B, Maheshwari DK, Dubey RC et al (2008) Beneficial effects of fluorescent pseudomonads on seed germination, growth promotion, and suppression of charcoal rot in groundnut (*Arachis hypogea* L.) J Microbiol Biotechnol 18:1578–1583

Siddiqui IA, Shaukat SS, Sheikh IH et al (2006) Role of cyanide production by *Pseudomonas fluorescens* CHA0 in the suppression of root-knot nematode, *Meloidogyne javanica* in tomato. World J Microbiol Biotechnol 22:641–650

Silveira A, Cardoso EJB (2004) Arbuscular mycorrhiza and kinetic parameters of phosphorus absorption by bean plants. Agric Sci 61:203–209

Singh S, Gupta G, Khare E et al (2014) Phosphate solubilizing rhizobia promote the growth of chickpea under buffering conditions. Int J Pure Appl Biosci 2:97–106

Song OR, Lee SJ, Lee YS et al (2008) Solubilization of insoluble inorganic phosphate by *Burkholderia cepacia* DA23 isolated from cultivated soil. Braz J Microbiol 39:151–156

Sridevi M, Mallaiah KV (2009) Phosphate solubilization by *Rhizobium* strains. Ind J Microbiol 49:98–102

Stajković O, Delić D, Jošić D et al (2011) Improvement of common bean growth by co-inoculation with *Rhizobium* and plant growth-promoting bacteria. Rom Biotechnol Lett 16:5919–5926

Tagore GS, Namdeo SL, Sharma SK et al (2013) Effect of *Rhizobium* and phosphate solubilizing bacterial inoculants on symbiotic traits, nodule leghemoglobin, and yield of chickpea genotypes. Int J Agron. Article ID: 581627, 8p

Tahir M, Mirza MS, Zaheer A (2013) Isolation and identification of phosphate solubilizer *Azospirillum, Bacillus* and *Enterobacter* strains by 16SrRNA sequence analysis and their effect on growth of wheat (*Triticum aestivum* L.) Aust J Crop Sci 7:1284–1292

Takahashi S, Anwar MR (2007) Wheat grain yield, phosphorus uptake and soil phosphorus fraction after 23 years of annual fertilizer application to an Andosol. Field Crops Res 101:160–171

Taurian T, Anzuay MS, Angelini JG et al (2010) Phosphate-solubilizing peanut associated bacteria: screening for plant growth-promoting activities. Plant Soil 329:421–431

Toro M, Azcón R, Barea JM (2008) The use of isotopic dilution techniques to evaluate the interactive effects of *Rhizobium* genotype, mycorrhizal fungi, phosphate-solubilizing rhizobacteria and rock phosphate on nitrogen and phosphorus acquisition by *Medicago sativa*. New Phytol 138:265–273

Turnau K, Orlowska E, Ryszka P et al (2006) Role of arbuscular mycorrhiza in phytoremediation toxicity monitoring of heavy metal rich industrial wastes in southern poland. In: Twardowska I, Herbert E, Allen (eds) Soil and water pollution monitoring, protection and remediation. Springer, Netherlands, pp 533–551

Valverde A, Burgos A, Fiscella T et al (2006) Differential effects of co inoculations with *Pseudomonas jessenii* PS06 (a phosphate-solubilizing bacterium) and *Mesorhizobium ciceri* C-2/2 strains on the growth and seed yield of chickpea under greenhouse and field conditions. Plant Soil 287:43–50

Vazquez P, Holguin G, Puente ME et al (2000) Phosphate-solubilizing microorganisms associated with the rhizosphere of mangroves in a semiarid coastal lagoon. Biol Fertil Soils 30:460–468

Vessey JK (2003) Plant growth promoting rhizobacteria as biofertilizers. Plant Soil 255:571–586

Vikram A, Hamzehzarghani H (2008) Effect of phosphate solubilizing bacteria on nodulation and growth parameters of greengram (*Vigna radiate* L. Wilczek). Res J Microbiol 3:62–72

Vikram A, Hamzehzarghani H, Alagawadi AR et al (2007) Production of plant growth promoting substances by phosphate solubilizing bacteria isolated from vertisols. J Plant Sci 2:326–333

Vyas P, Gulati A (2009) Organic acid production in vitro and plant growth promotion in maize under controlled environment by phosphate-solubilizing fluorescent *Pseudomonas*. BMC Microbiol 9:174

Walpola BC, Yoon MH (2012) Prospectus of phosphate solubilizing microorganisms and phosphorus availability in agricultural soils: a review. Afr J Microbiol Res 6:6600–6605

Walpola BC, Song JS, Keum MJ et al (2012) Evaluation of phosphate solubilizing potential of three *Burkholderia* species isolated from green house soils. Kor J Soil Sci Fertil 45:602–609

Wani PA, Khan MS, Zaidi A (2007a) Co-inoculation of nitrogen-fixing and phosphate-solubilizing bacteria to promote growth, yield and nutrient uptake in chickpea. Acta Agron Hung 55:315–323

Yazdani M, Bahmanyar MA, Pirdashti H et al (2009) Effect of phosphate solubilization microorganisms (PSM) and plant growth promoting rhizobacteria (PGPR) on yield and yield components of Corn (*Zea mays L.*) Proc World Acad Sci Eng Technol 37:90–92

Yi Y, Huang W, Ge Y (2008) Exopolysaccharide: a novel important factor in the microbial dissolution of tricalcium phosphate. World J Microbiol Biotechnol 24:1059–1065

Zabihi HR, Savaghebi GR, Khavazi K et al (2011) *Pseudomonas* bacteria and phosphorous fertilization, affecting wheat (*Triticum aestivum* L.) yield and P uptake under greenhouse and field conditions. Acta Physiol Plant 33:145–152

Zaidi A, Khan MS (2006) Co-inoculation effects of phosphate solubilizing microorganisms and glomus fasciculatum on green gram-Bradyrhizobium symbiosis. Turk J Agric For 30:223–230

Zaidi A, Khan MS, Amil M (2003) Interactive effect of rhizotrophic microorganisms on yield and nutrient uptake of chickpea (*Cicer arietinum* L.) Eur J Agron 19:15–21

Zaidi A, Khan MS, Aamil M (2004) Bioassociative effect of rhizospheric microorganisms on growth, yield, and nutrient uptake of greengram. J Plant Nutr 27:601–612

Zaidi S, Usmani S, Singh BR et al (2006) Significance of *Bacillus subtilis* strain SJ-101 as a bioinoculant for concurrent plant growth promotion and nickel accumulation in *Brassica juncea*. Chemosphere 64:991–997

Zaidi A, Khan MS, Ahemad M, Oves M (2009) Plant growth promotion by phosphate solubilizing bacteria. Acta Microbiol Immunol Hung 56:283–284

Zou X, Binkley D, Doxtader KG (1992) A new method for estimating gross phosphorus mineralization and immobilization rates in soils. Plant Soil 147:243–250

Mycorrhizosphere Interactions to Improve a Sustainable Production of Legumes

9

José-Miguel Barea, Rosario Azcón, and Concepción Azcón-Aguilar

Abstract

The sustainability and productivity of agroecosystems depends exquisitely on the functionality of a framework of plant–soil interactions where microbial populations, including both mutualistic symbionts and saprophytic microorganisms, living at the root–soil interfaces, the rhizosphere, are involved. Among various beneficial and consumable plant species, legumes form useful symbiotic relationships with two types of soil microbiota: N_2-fixing bacteria, often called rhizobia, and arbuscular mycorrhizal (AM) fungi. Also, the legume rhizosphere inhabits other valuable microbes such as plant growth-promoting rhizobacteria (PGPR). These microorganisms interact intensely among themselves, and with legume roots, to develop the multifunctional legume mycorrhizosphere, a microcosm environment of variable activities, appropriate for legume productivity. This chapter highlights (1) the types of microorganisms and processes involved in the establishment and functioning of the mycorrhizosphere, (2) the impact of the mycorrhizosphere activities on legume production, and (3) the possibilities to tailor an efficient mycorrhizosphere as a biotechnological tool to improve legume performance in different production systems following efficient rhizobial, PGPR, and AM fungal inoculants.

J.-M. Barea (✉) • R. Azcón • C. Azcón-Aguilar
Departamento de Microbiología del Suelo y Sistemas Simbióticos, Estación, Experimental del Zaidín, Prof. Albareda 1, 18008 Granada, Spain
e-mail: jmbarea@eez.csic.es

9.1   Introduction

The great challenge that science and society are currently facing is how to satisfy the increasing demands of healthy foods for the constantly growing human populations. Considering these challenges, there is urgent need to increase agricultural/ food production globally. In this regard, intensive agriculture appears, theoretically, fundamental to respond to this challenge. However, high-input agricultural practices are concomitant with the mass consumption of nonrenewable natural resources particularly rock phosphate (RP) reserves (George et al. 2016). Also, the excessive use of agrochemicals in modern farming practices causes stress to the plants and may cause climatic change (Barea 2015). Due to these problems, society and science both are becoming aware on the necessity to follow sustainable agricultural production models which could provide a healthy and nutritious food supply without compromising on yields (Altieri 2004). Similarly, a sustainable management (restoration practices) of natural soil–plant ecosystems is peremptory to preserve the biodiversity and environmental quality (Barea et al. 2011). Concerning quality food production, diverse research approaches have been proposed and practiced to achieve environmental and economical sustainability. One of these approaches, exploitation of soil microbial communities (Barea 2015), has been considered as safe, inexpensive, and environmentally friendly. Indeed, diverse genetic and functional groups of soil microorganisms are known to play decisive roles (microbial services) in agriculture, mainly by propelling nutrient cycling, enhancing plant nutrition, and promoting plant health and soil quality (Lugtenberg 2015).

Microbial activities are particularly relevant at the root–soil interface microhabitats known as the rhizosphere, where microorganisms are stimulated by carbon substrates provided by plant rhizodeposits (Hirsch et al. 2013b). Formation, development, significance, functioning, and managing of the rhizosphere have been reviewed (Barea et al. 2013a). Currently, much attention is given to optimize the functions of root-associated microbiome in enhancing plant nutrient capture and for increasing plant resistance/tolerance to either biotic or abiotic stress factors. Accordingly, several strategies for identifying and utilizing beneficial microbial services have been proposed to promote a sustainable and environmentally friendly agricultural production (Raaijmakers and Lugtenberg 2013; Barea 2015). It is noteworthy to point out that the use of molecular techniques has evidenced that only 1% of microorganisms living in the bulk soil, and 10% of those from the rhizosphere microbiome, are able to grow in standard in vitro culture media and can therefore be isolated and multiplied (Hirsch et al. 2013a). The rest of soil microorganisms are considered as unculturable but can be detected and analyzed for their effectiveness using culture-independent molecular approaches (Barret et al. 2013; Schreiter et al. 2015). Most studies on the plant-associated microbiome focus on bacteria and fungi and both of them can establish either saprophytic or symbiotic relationships with the plant which could either be detrimental or beneficial (Spence and Bais 2013; Lugtenberg et al. 2013a, b). Beneficial plant mutualists are both the N_2-fixing bacteria (Olivares et al. 2013) and the multifunctional arbuscular mycorrhizal (AM) fungi (van der Heijden et al. 2015).

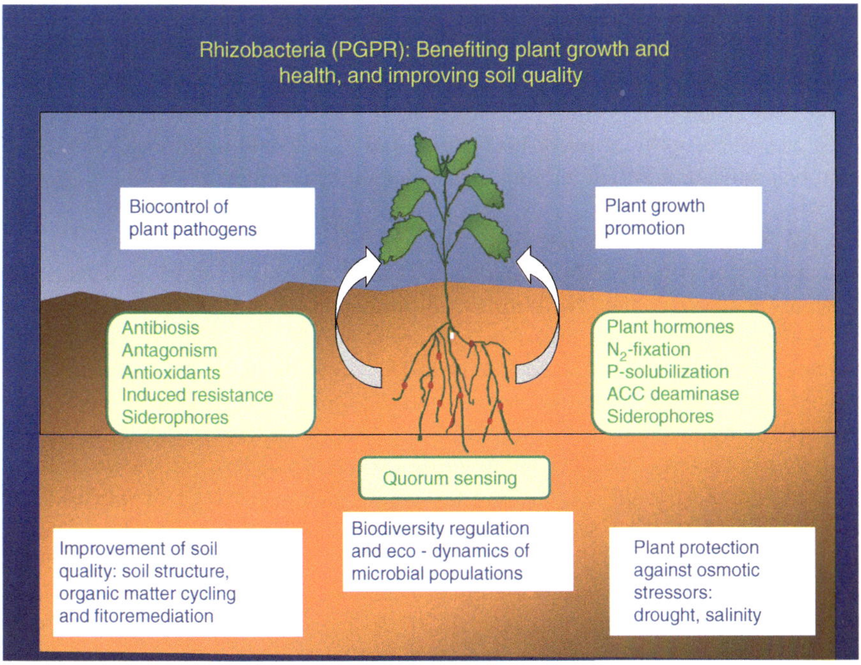

Fig. 9.1 Plant growth promoting activities of rhizobacteria (modified from Barea et al. (2013a))

the reduction in the saprophytic growth of the pathogens which occurs mainly via antagonistic activities, as mediated by the production of antibiotics, (2) the reduction of the virulence of the pathogen, and (3) the induction of systemic resistance in the host plants (Pieterse et al. 2014). *Trichoderma* spp. are rhizosphere fungi which promote plant growth and act as pathogen antagonists following the above indicated mechanisms and, additionally, exert as mycoparasitic agent (Hermosa et al. 2012).

Another fundamental activity of PGPR is nutrient cycling, particularly nitrogen fixation (Azcón-Aguilar and Barea 2015) and phosphate mobilization (Khan et al. 2010; Barea and Richardson 2015). The N_2 fixation process is the first step in cycling N from the atmosphere to the biosphere, a key N input to plant productivity (Arrese-Igor 2010). Many free-living diazotrophic bacteria are recognized to be able to fix N_2 that these microbes use as a source of N for themselves, with a low direct N transfer to the plant, having therefore a limited agronomic significance (Ramos-Solano et al. 2009; Olivares et al. 2013).

Diverse rhizobacteria and rhizofungi have the capacity to mobilize P from poorly available sources of this element and supply soluble P to plants (Zaidi et al. 2010; Barea and Richardson 2015). The mechanisms whereby P-mobilizing microorganisms release available P from sparingly soluble soil P forms, either inorganic (solubilization) or organic (mineralization), by means of activities largely based on producing specific enzymes and/or chelating organic acids, have recently been discussed (Barea and Richardson 2015). The effect of phosphate-mobilizing microorganisms, mostly

PGPR, has been tested under field conditions, but their effectiveness in the soil–plant system is variable (Antoun 2012). One of the reasons for a lack in realizing benefit to the plant is that the P ions made available could be refixed by the soil constituents before they reach the root surface. However, if phosphate ions, as released by the PMB, are taken up by a mycorrhizal mycelium, this would result in a synergistic microbial interaction which improves P acquisition by the plant.

9.2.1.2 N_2-Fixing Symbiotic Bacteria

Associative and symbiotic N_2-fixing bacteria are fundamental in plant N nutrition (Olivares et al. 2013). The N_2-fixing symbiotic bacteria belonging to different genera, collectively termed as "rhizobia," are able to fix N_2 in symbiosis with legume plants (Olivares et al. 2013; de Bruijn 2015). How these bacteria interact with legume roots to form N_2-fixing nodules and the molecular aspect determinants of host specificity in the rhizobia–legume symbiosis are described elsewhere in this book. Other bacteria, from the genus *Frankia* (actinomycetes) are known to form N_2-fixing nodules on the roots of the so-called "actinorrhizal" plant species. The associative bacteria, like *Azospirillum*, colonize root surfaces and establish diazotrophic rhizocenosis with the plant and can even invade intercellular tissues; however N_2-fixing structures are not formed (Gutiérrez-Mañero and Ramos-Solano 2010; Bashan et al. 2011; Olivares et al. 2013). *Azospirillum* enhance N supply to the plant but act mainly by increasing the production of auxin-type phytohormones, which affect the rooting patterns thereby benefiting plant nutrient uptake from soil rather than as N_2-fixing bacteria (Dobbelaere et al. 2001).

9.2.1.3 Arbuscular Mycorrhizal (AM) Fungi

Some 50,000 fungal species are recognized to form mycorrhizal symbiosis with about 250,000 plant species and thereby to carry out fundamental roles in terrestrial ecosystems, particularly by propelling nutrient and carbon cycles and other agroecosystem services. Actually, mycorrhizal fungi provide up to 80% of plant N and P (van der Heijden et al. 2015). Several types of mycorrhiza are recognized, but the most widespread and agronomically important type is constituted by the arbuscular mycorrhizal (AM) associations, which colonize approximately 80% of terrestrial plant species growing in almost all terrestrial agroecosystems worldwide (Brundrett 2009).

AM fungi are ubiquitous soilborne microscopic fungi whose SSU rDNA phylogeny revealed that they have a monophyletic origin constituting the phylum *Glomeromycota* (Schüßler et al. 2001). There are both fossil and phylogenetic evidences that the terrestrial AM fungi are about 460 million years old, suggesting that they existed before the land flora, consisting on bryophyte-like plants, at 450 million years ago (Honrubia 2009; Schüßler and Walker 2011; Barea and Azcón-Aguilar 2013; Selosse et al. 2015). However, molecular clock analyses further revealed its origin of land plant to around 477 million years and that the origin of AM fungi took place 50–200 million years earlier than the land plants (Schüßler and Walker 2011; Shtark et al. 2012; Barea and Azcón-Aguilar 2013). Morphological (fossil records) and phylogenetic (molecular) studies support that AM fungi facilitated plant terrestrialization and that roots coevolved in association with AM fungi in such a way that the majority of the extant vascular plants live associated with AM fungi (Field et al. 2015).

The AM fungal community diversity studies were based initially on the morphological characterization of their large multinucleate spores, until the molecular tools became available (see Robinson-Boyer et al. 2009 for references of pioneering studies). Essentially, molecular-based identification processes include the PCR-amplified rDNA fragments of the spores and/or the mycelia from AM fungi followed by cloning, fingerprinting, and sequencing (Krüger et al. 2011; Brearley et al. 2016). New molecular approaches allow the quantitatively analyses of the effect of environment, geographical location, or management on the AM fungal communities. The Q-PCR technique can be used for simultaneous specific and quantitative investigations of particular taxa of AM fungi in roots and soils colonized by several taxa (König et al. 2010; Redecker et al. 2013). In addition, new techniques of high-throughput sequencing are available which have improved our understanding on the biology, evolution, and diversity of AM fungi (Lumini et al. 2010; Drumbell 2013; Öpik et al. 2013). The morphological characterization of AM fungi is also currently used, as complementary to the molecular methods (Oehl et al. 2011a, b; Redecker et al. 2013).

An important aspect in AM fungal diversity studies is to consider that AM fungi use three types of propagules for root colonization: spores, fragments of AM roots, and internal mycelium (IRM) and the external AM mycelium (ERM) developing in the root-associated soil (Smith and Read 2008). Most of AM fungal diversity surveys commonly focused on a single propagule compartment, traditionally the spore community in soil. However, with the spread of sequence-based identification methods, many studies have now addressed to the AM fungi colonizing roots, the IRM, or mycorrhizospheric soil samples that include both ERM and spores. The related information on this aspect was reviewed recently by Varela-Cervero et al. (2015, 2016), where the AM fungal diversity was analyzed in the three different propagule types. These studies suggest that AM fungal taxa are differentially allocated among soil mycelium, soil spores, and colonized root propagules in a natural environment. Obviously, these results are relevant for exploiting AM fungal diversity and designing vegetation restoration programs employing AM inoculation. The analysis of the genome of the AM fungi has been addressed in several studies based on functional molecular approaches (Gianinazzi-Pearson et al. 2012). The complete genome of the model AM fungus *Rhizophagus irregularis* (formerly *Glomus intraradices*) has been sequenced (Tisserant et al. 2013; Lin et al. 2014).

The most significant biological characteristic of AM fungi is their ability to establish association with members of almost all phyla of land plants, regardless of their taxonomic position, life form, or geographical distribution (Smith and Read 2008; Brundrett 2009). In this context, it is relevant to note that AM fungi exhibit little host specificity, while a single plant root can be colonized by many different AM fungi. However, a certain degree of host preference (functional compatibility) has been demonstrated, a fact having ecological and agronomical consequences. From the ecological point of view, a diverse AM fungal population is needed for maintaining the diversity, stability, and productivity of natural ecosystems, while an appropriate selection of AM fungal inocula is fundamental for the effectiveness of AM symbioses in plant production in agricultural systems (Barea and Azcón-Aguilar 2013).

9.2.2   The AM Symbiosis: Characteristic, Establishment, and Functions

This review will focus only on AM symbiosis and fungi, but a brief reference to the other mycorrhizal types is given. There are two main types of mycorrhiza: ecto- and endomycorrhiza (Smith and Read 2008; van der Heijden et al. 2015). In ectomycorrhizas, the fungus forms a mantle of hyphae around the feeder roots. The mycelium penetrates the root and grows between the cortical cells forming the so-called Hartig net where nutrient exchange between partners takes place. About 2% of higher plant species, mainly forest trees in the Fagaceae, Betulaceae, Pinaceae, and some woody legumes form ectomycorrhiza. The fungi involved belong mostly to class *Basidiomycetes* and *Ascomycetes* (mushrooms and truffles). In endomycorrhizas, the fungi colonize the root cortex both intercellularly and intracellularly and develop an extraradical mycelium growing in soil, a network of hyphae without extructuring a mantle around the root. There are three types of endomycorrhizal fungi: "ericoid," "orchid," and "arbuscular." Of these, arbuscular fungi are the most common type and widely distributed throughout the plant kingdom. The widespread and ubiquitous AM symbiosis is characterized by the treelike symbiotic structures, termed "arbuscules" that the fungus develops within the root cortical cells and where most of the nutrient exchange between the fungus and the plant occurs. An intermediate mycorrhizal type, the ectendomycorrhiza, is formed by plants in families in the Ericales and in the Monotropaceae and Cistaceae. In these mycorrhizal associations, the fungi form both a hyphal mantle and intracellular penetrations (Smith and Read 2008). The AM symbiosis is that established by plant of agronomic interest and by herbaceous, arbustive, and some trees in natural ecosystems (Barea and Azcón-Aguilar 2013; van der Heijden et al. 2015). The AM associations are known to benefit plant fitness and soil quality, mainly by improving nutrient acquisition and plant health (Barea 1991). As the AM symbiosis is the mycorrhizal type formed by legume plants (all but *Lupinus* spp.), this review will focus only on the AM symbiosis and their role at improving these species, the target of this chapter, both in agricultural and natural systems.

The cellular and developmental programs controlling the processes of AM formation, from propagule activation until the intracellular accommodation of the fungal symbiont, have recently been reviewed (Gutjahr and Parniske 2013; Bonfante and Desirò 2015). This molecular cross talk prior to physical contact is the recognition by the fungus of plant signaling molecules, the strigolactones, which stimulate the fungus to ramify (López-Ráez et al. 2011). On the other side, plants perceive diffusible fungal signals, called "Myc factors" (lipochitooligosaccharides), analogous to the nodulation (NOD) factor of nitrogen-fixing rhizobia, which induce symbiosis-specific responses in the host root (Genre and Bonfante 2010; Maillet et al. 2011; Bonfante and Desirò 2015). After contacting the root epidermal cells, the fungal hyphae form an appressorium from which the fungus penetrates the epidermal cells to develop an intraradical mycelium until the formation of the tree-shaped structures, the arbuscules. Each fungal branch within a plant cell is surrounded by a plant-derived periarbuscular membrane and an apoplastic interface between the plant and fungal plasma membranes. The resulting structure is fundamental for the exchange of symbiotic signals and nutrient between symbionts (Gutjahr and Parniske 2013; Bonfante and Desirò 2015).

The AM fungi contribute to nutrient, mainly P but also N, acquisition, and supply to plants by linking the geochemical and biotic portions of the soil ecosystem, thereby affecting rates and patterns of nutrient cycling in both agricultural and natural ecosystems. The extraradical mycelium of AM fungi is profusely branched and provides a very efficient nutrient-absorbing system beyond the Pi-depletion zone surrounding the plant roots, thereby reducing the distance that Pi must diffuse through the soil prior to its interception. Actually, the AM fungal mycelium can spread through the soil over considerably longer distances (up to 25 cm) than root hairs, and the hyphal length densities in field soils range from 3 to 14 m/g (Smith and Smith 2011, 2012). The ability of the AM hyphae to grow beyond the root Pi-depletion zone and deliver the intercepted Pi to the plant is thought to be the reason why AM associations increase Pi accumulation and plant growth in soils with low P availability. In addition, functionality of P and N transporters is fundamental in nutrient acquisition by AM plants (Gianinazzi-Pearson et al. 2012; Pérez-Tienda et al. 2014; Bonfante and Desirò 2015). The AM symbiosis also improves plant health through increased protection against environmental factors causing plant stress, including biotic (e.g., pathogen or insect attack and parasitic plants) or abiotic (e.g., drought, salinity, heavy metals, organic pollutants) stressors, and enhancing soil structure through the formation of the aggregates necessary for good soil tilth. The AM effects on plants (Tamayo et al. 2014; Pozo et al. 2013, 2015) including legumes (Azcón et al. 2013; Sánchez-Romera et al. 2016; Hidri et al. 2016) have been reported and presented in Fig. 9.2. The effects of mycorrhizal fungi at maintaining plant community structure and functions are well known, but the information is scarce concerning their relative influence in a wider context considering the multiple abiotic and biotic interactions occurring in plant communities (van der Heijden et al. 2015). A deep understanding of these aspects is important because the interactions among plants and mycorrhizal fungi can affect plant recruitment dynamics and the final result of plant competition, effects which are mediated by the mycelia network of mycorrhizal fungal activity.

9.2.3   Mycorrhizosphere Establishment

The concept of mycorrhizosphere, pragmatically the rhizosphere of a mycorrhizal plant, was recognized long time ago as a scenario of interactions among AM fungi and other members of the root-associate microbiome resulting in activities promoting plant nutrition and plant health (Linderman 1988). How the mycorrhizosphere is established has recently been reviewed (Azcón-Aguilar and Barea 2015). Briefly, AM fungal colonization affects the structure and diversity of the rhizosphere

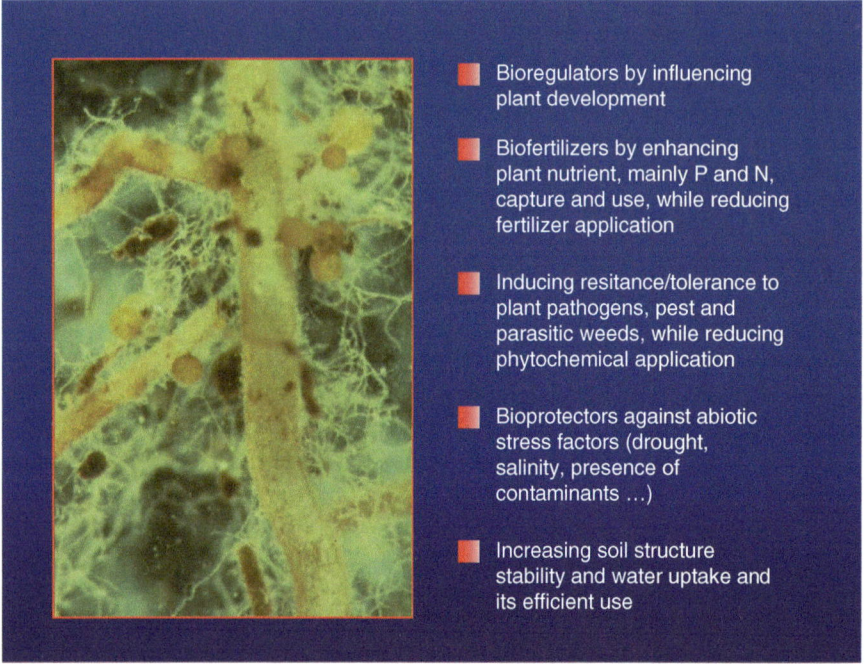

Fig. 9.2 Arbuscular mycorrhiza services in agro-ecology: biostimulants, biofertilizers, biofortifi-cants, bioprotectors and soil quality improvers (photography courtesy of Prof. S. Rosendahl, University of Copenhagen)

microbial communities. Conversely, some rhizosphere bacteria are known to help mycorrhiza formation, so termed as "mycorrhiza-helper-bacteria" (Frey-Klett et al. 2007; Fernández-Bidondo et al. 2011). The AM mycelium developing in soil itself creates a new physical environment for soil microorganisms; however, the main reason for the mycorrhizosphere effects is that root colonization by AM fungi changes mineral nutrient composition and other fundamental processes of plant physiology, such as the hormonal balance and C allocation patterns. These changes in turn alter the chemical composition of root exudates and the exudation rates (Barea et al. 2013a). The resulting situations of microbial interactions in the mycor-rhizosphere are fundamental for plant growth and plant protection (Fig. 9.3).

9.3   Tripartite Symbiosis in Legumes: Establishment and Functioning

The establishment of a functional and effective mycorrhizosphere is a key issue for legume productivity improvement (Azcón and Barea 2010; Muleta 2010). Here, the interaction between rhizobial bacteria and AM fungi and their interaction with the roots of their common legume host is discussed considering (1) a shared signaling

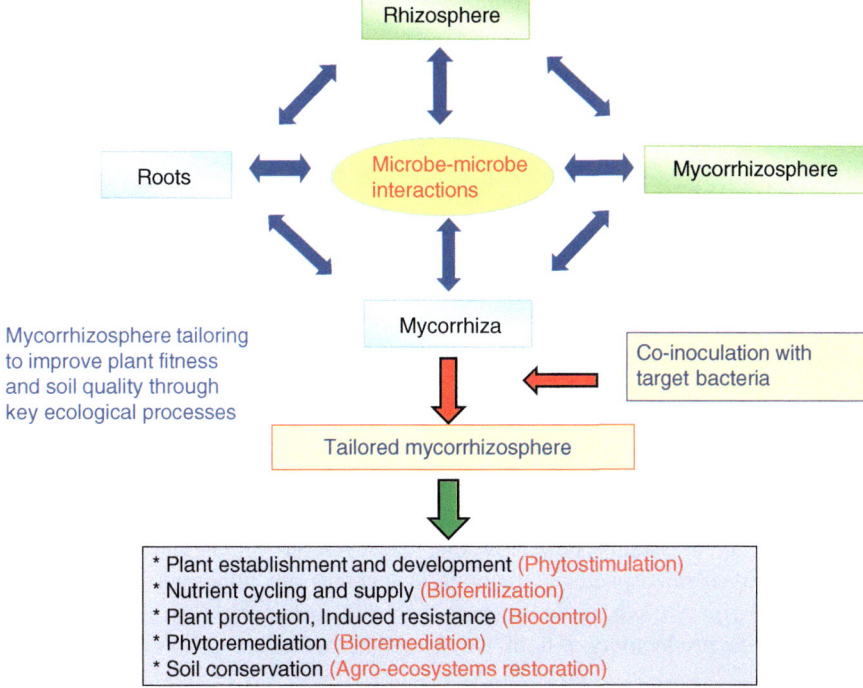

Fig. 9.3 Mycorrhizosphere tailoring: microbial interactions involving selected AM fungi and PGPR to improve nutrition and health, and soil quality (modified from Barea et al. (2013a))

pathway for AM fungi and root nodule symbiosis establishment, (2) physiological interactions related to the formation and functioning of the tripartite symbiosis, and (3) the use of ^{15}N to ascertain the role of AM fungi in N_2 fixation.

9.3.1 Establishment of Arbuscular Mycorrhiza and Root Nodule Symbiosis: A Shared Signaling Pathway

Root colonization by AM fungi depends on a number of genes termed symbiosis (SYM) genes (Gutjahr and Parniske 2013; Bonfante and Desirò 2015). Such a plant–microbe symbiotic toolkit evolved with the AM symbiosis from basal land plants to extant flora (Delaux et al. 2013). Even more, ancestors of land plants, the green algae in the Charophytes, appeared preadapted for symbiosis as they possessed SYM genes, a symbiotic toolkit which was later recruited and further developed alongside AM fungi and plant coevolution (Delaux et al. 2015). In addition, during plant evolution, the SYM genes were again recruited for other plant root symbioses, like the N_2-fixing rhizobial root nodules; thus a similar plant gene toolkit can modulate both types of legume symbioses. Actually, the legume–rhizobia symbiosis evolved much later than the AM symbiosis from a set of preadaptations during coevolution with AM fungi; thus the legume root symbioses may be considered as a component of an "evolutionary plant–microbial continuum" (Shtark et al. 2012). The genes required for AM establishment were identified first in legumes but were then found in many AM host plants (Parniske 2008; Bonfante and Genre 2010). The SYM genes, common to both symbioses, are known to encode proteins involved in a signaling transduction pathway starting with the perception of microbial signals at the plant plasma membrane, by means of a receptor kinase, and ending with the intracellular accommodation of both symbionts, AM fungi and rhizobial bacteria, into the host cell (Genre et al. 2013; Lagunas et al. 2015; Bonfante and Desirò 2015; Sun et al. 2015).

9.3.2 Formation and Functioning of the Tripartite Symbioses: Physiological Interactions

The AM associations are recognized as an adaptive strategy for P acquisition in soils with low P availability which is an important nutrient for legumes, required growth, nodulation, and N_2 fixation (Olivares et al. 2013; Azcón-Aguilar and Barea 2015). Since the mycorrhizosphere interactions in legumes are important for N and P cycling, the tripartite symbiosis becomes special in sustainable agriculture in order to improve the productivity of both, woody and herbaceous, legumes (Courty et al. 2015).

Numerous experiments have been conducted to study the physiological and biochemical basis of AM fungal x rhizobia interactions. The information, reviewed by Azcón and Barea (2010), reinforce the idea that the main cause of such interactions is the supply of P by the AM fungi to satisfy the high P demand for nodule formation and N_2 fixation, leading to an increased fixation rates.

A topic of research interest is whether AM fungi and rhizobia compete for photosynthates. Since legumes use up to 4–16% of photosynthesis products to satisfy the demands of their heterotrophic rhizobial and AM fungal symbionts, a reduction in productivity, therefore, results under photosynthate limitation. The main conclusions of former studies, as reviewed by Ha and Gray (2008), are that when host photosynthesis is limited, AM fungi usually show a competitive advantage for carbohydrates over the rhizobia, but under normal situations, the photosynthetic capacity of plants exceeds the carbon demand of the tripartite symbiosis. Pioneering results show that AM plants have developed a mechanism for enhancing photosynthesis to compensate for the C cost of the symbioses, as further corroborated (Mortimer et al. 2008). Accordingly, a comprehensive meta-analysis of potential photosynthate limitation of the symbiotic responses of legumes to rhizobia and AM fungi was carried out (Kaschuk et al. 2010). These authors analyzed 348 data points from published studies with 12 legume species using response variables plant yield, harvest index, and seed protein and certain lipid production. They found an increase in the target parameters supporting that legumes are not C limited under symbiotic conditions.

9.3.3 Evaluation of N_2 Fixation and Role of AM Fungi in This Process: A ^{15}N Approach

Plant-available N occurs in six isotopic forms; but only two of them are stable: ^{14}N with a natural abundance of 99.634% and ^{15}N with a natural abundance of 0.366%. The ratio of $^{15}N/^{14}N$ remains almost constant in the atmosphere, plants, and soils (Zapata 1990). However, addition of a small amount of ^{15}N-enriched inorganic fertilizer to soil increases the $^{15}N/^{14}N$ ratio and consequently the soil N pool. In this context, a basic concept in agronomy must be considered: "when a plant is confronted with two or more sources of a nutrient, the nutrient uptake from each source is proportional to the amounts available in each source" (Zapata 1990). Consequently, plants, after growing in a ^{15}N-labeled soil, will take N from soil at the isotopic proportion provided by the new $^{15}N/^{14}N$ ratio after the ^{15}N-enriched inorganic fertilizer addition, according to the known amount and richness of the added ^{15}N. Simple calculations allow us to quantify the amount of N in plant derived from soil or from the fertilizer.

These approaches and concepts form the basis to measure N_2 fixation rates of nodulated legumes. For quantitative measurements, an appropriate "non-fixing" reference crop is needed (Danso 1986). Furthermore, both N_2-fixing and non-fixing plants are grown on a ^{15}N-labeled soil, and both type of plants will take up ^{15}N and ^{14}N at a similar rate. However, since the N_2-fixing plants use atmospheric N as an additional source of available N, the ratio $^{15}N/^{14}N$ will be lower in the N_2-fixing plant (Danso 1986). The technique can be used to select the more efficient rhizobial strain that shows the highest reduction in $^{15}N/^{14}N$ ratio. These methodologies are used to measure N_2 fixation by rhizobia–legume symbioses under field conditions (Azcón-Aguilar and Barea 2015).

These techniques have also been applied to ascertain the role AM fungi in N_2 fixation by nodulated legumes, as reported in the first edition of this book (Azcón and Barea 2010), later discussed thoroughly (Azcón-Aguilar and Barea 2015). Briefly, the effect of the co-inoculation of rhizobia and AM fungi on N_2 fixation by legumes growing under field conditions using ^{15}N-aided methodologies was investigated first time by Barea et al. (1987). A lower $^{15}N/^{14}N$ ratio was recorded for the shoots of rhizobia-inoculated AM plants compared to those obtained in non-mycorrhizal plants. This finding indicated an enhancement of the N_2 fixation rates which was induced by the AM activity on the rhizobium–legumes symbiosis. Other studies further corroborated the contribution of the AM symbiosis to N_2 fixation by rhizobia-inoculated legumes both under greenhouse and field conditions (Toro et al. 1998; Barea et al. 1989a, b; Chalk et al. 2006).

In addition, the isotopic techniques have been also used to measure N transfer in mixed cropping or natural plant communities where legumes are usually involved (Zapata et al. 1987). Actually, the root exudates of legume plants are enriched in N compounds derived from fixation, and when nonlegume plants are growing nearby, these can capture N from the intermixed rhizospheres. This fact, known as "N transfer from fixation," can be measured by using ^{15}N-aided techniques (Danso 1986). As

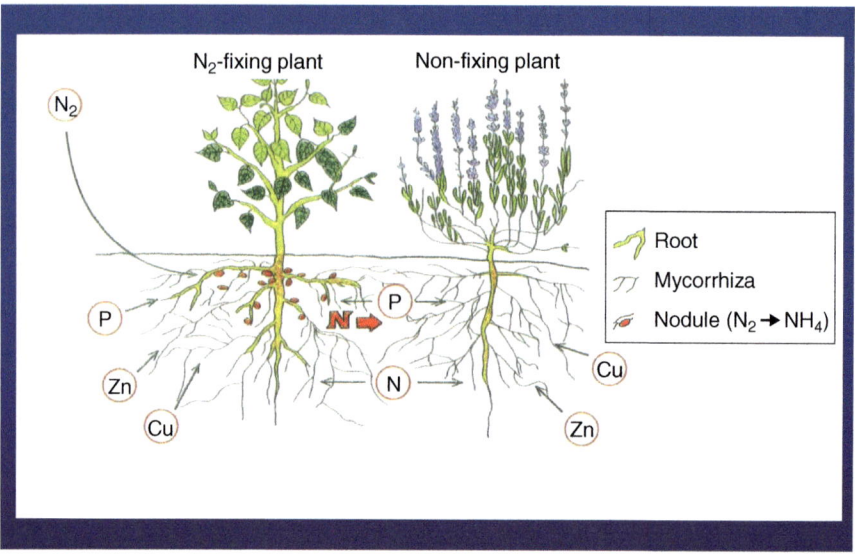

Fig. 9.4 Role of arbuscular mycorrhiza in nutrient acquisition by plant communities including legume and non-legume plants (drawing by Esperanza Campos: reprinted from Azcón and Barea (2010))

a part of the N in the exudates of the legume derived from N_2 fixation, which is at the natural abundance level ($^{15}N/^{14}N$ ratio), this N lowers the $^{15}N/^{14}N$ ratio in a ^{15}N-labeled soil. Consequently, a non-fixing plants growing near legumes will have a lower $^{15}N/^{14}N$ ratio than those growing near other nonlegume species. The effect of AM fungi in helping N transfer from N_2-fixing to non-fixing plants can also be measured, and it was found that the AM symbiosis increased "N transfer from fixation" to nonlegume plants in intermixed mycorrhizospheres both in the greenhouse (Barea et al. 1989a) and in the field conditions (Barea et al. 1989b; Requena et al. 2001). The role of AM symbiosis at improving nodulation and N_2 fixation in legumes, and helping "N transfer" from the rhizosphere of a N_2-fixing legume to a non-fixing plant growing nearby, is presented in Fig. 9.4.

9.4 Mycorrhizosphere Managing for Improving Legume Productivity

Managing mycorrhizosphere interactions (mycorrhizosphere tailoring) is a feasible biotechnological tool to improve plant growth and health and soil quality (Azcón-Aguilar and Barea 2015). The impact of mycorrhizosphere tailoring on legume performance has been tested under field conditions and involves the common host AM plant, N_2-fixing nodulating rhizobia, and phosphate-mobilizing bacteria (Azcón and Barea 2010; Shtark et al. 2012, 2015a; Zhukov et al. 2013; Larimer et al. 2014). However, the success of these organisms under field environment has been variable.

The inoculum production technologies, as a basis for tailoring legume mycorrhizosphere, are discussed in the following section.

9.4.1   Field Strategies for Testing a Managed Legume Mycorrhizosphere

Only few studies involving AM fungi–rhizobia interactions have been carried out under field conditions for a sustainable agricultural production. Since the pioneering work on dual inoculation of *Medicago sativa* grown in normal cultivation systems in an arable soil (Azcón-Aguilar et al. 1979), some experiments aimed at evaluating the role of AM fungi in improving N_2 fixation, either in controlled or in real field conditions, have been carried out all over the world. The accumulating data, however, suggest that several factors must be considered so that the inoculation effects of microbial symbionts on legumes become successful: (1) selection of appropriate rhizobial strain/AM fungus combination (Azcón et al. 1991; Ahmad 1995) and (2) fertility level of soils. In this context, a beneficial impact of AM fungi on legume symbiotic performance was corroborated mainly under low soil P levels (Chalk et al. 2006; Uyanoz et al. 2007; Pagano et al. 2008). Conversely, the AM symbiosis was not effective and did not promote N_2 fixation in soils with high levels of available P (Zaidi and Khan 2007; Lesueur and Sarr 2008).

Several studies based on managing the mycorrhizosphere of legumes for the revegetation of degraded areas suffering disturbance of their plant cover have been carried out (Azcón and Barea 2010). The information on the interactions of AM fungi and rhizobia, obtained from restoration (by revegetation) under field experiments, most of them concerning with Mediterranean desertification-threatened areas, was further analyzed (Barea et al. 2011). Two model experiments carried out in semiarid Mediterranean ecosystems of southeast Spain are briefly discussed. In one of them six species of woody legumes, adapted to the drought and nutrient-deficient condition of the target environment, were inoculated with both N_2-fixing bacteria and AM fungi (Herrera et al. 1993). The target shrub legumes included both native (*Anthyllis cytisoides* and *Spartium junceum*) and allochthonous (*Robinia pseudoacacia*, *Acacia caven*, *Prosopis chilensis*, and *Medicago arborea*) species. After 4 years of field grown, only the native shrub legumes were able to survive and thrive under the experimental conditions. It was also shown that the tailored mycorrhizosphere improved plant survival, outplanting performance, and biomass production. *Anthyllis cytisoides*, a highly mycotrophic legume species from the natural succession, adapted to drought and nutrient-deficient soils, was selected for further revegetation studies. The idea was to promote an integral restoration of the target degraded ecosystem. For that, seedlings with an optimized mycorrhizosphere were transplanted to facilitate plant establishment and nutrient acquisition and to improve physicochemical properties of soil. The established mycorrhizosphere-tailored plants acted as "resource islands," while inoculum supplied nutrient to the surrounding vegetation (Barea et al. 2011). In a follow-up study, Requena et al. (2001) carried out a time course (3 years) field experiment. During this study, *A. cytisoides* seedlings were inoculated with a mixture of five taxa of AM fungi representing the natural abundance and diversity and rhizobial symbionts, and the inoculated plants were transplanted into the target degraded area in semiarid Mediterranean ecosystem of southeast Spain. Control seedlings only got AM fungal and rhizobial inocula from the field soil along the time course trial. Tailoring *A. cytisoides* mycorrhizosphere at pre-transplanting assisted the establishment and nutrient acquisition by the test plant and also increased N content and organic matter accumulation in soil and the formation of hydrostable soil aggregates around plants. *Lavandula* seedlings (a small shrub species from the natural succession of the site) were also transplanted in some plots in the same experimental area, to be grown either alone or near *A. cytisoides* plants. The soil in these plots was labeled with ^{15}N to measure N_2 fixation and N transfer from the fixing *A. cytisoides* to the non-fixing *Lavandula* plants (Requena et al. 2001). It was demonstrated that AM inoculation improved N_2 fixation and N transfer.

The role of AM fungi to promote an integral restoration of target degraded ecosystems based on using nodulated legumes from the natural succession of the site was tested in other experiments (Medina et al. 2004; Alguacil et al. 2005, 2011). These authors found that AM inoculation enhanced plant establishment and growth and improved physico-biochemical properties of soil, including enzymatic activities related to C, N, and P cycles and hydrostable soil water-stable aggregates. An increase in the amount and diversity of AM propagules was also found (Alguacil et al. 2011; Martínez-García et al. 2011). These findings supported the hypothesis that the AM management strategy are extremely important in improving both plant development and soil quality and can be considered as a successful biotechnological tool to aid the restoration of self-sustaining ecosystems.

9.4.2 Interactions Between AM Fungi and Phosphate-Solubilizing Microbes: Importance in Legume Improvement

Multitrophic interactions involving AM fungi and phosphate-solubilizing bacteria (PSB) and their consequential impact on legumes (Azcón and Barea 2010; Zaidi et al. 2010) are discussed in the following section.

During the interactions between AM fungi and PSB, PSB solubilize phosphate ions that AM fungi can capture and transport to the plant (Zaidi et al. 2010; Azcón-Aguilar and Barea 2015). The interactions between these two groups of microorganisms concerning on how they affect the development of each other have been analyzed recently (Ordoñez et al. 2016). The mechanisms whereby P-mobilizing microorganisms release available P from sparingly soluble soil P forms involve solubilization (inorganic P) or mineralization (organic P) (which are activated by specific enzymes and/or chelating organic acids, respectively, released by PSB into the surrounding environment (Richardson et al. 2009; Zaidi et al. 2009, 2010; Barea and Richardson 2015)). However, the Pi made available by PSB acting on sparingly soluble P sources may not reach to the root surface due to limited diffusion of this ion in soil solution (Barea and Richardson 2015). This connects with the well-known fact that the external mycelium of the AM fungi is involved in plant uptake of solubilized P (Smith and Smith 2011, 2012). Therefore, it was proposed that if P is solubilized by PSB, AM fungi can tap these Pi ions and translocate them to plants suggesting a microbial interaction, which could improve P supply to the host plants synergistically or additively, as first suggested by Azcón et al. (1976). The ^{32}P-based methodologies have been applied to assess how the interaction among AM fungi with PSB contributed to plant P nutrition, using RP as a source of sparingly available P, and different plant species, mainly legumes (Azcón-Aguilar and Barea 2015). The interactions between AM fungi and PSB are important for P acquisition by the legume plants, and several experiments have investigated this mycorrhizosphere activity. In general, these studies found that dual inoculation produced an increased biomass and P content of plants co-inoculated with AM and PSB and reduced the specific activity (SA = ^{32}P/^{31}P quotient) of the host plant. This lowering of SA indicates that the plant used an extra amount of ^{31}P, solubilized by the PSB from either endogenous or added as RP and that solubilized P can be captured by the AM mycelium from the soil microhabitat where PSB demonstrated P solubilizing activity (Barea et al. 2007; Barea 2010). Conclusively, plants dually inoculated with both AM fungi and PSB appear to be more efficient in terms of P supply compared to non-inoculated or singly inoculated plants.

To validate these concepts, some experiments were carried out in the field (Barea et al. 2002) using *Rhizobium*-inoculated alfalfa plants to investigate the interactive effects of PSB and AM fungi on P capture, cycling, and supply either from naturally existing P sources or from added RP. Results from this field trial suggested that interactions between AM fungi, rhizobia, and PSB can have a cooperative fundamental role in increasing plant biomass and P and N nutrition in the target-tailored legume mycorrhizosphere. The interactions in the mycorrhizosphere involving AM fungi, PSB, and rhizobia to benefit plant nutrition are illustrated in Fig. 9.5.

9.4.3 Implementing Inoculum Production Technology: Tailoring Legume Mycorrhizosphere

The technology for production of rhizobial, free-living PGPR, and AM fungal inoculants was described in the first edition of this book (Azcón and Barea 2010; Patil and Alagawady 2010). Since then some recent advances in this area have been made which are discussed in the following section.

A comprehensive review on the formulation and practical perspectives of inoculant technology for *Rhizobium* and PGPR has recently been published (Bashan et al. 2014). The authors pointed out a number of top priorities of research to implement the production steps and delivery systems for the bacterial inocula. Special emphasis must be given to the evaluation of carriers and to improve the survival of microorganisms in the inoculants and their shelf life. They encourage the implementation of polymeric/encapsulated formulations. Several companies are producing PGPR inoculum products worldwide (Ravensberg 2015; Kamilova et al. 2015; Borriss 2015).

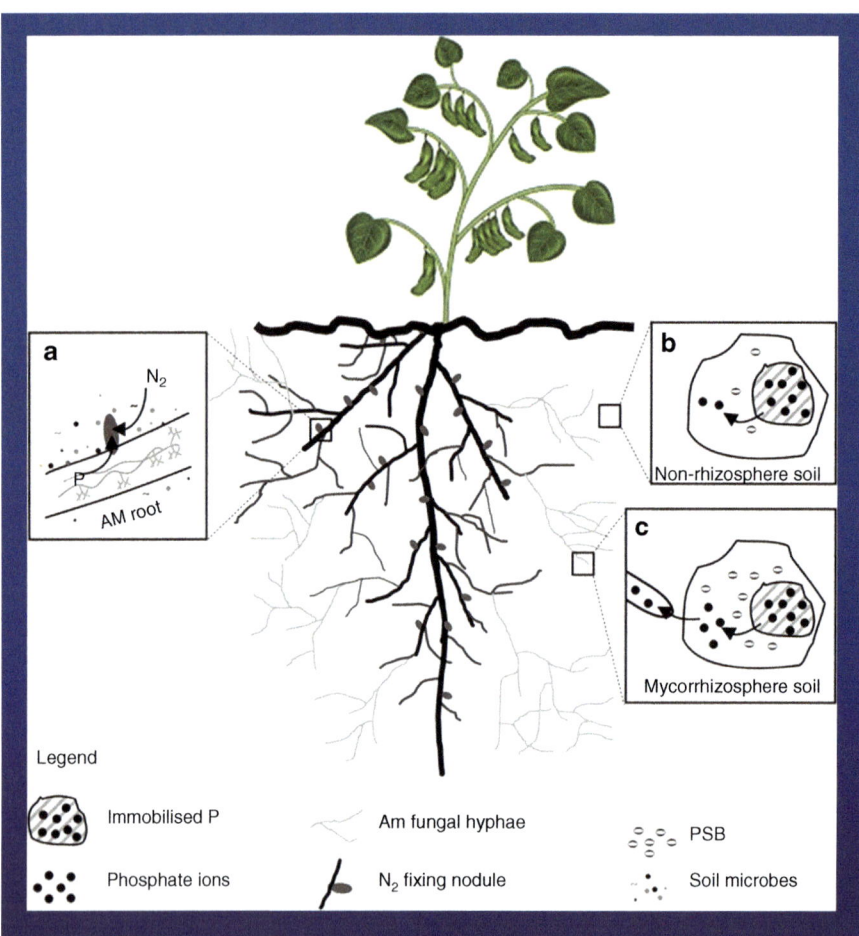

Fig. 9.5 Interactions in the mycorrhizosphere of legumes involving AM fungi, PSB and rhizobia where- (**a**) AM fungi provides P for N_2-fixation (**b**) limited plant access to the Pi solubilized by PSB in non-rhizosphere soil and (**c**) AM fungal uptake and transport of P solubilized by PSB in the mycorrhizosphere [reprint from Azcón-Aguilar and Barea (2015) with permission of the publisher]

Concerning AM fungi-based inocula, the difficulty to culture these obligate symbionts fungi in the absence of their host plant is a major obstacle to produce AM inoculants and for the development of inoculation techniques. Despite these problems, several procedures have been developed to multiply AM fungi and to produce high-quality inocula either on-farm, ex vitro in greenhouses, or in vitro monoxenic root organ cultures (Ijdo et al. 2011; Rouphael et al. 2015). The resulting materials (spores, hyphae, root fragments, etc.), from "culturing" AM fungi, are incorporated into different carriers to produce several formulations, including encapsulation, to be applied at an agronomical scale using different techniques. Inoculation of nursery-produced seedlings is a recommendable method for establishing selected fungi in the roots before planting out into the field, as is the case with horticulture and plantation crops. Mixed microbial inoculants, including PGPR, are recommended (Rouphael et al. 2015). This is particularly relevant in the case of legumes because they associate with rhizobia, PGPR, and AM fungi (Azcón and Barea 2010). In this context, Shtark et al. (2015a, b) suggested the development of multi-component microbial inoculants for legume improvement and to decrease the use of mineral fertilizers and pesticides. However, these authors point out constraints for the certification of these multicomponent inoculants due to the current procedures imposed by governmental registration of inoculants. Several companies worldwide are producing plant-based AM inoculum products which are now commercially available to be applied in forestry, agriculture, and horticulture (Vosátka et al. 2008; Gianinazzi et al. 2010; Singh et al. 2014). A key point is, however, to develop appropriate methodologies for assaying the establishment, persistence, and effectiveness of AM fungal inoculants in the field (Verbruggen et al. 2013). In this context, Pellegrino et al. (2012) used molecular tracing techniques for such purposes, using alfalfa as a test plant. The results revealed the success of AM fungal inoculum establishment and its effectiveness.

Apart from microbial inoculations, other opportunities to exploit the beneficial activities of soil microorganisms are now emerging. Diverse research approaches are currently challenged to ascertain whether plant rhizosphere can be engineered to encourage beneficial organisms while preventing the presence/emergence of pathogens (Achouak and Haichar 2013; Spence and Bais 2013). To address these issues, some approaches are used which are based on combining molecular microbial ecology with ecophysiology and plant genetics which will allow a better understanding of plant–microbiome interactions in the rhizosphere (Zancarini et al. 2013). This "biased rhizosphere" concept/action is a challenge for the future, and possible approaches have recently been discussed (Savka et al. 2013). According to Bakker et al. (2012), one strategy to manipulate the plant to recruit beneficial microorganisms in its rhizosphere relies on developing plants able to shape their microbiome by targeting particular taxa for specific functions such as N_2 fixation, P mobilization, biocontrol, etc. Undoubtedly application of biased rhizosphere to foster beneficial microbial services opens new opportunities for future agricultural developments based on exploiting the beneficial microbial services to reduce the agrochemicals inputs thereby achieving environmental sustainability and economic objectives.

Conclusion

Both plant mutualistic symbionts and saprophytic microorganisms living at the root–soil interfaces, the rhizosphere, are essential for plant nutrition and health. Legumes are plant species of great agricultural/environmental importance, known to establish beneficial symbiotic relationships with N_2-fixing bacteria and AM fungi, associated with many saprophytic microorganisms developing the so-called mycorrhizosphere. Managing the microbial symbionts and saprobes, including PGPR, involved in legume mycorrhizosphere has a great relevance to improve legume productivity either in sustainable agriculture or in the maintenance of natural plant communities. Consolidated information supports that opportunity exists to exploit the interactive effects of phosphate-mobilizing PGPR, N_2-fixing rhizobia, and AM fungi, through tailored management of the mycorrhizosphere, thereby benefiting P and N cycling and plant nutrition. In addition, the interactions among AM fungi and certain PGPR help plants to tolerate the negative impact of biotic (pathogens, insect, parasitic plants) or abiotic (salinity, drought, contaminants) stressors. The technologies for the production of efficient rhizobial, PGPR, and AM fungal inoculants nowadays are commercially available and are used in the field of agriculture, horticulture, and revegetation of degraded ecosystems. The production of legumes employing selected microbial inoculants is likely to become even more important in future due to the agroecological threats of agrochemicals, which urgently requires to be reduced, and even avoided, to increase food quality, sustainable food production, and environmental protection. Therefore, to popularize and improve the use of tailored mycorrhizospheres in legume plants is a major challenge for the scientists, farming communities, and industry.

Acknowledgments This research was supported by the Andalusian Research Programme (Project CVI-7640) and the Spanish National Research Programme (R & D)-European Union (Feder) (Project CGL2015-69118-C2-2-P).

References

Achouak W, Haichar FZ (2013) Shaping of microbial community structure and function in the rhizosphere by four diverse plant species. In: de Bruijn FJ (ed) Molecular microbial ecology of the rhizosphere, vol 1. Wiley Blackwell, Hoboken, NJ, pp 161–167

Ahmad MH (1995) Compatibility and coselection of vesicular-arbuscular mycorrhizal fungi and rhizobia for tropical legumes. Crit Rev Biotechnol 15:229–239

Alguacil MM, Caravaca E, Roldán A (2005) Changes in rhizosphere microbial activity mediated by native or allochthonous AM fungi in the reafforestation of a Mediterranean degraded environment. Biol Fertil Soils 41:59–68

Alguacil MM, Torres MP, Torrecillas E, Díaz G, Roldán A (2011) Plant type differently promote the arbuscular mycorrhizal fungi biodiversity in the rhizosphere after revegetation of a degraded, semiarid land. Soil Biol Biochem 43:167–173

Altieri MA (2004) Linking ecologists and traditional farmers in the search for sustainable agriculture. Front Ecol Environ 2:35–42

Antoun H (2012) Beneficial microorganisms for the sustainable use of phosphates in agriculture. Proc Eng 46:62–67

Arrese-Igor C (2010) Biological nitrogen fixation. In: González-Fontes A, Gárate A, Bonilla I (eds) Agricultural sciences: topics in modern agriculture. Stadium Press, Houston, TX, pp 233–255

Azcón R, Barea JM (2010) Mycorrhizosphere interactions for legume improvement. In: Khan MS, Zaidi A, Musarrat J (eds) Microbes for legume improvement, 1st edn. Springer, New York, pp 237–271

Azcón R, Barea JM, Hayman DS (1976) Utilization of rock phosphate in alkaline soils by plant inoculated with mycorrhizal fungi and phosphate-solubilizing bacteria. Soil Biol Biochem 8:135–138

Azcón R, Rubio R, Barea JM (1991) Selective interactions between different species of mycorrhizal fungi and *Rhizobium meliloti* strains, and their effects on growth, N_2 fixation (N^{15}) in *Medicago sativa* at four salinity levels. New Phytol 117:399–404

Azcón R, Medina A, Aroca R, Ruíz-Lozano JM (2013) Abiotic stress remediation by the arbuscular mycorrhizal symbiosis and rhizosphere bacteria/yeast interactions. In: de Bruijn FJ (ed) Molecular microbial ecology of the rhizosphere. Wiley, Hoboken, NJ, pp 991–1002

Azcón-Aguilar C, Barea JM (2015) Nutrient cycling in the mycorrhizosphere. (In: Gianfreda, L. (Guest Editor) Biogeochemical processes in the rhizosphere and their influence on plant nutrition. Special issue). J Soil Sci Plant Nutr 15:372–396

Azcón-Aguilar C, Azcón R, Barea JM (1979) Endomycorrhizal fungi and *Rhizobium* as biological fertilizers for *Medicago sativa* in normal cultivation. Nature 279:325–327

Bakker MG, Manter DK, Sheflin AM, Weir TL, Vivanco JM (2012) Harnessing the rhizosphere microbiome through plant breeding and agricultural management. Plant Soil 360:1–13

Barea JM (1991) Vesicular-arbuscular mycorrhizae as modifiers of soil fertility. In: Stewart BA (ed) Advances in soil science, vol 7. Springer, New York, pp 1–40

Barea JM (2010) Mycorrhizas and agricultural fertility. In: González-Fontes A, Gárate A, Bonilla I (eds) Agricultural sciences: topics in modern agriculture. Stadium Press, Houston, TX, pp 257–274

Barea JM (2015) Future challenges and perspectives for applying microbial biotechnology in sustainable agriculture based on a better understanding of plant-microbiome interactions. (In: Gianfreda, L. (Guest Editor) Biogeochemical processes in the rhizosphere and their influence on plant nutrition. Special issue). J Soil Sci Plant Nutr 15:261–282

Barea JM, Azcón-Aguilar C (1983) Mycorrhizas and their significance in nodulating nitrogen-fixing plants. In: Brady N (ed) Advances in agronomy, vol 36. Academic, New York, pp 1–54

Barea JM, Azcón-Aguilar C (2013) Evolution, biology and ecological effects of arbuscular mycorrhiza. In: Camisao AF, Pedroso CC (eds) Symbiosis: evolution biology and ecological effects. Nova Science, New York, pp 1–34

Barea JM, Richardson AE (2015) Phosphate mobilisation by soil microorganisms. In: Lugtenberg B (ed) Principles of plant-microbe interactions. Microbes for sustainable agriculture. Springer International, Heidelberg, pp 225–234

Barea JM, Azcón-Aguilar C, Azcón R (1987) Vesicular-arbuscular mycorrhiza improve both symbiotic N_2-fixation and N uptake from soil as assessed with a ^{15}N technique under field conditions. New Phytol 106:717–725

Barea JM, Azcón R, Azcón-Aguilar C (1989a) Time-course of N_2 fixation (^{15}N) in the field by clover growing alone or in mixture with ryegrass to improve pasture productivity, and inoculated with vesicular-arbuscular mycorrhizal fungi. New Phytol 112:399–404

Barea JM, El-Atrach F, Azcón R (1989b) Mycorrhiza and phosphate interactions as affecting plant development, N_2 fixation, N-transfer and N-uptake from soil in legume grass mixtures by using a N^{15} dilution technique. Soil Biol Biochem 21:581–589

Barea JM, Toro M, Orozco MO, Campos E, Azcón R (2002) The application of isotopic (P^{32} and N^{15}) dilution techniques to evaluate the interactive effect of phosphate-solubilizing rhizobacteria, mycorrhizal fungi and *Rhizobium* to improve the agronomic efficiency of rock phosphate for legume crops. Nutr Cycl Agroecosyst 63:35–42

Barea JM, Toro M, Azcón R (2007) The use of 32P isotopic dilution techniques to evaluate the interactive effects of phosphate-solubilizing bacteria and mycorrhizal fungi at increasing plant P availability. In: Velázquez E, Rodríguez-Barrueco C (eds) First international meeting on microbial phosphate solubilization, Developments in plant and soil sciences. Springer, Dordrecht, pp 223–227

Barea JM, Palenzuela J, Cornejo P, Sánchez-Castro I, Navarro-Fernández C, Lopéz-García A, Estrada B, Azcón R, Ferrol N, Azcón-Aguilar C (2011) Ecological and functional roles of mycorrhizas in semi-arid ecosystems of Southeast Spain. J Arid Environ 75:1292–1301

Barea JM, Pozo MJ, Azcón R, Azcón-Aguilar C (2013a) Microbial interactions in the rhizosphere. In: de Bruijn FJ (ed) Molecular microbial ecology of the rhizosphere, vol 1. Wiley, Hoboken, NJ, pp 29–44

Barret M, Tan H, Egan F, Morrissey JP, Reen J, O'Gara F (2013) Exploiting new systems-based strategies to elucidate plant–bacterial interactions in the rhizosphere. In: de Bruijn FJ (ed) Molecular microbial ecology of the rhizosphere, vol 1. Wiley Blackwell, Hoboken, NJ, pp 57–68

Bashan Y, Trejo A, de Bashan LE (2011) Development of two culture media for mass cultivation of *Azospirillum* spp. and for production of inoculants to enhance plant growth. Biol Fertil Soils 47:963–969

Bashan Y, de Bashan LE, Prabhu SR, Hernandez J-P (2014) Advances in plant growth-promoting bacterial inoculant technology: formulations and practical perspectives (1998-2013). Plant Soil 378:1–33

Bonfante P, Desirò A (2015) Arbuscular mycorrhizas: the lives of beneficial fungi and their plant host. In: Lugtenberg B (ed) Principles of plant-microbe interactions. Microbes for sustainable agriculture. Springer International, Heidelberg, pp 235–245

Bonfante P, Genre A (2010) Mechanisms underlying beneficial plant-fungus interactions in mycorrhizal symbiosis. Nat Commun 1:48

Borriss R (2015) Towards a new generation of commercial microbial disease control and plant growth promotion products. In: Lugtenberg B (ed) Principles of plant-microbe interactions. Microbes for sustainable agriculture. Springer International, Heidelberg, pp 329–337

Brearley FQ, Elliott DR, Iribar A, Sen R (2016) Arbuscular mycorrhizal community structure on co-existing tropical legume trees in French Guiana. Plant Soil 403:253–265

Brundrett MC (2009) Mycorrhizal associations and other means of nutrition of vascular plants: understanding the global diversity of host plants by resolving conflicting information and developing reliable means of diagnosis. Plant Soil 320:37–77

Chalk PM, Souza RD, Urquiaga S, Alves BJR, Boddey RM (2006) The role of arbuscular mycorrhiza in legume symbiotic performance. Soil Biol Biochem 38:2944–2951

Courty PE, Smith P, Koegel S, Redecker D, Wipf D (2015) Inorganic nitrogen uptake and transport in beneficial plant root-microbe interactions. Crit Rev Plant Sci 34:4–16

Danso SKA (1986) Review, estimation of N_2 fixation by isotope dilution: an appraisal of techniques involving ^{15}N enrichment and their application. Soil Biol Biochem 18:243–244

de Bruijn FJ (2015) Biological nitrogen fixation. In: Lugtenberg B (ed) Principles of plant-microbe interactions. Microbes for sustainable agriculture. Springer International, Heidelberg, pp 215–224

Delaux P-M, Sejalon-Delmas N, Becard G, Ane J-M (2013) Evolution of the plant-microbe symbiotic 'toolkit'. Trends Plant Sci 18:298–304

Delaux P-M, Radhakrishnan GV, Jayaraman D et al (2015) Algal ancestor of land plants was pre-adapted for symbiosis. Proc Natl Acad Sci U S A 112:13390–13395

Dobbelaere S, Croonenborghs A, Thys A, Ptacek D, Vanderleyden J, Dutto P, Labandera-González C, Caballero-Mellado J, Aguirre JF, Kapulnik Y, Brener S, Burdman S, Kadouri D, Sarig S, Okon Y (2001) Responses of agronomically important crops to inoculation with *Azospirillum*. Aust J Plant Physiol 28:871–879

Drumbell AJ (2013) Arbuscular mycorrhizal fungi throughout the year: using massively parallel pyrosequencing to quantify spatiotemporal seasonal dynamics. In: de Bruijn FJ (ed) Molecular microbial ecology of the rhizosphere. Wiley, Hoboken, NJ, pp 1113–1122

Fernández-Bidondo L, Silvani V, Colombo R, Pérgola M, Bompadre J, Godeas A (2011) Pre-symbiotic and symbiotic interactions between *Glomus intraradices* and two *Paenibacillus* species isolated from AM propagules. *In vitro* and *in vivo* assays with soybean (AG043RG) as plant host. Soil Biol Biochem 43:1866–1872

Field KJ, Pressel S, Duckett JG, Rimington WR, Bidartondo MI (2015) Symbiotic options for the conquest of land. Trends Ecol Evol 30:477–486

Frey-Klett P, Garbaye J, Tarkka M (2007) The mycorrhiza helper bacteria revisited. New Phytol 176:22–36

Genre A, Bonfante P (2010) The making of symbiotic cells in arbuscular mycorrhizal roots. In: Koltai H, Kapulnik Y (eds) Arbuscular mycorrhizas: physiology and function. Springer, Dordrecht, pp 57–71

Genre A, Chabaud M, Balzergue C, Puech-Pages V, Novero M, Rey T, Fournier J, Rochange S, Becard G, Bonfante P, Barker DG (2013) Short-chain chitin oligomers from arbuscular mycorrhizal fungi trigger nuclear Ca^{2+} spiking in *Medicago truncatula* roots and their production is enhanced by strigolactone. New Phytol 198:179–189

George TS, Hinsinger P, Turner BL (2016) Phosphorus in soils and plants—facing phosphorus scarcity. Plant Soil 401:1–6

Gianinazzi S, Gollotte A, Binet MN, van Tuinen D, Redecker D, Wipf D (2010) Agroecology: the key role of arbuscular mycorrhizas in ecosystem services. Mycorrhiza 20:519–530

Gianinazzi-Pearson V, van Tuinen D, Wipf D, Dumas-Gaudot E, Recorbet G, Lyu Y, Doidy J, Redecker D, Ferrol N (2012) Exploring the genome of Glomeromycotan fungi. In: Hock B (ed) The Mycota, a comprehensive treatise on fungi as experimental systems for basic and applied research. Springer, Berlin, pp 1–21

Gutiérrez-Mañero J, Ramos-Solano B (2010) Bacteria and agriculture. In: González-Fontes A, Gárate A, Bonilla I (eds) Agricultural sciences: topics in modern agriculture. Stadium Press, Houston, TX, pp 275–289

Gutjahr C, Parniske M (2013) Cell and developmental biology of arbuscular mycorrhiza symbiosis. In: Schekman R (ed) Annual review of cell and developmental biology, vol 29, pp 593–617

Ha Y, Gray VM (2008) Growth yield of *Vicia faba* L in response to microbial symbiotic associations. S Afr J Bot 74:25–32

Hermosa R, Viterbo A, Chet I, Monte E (2012) Plant-beneficial effects of *Trichoderma* and of its genes. Microbiology 158:17–25

Herrera MA, Salamanca CP, Barea JM (1993) Inoculation of woody legumes with selected arbuscular mycorrhizal fungi and rhizobia to recover desertified mediterranean ecosystems. Appl Environ Microbiol 59:129–133

Hidri R, Barea JM, Metoui-Ben Mahmoud O, Abdellya C, Azcón R (2016) Impact of microbial inoculations on biomass accumulation by *Sulla carnosa* provenances, and in regulating nutrition, physiological and antioxidant activities of this species under non-saline and saline conditions. J Plant Physiol 201:28–41

Hirsch PR, Mauchline TH, Clark IM (2013a) Culture-independent molecular approaches to microbial ecology in soil and the rhizosphere. In: de Bruijn FJ (ed) Molecular microbial ecology of the rhizosphere, vol 1. Wiley Blackwell, Hoboken, NJ, pp 45–55

Hirsch PR, Miller AJ, Dennis PG (2013b) Do root exudates exert more influence on rhizosphere bacterial community structure than other rhizodeposits? In: de Bruijn FJ (ed) Molecular microbial ecology of the rhizosphere, vol 1. Wiley Blackwell, Hoboken, NJ, pp 229–242

Honrubia M (2009) The Mycorrhizae: a plant-fungus relation that has existed for more than 400 million years. Anal Jardin Bot Madrid 66:133–144

Ijdo M, Cranenbrouck S, Declerck S (2011) Methods for large-scale production of AM fungi: past, present, and future. Mycorrhiza 21:1–16

Kamilova F, Okon Y, de Weert S, Hora K (2015) Commercialization of microbes: manufacturing, inoculation, best practice for objective field testing, and registration. In: Lugtenberg B (ed) Principles of plant-microbe interactions. Microbes for sustainable agriculture. Springer International, Heidelberg, pp 319–327

Kaschuk G, Leffelaar PA, Giller KE, Alberton O, Hungria M, Kuyper TW (2010) Responses of legumes to rhizobia and arbuscular mycorrhizal fungi: a meta-analysis of potential photosynthate limitation of symbioses. Soil Biol Biochem 42:125–127

Khan M, Zaidi A, Ahemad M, Oves M, Wani P (2010) Plant growth promotion by phosphate solubilizing fungi—current perspective. Arch Agron Soil Sci 56:73–98

Kloepper JW (1994) Plant growth-promoting rhizobacteria (other systems). In: Okon Y (ed) *Azospirillum*/plant associations. CRC, Boca Raton, FL, pp 111–118

Kloepper JW, Zablotowicz RM, Tipping EM, Lifshitz R (1991) Plant growth promotion mediated by bacterial rhizosphere colonizers. In: Keister DL, Cregan PB (eds) The rhizosphere and plant growth. Kluwer Academic, Dordrecht, pp 315–326

König S, Wubet T, Dormann CF, Hempel S, Renker C, Buscot F (2010) Taqman real-time PCR assays to assess arbuscular mycorrhizal responses to field manipulation of grassland biodiversity: effects of soil characteristics, plant species richness, and functional traits. Appl Environ Microbiol 76:3765–3775

Krüger M, Walker C, Schüßer A (2011) *Acaulospora brasiliensis* comb. Nov. and *Acaulospora alpina* (Glomeromycota) from upland Scotland: morphology, molecular phylogeny and DNA-based detection in roots. Mycorrhiza 21:577–587

Lagunas B, Schaefer P, Gifford ML (2015) Housing helpful invaders: the evolutionary and molecular architecture underlying plant root-mutualist microbe interactions. J Exp Bot 66:2177–2186

Larimer AL, Clay K, Bever JD (2014) Synergism and context dependency of interactions between arbuscular mycorrhizal fungi and rhizobia with a prairie legume. Ecology 95:1045–1054

Lesueur D, Sarr A (2008) Effects of single and dual inoculation with selected microsymbionts (rhizobia and arbuscular mycorrhizal fungi) on field growth and nitrogen fixation of *Calliandra calothyrsus* Meissn. Agric Syst 73:37–45

Lin K, Limpens E, Zhang Z et al (2014) Single nucleous genome sequencing reveals high similarity among nuclei of an endomycorrhizal fungus. PLoS Genet 10:e1004078

Linderman RG (1988) Mycorrhizal interactions with the rhizosphere microflora. The mycorrhizosphere effects. Phytopathology 78:366–371

López-Ráez JA, Pozo MJ, García-Garrido JM (2011) Strigolactones: a cry for help in the rhizosphere. Botany 89:513–522

Lucy M, Reed E, Glick BR (2004) Applications of free living plant growth-promoting rhizobacteria. Antonie Van Leeuwenhoek 86:1–25

Lugtenberg B (2015) Life of microbes in the rhizosphere. In: Lugtenberg B (ed) Principles of plant-microbe interactions. Microbes for sustainable agriculture. Springer International, Heidelberg, pp 7–15

Lugtenberg BJJ, Malfanova N, Kamilova F, Berg G (2013a) Microbial control of plant root diseases. In: de Bruijn FJ (ed) Molecular microbial ecology of the rhizosphere, vol 2. Wiley Blackwell, Hoboken, NJ, pp 575–586

Lugtenberg BJJ, Malfanova N, Kamilova F, Berg G (2013b) Plant growth promotion by microbes. In: de Bruijn FJ (ed) Molecular microbial ecology of the rhizosphere, vol 2. Wiley Blackwell, Hoboken, NJ, pp 561–573

Lumini E, Orgiazzi A, Borriello R, Bonfante P, Bianciotto V (2010) Disclosing arbuscular mycorrhizal fungal biodiversity in soil through a land-use gradient using a pyrosequencing approach. Environ Microbiol 12:2165–2179

Maillet F, Poinsot V, Andre O, Puech-Pages V, Haouy A, Gueunier M, Cromer L, Giraudet D, Formey D, Niebel A, Martínez EA, Driguez H, Bécard G, Denarie J (2011) Fungal lipochitooligosaccharide symbiotic signals in arbuscular mycorrhiza. Nature 469:58–64

Martínez-García LB, Armas C, Padilla FM, Miranda JD, Pugnaire FI (2011) Shrubs influence arbuscular mycorrhizal fungi communities in a semiarid environment. Soil Biol Biochem 43:682–689

Martinez-Viveros O, Jorquera MA, Crowley DE, Gajardo G, Mora ML (2010) Mechanisms and practical considerations involved in plant growth promotion by rhizobacteria. J Soil Sci Plant Nutr 10:293–319

Medina A, Vassileva M, Caravaca F, Roldán A, Azcón R (2004) Improvement of soil characteristics and growth of *Dorycnium pentaphyllum* by amendment with agrowastes and inoculation with AM fungi and/or the yeast *Yarrowia lipolytica*. Chemosphere 56:449–456

Mendes R, Kruijt M, de Bruijn I, Dekkers E, van der Voort M, Schneider JHM, Piceno YM, DeSantis TZ, Andersen GL, Bakker PAHM, Raaijmakers JM (2011) Deciphering the rhizosphere microbiome for disease-suppressive bacteria. Science 332:1097–1100

Mercado-Blanco J (2015) Life of microbes inside the plant. In: Lugtenberg B (ed) Principles of plant-microbe interactions. Springer International, Heidelberg, pp 25–32

Mortimer PE, Pérez-Fernández MA, Valentine AJ (2008) The role of arbuscular mycorrhizal colonization in the carbon and nutrient economy of the tripartite symbiosis with nodulated *Phaseolus vulgaris*. Soil Biol Biochem 40:1019–1027

Muleta D (2010) Legume responses to arbuscular mycorrhizal fungi inoculation in sustainable agriculture. In: Khan MS, Musarrat J, Zaidi A (eds) Microbes for legume improvement, 1st edn. Springer, Vienna, pp 293–323

Oehl F, Sieverding E, Palenzuela J, Ineichen K, da Silva GA (2011a) Advances in *Glomeromycota* taxonomy and classification. IMA Fungus 2:191–199

Oehl F, da Silva GA, Sánchez-Castro I, Goto BT, Maia LC, Evangelista Vieira HE, Barea JM, Sieverding E, Palenzuela J (2011b) Revision of *Glomeromycetes* with entrophosporoid and glomoid spore formation with three new genera. Mycotaxon 117:297–316

Olivares J, Bedmar EJ, Sanjuan J (2013) Biological nitrogen fixation in the context of global change. Mol Plant Microb Interact 26:486–494

Öpik M, Zobel M, Cantero JJ, Davison J, Facelli JM, Hiiesalu I, Jairus T, Kalwij JM, Koorem K, Leal ME, Liira J, Metsis M, Neshataeva V, Paal J, Phosri C, Polme S, Reier U, Saks U, Schimann H, Thiery O, Vasar M, Moora M (2013) Global sampling of plant roots expands the described molecular diversity of arbuscular mycorrhizal fungi. Mycorrhiza 23:411–430

Ordoñez YM, Fernández BR, Lara LS, Rodríguez A, Uribe-Vélez D, Sanders IR (2016) Bacteria with phosphate solubilizing capacity alter mycorrhizal fungal growth both inside and outside the root and in the presence of native microbial communities. PLoS One 11:e0154438

Pagano MC, Cabello MN, Bellote AF, Sa NM, Scotti MR (2008) Intercropping system of tropical leguminous species and *Eucalyptus camaldulensis*, inoculated with rhizobia and/or mycorrhizal fungi in semiarid Brazil. Agric Syst 74:231–242

Parniske M (2008) Arbuscular mycorrhiza: the mother of plant root endosymbioses. Nat Rev Microbiol 6:763–775

Patil CR, Alagawady AR (2010) Microbial inoculants for sustainable legume production. In: Khan MS, Zaidi A, Musarrat J (eds) Microbes for legume improvement, 1st edn. Springer, New York, pp 515–535

Pellegrino E, Turrini A, Gamper HA, Cafa G, Bonari E, Young JPW, Giovannetti M (2012) Establishment, persistence and effectiveness of arbuscular mycorrhizal fungal inoculants in the field revealed using molecular genetic tracing and measurement of yield components. New Phytol 194:810–822

Pérez-Tienda J, Correa A, Azcón-Aguilar C, Ferrol N (2014) Transcriptional regulation of host NH_4^+ transporters and GS/GOGAT pathway in arbuscular mycorrhizal rice roots. Plant Physiol Biochem 75:1–8

Pieterse CMJ, Zamioudis C, Berendsen RL, Weller DM, Van Wees SCM, Bakker PAHM (2014) Induced systemic resistance by beneficial microbes. Annu Rev Phytopathol 52:347–375

Pozo MJ, Jung SC, Martínez-Medina A, López-Ráez JA, Azcón-Aguilar C, Barea JM (2013) Root allies: arbuscular mycorrhizal fungi help plants to cope with biotic stresses. In: Aroca R (ed) Symbiotic endophytes. Springer, Berlin, pp 289–307

Pozo MJ, López-Ráez JA, Azcón-Aguilar C, García-Garrido JM (2015) Phytohormones as integrators of environmental signals in the regulation of mycorrhizal symbioses. New Phytol 205:1431–1436

Raaijmakers JM, Lugtenberg BJJ (2013) Perspectives for rhizosphere research. In: de Bruijn FJ (ed) Molecular microbial ecology of the rhizosphere, vol 2. Wiley Blackwell, Hoboken, NJ, pp 1227–1232

Ramos-Solano B, Barriuso Maicas J, Pereyra de la Iglesia MT, Domenech J, Gutiérrez Mañero FJ (2008) Systemic disease protection elicited by plant growth promoting rhizobacteria strains: relationship between metabolic responses, systemic disease protection, and biotic elicitors. Phytopathology 98:451–457

Ramos-Solano B, Barriuso J, Gutiérrez Mañero J (2009) Biotechnology of the rhizosphere. In: Kirakosyan A, Kaufman P (eds) Recent advances in plant biotechnology. Springer, New York, pp 137–162

Ravensberg WJ (2015) Commercialisation of microbes: present situation and future prospects. In: Lugtenberg B (ed) Principles of plant-microbe interactions. Microbes for sustainable agriculture. Springer International, Heidelberg, pp 309–317

Redecker D, Schuessler A, Stockinger H, Stuermer SL, Morton JB, Walker C (2013) An evidence-based consensus for the classification of arbuscular mycorrhizal fungi (*Glomeromycota*). Mycorrhiza 23:515–531

Requena N, Pérez-Solis E, Azcón-Aguilar C, Jeffries P, Barea JM (2001) Management of indigenous plant-microbe symbioses aids restoration of desertified ecosystems. Appl Environ Microbiol 67:495–498

Richardson AE, Barea JM, McNeill AM, Prigent-Combaret C (2009) Acquisition of phosphorus and nitrogen in the rhizosphere and plant growth promotion by microorganisms. Plant Soil 321:305–339

Robinson-Boyer L, Grzyb I, Jeffries P (2009) Shifting the balance from qualitative to quantitative analysis of arbuscular mycorrhizal communities in field soils. Fungal Ecol 2:1–9

Rouphael Y, Franken P, Schneider C, Schwarz D, Giovannetti M, Agnolucci M, De Pascale S, Bonini P, Colla G (2015) Arbuscular mycorrhizal fungi act as biostimulants in horticultural crops. Sci Hortic 196:91–108

Sánchez-Romera B, Ruiz-Lozano JM, Zamarreno AM, Garcia-Mina JM, Aroca R (2016) Arbuscular mycorrhizal symbiosis and methyl jasmonate avoid the inhibition of root hydraulic conductivity caused by drought. Mycorrhiza 26:111–122

Savka MA, Dessaux Y, McSpadden Gardener BB, Mondy S, Kohler PRA, de Bruijn FJ, Rossbach S (2013) The "biased rhizosphere" concept and advances in the omics era to study bacterial competitiveness and persistence in the phytosphere. In: de Bruijn FJ (ed) Molecular microbial ecology of the rhizosphere, vol 2. Wiley Blackwell, Hoboken, NJ, pp 1147–1161

Schreiter S, Eltlbany N, Smalla K (2015) Microbial communities in the rhizosphere analyzed by cultivation-independent DNA-based methods. In: Lugtenberg B (ed) Principles of plant-microbe interactions. Microbes for sustainable agriculture. Springer International, Heidelberg, pp 289–298

Schüßler A, Walker C (2011) Evolution of the 'plant-symbiotic' fungal phylum, Glomeromycota. In: Pöggeler S, Wöstemeyer J (eds) Evolution of fungi and fungal-like organisms. Springer, Berlin, pp 163–185

Schüßler A, Schwarzott D, Walker C (2001) A new fungal phylum, the *Glomeromycota*, phylogeny and evolution. Mycol Res 105:1413–1421

Selosse M-A, Strullu-Derrien C, Martin FM, Kamoun S, Kenrick P (2015) Plants, fungi and oomycetes: a 400-million year affair that shapes the biosphere. New Phytol 206:501–506

Shtark O, Provorov N, Mikić A, Borisov A, Ćupina B, Tikhonovich I (2011) Legume root symbioses: natural history and prospects for improvement. Field Veg Crop Res 48:291–304

Shtark OY, Borisov AY, Zhukov VA, Tikhonovich IA (2012) Mutually beneficial legume symbioses with soil microbes and their potential for plant production. Symbiosis 58:51–62

Shtark O, Kumari S, Singh R, Sulima A, Akhtemova G, Zhukov V, Shcherbakov A, Shcherbakova E, Adholeya A, Borisov A (2015a) Advances and prospects for development of multi-component microbial inoculant for legumes. Legum Perspect 8:40–44

Shtark O, Zhukov V, Sulima A, Singh R, Naumkina T, Akhtemova G, AY B (2015b) Prospects for the use of multi-component symbiotic systems of the legumes. Ecol Genet 13:33–46

Singh S, Srivastava K, Sharma S, Sharma AK (2014) Mycorrhizal inoculum production. In: Solaiman ZM, Abbott LK, Varma A (eds) Mycorrhizal fungi: use in sustainable agriculture and land restoration, Soil biology, vol 41. Springer, Berlin, pp 67–80

Smith SE, Read DJ (2008) Mycorrhizal symbiosis, 3rd edn. Elsevier, Academic, New York

Smith SE, Smith FA (2011) Roles of arbuscular mycorrhizas in plant nutrition and growth: new paradigms from cellular to ecosystem scales. Annu Rev Plant Biol 62:227–250

Smith SE, Smith FA (2012) Fresh perspectives on the roles of arbuscular mycorrhizal fungi in plant nutrition and growth. Mycologia 104:1–13

Spence C, Bais H (2013) Probiotics for plants: rhizospheric microbiome and plant fitness. In: de Bruijn FJ (ed) Molecular microbial ecology of the rhizosphere, vol 2. Wiley Blackwell, Hoboken, NJ, pp 713–721

Sun J, Miller JB, Granqvist E, Wiley-Kalil A, Gobbato E, Maillet F, Cottaz S, Samain E, Venkateshwaran M, Fort S, Morris RJ, Ane J-M, Denarie J, Oldroyd GED (2015) Activation of symbiosis signaling by arbuscular mycorrhizal fungi in legumes and rice. Plant Cell 27:823–838

Tamayo E, Gómez-Gallego T, Azcón-Aguilar C, Ferrol N (2014) Genome-wide analysis of copper, iron and zinc transporters in the arbuscular mycorrhizal fungus *Rhizophagus irregularis*. Front Plant Sci 5:547

Tisserant E, Malbreil M, Kuo A, Kohler A et al (2013) Genome of an arbuscular mycorrhizal fungus provides insight into the oldest plant symbiosis. Proc Natl Acad Sci U S A 110:20117–20122

Toro M, Azcón R, Barea JM (1998) The use of isotopic dilution techniques to evaluate the interactive effects of *Rhizobium* genotype, mycorrhizal fungi, phosphate-solubilizing rhizobacteria and rock phosphate on nitrogen and phosphorus acquisition by *Medicago sativa*. New Phytol 138:265–273

Uyanoz R, Akbulut M, Cetin U, Gultepe N (2007) Effects of microbial inoculation, organic and chemical fertilizer on yield and physicochemical and cookability properties of bean (*Phaseolus vulgaris* L.) seeds. Philipp Agric Sci 90:168–172

van der Heijden MGA, Martin FM, Selosse M-A, Sanders IR (2015) Mycorrhizal ecology and evolution: the past, the present, and the future. New Phytol 205:1406–1423

Varela-Cervero S, Vasar M, Davison J, Barea JM, Opik M, Azcon-Aguilar C (2015) The composition of arbuscular mycorrhizal fungal communities differs among the roots, spores and extraradical mycelia associated with five Mediterranean plant species. Environ Microbiol 17:2882–2895

Varela-Cervero S, López-García A, Barea JM, Azcón-Aguilar C (2016) Spring to autumn changes in the arbuscular mycorrhizal fungal community composition in the different propagule types associated to a Mediterranean shrubland. Plant Soil 403:1–14

Verbruggen E, van der Heijden MGA, Rillig MC, Kiers ET (2013) Mycorrhizal fungal establishment in agricultural soils: factors determining inoculation success. New Phytol 197:1104–1109

Vosátka M, Albrechtová J, Patten R (2008) The international marked development for mycorrhizal technology. In: Varma A (ed) Mycorrhiza: state of the art, genetics and molecular biology, eco-function, biotechnology, eco-physiology, structure and systematics, 3rd edn. Springer, Berlin, pp 419–438

Zaidi A, Khan MS (2007) Stimulatory effects of dual inoculation with phosphate solubilising microorganisms and arbuscular mycorrhizal fungus on chickpea. Aust J Exp Agric 47:1016–1022

Zaidi A, Khan MS, Ahemad M, Oves M (2009) Plant growth promotion by phosphate solubilizing bacteria. Acta Microbiol Inmunol Hung 56:263–284

Zaidi A, Ahemad M, Oves M, Ahmad E, Khan MS (2010) Role of phosphate-solubilizing bacteria in legume improvement. In: Khan MS, Musarrat J, Zaidi A (eds) Microbes for legume improvement, 1st edn. Springer, Vienna, pp 273–292

Zancarini A, Lépinay C, Burstin J, Duc G, Lemanceau P, Moreau D, Munier-Jolain N, Pivato B, Rigaud T, Salon C, Mougel C (2013) Combining molecular microbial ecology with ecophysiology and plant genetics for a better understanding of plant-microbial communities' interactions in the rhizosphere. In: de Bruijn FJ (ed) Molecular microbial ecology of the rhizosphere, vol 1. Wiley Blackwell, Hoboken, NJ, pp 69–86

Zapata F (1990) Isotope techniques in soil fertility and plant nutrition studies. In: Hardarson G (ed) Use of nuclear techniques in studies of soil-plant relationships. IAEA, Vienna, pp 61–128

Zapata F, Danso SKA, Hardanson G, Fried M (1987) Nitrogen-fixation and translocation in field grown fababean. Agron J 79:505–509

Zhukov VA, Shtark OY, Borisov AY, Tikhonovich IA (2013) Breeding to improve symbiotic effectiveness of legumes. In: Andersen SB (ed) Plant breeding from laboratories to fields. InTech, Rijeka, Croatia. doi:10.5772/53003

Legume Response to Arbuscular Mycorrhizal Fungi Inoculation in Sustainable Agriculture

10

Diriba Muleta

Abstract

Globally, there is a widespread interest in the use of legumes due to their multi-faceted functions. Also, legumes (Fabaceae, Syn. Leguminosae) are essential components in natural and managed terrestrial ecosystems due to their ability to intimately interact with different rhizosphere microorganisms. Among soil microbiota, the arbuscular mycorrhizal fungi (AMF) are universal and ubiquitous rhizosphere microflora forging symbiosis with plethora of plant species roots and acting as biofertilizers, bioprotectants, mycoremediators, and biodegraders. The arbuscular mycorrhizal-legume (herb or tree) symbiosis is viewed as a better alternative for enhancing soil fertility and the rehabilitation of arid lands and, therefore, provides an important direction for future agricultural research. The sole application of AMF has been found to improve the overall performance of leguminous plants growing under diverse farming practices. In addition, the interaction of AM fungi with other plant growth-promoting rhizobacteria has shown considerable increase in growth and yield of legumes. Here, legume growth responses to single or composite inoculation of AMF for sustainable production of legumes cultivated in different agroecological niches are highlighted. Furthermore, mycorrhizal dependency of legumes and effects of arbuscular mycorrhizal fungi on productivity of legumes grown under stressed environment are described.

D. Muleta
Institute of Biotechnology, College of Natural Sciences, Addis Ababa University,
P.O. Box 1176, Addis Ababa, Ethiopia
e-mail: dmuleta@gmail.com; diriba.muleta@aau.edu.et

© Springer International Publishing AG 2017
A. Zaidi et al. (eds.), *Microbes for Legume Improvement*,
DOI 10.1007/978-3-319-59174-2_10

10.1   Introduction

The constantly declining cultivable lands and consistently rising human populations require that the production of crops be increased substantially at global level. Due to changes in abiotic and biotic soil properties, reestablishment of proper vegetation cover has been adversely affected (Miller 1987). Due to these factors and increased pressure for food production, there is urgent need to upgrade farming practices so that food demands are fulfilled. In this regard, farming communities are adopting intensive agricultural systems that involve the use of significant quantities of agro-chemicals to optimize crop production (Hooker and Black 1995). However, the cost and environmental threats associated with the use of high-input strategy demands that agricultural systems should be modified in order to make them more productive and sustainable (van der Vossen 2005). To circumvent such problems, the use of microorganisms especially symbiotic fungi opens up the new possibility of a more sustainable and low-cost agricultural practices.

Legumes form symbioses with both rhizobia (Spaink 1996) and mycorrhizal fungi (Harrison 1999; Lodwig et al. 2003). Legumes have been grown for food, as fodder, fiber, industrial and medicinal compounds, flowers, and other end uses. Leguminous plants are also highly suitable for agroforestry system, the area that receives due attention for sustainable agriculture. Nutrient-acquisition symbioses between plants and soil microbes are important to plant evolution and ecosystem function (Simms and Taylor 2002). A complex yet positive interactions between plants and soil microbes determine the soil fertility and consequently the plant health (Jeffries et al. 2003). Among numerous useful soil microflora, arbuscular mycorrhizas are the most important organisms that form symbioses with majority of plants including legumes (Barea and Azcon-Aguilar 1983) grown under P-deficient soils and influence plant community development, nutrient uptake, water relations, and aboveground productivity (van der Heijden et al. 2008). Arbuscular mycorrhizas also act as bioprotectants and protect plants from pathogens and toxic stresses (van der Heijden et al. 2008). However, in order to optimize their beneficial impacts, it is important to ensure that management practices such as minimum tillage, reduced use of inappropriate fertilizer, appropriate crop rotations with minimal fallow, and rationalized pesticide use be adopted regularly.

Arbuscular mycorrhizal fungi and rhizobia play a key role in enhancing plant productivity, plant nutrition, and plant resistance (Demir and Akköprü 2007). The activities of nitrogen-fixing bacteria are enhanced in the rhizosphere of mycorrhizal plants where synergistic interactions of such microorganisms with mycorrhizal fungi have been demonstrated (Barea et al. 2002), and hence, the symbiosis of AMF with rhizobia is considered crucial for legumes (van der Heijden et al. 2006). Realizing the importance of rhizobia and AMF symbiosis, pot and field experiments were conducted where both symbionts showed higher plant biomass and better N and P acquisition, although these effects were also dependent on the specific symbiont combination (Azcón et al. 1991; Requena et al. 2001; Xavier and Germida 2002). Similarly, the tripartite symbiosis of legume-mycorrhiza-rhizobium has conclusively shown improvements in overall growth of leguminous plants (Babajide et al. 2008; Wu et al. 2009). For some plant species, the association with mycorrhizal fungi is indispensable. The degree of dependence however varies with plant species, particularly the root morphology, and conditions of soil and climate. Mycorrhizal dependencies of leguminous plants such as *Acacia* and *Albizia* have been well demonstrated (Plenchette et al. 2005; Ghosh and Verma 2006). Several research findings have also indicated the remarkable roles of AM fungi in amelioration of various types of stresses (abiotic/biotic) in leguminous plants (Rabie and Almadini 2005; Khan 2006; Aysan and Demir 2009). Mycorrhizal legumes are also well known for rehabilitation of badly degraded lands and/or desertified habitats emphasizing the ecological significance of this special association (Requena et al. 2001; Quatrini et al. 2003). Under conditions of low N and P availability which occur in many tropical soils, the possible transfer of nutrients from the host plant to another plant by AMF active hyphae may take place. Hyphae of mycorrhizas may spread from one infected plant and enter the roots of one or more other plants (Heap and Newman 1980). It has been shown that plant assimilates may be transported from one plant to another through AM hyphal connections. In a study, the transfer of ^{14}C photosynthate from one plant to another was found primarily through AM hyphae rather than leakage from the roots of the donor plants (Francis and Read 1984). More specifically, different experimental results (Snoeck et al. 2000; Li et al. 2009) have verified the transfer of fixed N from legume mycorrhizal plants to nearby/adjacent nonleguminous plants via active hyphal connections. Diverse experimental results show that AMF differ in their capacity to supply plant nutrients such as P (van der Heijden et al. 2003; Ghosh and Verma 2006) suggesting mass production of suitable strains for sustainable inoculum development. Although the technology for the production of rhizobial and free-living PGPR inoculants are commercially available, the production of AM fungi inocula and the development of inoculation techniques have restricted the manipulation of AM fungi. An appropriate management of selected AM fungi is now available for exploiting the benefits of these microorganisms in agriculture, horticulture, and revegetation of degraded ecosystems (Barea et al., 2005). And large quantities of AMF inoculum can be produced by pot culture technique (Nopamornbodi et al. 1988). The traditional and most widely used approach has been to grow the fungus with suitable host plants in solid growth medium individually or in combination on the solid growth media (Tiwari and Adholeya 2002). However, the current biotechnology practices now allow the production of efficient AM fungal inoculants to mass propagate them for large-scale production systems (Gianinazzi and Vosátka 2004).

10.2 Mycorrhizal Association with Legumes

Legumes are an important plant group which can form symbiosis with P-acquiring arbuscular mycorrhizal fungi (AMF) (Pagano et al. 2007; Valsalakumar et al. 2007; Molla and Solaiman 2009). Scheublin et al. (2004) have analyzed the AMF community composition in the roots of three nonlegumes and in the roots and root nodules of three legumes growing in a natural dune grassland and found differences in

AMF communities between legumes and nonlegumes and between legume roots and root nodules. One AMF sequence type was much more abundant in legumes than in nonlegumes (39% and 13%, respectively). Root nodules contained characteristic AMF communities that were different from those in legume roots, even though the communities were similar in nodules from different legume species. Legumes and root nodules have relatively high N concentrations and high P demands. Accordingly, the presence of legume- and nodule-related AMF can be explained by the specific nutritional requirements of legumes or by host-specific interactions among legumes, root nodules, and AMF. In other experiments, Muleta et al. (2007, 2008) have reported more AMF spore counts under *Acacia abyssinica*, *Albizia gummifera*, and *Millettia ferruginea* shade trees than under nonleguminous shade trees in both natural coffee forest and in soils of smallholder agroforestry coffee system in southwestern Ethiopia. Similar observations have also been reported elsewhere under canopies of legume plants (He et al. 2004). Colozzi and Cardoso (2000) have also demonstrated that legume intercropping cultivation increased spores concentration of AMF in the soil.

Valsalakumar et al. (2007) in a field study identified the AM fungi associated with greengram [*Phaseolus aureus* Roxb. (=*Vigna radiata* var. *radiata*)]. The findings show that *Glomus mosseae*, *G. microcarpum*, *Gigaspora margarita*, and *Scutellospora* sp. colonized the greengram. *Glomus mosseae* was the most abundant AM fungal associate (81%) followed by *G. microcarpum* (24%) and *G. margarita* (24%) and *Scutellospora* sp. (5%) identified in soils studied. The range of distribution varied from a single species of AM fungus to three species belonging to two genera in one sample. Similarly, Bakarr and Janos (1996) examined the fine roots of 27 forest tree species for mycorrhizal colonization, a forestry plantation and a reforestation site in Sierra Leone, West Africa. Twenty tree species had arbuscular mycorrhizas, of which seven species were ectomycorrhiza colonizing six legume species belonging to Caesalpinioideae. Three species of Australian *Acacia* used widely in reforestation in Sierra Leone had arbuscular mycorrhizas. The effects of AMF, P addition, and their interaction on the growth and P uptake of three facultative mycotrophic legume trees (*Anadenanthera peregrina*, *Enterolobium contortisiliquum*, and *Plathymenia reticulata*) were investigated (Pagano et al. 2007). Phosphorus fertilization improved the growth of all the legume tree species. In turn, P enhanced the positive effects of AMF on the three studied species. Tissue nutrient concentrations showed slight variation among species and were influenced by both AMF inoculation and P. Plants inoculated with higher doses of KH_2PO_4 showed more vigorous seedlings. Results suggest that in low fertility soils, *A. peregrina*, *E. contortisiliquum*, and *P. reticulata* seedlings should be inoculated with AMF to enhance plant growth.

The application of AMF in soils has shown a tremendous improvement in growth and yields of diverse leguminous plants raised under both greenhouse and field conditions. For instance, inoculation with AMF improved growth of chickpea and doubled P uptake at low and intermediate levels of P fertilization in a pot experiment on sterilized low P calcareous soil (Weber et al. 1992). In a follow-up study, Ndiaye et al. (2009) evaluated the effects of different indigenous AM fungi on the mobilization of P from Senegalese natural rock phosphate (NRP) for growth of *Gliricidia sepium* and *Sesbania sesban* seedlings. In this study, the levels of NRP were found compatible with high AM fungal proliferation but changed the pattern of root colonization which varied with plant cultivars and fungal species. The mixed applications of NRP and AM fungi facilitated the measured growth parameters of *G. sepium* and *S. sesban* after 4 months cultivation. AM fungi in the presence of 600 or 800 mg NRP enhanced the weight of *S. sesban* by 200%. For *Gliricidia*, only *G. aggregatum* in the presence of high NRP levels showed similar growth promotory effects. On the other hand, *G. fasciculatum* enhanced the height of *Sesbania* by twofolds when grown in the presence of 400, 600, and 800 mg NRP. Generally, the impact of composite application of AM fungi and NRP on nutritional content was more obvious for *Sesbania* than for *Gliricidia* seedlings.

10.3 Mycorrhizal Status of Legumes That Do Not Form Nodules

It is interesting that certain leguminous tribes that cannot form nodules may be colonized by AM fungi. Cárdenas et al. (2006) investigated early responses to Nod factors and mycorrhizal colonization in a non-nodulating *Phaseolus vulgaris* mutant. The results indicate that even though *P. vulgaris* non-nodulating mutant (NN-mutant) is deficient in early nodulin gene expression when inoculated with *Rhizobium etli*, it can be effectively colonized by AM fungus, *G. intraradices*. Sometimes Nod mutants of other legumes fail to establish a mycorrhizal symbiosis (Bradbury et al. 1991) indicating that common elements of the infection process may exist in both associations.

10.4 Dual Inoculation of Legume Plants with Mycorrhizal Fungi and Rhizobia

The majority of legumes have the capacity to form a dual symbiotic interaction with N_2-fixing rhizobia and P-acquiring AM fungi (Lodwig et al. 2003; Navazio et al. 2007). Arbuscular mycorrhizal fungi and rhizobia together play a key role in natural ecosystems and influence plant productivity, nutrition, resistance, and plant community structure (van der Heijden et al. 2006; Demir and Akköprü 2007). The bioavailability of N and P is enhanced in the rhizosphere of mycorrhizal plants following synergistic interactions between the two groups of microorganisms (Barea et al. 2002). The authors further suggested that the inoculation of such phytobeneficial microbes has been shown to improve the overall performance of legumes indicating the importance of the tripartite symbiosis between legume-mycorrhiza and rhizobia in a given ecosystem. Studies have demonstrated that the two symbioses share some components of their developmental programs (Harrison 2005; Navazio et al. 2007). Synergistic effect of dual colonization of roots with AMF and *Rhizobium* on growth, nutrient uptake, and N_2 fixation in many legume plants has been reported (Xavier and Germida 2002; Stancheva et al. 2008) and discussed in the following section.

10.4.1 Dual Inoculation of AM Fungi and Rhizobia Under Greenhouse Conditions

Response of *Leucaena leucocephala* to inoculation with *Glomus fasciculatum* and/ or *Rhizobium* was studied in a P-deficient unsterile soil (Manjunath et al. 1984). The findings show that *G. fasciculatum* inoculation alone improved nodulation by native rhizobia and *Rhizobium* only treatment increased colonization of roots by native mycorrhizal fungi. However, when AM fungi and *Rhizobium* were used together, it improved nodulation, mycorrhizal colonization, dry weight, and N and P contents of the plants compared to single inoculation of each organism in a similar study. Eom et al. (1994) evaluated two wild legume plants, *Glycine soja* and *Cassia mimosoides* var. *nomame*, and a cultivated plant, soybean, inoculated with *Scutellospora heterogama*, isolated from natural soils and rhizobial cells. The AMF-colonized wild legume plants showed greater growth compared to soybean, whereas the soybean showed more nodulation than AM-colonized *Cassia mimosoides* plants. Moreover, *S. heterogama* appeared to stimulate the triple symbiosis of the wild legume plants. In addition, Babajide et al. (2008) in a greenhouse experiments determined the effect of different rhizobial and mycorrhizal species (*G. clarum*) on growth, nodulation, and biomass yield of soybean grown under low fertile eroded soil condition. Plant growth and biomass yield were significantly enhanced by AM fungus in both sterile and non-sterile soils compared to the control. However, combined inoculation of mycorrhiza with any of the rhizobial strains further improved plant growth and biomass production. The effect of composite inoculation of mycorrhiza + R25B *Rhizobium* was more pronounced, which substantially increased the plant height (68.8 cm), stem circumference (2.94 cm), number of leaves (39.0), shoot dry weight (16.1 g), and root dry weight (4.6 g), relative to control values of 33.2, 0.60 cm, 15, 4.4, and 1.6 g, respectively. Nodulation was equally enhanced by mycorrhizal and rhizobial inoculations under sterile and non-sterile soils. The percentage of mycorrhizal root colonization ranged from 4 to 42%, and root colonization was highest for mycorrhizal inoculated plants grown in sterile soil. From these findings authors concluded that dual inoculation of mycorrhiza and *Rhizobium* may be beneficial to soybean production in the tropics, where nutrients particularly available P and total N are very low. Ahmad (1995) studied the effect of dual inoculation on three local cultivars (Miss Kelly, Portland Red, Round Red) of red kidney bean with four strains of *R. phaseoli* (B36, B17, T2, and CIAT652) and three species of AM fungi (*G. pallidum*, *G. aggregatum*, and *Sclerocystis microcarpa*) in sterilized and non-sterilized soil. Symbiotic efficiency including improved plant growth and enhanced N and P was dependent on the specific combinations of *Rhizobium* strain, AM fungus, and cultivars of kidney bean. The rhizobial strains B36 and B17 co-inoculated with *G. pallidum* or *G. aggregatum* increased the growth of Miss Kelly and Portland Red, while rhizobial strain T2 paired with any of the three AM fungi was found as the best compatible pairing for the Round Red kidney beans. From these results, the author suggested that even though dual inoculation significantly improved the growth of the bean plants, the best performing combination of AM fungus and rhizobia requires further trials so that it is recommended for legume promotion in different geographical regions. Tajini et al. (2012) have also investigated the effect of dual inoculation of common bean with *G. intraradices* and *Rhizobium tropici* CIAT899 under glasshouse conditions. Two common bean genotypes (i.e., CocoT and Flamingo) varying in their effectiveness for nitrogen fixation were inoculated with *G. intraradices* and *R. tropici* CIAT899 and grown for 50 days in soil–sand substrate. Inoculation of common bean plants with the AM fungi resulted in a significant increase in nodulation compared to plants without inoculation. The combined inoculation of AM fungi and rhizobia significantly increased various plant growth parameters compared to simple inoculated plants. In addition, the combined inoculation of AM fungi and rhizobia resulted in significantly higher nitrogen and phosphorus accumulation in the shoots of common bean plants and improved phosphorus use efficiency compared to their controls, which were not dually inoculated. It is concluded that inoculation with rhizobia and AM fungi could improve the efficiency in P use for symbiotic nitrogen fixation especially under phosphorus deficiency. Combined inoculation with *G. intraradices* and *R. tropici* CIAT899 increases P use efficiency for symbiotic nitrogen fixation in common bean. Similarly, potted bean plants were grown in a glasshouse with and without organic and chemical fertilizers, uninoculated or inoculated with rhizobia (a mixed culture of *R. leguminosarum* bv. phaseoli and *R. tropici*) and AMF (*Glomus* spp.), singly or in combination (Aryal et al. 2003). Treatment effects on growth, nodulation, AMF colonization, and nutrient uptake of plants were evaluated. Rhizobial inoculation positively influenced root dry weight and nodulation of plants. Shoot and root dry weights and nodulation were again higher in dually inoculated plants compared to singly inoculated plants. Compared to control, single inoculation either with rhizobia or AMF did not increase pod yield. But, dual inoculation significantly increased pod yield compared to control or singly inoculated plants. Inoculation also significantly increased pod yield in organic fertilization treatment, but not in chemical fertilization treatment. AMF colonization, spore population, and shoot N and P were also significantly higher in dually inoculated plants. Under fertilized conditions, nodulation, AMF colonization, and spore population were generally more pronounced in dually inoculated organic plants than in chemical plants. Shoot Ca and K remained unaffected by inoculation either in fertilized or unfertilized conditions. Dual inoculation significantly increased the concentration of shoot Mg in organic plants, but not in chemical. In general, better positive effects of inoculations were observed in organic plants than in chemical suggesting higher dependency of organic plants on these symbionts for better growth and development. A similar study was conducted by Jia et al. (2004) to investigate the effects of the interactions between *Rhizobium* and AMF on N and P accumulation by broad bean and how increased N and P content influence biomass production, leaf area, and net photosynthetic rate. The AM fungus increased biomass production and photosynthetic rates by stimulating the ratio of P to N accumulation, and an increase in P was consistently correlated with an increase in N accumulation and N productivity, expressed in terms of biomass and leaf area. Photosynthetic N use efficiency, irrespective of the inorganic source of N (e.g., NO^{3-} or N_2) was enhanced by increased P supply due to AMF colonization. However, *Rhizobium* significantly declined AMF colonization irrespective of N supply and without *Rhizobium*; AMF colonization was higher in low N treatments. Presence or absence of AMF did not have a significant effect on nodule mass but high N with or without AMF led to a significant decline in nodule biomass. Furthermore, plants with the *Rhizobium* and AMF had higher photosynthetic rates per unit leaf area. Geneva et al. (2006) reported that the dual inoculation of pea plants with *G. mosseae* or *G. intraradices* and *R. leguminosarum* bv. *viciae*, strain D 293, significantly increased the plant biomass, photosynthetic rate, nodulation, and N_2 fixing activity in comparison to single inoculation of *R. leguminosarum* bv. *viciae* strain D 293. In addition, the co-inoculation significantly increased the total P content in plant tissues, acid phosphatase activity, and percentage of root colonization. Among all the microbial pairings, the co-inoculation of *R. leguminosarum* with *G. mosseae* was most effective at low P level, while *G. intraradices* inoculated with *R. leguminosarum* was most effective at higher P level. Xavier and Germida (2002) investigated also the effect of synergism between AMF and *R. leguminosarum* bv. *viciae* strains on lentil (*Lens culinaris* cv. Laird). Plants were inoculated with the AMF species *G. clarum* NT4 or *G. mosseae* NT6 and/or nine *Rhizobium* strains varying in efficacy and grown for 110 days in soil containing indigenous AMF and rhizobia. The results suggest that synergistic interactions between AMF and *Rhizobium* strains can enhance lentil productivity. In another study, Wu et al. (2009) have determined the single and combined effects of *G. mosseae* and *Rhizobium* on *Medicago sativa* grown on three types of coal mine substrates, namely, a mixture of coal wastes and sands (CS), coal wastes and fly ash (CF), and fly ash (FA) in pot experiment. When *Rhizobium* was used alone, it did not result in any growth response but sole application of *G. mosseae* had a significant effect on plant growth. Inoculation of G. *mosseae* also increased the survival rate of *M. sativa* in CS substrate. When *G. mosseae* inoculated *M. sativa* plant was grown with CF and FA substrates, the dry matter accumulation in the test plants was 1.8 and 5.1 times higher than those without inoculation. However, when *M. sativa* was inoculated with *G. mosseae* and *Rhizobium* together and grown in CS and CF substrates, the N, P, and K uptake by the test plant increased substantially suggesting a synergistic effect of the two phylogenetically distinct organisms which could be exploited for revegetation of coal mine substrates. In another greenhouse trial, Mehdi et al. (2006) reported that the effects of AM fungi (*G. mosseae* and *G. intraradices*), rhizobial (*R. leguminosarum* bv. *viciae*) strains, and P (superphosphate and phosphate rock) fertilizers significantly increased the dry biomass of shoots and seeds, P and N contents (shoots and seeds) of lentil plants, and percent of root colonized by AM fungus. The rhizobial strain possessing P-solubilizing ability showed a more beneficial effect on plant growth and nutrient uptake than the strain without this activity, although both strains had similar N_2-fixing efficiency. Synergistic relationships were observed between AM fungi and some rhizobial strains that related to the compatible pairing of these two microsymbionts. Moreover, the P uptake efficiency was increased when P fertilizers were applied along with AM fungi and/or P-solubilizing rhizobial strains emphasizing the remarkable importance of dual inoculation in the improvement of plant growth responses. Likewise, Meghvansi et al. (2008) observed the comparative efficacy of three AMF combined with cultivar-specific *B. japonicum* (CSBJ) in soybean under greenhouse conditions. Soybean seeds of four cultivars, namely, JS 335, JS 71-05, NRC 2, and NRC 7, were inoculated with three AMF (*G. intraradices*, *Acaulospora tuberculata*, and *Gigaspora gigantea*) and CSBJ isolates, individually or in combination, and were grown in pots using autoclaved alluvial soil of a nonlegume cultivated field of Ajmer (Rajasthan). Their findings indicate that among the single inoculations of three AMF, *G. intraradices* produced the largest increases in the parameters (nodulation, plant growth, and seed yield) studied followed by *A. tuberculata* and *G. gigantea* indicating that plant acted selectively on AMF symbiosis. The dual inoculation with AMF + *B. japonicum* CSBJ further improved these parameters demonstrating synergism between the two microsymbionts. Among all the dual treatments, *G. intraradices* + *B. japonicum* showed the greatest increase (115.19%), in seed weight per plant suggesting a strong selective synergistic relationship between AMF and *B. japonicum*. The cv. JS 335 exhibited maximum positive response toward inoculation. The variations in efficacy of different treatments with soybean cultivars, however, indicated the specificity of the inoculants. These results provide a basis for selection of an appropriate combination of specific AMF and *Bradyrhizobium* which could further be utilized for identifying the symbiotic effectiveness and competitive ability of microsymbionts under field conditions. Likewise, a pot trial was set up (Stancheva et al. 2008) to evaluate the response of alfalfa (*Medicago sativa* L.) to AMF species *G. intraradices* and *S. meliloti*, strain 1021, regarding the dry biomass accumulation, mycorrhizal fungi colonization, nodulation, and nitrogen fixation activity. Alfalfa plants were grown in a glasshouse until the flowering stage (58 days), in 4 kg plastic pots using leached cinnamon forest soil (Chromic luvisols—FAO) at P levels 42 mg P_2O_5 kg^{-1} soil (applied as 133 mg kg soil^{-1} tunisian phosphorite). The results demonstrated that the dual inoculation of alfalfa plants with *G. intraradices* and *S. meliloti*, strain 1021, significantly increased the percent of root colonization and acid phosphatase activities in the root tissue and in soil in comparison to a single inoculation with *G. intraradices*. Co-inoculation also significantly increased the plant biomass, total P and N content in plant tissues. Under conditions of dual inoculation, high nitrogenase activity was established, especially at the floral budding stage compared to the single inoculation of *S. meliloti* strain 1021. In addition, the interaction between AMF, *S. meliloti*, and *Medicago truncatula* Gaertn was investigated (de Varennes and Goss 2007). To generate a differential inoculum potential of indigenous AMF, five cycles of wheat, each of 1 month, were grown in sieved or undisturbed soil before *M. truncatula* was sown. The early colonization of *M. truncatula* roots by indigenous AMF was faster in undisturbed soil compared to sieved soil. *M. truncatula* grown in undisturbed soil had accumulated a greater biomass in aboveground tissues, had a greater P concentration, and derived more N from the atmosphere than plants grown in disturbed soil, although soil compaction resulted in plants having a smaller root system than those from disturbed soil. The difference in plant P content could not be explained by modifications in hydrolytic soil enzymes related to the P cycle as the activity of acid phosphatase was greater in sieved than in undisturbed soil, and the activity of alkaline phosphatase was unaffected by the treatment. Thus, the results observed were a consequence of the different rates of AMF colonization caused by soil disturbance. This study confirms that soil disturbance modifies the interaction between indigenous AMF, rhizobia, and legumes leading to a reduced efficacy of the bacterial symbiont.

Chickpea plants were also inoculated with six strains of *M. ciceri* and three AMF species, *G. intraradices* (GI), *G. mosseae* (GM), and *G. etunicatum* (GE), under pot experiments (Tavasolee et al. 2011). The plants inoculated with a number of AMF species, and bacterial strains increased overall plant dry mass compared to non-inoculated plants. GE was the most efficient in increasing plant dry matter. Individual AMF species were more effective than when mixed (GI + GM + GE). Bacterial treatments had increasing effect on root colonization by GI, GM, and GI + GM + GE. The results revealed that dual inoculation with AMF and rhizobia enhanced N, P, Zn, Fe, and Cu content in plants, but these increasing effects were different between fungal and bacterial treatments. Chaitra and Lakshman (2016) have also investigated the interaction between AM fungus, *Rhizobium* and *Azospirillum* on three leguminous crop plants (*Cicer arietinum* L., *Vigna unguiculata* L., and *Vigna radiata* L.) under greenhouse condition. Results revealed that triple inoculation of *Rhizobium*, AM fungus (*G. geosporum*) with *Azospirillum*, showed a significant plant growth biomass yield, percent root colonization, spore number, nodule number, nitrogen, and phosphorus content in shoots of *C. arietinum* L. and *V. unguiculata* L. compared to dual inoculation or single inoculation. However no improvement was observed in control/non-inoculated plants. The *V. radiata* responded positively with dual inoculation of *Rhizobiuam* with AM fungus (*G. geosporum*). This change has not been recorded in control plants compared to single/triple inoculation. Response to mineral fertilization and inoculation with rhizobia and/or arbuscular mycorrhiza fungi (AMF) of the *Anadenanthera colubrina*, *Mimosa bimucronata*, and *Parapiptadenia rigida* (Leguminosae–Mimosoideae) native trees from Brazilian riparian forests were studied in nursery conditions (Patreze and Cordeiro 2004). There were seven treatments varying in N, P fertilization, and inoculation with rhizobia (r), mycorrhiza (m), or both (rm): NP, P, P + r, P + rm, N, N + m, and N + rm. Results showed that AMF inoculations did not enhance the mycorrhizal colonization, and P uptake was not sufficient to sustain good growth of plants. The level of P mineral added affected negatively the AMF colonization in *A. colubrina* and *M. bimucronata*, but not in *P. rigida*. Native fungi infected the three legume hosts. The absence of mineral N limited growth of *A. colubrina* and *P. rigida*, but in *M. bimucronata* the lack of N was corrected by biological nitrogen fixation. N mineral added inhibited the nodulation, although spontaneous nodulation had occurred in *A. colubrina* and *M. bimucronata*. Rhizobia inoculation enhanced the number of nodules, nitrogenase activity, and leghemoglobin content of these two species. Thus, the extent of rhizobial and mycorrhizal symbiosis in these species under nursery conditions can affect growth and consequently the post-planting success.

10.4.2 Dual Inoculation with AMF and Rhizobia Under Field Conditions

Field investigations were conducted to study the effects of AM inoculation and triple superphosphate fertilization on nodulation, dry matter yield, and tissue N and P contents of *Bradyrhizobium*-inoculated soybean and lablab bean (Mahdi and Atabani 1992). Inoculation of both legumes with any of four AM fungi enhanced nodulation, dry matter yield, and plant N and P contents more than did triple super-phosphate. *Gigaspora margarita* and *G. mosseae* were superior to *G. calospora* and *Acaulospora* species and resulted in more extensive root infection, especially in soyabean. The integration of N_2 fixing trees into stable agroforestry systems in the tropics is being tested due to their ability to produce high biomass N and P yields, when symbiotically associated with rhizobia and mycorrhizal fungi (Marques et al. 2001; Kayode and Franco 2002). Accordingly, in a field trial, Marques et al. (2001) evaluated the effect of dual inoculation of *Rhizobium* spp. and mycorrhizal fungi on the growth of *Centrolobium tomentosum* Guill. ex Benth, a native leguminous tree of the Brazilian Atlantic Forest. Complete fertilization was compared to inoculation treatments of selected rhizobia strains BHICB-Ab1 or BHICB-Ab3, associated or not to AM fungi. Plants inoculated with strain BHICB-Ab1 and AMF increased the dry matter by 56% over uninoculated control, and N accumulation was greater than those observed for BHICB-Ab3 inoculated plants. Strain BHICB-Ab1 formed a synergetic relationship with mycorrhizal fungi as the combined inoculation enhanced plant height and dry weight more than single inoculation, while the growth of BHICB-Ab3 plants was not modified by AMF inoculation. Arbuscular mycorrhizal fungi also improved plants survival and possibly favored the nodule occupation by rhizobial strains as compared to the non-mycorrhizal plants. Similarly, *Acacia mangium* inoculated with rhizobial strains (BR 3609 and BR 3617) and three AM fungi, *G. clarium*, *Gigaspora margarita*, and *Scutellospora heterogama*, grew better than seeds planted without rhizobia and AMF inoculants (Kayode and Franco 2002). The authors observed that *S. heterogama* facilitated the growth better in both fallow and degraded soils. Seeds inoculated with rhizobia strains and AMF, however, produced more nodules and had higher AMF infection rates than seeds inoculated with rhizobia or AMF inoculants alone (Marques et al. 2001; Kayode and Franco 2002). Singh et al. (1991) evaluated the effect of live yeast cells (*Saccharomyces cerevisiae*) on nodulation and dry biomass of shoot and roots of legumes like *Leucaena leucocephala*, *Glycine max*, *Cajanus cajan*, *Phaseolus mungo*, *Phaseolus aureus*, and *Vigna unguiculata* in the presence of both AMF and *Rhizobium* strains. The results indicate that inoculation with live yeast cells remarkably enhanced the measured plant parameters. Root infection (native AMF) and the formation of vesicles, arbuscules, and spores were also increased with yeast inoculation. The increase in the parameters, however, varies with legumes and the type of yeast culture. On the other hand, the effect of whey application, the inoculation of *Glomus intraradices* Schenck & Smith and *Mesorhizobium ciceri* on root colonization, nodulation, yield, and the components of yield in chickpea (cv. Aziziye-94) were studied under rain-fed and irrigation management (Erman et al. 2011). Experiments were carried out in a split plot design with four replications in 2003 and 2004. The abovementioned factors were all applied to plants in single, double, and triple combinations. Arbuscular mycorrhizal fungus (AMF) inoculation, alone or in combination with other treatments, was very effective under rain-fed conditions, resulting in large increases in yield, root colonization, and phosphorus content of the seed and shoot. On the other hand, rhizobial inoculation increased significantly all traits examined, particularly root nodulation and the nitrogen content of seeds and shoots under irrigated conditions. Whey combined with AMF significantly increased root colonization, while its combination with *Rhizobium* increased the number of nodules. Combinations of two or three treatments were more effective than individual applications. The greatest yield, root colonization, and nodulation were obtained from the combination of all three treatments under irrigation. Although voluminous literature reports show superiority of plant performances under dual inoculation, sometimes the usual synergism was found to be less effective. For example, Nambiar and Anjaiah (1989), in a field experiment, reported that the effects of AMF on competition among inoculated bradyrhizobia were less evident, but inoculation with *Bradyrhizobium* strains increased root colonization by AMF and certain AMF/*Bradyrhizobium* inoculum strain combinations produced higher nodule numbers. Plants grown without *Bradyrhizobium* and AMF, but supplied with ammonium nitrate (300 g mL^{-1}) and potassium phosphate (16 g mL^{-1}), produced higher dry matter yields than those inoculated with both symbionts in the pot experiment. Inoculation with either symbiont in the field, however, did not result in higher pod yields at harvest. In a similar trial, Camila and Lazara (2004) have tested response to mineral fertilization and inoculation with rhizobia and/or AMF of the *Anadenanthera colubrina*, *Mimosa bimucronata*, and *Parapiptadenia rigida* (Leguminosae–Mimosoideae) native trees from Brazilian riparian forests, in nursery conditions. The findings showed that AMF inoculations did not enhance the mycorrhizal colonization, and P uptake was not sufficient to sustain good growth of plants. The level of P mineral added affected negatively the AMF colonization in *A. colubrina* and *M. bimucronata*, but not in *P. rigida*. Native fungi infected the three legume hosts. The absence of mineral N limited the growth of *A. colubrina* and *P. rigida*, but in *M. bimucronata* the lack of N was corrected by BNF. The applied N mineral, however, inhibited nodulation, although spontaneous nodulation occurred in *A. colubrina* and *M. bimucronata*. Rhizobia inoculation enhanced the number of nodules, nitrogenase activity, and leghemoglobin content of these two species. Thus, the extent of rhizobial and mycorrhizal symbiosis in these species under nursery conditions affected growth and consequently the post-planting success. Evidence is also available that improved formation of AM can inhibit nodulation, possibly due to inter-endophyte incompatibility of competition (Behlenfalvay et al. 1985). On the contrary, (Pacovsky et al. 1986) revealed that even though nodule numbers may not significantly be increased by AM colonization, yet the size and nitrogen-fixing activity may be increased. However, there is a report that suggests that symbiotic N$_2$ fixation is clearly accelerated in legume following AMF inoculation, but the response of *Rhizobium* symbiosis may vary according to the strains of the AM fungus involved (Linderman and

Paulitz 1990). These and other associated data thus indicate that the *Rhizobium*–AMF partnership nearly always exists but may not necessarily be optimal with the best combination of symbionts for the host legumes.

10.5   Mycorrhizal Dependency of Legumes

For some plant species, the association with mycorrhizal fungi is indispensable. The degree of dependence, however, varies with plant species, particularly the root morphology and conditions of soil and climate (Hayman 1986). Plants with thick roots poorly branched and with few root hairs are usually more dependent on mycorrhizas for normal growth and development. These species include onions, grapes, citrus, cassava, coffee, and tropical legumes. When the level of soil fertility and humidity are increased, the dependence on the mycorrhizal condition decreases to a point where the plant becomes immune to colonization (Khaliel et al. 1999). Furthermore, mycorrhizal dependencies of leguminous plants grown in stressed situations have also been well documented (Plenchette et al. 2005; Ghosh and Verma 2006). Growth and mineral uptake of 24 tropical forage legumes and grasses were compared under glasshouse conditions in a sterile low P oxisol, one part inoculated and the other not inoculated with mycorrhizal fungi (Duponnois et al. 2001). Shoot and root dry weights and total uptake of P, N, K, Ca, and Mg of the entire test plants were significantly increased by mycorrhizal inoculation. Mycorrhizal inoculation, with few exceptions, decreased the root/shoot ratio. Non-mycorrhizal plants, on the other hand, had lower quantities of mineral elements than mycorrhizal plants. Plant species, however, did not show any correlation between percentage mycorrhizal infection and growth. A great variation in dependence on mycorrhiza was observed among forage species. Total uptake of all elements by non-mycorrhizal legumes and uptake of P, N, and K by non-mycorrhizal grasses correlated inversely with mycorrhizal dependency. Mycorrhizal plants of all species used significantly greater quantities of soil P than the non-mycorrhizal plants, and utilization of soil P by non-mycorrhizal plants was correlated inversely with mycorrhizal dependency. As the production of grain and herbaceous legumes is often limited by low levels of available P in most savanna soils, the potential for managing AMF by selecting lines or accessions dependent on AMF as a strategy to improve plant P nutrition and productivity is required (Plenchette et al. 2005; Ghosh and Verma 2006). Accordingly, Nwoko and Sanginga (1999) evaluated the interactions between AMF and *Bradyrhizobium* species and their effects on growth and mycorrhizal colonization of ten recent selections of promiscuous soybean breeding lines and two herbaceous legumes (*Lablab purpureus* and *Mucuna pruriens*). Mycorrhizal colonization differed among promiscuous soybean lines (ranging from 16 to 33%) and was on average 20% for mucuna and lablab. Three groups of plants were identified according to mycorrhizal dependency (MD): (1) the highly dependent plants with MD >30% (e.g., soybean line 1039 and mucuna), (2) the intermediate group, with MD between 10 and 30% (e.g., soybean line 1576 and lablab), and (3) the majority of soybean lines (five lines out of ten) that were not mycorrhizal dependent. This great variability in MD and response to P application among promiscuous soybean and herbaceous legumes offers a potential for the selection of plant germplasm able to grow in P-deficient soil. Similar results have also been reported for different species of woody leguminous trees. For instance, Ghosh and Verma (2006) evaluated the effects of three AMF species (*G. occultum*, *G. aggregatum*, and *G. mosseae*) inoculations on growth responses of *Acacia mangium* in lateritic soil. All inoculations significantly enhanced growth with respect to shoot height, root diameter, leaf area, chlorophyll content, and biomass of *A. mangium* compared to uninoculated control seedlings. The mycorrhizal dependency factor indicated that the growth of *A. mangium* was 57% dependent on *G. occultum*, 47% on *G. mosseae*, and 46% on *G. aggregatum*. The findings indicate the presence of disparity among AMF species with regard to their growth enhancement in a particular mycorrhizal legume. It has also been demonstrated that mycorrhizal dependence and responsiveness of legumes declines with an increase in P added to the soil (Khaliel et al. 1999).

10.6 How Arbuscular Mycorrhizal Fungi Enhance Legumes' Performance

The AM fungi affect the growth and development of plants both directly and indirectly (Table 10.1). However, broadly, the principal contribution of AM fungi to plant growth is due to uptake of nutrients by extraradical mycorrhizal hyphae (Marschner 1998; Hodge and Campbell 2001; van der Heijden et al. 2006). The most prominent effect of AMF is to improve P nutrition of the host plant in soils with low P levels due to the large surface area of their hyphae and their high affinity P uptake mechanisms (Muchovej 2001). To substantiate this concept of plant growth promotion by AM fungi, several studies have shown that AM fungi contribute to up to 90% of plant P demand (Jakobsen et al. 1992; van der Heijden et al. 2006). For instance, the P depletion zone around non-mycorrhizal roots extends to only 1–2 mm, nearly the length of a root hair, whereas extraradical hyphae of AMF extend 8 cm or more beyond the root making the P in this greater volume of soil available to the host

Table 10.1 Direct and indirect effects of mycorrhizal fungi on crop productivity in organic farming systems

Direct effects	Indirect effects
Stimulation of crop productivity	Weed suppression
Nutrient acquisition (P, N, Cu, Fe, Zn)	Stimulation of nitrogen fixation
Enhanced seedling establishment	Stimulation of soil aggregation and soil structure
Drought resistance	Suppression of soil pathogens
Heavy metal/salt resistance	Soil biological activity stimulation
	Increased soil carbon storage
	Reduction of nutrient leaching

Adapted from van der Heijden et al. (2008)

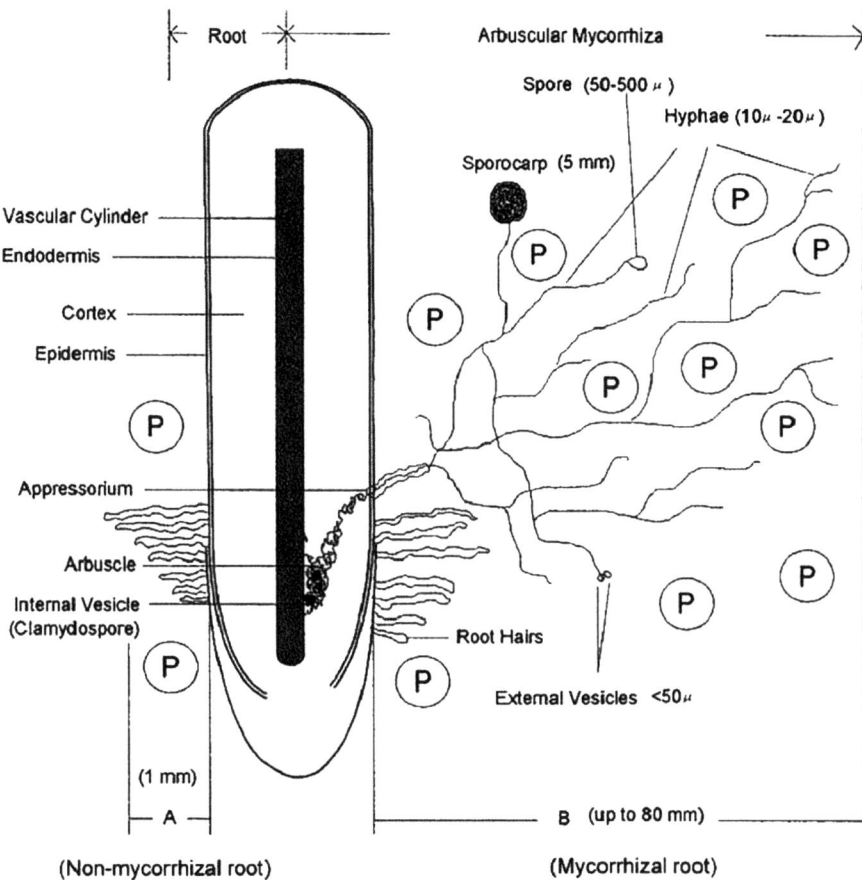

Fig. 10.1   Root colonized by endomycorrhizal fungus. Zone of P (or other nutrient) absorption by a non-mycorrhizal root (**a**) and by a mycorrhizal root (**b**) P phosphate ion (adapted from Muchovej (2001))

(Fig. 10.1). There are also reports of production of organic acids by AMF that could solubilize the insoluble mineral P (Lapeyrie 1988), an added advantage in terms of improvement of P uptake by host plants. In addition, AMF mycelia have also been shown to increase uptake of many other nutrients, including N, S, B, Cu, K, Zn, Ca, Mg, Na, Mn, Fe, Al, and Si (Clark and Zeto 2000). Apparently, besides providing P to their host plants, AM fungi can facilitate N_2 fixation by providing legumes with P and other immobile nutrients such as Cu and Zn, essential for N_2 fixation (Li et al. 1991; Kothari et al. 1991; Clark and Zeto 2000). There are reports that N fixation can be reduced or even completely inhibited in the absence of AMF at low nutrient availability (Azcón et al. 1991). The improvement in plant growth under both greenhouse and field conditions has also been suggested due to increased photosynthesis and improved carbon flow to the nodule and to AM sinks, giving rise to more and larger nodules that fix more nitrogen for the plant (Linderman and Paulitz 1990). In some cases, AMF may be responsible for acquiring 100% of host nutrients (Smith et al. 2004). Thus, Marschner (1998) and Hodge and Campbell (2001) have suggested that the overall improvement in plant nutrition following AM inoculation is due to (1) increased root surface through extraradical hyphae, which can extend beyond root depletion zone, (2) degradation of organic material, and (3) alteration of the microbial composition in the rhizosphere. More specifically, mechanisms as to how AMF contribute to plant health have been extensively studied leading to development of several hypotheses (Linderman 1994). The most important are (a) increased nutrient uptake that results in higher resistance of the plant to pathogen invasion or a compensation of the symptoms, (b) competition for photosynthates or space, (c) plant morphological changes and barrier formation, (d) changes in biochemical compounds related with plant defense, and (e) increased percentage of microbial antagonists in the rhizosphere. Under conditions of low N and P availability which exist in many tropical soils, the possible transfer of nutrients from the mycorrhizal plant to another plant via AMF hyphal network may occur. Underground hyphal links can be formed when hyphae of mycorrhizal fungi spread from one infected plant and enter the roots of one or more other plants (Heap and Newman 1980). Studies have ascertained that AM fungi did enhance N transfer from mycorrhizal legumes to another nonleguminous plant (Vankessel et al. 1985). Similarly, Snoeck et al. (2000) demonstrated that nearly 30% of the nitrogen fixed by legumes like *Desmodium* and *Leucaena* was transferred to associated coffee trees.

10.6.1 Alleviation of Environmental Stresses in Mycorrhizal Legumes

Currently, wide arrays of environmental stresses (abiotic/biotic) are increasing worldwide due to various types of anthropological activities that have seriously threatened plant distribution and function in a given ecosystem. Although plants have evolved mechanisms to cope such unfavorable factors, but they can perform better if grown with beneficial rhizosphere microbes (Aroca and Ruiz-Lozano 2009). Generally, phytobeneficial microbes greatly enhance tolerances of plants to a wide array of stresses (Fig. 10.2). The role of AM fungi and other phytobeneficial microbes in the promotion of biological and chemical properties of legumes under stressed environment is briefly discussed in the following section.

10.6.1.1 Tolerance to Salt/Alkaline and Acidity
Salinity is one of the most important abiotic stresses that limit crop growth and productivity across the globe. Soil salinity also decreases nodulation and N_2 fixation and nitrogenase activity of nodulated legumes (Karmakar et al. 2015). Thus, the development of salt-tolerant symbioses is an absolute necessity to enable cultivation of leguminous crops in salt-affected soils. For example, Rabie and Almadini (2005) while investigating the effects of dual inoculation of *Azospirillum brasilense* nitrogen fixing bacterium (NFB) and AMF (*G. clarum*) on *Vicia faba* grown with five levels of NaCl (0.0–6.0 dS m^{-1}) observed that AM-inoculated faba plants showed

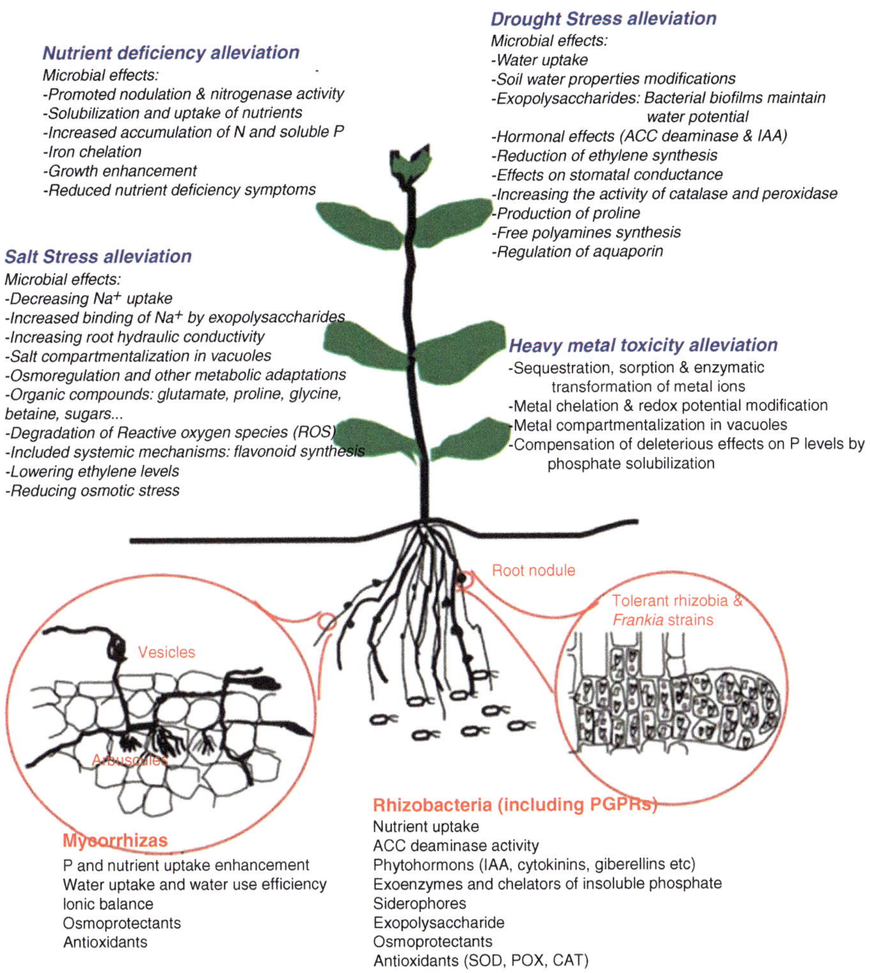

Drought Stress alleviation
Microbial effects:
-Water uptake
-Soil water properties modifications
-Exopolysaccharides: Bacterial biofilms maintain water potential
-Hormonal effects (ACC deaminase & IAA)
-Reduction of ethylene synthesis
-Effects on stomatal conductance
-Increasing the activity of catalase and peroxidase
-Production of proline
-Free polyamines synthesis
-Regulation of aquaporin

Nutrient deficiency alleviation
Microbial effects:
-Promoted nodulation & nitrogenase activity
-Solubilization and uptake of nutrients
-Increased accumulation of N and soluble P
-Iron chelation
-Growth enhancement
-Reduced nutrient deficiency symptoms

Salt Stress alleviation
Microbial effects:
-Decreasing Na^+ uptake
-Increased binding of Na^+ by exopolysaccharides
-Increasing root hydraulic conductivity
-Salt compartmentalization in vacuoles
-Osmoregulation and other metabolic adaptations
-Organic compounds: glutamate, proline, glycine, betaine, sugars...
-Degradation of Reactive oxygen species (ROS)
-Included systemic mechanisms: flavonoid synthesis
-Lowering ethylene levels
-Reducing osmotic stress

Heavy metal toxicity alleviation
-Sequestration, sorption & enzymatic transformation of metal ions
-Metal chelation & redox potential modification
-Metal compartmentalization in vacuoles
-Compensation of deleterious effects on P levels by phosphate solubilization

Root nodule

Tolerant rhizobia & *Frankia* strains

Vesicles

Arbuscules

Mycorrhizas
P and nutrient uptake enhancement
Water uptake and water use efficiency
Ionic balance
Osmoprotectants
Antioxidants

Rhizobacteria (including PGPRs)
Nutrient uptake
ACC deaminase activity
Phytohormons (IAA, cytokinins, giberellins etc)
Exoenzymes and chelators of insoluble phosphate
Siderophores
Exopolysaccharide
Osmoprotectants
Antioxidants (SOD, POX, CAT)

Fig. 10.2 Mechanisms adopted by N_2-fixing bacteria, PGPR, and AM fungi to alleviate abiotic stresses; *CAT* catalase, *IAA* indoleacetic acid, *PGPR* plant growth-promoting rhizobacteria, *POX* peroxidase, *SOD* superoxide dismutase (adapted from Bouizgarne et al. (2014))

decreases in salinity tolerance, % of mycorrhizal infection, and higher accumulation of proline with increasing levels of salinity. In addition, AMF infection significantly increased mycorrhizal dependency, N and P level, phosphatase enzymes, nodule numbers, protein content, and nitrogenase enzymes of all salinized faba plants compared to control and non-AM plants either in the absence or presence of NFB. In shoots of non-AM plants, Na^+ concentration was increased, while the concentrations of K^+, Mg^{2+}, and Ca^{2+} were decreased with increasing salinity. AM-colonized plants, on the other hand, had greater K^+/Na^+, Mg^{2+}/Na^+, and Ca^{2+}/Na^+ ratios relative to non-AM plants at all salinity levels. The Na^+ level in shoots of AM plants showed slight increase with gradual increase in salinity, while a noticeable increase was

observed in K⁺ and Ca²⁺ concentrations especially at higher salinity levels. The
results clearly showed that the inoculation of NFB along with AM plants synergistically increased the performance of the test legume under salinity stress providing
evidence for reducing the salt affected negative impact on legumes as also reported
for *Trifolium alexandrinum* plants grown under different salinity levels (2.2, 5, and
10 dS m⁻¹) in a pot experiment under glasshouse conditions (Shokri and Maadi
2009). Another study in Egypt (Abd-Alla et al. 2014) was devoted to investigating
the synergistic interaction of *Rhizobium* and arbuscular mycorrhizal fungi for
improving growth of faba bean grown in alkaline soil. Out of 20 isolates 3 of them
were selected as tolerant isolates and named as *Rhizobium* sp. Egypt 16 (HM622137),
Rhizobium sp. Egypt27 (HM622138), and *R. leguminosarum* bv. *viciae* STDF-
Egypt 19 (HM587713). The best alkaline tolerant was *R. leguminosarum* bv. *viciae*
STDF-Egypt 19 (HM587713). The effect of *R. leguminosarum* bv. *viciae* STDF-
Egypt 19 and mixture of AMF (*Acaulospora laevis, G. geosporum, G. mosseae,* and
Scutellospora armeniaca) both individually and in combination on nodulation,
nitrogen fixation, and growth of *Vicia faba* under alkalinity stress were assessed. A
significant increase over control in number and mass of nodules, nitrogenase activity, leghemoglobin content of nodule, mycorrhizal colonization, and more dry mass
of root and shoot was recorded in dual inoculated plants than plants with individual
inoculation. The enhancement of nitrogen fixation of faba bean could be attributed
to AMF facilitating the mobilization of certain elements such as P, Fe, K, and other
minerals that involve in synthesis of nitrogenase and leghemoglobin. Thus it is clear
that the dual inoculation with *Rhizobium* and AMF biofertilizer is more effective for
promoting growth of faba bean grown in alkaline soils than the individual treatment,
reflecting the existence of synergistic relationships among the inoculants.

The ability of crop plants to tolerate low soil pH has become extremely important
in the agricultural production systems of the humid tropics with soils of low pH
(Kamprath and Foy 1985). Studies by Dodd et al. (1990) and Sieverding (1991)
show that over 50 field trials with effective AMF in acid soils of varying fertility
resulted in an average increase of 20–25% in yields (3 tons ha⁻¹) and a greater stability in production year after year. Later on, the influence of soil acidity on the
levels of colonization by the microsymbionts and the dependency of pioneer plants
on the microsymbionts was investigated in an abandoned quarry of acid sulfate soil
(Maki et al. 2008). The levels of AM colonization in pioneer grass, forbs, and
legume shrubs grown in the field were assessed, and no significant decline in the
levels with an increase in soil acidity was observed. Most of the legume shrubs
formed root nodules. Several AM fungi and bradyrhizobia were cultured from the
rhizosphere soils of pioneer plants grown in the quarry. Pot experiments revealed
that the microsymbionts isolated from the field significantly promoted the growths
of pioneer grasses and legume shrubs in acid sulfate soil at pH 3.4. On the other
hand, Dodd et al. (1990) supported the idea that increasing the AMF inoculum
potential of acid-infertile soils by inoculation or pre-crops can greatly increase the
rate of establishment of mycorrhiza-dependent host plants. Thus, from these and
other studies, it was suggested that bacterial-AM-legume tripartite symbioses could
be a new approach to increase the tolerance of legume plants under stressed

environment. Integration of microbial inoculants with NPK application in acidic soils has showed promising results. For example, Bai et al. (2016) in a recent experiment quantified the influence of integrated use of AMF, *Rhizobium*, and N and P on growth, productivity, profitability, and nutrient use efficiencies of garden pea grown under acid Alfisol field. The experiment was laid out in randomized block design (RBD) replicated thrice comprising 13 treatments involving AMF (*G. mosseae*), *Rhizobium* (*R. leguminosarum*), and inorganic N and P fertilizers. The results revealed that dual inoculation of pea seed with AMF and *Rhizobium* enhanced the plant height, leaf area index, and dry matter accumulation significantly by 19.4 and 13.1, 10.7 and 10.7, and 16.6 and 16.7%, respectively, at 60 and 120 days after sowing (DAS). Similarly, dual inoculation exhibited significant respective increases of 9.5 and 14.6% in absolute and crop growth rates over generalized recommended NPK dose (GRD) during 60–120 DAS. The dual inoculation led to significant respective increases of 1- and 2.2-, 1.06- and 1.74-, 0.21- and 1.5-, and 1.05- and 1.60-folds in partial factor productivity, crop recovery efficiency, physiological efficiency, and % recovery of applied N and P, respectively, over GRD. The magnitude of increase in pea productivity, net returns, and boron to carbon (B/C) ratio following dual inoculation was to the tune of 20, 54.4, and 104.1%, respectively, over GRD. Dual inoculation also exhibited significant increases of 19.4 and 53% in production and monetary efficiencies of pea over GRD. Overall, dual inoculation of AMF and *Rhizobium* with 75% soil-test-based N and P dose in pea has great potential in enhancing pea productivity, profitability, and nutrient use efficiency besides saving about 25% fertilizer N and P without impairing pea productivity in Himalayan acid Alfisol.

10.6.1.2 Heavy Metals and Drought Tolerance

Working with *Trifolium repens*, Vivas et al. (2003) studied the effect of inoculation with naturally occurring microorganisms (an AM fungus and rhizosphere bacteria) isolated from a Cd-polluted soil. One of the bacterial isolate identified as a *Brevibacillus* sp. showed a marked PGPR activity. Mycorrhizal colonization also enhanced *Trifolium* growth and N, P, Zn, and Ni content, and the dual inoculation of AM fungus and *Brevibacillus* sp. further enhanced growth and nutrition and reduced Cd concentration, particularly at the highest Cd level. Interestingly, increasing Cd level in soil decreased Zn and Pb accumulation in shoot. Co-inoculation of *Brevibacillus* sp. and AM fungus increased shoot biomass over single mycorrhizal plants by 18% (at 13.6 mg Cd kg^{-1}), 26% (at 33.0 mg Cd kg^{-1}), and 35% (at 85.1 mg Cd kg^{-1}). In contrast, Cd transport from soil to plants was substantially reduced and at the highest Cd level; *Brevibacillus* sp. lowered this value by 37.5% in AM-colonized plants. However, the increase in Cd level highly reduced plant mycorrhization and nodulation. On the contrary, strong positive effect of this bacterium was observed for nodule formation in all treatments. In a similar study conducted by Al-Garni (2006), the composite inoculation of AM fungus and *Rhizobium* significantly increased dry weight, root/shoot ratios, leaf number and area, plant length, leaf pigments, total carbohydrates, and N and P content of cowpea plants grown in pots treated with 6 concentrations of Zn (0–1000 mg/kg dry soil) and Cd (0–100 mg/

kg dry soil) compared to non-inoculated controls. Moreover, tolerance index of inoculated cowpea plants was greater than uninoculated plants. And microsymbionts dependencies of test plants increased at higher levels of Zn and Cd in polluted soil. Metals accumulated by microsymbionts-infected cowpea plant were mostly distributed in root tissues, suggesting that an exclusion strategy for metal tolerance exists in such organisms. Yet in another study, the influence of AM fungus *G. macrocarpum* Tul. and Tul on growth, nutrients, and Pb uptake by *Bradyrhizobium*-inoculated soybean (var. IAC-14) was assessed in soils treated with different levels of Pb (Andrade et al. 2004). The results revealed that soybean shoot dry biomass was not affected by increasing doses of Pb, but the number of pods decreased significantly. Nodule dry weights of mycorrhizal roots were reduced by soil Pb additions, although the mycorrhizas stimulated plant nodulation significantly. The inoculation of AMF in soybeans provided higher rates of nutrients uptake, mainly P, inducing greater mycorrhizal-soybean growth. Thus, mycorrhizas improved Pb uptake and produced shoots with Pb concentrations 30% lower than those of non-mycorrhizal plants, at the highest Pb concentration added to the soil. AM fungus was, however, more susceptible to the higher Pb rates added to the soil than the soybean plants, decreasing both root AM colonization and spore production. This work indicated that a concentration of 600 mg dm^{-3} of Pb in the soil interfered with the establishment of double symbioses between AMF and *Bradyrhizobium* and with the fungus perpetuation in soil. Recent surveys indicate that ecosystem restoration of heavy metal contaminated soils practices need to incorporate microbial biotechnology research and development in order to harness the optimum benefits of bacterial-AM-legume tripartite symbiosis under heavy metal contaminated soils (Al-Garni 2006; Khan 2006).

Water deficit is considered one of the most important abiotic factors limiting plant growth and yields. Several eco-physiological investigations have shown that the AM symbiosis often alters the rates of water movement into, through, and out of the host plants, with consequent effects on tissue hydration and plant physiology (Ruiz-Lazano 2003) and consequently improve water uptake by plants (Aliasgharzad et al. 2006). AM fungi in combination with rhizobia or PGPR usually have an accumulative beneficial effect on plant drought tolerance (Aroca and Ruiz-Lozano 2009). For instance, in a controlled pot culture experiment performed by Aliasgharzad et al. (2006), soybean plants were inoculated with two species of AM fungi, *G. mosseae* (Gm) or *G. etunicatum* (Ge), or left non-inoculated (NM) as control in a sterile soil. Four levels of soil moisture (field capacity, 0.85 FC, 0.7 FC, 0.6 FC) in the presence or absence of *B. japonicum* were applied to the pots. Relative water content (RWC) of leaf at both plant growth stages (flowering and seed maturation) decreased with the dryness of soil; RWC was higher in all mycorrhizal than non-mycorrhizal plants irrespective of soil moisture level. At the lowest moisture level (0.6 FC), Ge was more efficient than Gm in maintaining high leaf RWC. Leaf water potential (LWP) had the same trend as RWC at flowering stage, but it was not significantly influenced by decrease in soil moisture to 0.7 FC during seed maturation stage. Seed and shoot dry weights were affected negatively by drought stress. Mycorrhizal plants, however, had significantly higher seed and shoot dry weights than non-mycorrhizal plants at all moisture levels except for seed weight at 0.6 FC. Root mycorrhizal colonization was positively correlated with RWC, LWP, shoot N and K, and seed weight, implying improvement of plant water and nutritional status as a result of colonization. Shoot K was enhanced considerably by both bacterial and fungal inoculations, particularly in plants with dual inoculations where the highest shoot K levels were found. The relatively higher shoot and seed dry weights in plants inoculated with both *G. etunicatum* and *B. japonicum* could be ascribed to their higher RWC and LWP, suggesting that drought avoidance is main mechanism of this plant-microbe association in alleviation of water stress in soybean. Aroca and Ruiz-Lozano (2009) also emphasized that phytobeneficial soil microorganisms enhance plant drought tolerance by different mechanisms including decreased oxidative stress, improved water status, or regulation of aquaporins. In addition, the authors further suggested that AM symbiosis improves almost every physiological parameter, like water status, leaf transpiration, photosynthesis, or root water uptake of the host plant under drought stress. At the same time, AMF in combination with rhizobia or other PGPR results in additive or synergistive effect on plant drought tolerance, although this depends on the compatibility of strains used for inoculation. Therefore, although there is evidence which help to understand as to soil microorganisms induce plant drought tolerance at physiological level, the mechanistic basis of drought tolerance at molecular level is inadequate. Currently, it is well documented that desertification is a complex and dynamic process which obviously has a negative environmental impact, particularly in arid, semiarid, and subhumid areas of the world, where the process is claiming several million hectares per annum (Herrera et al. 1993; Aroca and Ruiz-Lozano 2009). Consequently, the proportion of plants living under water shortage conditions is increasing. Thus, management of indigenous plant-microbes symbioses assists in restoration of desertified ecosystems (Requena et al. 2001). Legumes are the most appropriate candidates for revegetation of water-deficient, low-nutrient environments/disturbed ecosystems because of their ability to establish tripartite symbiotic associations with nitrogen-fixing rhizobia and AMF which improve nutrient acquisition and help plants to become established and cope with stress situations (Herrera et al. 1993). Studies show that useful legume tree species may contribute around 12 tons of dry litter and 190 kg of N ha^{-1} y^{-1} to renovate degraded soil (Franco and De Faria 1997). Sometimes, the combined effect of microsymbionts may, however, cause deleterious effect to legume host under moderate water stress condition. For instance, four *Phaseolus vulgaris* varieties were single or dual inoculated with two different AM fungus and/or two different *Rhizobium* strains (Franzini et al. 2010). All plants were grown under moderate drought conditions. Surprisingly, most of the biological treatments involving one fungus and one *Rhizobium* together caused a deleterious effect on plant growth. However, these negative effects were dependent on the *P. vulgaris* variety used as well as on the symbionts implicated. The results showed that AM symbiosis inhibited nodule development and N$_2$ fixation, causing diminution of plant growth. Therefore, under moderate drought conditions, the dual symbiosis formed by AM fungi and *Rhizobium* can be deleterious to *P. vulgaris* growth depending on the plant variety and the symbionts involved. Thus, under these common stress conditions, selection for the appropriate symbionts to each *P. vulgaris* variety is needed.

10.6.1.3 Tolerance of Soilborne Pathogens

The effects of AM fungi *G. mosseae* (Gm) and *G. fasciculatum* (Gf) and *R. legumi-nosarum* biovar *phaseoli* (Rlp) were examined on the patho-system of *Sclerotinia sclerotiorum* (Lib) de Bary (Ss) and common bean (Aysan and Demir 2009). The colonization and nodulation of two biological control agents exhibited differences as a result of reciprocal interactions of these items as well as the effect of Ss. Nodulation of Rlp decreased in triple inoculation. In addition, colonization of AMF significantly decreased in treatment of Ss + AMF than control AMF. Treatments of single inoculation of AMF and Rlp isolates reduced disease severity by 10.3–24.1%. It was found that single biological control agent's inoculations were more effective than dual inoculations (AMF + Rlp). While comparing the morphological parameters of common beans, all measured morphological parameters were decreased in treatments having pathogen isolate. Besides this, all biological control agents increased total content of P and N in treated plants compared to the controls. Root colonization by AMF can improve plant resistance/tolerance to biotic stresses. Studies indicate a range of mechanisms are involved in controlling the pathogen by mycorrhizal roots such as exclusion of pathogen, lignifications of cell wall, changed P nutrition, exudation of low molecular weight compounds, and others (Sharma et al. 2004). Sundaredan et al. (1993) investigated the interaction of *G. fasciculatum* with a wilt-causing soilborne pathogen, *Fusarium oxysporum,* against cowpea plants. It was found that pre-establishment by AM fungus reduced the colonization of the pathogen and the severity of the disease, as determined by reduction in vascular discoloration index. In mycorrhizal plants, the production of phytoalexin compounds was always higher than in the non-mycorrhizal plants, and a direct correlation between the concentration of the phytoalexins and the degree of mycorrhizal association was found. It is argued that the production of phytoalexin compounds in mycorrhizal plant could be one of the mechanisms imparting tolerance to the plants against wilt disease. Moreover, multiple lines of evidence reveal that AM fungi significantly reduced disease symptoms caused by fungal pathogens such as *Phytophthora, Gaeumannomyces, Fusarium, Pythium, Rhizoctonia, Verticillium,* and *Aphanomyces* (Demir and Akköprü 2007). In another study Gao et al. (2012) have investigated the disease incidence and index of soybean red crown rot under different P regimes in field and found that the natural inoculation of rhizobia and AMF could affect soybean red crown rot, particularly without P addition. Further studies in sand culture experiments showed that inoculation with rhizobia or AMF significantly decreased severity and incidence of soybean red crown rot, especially for co-inoculation with rhizobia and AMF at low P. The root colony forming unit (CFU) decreased over 50% when inoculated with rhizobia and/or AMF at low P. However, P addition only enhanced CFU when inoculated with AMF. Furthermore, root exudates of soybean inoculated with rhizobia and/or AMF significantly inhibited pathogen growth and reproduction. Quantitative RT-PCR results indicated that the transcripts of the most tested pathogen defense-related (PR) genes in roots were significantly increased by rhizobium and/or AMF inoculation. Among them, PR2, PR3, PR4, and PR10 reached the highest level with co-inoculation of rhizobium and AMF. The results indicated that inoculation with rhizobia and AMF could directly inhibit pathogen growth and reproduction and activate the plant overall defense system through increasing PR gene expressions. Combined with optimal P fertilization, inoculation with rhizobia and AMF could be considered as an efficient method to control soybean red crown rot in acid soils.

10.7   Inoculum Development and Formulations

Since AM fungi are an obligate biotrophs, the AM inoculum production on a commercial scale via a host plant is still an obstacle and hence limits its utility as inoculants in sustainable agricultural production systems. Despite the limitation in inoculums development, certain progress has been made in this direction, and some commercial inoculum is currently marketed in some countries in the world (Gianinazzi and Vosátka 2004). Currently, there has been a remarkable boom in enterprises producing mycorrhizal fungi inocula and related services for the retail sector, commercial plantations, horticulture, and, more recently, the developing agricultural market. There are number of reasons for increasing interest in developing mycorrhizal inocula by the mycorrhizal industry. Firstly, the positive effects of mycorrhizal fungi on plant health, growth, and yield have generated a greater interest among end users of mycorrhizal technology (Gianinazzi and Vosátka 2004). Secondly, it offers an environmentally friendly and economically attractive option in commercial cultivation (Toro et al. 1997). Therefore, AM fungi are gaining popularity as "biofertilizers"/efficient scavengers of nutrients, "bioprotectors," and "biocontrol" agents (Sylvia 1999), and hence, the industry of mycorrhizal inoculum production is expanding around the world (Todd 2004). However, extensive field trials are required to prove that bioagent indeed is effective and, hence, can be recommended for inoculant development and its consequent application over a wide range of soil, environmental conditions, and crop types (Leggett et al. 2007). The first consideration in inoculum production involves the selection of fungal isolates endowed with growth-promoting activity (Ryan and Graham 2002). Other factors to be considered in the production of inoculum include soil conditions, the host plant used to grow fungus (Sieverding 1991; Ryan and Graham 2002). Several host plants including Sudan grass (*Sorghum bicolor* var. Sudanese), bahia grass (*Paspalum notatum*), guinea grass (*Panicum maximum*), cenchrus grass (*Cenchrus ciliaris*), clover (*Trifolium subterraneum*), strawberry (*Fragaria* sp.), sorghum (*Sorghum vulgare*), maize (*Zea mays*), barley (*Hordeum vulgare*), and onion (*Allium cepa*) have been attempted to produce AM inoculum. However, mass propagation of AM fungi varies greatly with root structure and habitat of host plant (Bever et al. 1996). Furthermore, since there are greater variations in soils and climates around the world, the locally available materials for inoculum production should be tested (Sieverding 1991). The traditional and most widely used approach has been to grow the fungus with the host plant in solid growth medium individually or a combination of the solid growth media such as soil, sand, peat, vermiculite, perlite, clay, or various types of composted barks (Sylvia and Jarstfer 1992). For the commercial development of AM inoculants, numerous strategies have been adopted time to time with their own merits and demerits. Some of the recently followed techniques for the production of mycorrhizal inoculum including soil or soilless technologies, like nutrient film technique (NFT) (Mosse and Thompson 1984), circulation hydroponic culture system, aeroponic culture system (Sylvia and Hubbell 1986), root organ culture, and tissue culture (Nopamornbodi et al. 1988), are briefly discussed in the following section.

10.7.1 Inoculum Production Strategies

10.7.1.1   Soil-Based Pot Culture Method

Soil-based pot culture is a common method for production of AM fungal inoculum (Menge 1984). Soil inoculum contains all AM fungal structures; this inoculum source is highly infective (Sieverding 1991). The author further suggested that the success for good soil inoculum production depends on the selection of the host plant and the ambient conditions. The soil inoculum (containing AM-infected roots, AM spores, and mycelium) is chopped and homogenized before use. Soil may contain abiotic and biotic components which make it undesirable substrate in which to grow and subsequently to distribute the AM fungal inoculum. Soil inocula are considered impractical because of their bulk and the risk of contamination by insects, nematodes, and plant pathogens (Sylvia and Jarstfer 1992). However, chopped roots in peat blocks (Warner 1985) and spores within a clay matrix (Dehne and Backhaus 1986) have been proposed for field application. Soil-based pot culture method is cost-effective with low inputs, and thousands of infectious propagules can be extracted per gram of soil. However, the major disadvantage associated with this technique includes bulk amount, vulnerability of pest to infestation, and nutrient management (Sharma et al. 2000). To overcome these problems, soilless technologies were discovered and are discussed.

10.7.1.2   Nutrient Film Technique (NFT)

In this method, large volume of nutrient liquid in a film is recycled which flows over the roots of plants. However, any host in the NFT should be grown first in the soil substrate with AM inoculums in order to infect the roots. This technique eliminates the possibility of contamination and helps to produce large quantities of AM-infected roots. However, higher sporulation compared to soil system is not achieved. Yun-Jeong and Eckhard (2005) used NFT culture system for nursery production of arbuscular mycorrhizal horticultural crops. In the NFT system, a thin layer of glass beads was used to provide solid support for plant and fungus growth, and nutrient solution was supplied intermittently (15 min, six times per day). A modified nutrient solution (80 μM P) was used and was changed with fresh solution at 3-day intervals. The dry matter accumulation in *Glomus mosseae* (BEG 107)-colonized lettuce (*Lactuca sativa* var. capitata) was significantly higher than non-mycorrhizal lettuce in perlite during the precolonization period. The root colonization rate was also high at rates up to 80 μM P supply. On the NFT system, growth differences between mycorrhizal and non-mycorrhizal plants were less than in perlite. However, root colonization rate was not reduced during the NFT culture period. In this system, high amounts of fungal biomass were produced. The authors suggested that using this technique, metal and other nutrient concentrations in fungal hyphae can be determined. Furthermore, this modified NFT culture system would also be suitable for fungal biomass production on a large scale with a view to additional aeration by intermittent nutrient supply, optimum P supply, and a use of glass beads as support materials. Furthermore, bulk inoculum composition with a mixture of spores, colonized roots, and hyphae grown in soilless media by the modified NFT system might be a useful way to mass-produce mycorrhizal crops and inoculum for commercial horticultural purposes.

10.7.1.3 Aeroponic Method

A culture system which applies a fine mist of defined nutrient solution to the roots of trap plant is termed as aeroponic culture (Zobel et al. 1976). For this, plants are generally inoculated in sand or vermiculite before they are transferred to these systems. Plants have also been inoculated directly in the aeroponic system (Hung et al. 1991). Applying aeroponics higher number spores have been produced compared to soil-based pot cultures. Since no substrate is present with the inoculum with aeroponic culture of roots, it is possible to produce inoculum with hundreds of thousands of propagules per dry gram of roots (Sylvia and Jarstfer 1992). The aeroponics has distinct advantage over other AM-producing techniques, like the highly aerated rooting environment of aeroponics stimulates rapid and abundant sporulation of the AM fungi and this system reduces the risk of contamination but this technique is a costly affair. For example, in an aeroponic culture, root colonization and sporulation of *G. mosseae* (Nicol. & Gerd.) Gerd. & Trappe and *G. intraradices* Schenck & Smith with bahia grass was found superior relative to a soil-based pot culture (Sylvia and Hubbell 1986). Similarly, Martin-Laurent et al. (1999) designed an experiment to produce *Acacia mangium* saplings associated with AM fungi using aeroponics (a soilless plant culture method). *A. mangium* seedlings were first grown in multi-pots and inoculated with Endorize (a commercial AM fungal inoculums) followed by transferring to aeroponic systems or to soil. Aeroponics was found as a better system than soil and doubled the production of tree saplings compared to soil. Moreover, compared to plants grown in soil, aeroponically grown saplings inoculated with AM fungal inoculum exhibited significantly different rates of mycorrhization, leading to an increase in chlorophyll contents in plant tissues. The authors suggested that the aeroponic system is an innovative and appropriate technology which could be used to produce large quantities of tree saplings associated with soil microorganisms, such as AM fungi, for reforestation of degraded land in the humid tropics. Aeroponically produced *G. deserticola* and *G. etunicatum* inocula retained their infectivity after cold storage (4°C) in either sterile water or moist vermiculite for at least 4 and 9 months, respectively (Hung and Sylvia 1988).

10.7.1.4 Root Organ Culture System

The root organ culture system is the most attractive and advanced cultivation method for AMF development. This technique uses root-inducing transfer-DNA-transformed roots of a host plant to develop the symbiosis on a specific medium in vitro which provides pure, viable, contamination-free inoculum using less space. Systems utilizing excised roots of various host plants and different media formulations have been developed to culture glomalean fungi monoxenically (Mugnier and Mosse 1987). Less than 5% of currently known AM species have, however, been successfully cultivated using dual culture approach. *Gigaspora margarita* (Miller-Wideman and Watrud 1984); *G. fasciculatum*, *G. intraradices*, and *G. macrocarpum* (Declerck et al. 1998); and *G. versiforme* (Diop et al. 1994) have been maintained and sporulated in association with excised tomato roots or roots of carrot transformed by "hairy root" inducing T-DNA from *Agrobacterium rhizogenes*. Evidently, the rate of in vitro spore formation of the AM fungus *G. versiforme* was followed in Petri dishes, using mycorrhizal root-segment inoculum associated with Ri T-DNA transformed carrot roots (Declerck et al. 1996). Three phases of sporulation were observed: a lag phase, a period of intensive spore production, and a plateau phase. An average of 9500 spores/Petri dish was produced after 5 months of dual culture. The root organ culture system supported extensive root colonization with many arbuscules and vesicles being formed. The fungus, both within root segments and as spores produced, was viable and able to complete its life cycle in vitro. The mycorrhizal root segments, however, exhibited higher inoculum potential due to numerous vesicles and extensive intraradical mycelium. The in vitro propagation on root organ culture, however, may not change drastically the traditional process but will improve the quality of strain and the supply of spores (Dalpé and Monreal 2004).

10.7.2 Formulations

Different formulated products are available in the market, which creates the need for the establishment of standards for widely accepted quality control. In most cases, fresh AMF inoculum is applied (Fig. 10.3). In preparation and formulation of mycorrhizal inoculum, the most widely used methods are based on the entrapment of fungal materials in natural polysaccharide gels (Sieverding 1991; Vassilev et al. 2005). The potential of such inoculant preparations is illustrated by various studies which include immobilization of mycorrhized root pieces, vesicles, and spores, in some cases co-entrapped with other plant beneficial microorganisms (Vassilev et al. 2001).

In a study, Vassilev et al. (2001) assessed the applicability of microbial inoculants entrapped in alginate gel. For this, AM fungus *G. deserticola* enriched with rock phosphate, either in free form or entrapped in calcium alginate alone or in combination with P-solubilizing yeast (*Yarrowia lipolytica*), was inoculated into soil microcosms. Plant dry weight, soluble P acquisition, and mycorrhizal index were equal in treatments inoculated with free and alginate-entrapped AM fungus. Dual inoculation with entrapped *G. deserticola* and free cells of *Y. lipolytica* significantly increased all measured variables. The highest rates of the latter were obtained when both fungal microorganisms were applied co-entrapped in the carrier. The yeast culture behaved as a "mycorrhiza helper microorganism" enhancing mycorrhization of plant roots. These results indicate that dual inoculation with an AM

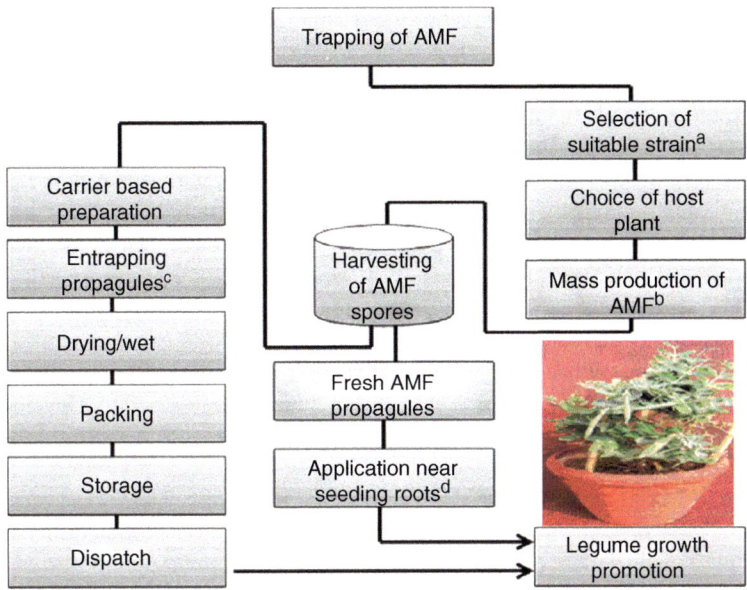

Fig. 10.3 Various stages involved in production and inoculation of AMF: [a]strains to be selected must be the best performer, [b]cultivation using suitable host by employing either conventional soil-based or soilless techniques, [c]include immobilization of mycorrhized root pieces, vesicles, and spores, in some cases co-entrapped with other plant beneficial microorganisms, [d]supplying propagules near seedlings in soil at appropriate rate (modified from Sieverding (1991))

fungus and a P-solubilizing microorganism co-entrapped in alginate can be an efficient technique for plant establishment and growth in nutrient-deficient soils. Likewise, Weber et al. (2005) studied dual inoculation of *Acacia mangium* grown in aeroponic culture using selected strains of *Bradyrhizobium* sp. and *G. intraradices*. A single-step technique with alginate as an embedding and sticking agent for an inoculum composed of AM-infected sheared roots was used to infect plants. This method resulted in the successful establishment of AM in 100% of the inoculated plants after 7 weeks. The results indicated that dual microbial inoculation with *G. intraradices* strain S-043 and *Bradyrhizobium* strain AUST 13C stimulated the growth of *A. mangium* in aeroponic culture. The effects of single and dual microbial inoculations were also evaluated at two levels of P in the nutrient medium. A concentration of 5 mg P kg^{-1} stimulated the development of AM without affecting plant development or establishment of *Bradyrhizobium* symbiosis. In contrast, saplings supplemented with a higher concentration of P (25 mg kg^{-1}) alone or co-inoculated with *Bradyrhizobium* had lower AM frequencies.

Conclusion

The great agricultural and environmental importance of legumes together with its ability to harbor conventional symbionts and other PGPR make legumes a target crop in sustainable agriculture. Accordingly, beneficial soil microbes have become one of the established, promising, and sustainable low-input soil management options. Moreover, legumes, in general, have the potential to form mycorrhizal symbiosis. Mycorrhizal fungi affect the ecophysiology of nodulated legumes, the microbiota of soil, and associated nonlegume plants. Concomitantly, the inoculation of both rhizobia and AMF has increased growth and development of plants under varying conditions. Furthermore, co-inoculation with rhizobia and mycorrhizal fungi is currently being suggested as a possible solution to reforestation and amendment of soil fertility. Also, AMF alleviate various types of abiotic/biotic stresses and have been reported to increase tolerance of legumes to salt, heavy metals, acidic soils, drought, soilborne pathogens, etc. Due to these, use of AM inoculum may provide solutions to ever-increasing costs of agrochemicals and other health-problem causing factors. However, further research is needed to better understand the prospect of AM inoculum in legume production. Development of suitable technology for mass production of inoculants, simple application methods, and assessment of the mycorrhized fields are urgently required to harness the full potential of mycorrhizas. Apart from these, factors deleterious to mycorrhizal diversity and their associated activities, such as pesticides, fertilizers, and poor management practices, need to be carefully monitored.

Acknowledgments The author would like to thank Prof. Md. Saghir Khan for his prompt and kind initiation to write this chapter and for his meticulous edition of the chapter. I would also like to express my earnest thanks to my wife Elfinesh Tolera for her unvarying encouragement and materials support.

References

Abd-Alla MH, El-Enany A-WE, Nafady NA, Khalaf DM, Morsy FM (2014) Synergistic interaction of *Rhizobium leguminosarum* bv. *viciae* and arbuscular mycorrhizal fungi as a plant growth promoting biofertilizers for faba bean (Vicia faba L.) in alkaline soil. Microbiol Res 169:49–58

Ahmad MH (1995) Compatibility and co-selection of vesicular-arbuscular mycorrhizal fungi and rhizobia for tropical legumes. Crit Rev Biotechnol 15:229–239

Al-Garni SMS (2006) Increased heavy metal tolerance of cowpea plants by dual inoculation of an arbuscular mycorrhizal fungi and nitrogen-fixer *Rhizobium* bacterium. Afr J Biotechnol 5:133–142

Aliasgharzad N, Neyshabouri MR, Salimi G (2006) Effects of arbuscular mycorrhizal fungi and *Bradyrhizobium japonicum* on drought stress of soybean. Biologia Bratislava 61(Suppl. 19):S324–S328

Andrade SAL, Abreu CA, de Abreu MF, Silveira APD (2004) Influence of lead additions on arbuscular mycorrhiza and *Rhizobium* symbioses under soybean plants. Appl Soil Ecol 26:123–131

Aroca R, Ruiz-Lozano JM (2009) Induction of plant tolerance to semi-arid environments by beneficial soil microorganisms–a review. In: Lichtfouse E (ed) Climate change, intercropping, pest control and beneficial microorganisms, sustainable agriculture reviews 2. Springer, Dordrecht, pp 121–135

Aryal UK, HL X, Fujita M (2003) Rhizobia and AM fungal inoculation improve growth and nutrient uptake of bean plants under organic fertilization. J Sustain Agric 21:27–39

Aysan E, Demir S (2009) Using arbuscular mycorrhizal fungi and *Rhizobium leguminosarum* biovar *phaseoli* against *Sclerotinia sclerotiorum* (Lib) de Bary in the common bean (*Phaseolus vulgaris* L.) Plant Pathol J 8:74–78

Azcón R, Rubio R, Barea JM (1991) Selective interactions between different species of mycorrhizal fungi and *Rhizobium meliloti* strains, and their effects on growth, N_2-fixation (^{15}N) and nutrition of *Medicago sativa* L. New Phytol 117:399–404

Bai B, Suri VK, Kumar A, Choudhary KA (2016) Influence of dual inoculation of AM fungi and *Rhizobium* on growth indices, production economics, and nutrient use efficiencies in Garden Pea (*Pisum sativum* L.) Commun Soil Sci Plant Anal 47:941–954

Babajide PA, Akanbi WB, Alamu LO, Ewetola EA, Olatunji OO (2008) Growth, nodulation and biomass yield of soybean (*Glycine max*) as influenced by biofertilizers under simulated eroded soil condition. Res J Agron 2:96–100

Bakarr MI, Janos DP (1996) Mycorrhizal associations of tropical legume trees in Sierra Leone, West Africa. Forest Ecol Manag 89:89–92

Barea JM, Azcon-Aguilar C (1983) Mycorrhizas and their significance in nodulating nitrogen-fixing plants. In: Brady NC (ed) Advances in agronomy, vol 36. Academic, New York, pp 1–54

Barea JM, Werner D, Azcón-Guilar C, Azcón R (2005) Interactions of arbuscular mycorrhiza and nitrogen-fixing symbiosis in sustainable agriculture. In: Werner D, Newton WE (eds) Nitrogen fixation in agriculture, forestry, ecology and the environment. Springer, Dordrecht, pp 199–222

Barea JM, Azcon R, Azcon-Aguilar C (2002) Mycorrhizosphere interactions to improve plant fitness and soil quality. Antonie Van Leeuwenhoek 81:343–351

Behlenfalvay GJ, Brown MS, Stafford AE (1985) *Glycine-Rhizobium*-symbiosis II. Antagonistic effects between mycorrhizal colonization and nodulation. Plant Physiol 79:1054–1058

Bever JD, Morton JB, Antonovics J, Schultz PA (1996) Host-dependent sporulation and species diversity of arbuscular mycorrhizal fungi in mown grassland. J Ecol 84:71–82

Bradbury SM, Peterson RL, Bowley SR (1991) Interactions between three alfalfa nodulation genotypes and two *Glomus* species. New Phytol 119:115–120

Camila MP, Lazara C (2004) Nitrogen-fixing and vesicular-arbuscular mycorrhizal symbioses in some tropical legume trees of tribe mimoseae. Forest Ecol Manag 196:275–285

Cárdenas L, Alemán E, Nava N, Santana O, Sánchez F, Quinto C (2006) Early responses to Nod factors and mycorrhizal colonization in a non-nodulating *Phaseolus vulgaris* mutant. Planta 223:746–754

Chaitra B, Negalur Lakshman HC (2016) Interaction between AMF (*Glomus geosporum*) *Rhizobium, Azospirillum* and their effect on three leguminous plant to improve growth and N, P, K, uptake. Sci Res Rep 6:68–74

Clark RB, Zeto SK (2000) Mineral acquisition by arbuscular mycorrhizal plants. J Plant Nutr 23:867–902

Colozzi A, Cardoso EJBN (2000) Detection of arbuscular mycorrhizal fungi in roots of coffee plants and *Crotalaria* cultivated between rows. Pesqui Agropecu Bras 35:2033–2042

Dalpé Y, Monreal M (2004) Arbuscular Mycorrhiza inoculum to support sustainable cropping systems. Proceedings of a symposium on the Great Plains Inoculant Forum, 27 and 28 March 2003, Saskatoon, Saskatchewan

Declerck S, Strullu DG, Plenchette C (1996) In vitro mass-production of the arbuscular mycorrhizal fungus, *Glomus versiforme*, associated with Ri T-DNA transformed carrot roots. Mycol Res 100:1237–1242

Declerck S, Strullu DG, Plenchette C (1998) Monoxenic culture of the intraradical forms of *Glomus* sp. isolated from a tropical ecosystem: a proposed methodology for germplasm collection. Mycologia 90:579–585

Dehne HW, Backhaus GF (1986) The use of vesicular-arbuscular mycorrhizal fungi in plant production. I Inoculum production. Z Pflanzenkr Pflanzenschutz 93:415–424

Demir S, Akköprü A (2007) Using of arbuscular mycorrhizal fungi (AMF) for biocontrol of soil borne fungal plant pathogens. In: Chincholkar SB, Mukerji KG (eds) Biological control of plant diseases. Haworth, Binghamton, NY, pp 17–37

de Varennes A, Goss MJ (2007) The tripartite symbiosis between legumes, rhizobia and indigenous mycorrhizal fungi is more efficient in undisturbed soil. Soil Biol Biochem 39: 2603–2607

Diop TA, Plenchette C, Strullu DG (1994) Dual axenic culture of sheared-root inocula of vesicular-arbuscular mycorrhizal fungi associated with tomato roots. Mycorrhiza 5:17–22

Dodd JC, Arias I, Koomen I, Hayman DS (1990) The management of populations of vesicular-arbuscular mycorrhizal fungi in acid-infertile soils of a savanna ecosystem. I. The effect of pre-cropping and inoculation with VAM-fungi on plant growth and nutrition in the field. Plant Soil 122:229–240

Duponnois R, Plenchette C, Bâ AM (2001) Growth stimulation of seventeen fallow leguminous plants inoculated with *Glomus aggregatum* in Senegal. Eur J Soil Biol 37:181–186

Eom AH, Lee SS, Ahn TK, Lee MW (1994) Ecological roles of arbuscular mycorrhizal fungi in two wild legume plants. Mycoscience 35:69–75

Erman M, Demir S, Ocak E, Tufenkci S, Oguz F, Akkopru A (2011) Effects of *Rhizobium*, arbuscular mycorrhiza and whey applications on some properties in chickpea (*Cicer arietinum* L.) under irrigated and rain-fed conditions 1—yield, yield components, nodulation and AMF colonization. Field Crops Res 122:14–24

Francis R, Read DJ (1984) Direct transfer of carbon between plants connected by vesicular-arbuscular mycorrhizal mycelium. Nature 307:53–56

Franco AA, De Faria SM (1997) The contribution of N_2-fixing tree legumes to land reclamation and sustainability in the tropics. Soil Biol Biochem 29:897–903

Franzini VI, Azco'n R, Mendes FL, Aroca R (2010) Interactionsbetween *Glomus* species and *Rhizobium* strains affect the nutritional physiology of drought-stressed legume hosts. J Plant Physiol 167:614–619

Gao X, Lu X, Wu M, Zhang H, Pan R, Tian J, Li S, Liao H (2012) Co-inoculation with Rhizobia and AMF inhibited soybean red crown rot: from field study to plant defense-related gene expression analysis. PLoS One 7(3):e33977. doi:10.1371/journal.pone.0033977

Geneva M, Zehirov G, Djonova E, Kaloyanova N, Georgiev G, Stancheva I (2006) The effect of inoculation of pea plants with mycorrhizal fungi and *Rhizobium* on nitrogen and phosphorus assimilation. Plant Soil Environ 52:435–440

Ghosh S, Verma NK (2006) Growth and mycorrhizal dependency of *Acacia mangium* Willd. Inoculated with three vesicular arbuscular mycorrhizal fungi in lateritic soil. New Forests 31:75–81

Gianinazzi S, Vosátka M (2004) Inoculum of arbuscular mycorrhizal fungi for production systems: science meets business. Can J Bot 82:1264–1271

Harrison MG (2005) Signaling in the arbuscular mycorrhizal symbiosis. Annu Rev Microbiol 59:19–42

Harrison MJ (1999) Molecular and cellular aspects of the arbuscular mycorrhizal symbiosis. Annu Rev Plant Physiol 50:361–389

Hayman DS (1986) Mycorrhizae of nitrogen-fixing legumes. World J Microbiol Biotechnol 2:121–145

He X, Pen-Mouratov S, Steinberger Y (2004) Research note: spatial variation of AM fungal spore numbers under canopies of *Acacia raddiana*. Arid Land Res Manag 18:295–299

Heap AJ, Newman EL (1980) Links between roots by hyphae of vesiculararbuscular mycorrhizas. New Phytol 85:169–171

Herrera MA, Salamanca CP, Barea JM (1993) Inoculation of woody legumes with selected arbuscular mycorrhizal fungi and rhizobia to recover desertified mediterranean ecosystems. Appl Environ Microbiol 59:129–133

Hodge A, Campbell CDFAH (2001) An arbuscular mycorrhizal fungus accelerates decomposition and acquires nitrogen directly from organic material. Nature 413:297–299

Hooker JE, Black KE (1995) Arbuscular mycorrhizal fungi as components of sustainable soil-plant systems. Crit Rev Biotechnol 15:201–212

Hung LL, O'Keefe DM, Sylvia DM (1991) Use of hydrogel as a sticking agent and carrier for vesicular–arbuscular mycorrhizal fungi. Mycol Res 95:427–429

Hung LLL, Sylvia DM (1988) Production of vesicular-arbuscular mycorrhizal fungus inoculum in aeroponic culturet. Appl Environ Microbiol 54:353–357

Jakobsen I, Abbott LK, Robson AD (1992) External hyphae of vesicular-arbuscular mycorrhizal fungi associated with *Trifolium subterraneum* L. I: spread of hyphae and phosphorus inflow into roots. New Phytol 120:371–380

Jeffries P, Gianinazzi S, Perotto S, Turnau K, Barea JM (2003) The contribution of arbuscular mycorrhizal fungi in sustainable maintenance of plant health and soil fertility. Biol Fertil Soils 37:1–16

Jia Y, Gray VM, Straker CJ (2004) The influence of *Rhizobium* and arbuscular mycorrhizal fungi on nitrogen and phosphorus accumulation by *Vicia faba*. Ann Bot 94:251–258

Kamprath EJ, Foy CD (1985) Lime-fertilizer-plant interactions in acid soils. In: Englestad O (ed) Fertilizer technology and use, 3rd edn. Soil Science Society of America, Madison, WI

Karmakar K, Rana A, Rajwar A, Sahgal M, Johri BN (2015) Legume-rhizobia symbiosis under stress. In: Arora NK (ed) Plant microbes symbiosis: applied facets. Springer, India, pp 241–258

Kayode J, Franco AA (2002) Response of *Acacia mangium* to rhizobia and arbuscular mycorrhizal fungi. Trop Sci 42:116–119

Khaliel AS, Elkhider KA, Bahkali AH (1999) Response and dependence of haricot bean to inoculation with arbuscular mycorrhiza. Saudi J Biol Sci 6:126–132

Khan A (2006) Mycorrhizoremediation–an enhanced form of phytoremediation. J Zhejiang University Sci B7:503–514

Kothari SK, Marschner H, Römheld V (1991) Contribution of the VA mycorrhizal hyphae in acquisition of phosphorus and zinc by maize grown in a calcareous soil. Plant Soil 131:177–185

Lapeyrie F (1988) Oxalate synthesis from soil bicarbonate by fungus *Paxillus involutus*. Plant Soil 110:3–8

Leggett M, Cross J, Hnatowich G, Holloway G (2007) Challenges in commercializing a phosphate-solubilizing microorganism: *Penicillium bilaiae*, a case history. In: Velázquez E, Rodríguez-Barrueco C (eds) First international meeting on miccrobial phosphate solubilization. Springer, Dordrecht, pp 215–222

Li XL, Marschner H, George E (1991) Acquisition of phosphorus and copper by VA–mycorrhizal hyphae and root-to-shoot transport in white clover. Plant Soil 136:49–57

Li Y, Ran W, Zhang R, Sun S, Xu G (2009) Facilitated legume nodulation, phosphate uptake and nitrogen transfer by arbuscular inoculation in an upland rice and mung bean intercropping system. Plant Soil 315:285–296

Linderman RG (1994) Role of VAM fungi in biocontrol. In: Pfleger FL, Linderman RG (eds) Mycorrhizae and plant health. APS, St Paul, MN, pp 1–26

Linderman RG, Paulitz TC (1990) Mycorrhizal rhizobacterial interactions. In: Hornby D (ed) Biological control of soilborne plant pathogens. CABI International, Wallingford

Lodwig EM, Hosie AHF, Bourdès A, Findlay K, Allaway D, Karunakaran R, Downie JA, Poole PS (2003) Amino-acid cycling drives nitrogen fixation in the legume–*Rhizobium* symbiosis. Nature 422:722–726

Mahdi AA, Atabani IMA (1992) Response of *Bradyrhizobium*-inoculated soyabean and lablab bean to inoculation with vesicular-arbuscular mycorrhizae. Exp Agric 28:399–408

Maki T, Nomachi M, Yoshida S, Ezawa T (2008) Plant symbiotic microorganisms in acid sulfate soil: significance in the growth of pioneer plants. Plant Soil 310:55–65

Manjunath A, Bagyaraj DJ, Gowda HSG (1984) Dual inoculation with VA mycorrhiza and Rhizobium is beneficial to *Leucaena*. Plant Soil 78:445–448

Marques MS, Pagano M, Scotti M (2001) Dual inoculation of a woody legume (*Centrolobium tomentosum*) with rhizobia and mycorrhizal fungi in south-eastern Brazil. Agrofor Syst 50:107–117

Marschner H (1998) Mineral Nutrition of higher plants. Academic, London

Martin-Laurent F, Lee SK, Tham FY, Jie H, Diem HG (1999) Aeroponic production of *Acacia mangium* saplings inoculated with AM fungi for reforestation in the tropics. Forest Ecol Manag 122:199–207

Meghvansi MK, Prasad K, Harwani D, Mahna SK (2008) Response of soybean cultivars toward inoculation with three arbuscular mycorrhizal fungi and *Bradyrhizobium japonicum* in the alluvial soil. Eur J Soil Biol 44:316–323

Mehdi Z, Nahid S-R, Alikhani HA, Nasser A (2006) Responses of lentil to co-inoculation with phosphate-solubilizing rhizobial strains and arbuscular mycorrhizal fungi. J Plant Nutr 29:1509–1522

Menge JA (1984) Inoculum production. In: Powell CL, Bagyaraj DJ (eds) VA mycorrhiza. CRC, Boca Raton, FL, pp 187–203

Miller RM (1987) The ecology of vesicular-arbuscular mycorrhizae in grass and shrublands. In: Safir GR (ed) Ecophysiology of VA mycorrhizal plants. CRC, Boca Raton, FL, pp 135–170

Miller-Wideman MA, Watrud LS (1984) Sporulation of *Gigaspora margarita* on root cultures of tomato. Can J Microbiol 30:642–646

Molla MN, Solaiman ARM (2009) Association of arbuscular mycorrhizal fungi with leguminous crops grown in different agro-ecological zones of Bangladesh. Arch Agron Soil Sci 55: 233–245

Mosse B, Thompson JP (1984) Vesicular-arbuscular endomycorrhizal inoculum production. I. Exploratory experiments with beans (*Phaseolus vulgaris*) in nutrient flow culture. Can J Bot 62:1523–1530

Muchovej RM (2001) Importance of mycorrhizae for agricultural crops. http://edis.ifas.ufl.edu/pdffiles/AG/AG11600.pdf. Accessed October 2009

Mugnier J, Mosse B (1987) Vesicular-arbuscular mycorrhizal infection in transformed root-inducing T-DNA roots grown axenically. Phytopathology 77:1045–1050

Muleta D, Assefa F, Nemomissa S, Granhall U (2007) Composition of coffee shade tree species and density of indigenous arbuscular mycorrhizal fungi (AMF) spores in Bonga natural coffee forest, southwestern Ethiopia. Forest Ecol Manag 241:145–154

Muleta D, Assefa F, Nemomissa S (2008) Granhall U (2008) Distribution of arbuscular mycorrhizal fungi spores in soils of smallholder agroforestry and monocultural coffee systems in southwestern Ethiopia. Biol Fertil Soils 44:653–659

Nambiar PTC, Anjaiah V (1989) Competition among strains of *Bradyrhizobium* and vesicular-arbuscular mycorrhizae for groundnut (*Arachis hypogaea* L.) root infection and their effect on plant growth and yield. Biol Fertil Soils 8:311–318

Navazio L, Moscatiello R, Genre A, Novero M, Baldan B, Bonfante P, Mariani P (2007) A diffusible signal from arbuscular mycorrhizal fungi elicits a transient cytosolic calcium elevation in host plant cells. Plant Physiol 144:673–681

Ndiaye F, Manga A, Diagne-Leye G, Samba SAN, Diop TA (2009) Effects of rock phosphate and arbuscular mycorrhizal fungi on growth and nutrition of *Sesbania sesban* and *Gliricidia sepium*. Afr J Microbiol Res 3:305–309

Nopamornbodi O, Rojanasiriwong W, Thomsurakul S (1988) Production of VAM fungi, *Glomus intraradices* and *G. mosseae* in tissue culture. In: Mahadevan A, Raman N, Natarajan K (eds) Mycorrhizae for green Asia. University of Madras, Madras, pp 315–316

Nwoko H, Sanginga N (1999) Dependence of promiscuous soybean and herbaceous legumes on arbuscular mycorrhizal fungi and their response to bradyrhizobial inoculation in low P soils. Appl Soil Ecol 13:251–258

Pacovsky RS, Fuller G, Stafford AE, Paul EA (1986) Nutrient and growth interactions in soybeans colonized with *Glomus fasciculatum* and *Rhizobium japonicum*. Plant Soil 92:37–45

Pagano MC, Cabello MN, Scotti MR (2007) Phosphorus response of three native Brazilian trees to inoculation with four arbuscular mycorrhizal fungi. J Agric Technol 3:231–240

Patreze CM, Cordeiro L (2004) Nitrogen-fixing and vesicular–arbuscular mycorrhizal symbioses in some tropical legume trees of tribe Mimoseae. Forest Ecol Manag 196:275–285

Plenchette C, Clermont-Dauphin C, Meynard JM, Fortin JA (2005) Managing arbuscular mycorrhizal fungi in cropping systems. Can J Plant Sci 85:31–40

Quatrini P, Scaglione G, Incannella G, Badalucco L, Puglia AM, Lamantia T (2003) Microbial inoculants on woody legumes to recover a municipal landfill site. Water Air Soil Pollut Focus 3:189–199

Rabie GH, Almadini AM (2005) Role of bioinoculants in development of salt-tolerance of *Vicia faba* plants under salinity stress. Afr J Biotechnol 4:210–222

Requena N, Pérez-Solis E, Azcón-Aguilar C, Jeffries P, Barea JM (2001) Management of indigenous plant-microbe symbiosis aids restoration of desertified ecosystems. Appl Environ Microbiol 67:495–498

Ruiz-Lazano JM (2003) Arbuscular mycorrhizal symbiosis and alleviation of osmotic stress. New perspectives for molecular studies. Mycorrhiza 13:309–317

Ryan MH, Graham JH (2002) Is there a role for arbuscular mycorrhizal fungi in production agriculture? Plant Soil 244:263–271

Scheublin TR, Ridgway KP, Young JPW, van der Heijden MGA (2004) Nonlegumes, legumes, and root nodules harbor different arbuscular mycorrhizal fungal communities. Appl Environ Microbiol 70:6240–6246

Sharma AK, Singh C, Akhauri P (2000) Mass culture of arbuscular mycorrhizal fungi and their role in biotechnology. Proc Indian Natl Sci Acad (PINSA) B66:223–238

Sharma MP, Tanu AG, Sharma OP (2004) Prospects of arbuscular mycorrhiza in sustainable management of root- and soil-borne diseases of vegetable crops. In: Mukerji KG (ed) Fruit and vegetable diseases. Kluwer Academic, Dordrecht, pp 501–539

Shokri S, Maadi B (2009) Effects of arbuscular mycorrhizal fungus on the mineral nutrition and yield of *Trifolium alexandrinum* plants under salinity stress. J Agron 8:79–83

Sieverding E (1991) Vesicular-arbuscular mycorrhizal management in tropical agrosystems. GTZ, Eschborn, Germany

Simms EL, Taylor DL (2002) Partner choice in nitrogen-fixation mutualisms of legumes and rhizobia. Integr Comp Biol 42:369–380

Singh CS, Kapoor A, Wange SS (1991) The enhancement of root colonisation of legumes by vesicular-arbuscular mycorrhizal (VAM) fungi through the inoculation of the legume seed with commercial yeast (*Saccharomyces cerevisiae*). Plant Soil 131:129–133

Smith SE, Smith FA, Jakobsen I (2004) Functional diversity in arbuscular mycorrhizal (AM) symbioses: the contribution of the mycorrhizal P uptake pathway is not correlated with mycorrhizal responses in growth or total P uptake. New Phytol 162:511–524

Snoeck D, Zapata F, Domenach A (2000) Isotopic evidence of the transfer of nitrogen fixed by legumes to coffee trees. Biotechnol Agron Soc Environ 4:95–100

Spaink HP (1996) Regulation of plant morphogenesis by lipochin oligosaccharides. Crit Rev Plant Sci 15:559–582

Stancheva I, Geneva M, Djonova E, Kaloyanova N, Sichanova M, Boychinova M, Georgiev G (2008) Response of alfalfa (*Medicago sativa* L) growth at low accessible phosphorus source to the dual inoculation with mycorrhizal fungi and nitrogen fixing bacteria. Gen Appl Plant Physiol 34:319–326

Sundaredan P, Raja NU, Gunasekaran P (1993) Induction and accumulation of phytoalexins in cowpea roots infected with a mycorrhizal fungus *Glomus fasciculatum* and their resistance to *Fusarium* wilt disease. J Biosci 18:291–301

Sylvia DM (1999) Fundamentals and applications of arbuscular mycorrhizae: a 'biofertilizer' perspective. In: Siqueira JO (ed) Soil fertility, biology, and plant nutrition interrelationships. SBCS, Viçosa, pp 705–723

Sylvia DM, Hubbell DH (1986) Growth and sporulation of vesicular-arbuscular mycorrhizal fungi in aeroponic and membrane systems. Symbiosis 1:259–267

Sylvia DM, Jarstfer AG (1992) Sheared roots inocula of vesicular mycorrhizal fungi. Appl Environ Microbiol 58:229–232

Tajini F, Trabelsi M, Drevon J-J (2012) Combined inoculation with *Glomus intraradices* and *Rhizobium tropici* CIAT899 increases phosphorus use efficiency for symbiotic nitrogen fixation in common bean (*Phaseolus vulgaris* L.) Saudi J Biol Sci 19:157–163

Tavasolee A, Aliasgharzad N, Salehi Jouzani G, Mardi M, Asgharzadeh A (2011) Interactive effects of Arbuscular mycorrhizal fungi and rhizobial strains on chickpea growth and nutrient content in plant. Afr J Biotechnol 10:7585–7591

Tiwari P, Adholeya A (2002) In vitro co-culture of two AMF isolates *Gigaspora margarita* and *Glomus intraradices* on Ri T-DNA transformed roots. FEMS Microbiol Lett 206:39–43

Todd C (2004) Mycorrhizal fungi, nature's key to plant survival and success. Pac Hort 65:8–12

Toro M, Azco'n R, Barea J (1997) Improvement of arbuscular mycorrhiza development by inoculation of soil with phosphate-solubilizing rhizobacteria to improve rock phosphate bioavailability [^{32}P] and nutrient cycling. Appl Environ Microbiol 63:4408–4412

Valsalakumar N, Ray JG, Potty VP (2007) Arbuscular mycorrhizal fungi associated with green gram in South India. Agron J 99:1260–1264

van der Heijden MGA, Wiemken A, Sanders IR (2003) Different arbuscular mycorrhizal fungi alter coexistence and resource distribution between co-occurring plant. New Phytol 157:569–578

van der Heijden MGA, Rinaudo V, Verbruggen E, Scherrer C, Bàrberi P, Giovannetti M (2008) The significance of mycorrhizal fungi for crop productivity and ecosystem sustainability in organic farming systems. 16th IFOAM Organic World Congress, Modena, Italy, 16–20 June 2008

van der Heijden MGA, Streitwolf-Engel R, Riedl R, Siegrist S, Neudecker A, Ineichen K, Boller T, Wiemken A, Sanders IR (2006) The mycorrhizal contribution to plant productivity, plant nutrition and soil structure in experimental grassland. New Phytol 172:739–752

van der Vossen HAM (2005) A critical analysis of the agronomic and economic sustainability of organic coffee production. Exp Agric 41:449–473

Vankessel C, Singleton PW, Hoben HJ (1985) Enhanced N-transfer from a soybean to maize by vesicular arbuscular mycorrhizal (VAM) fungi. Plant Physiol 79:562–563

Vassilev N, Nikolaeva I, Vassileva M (2005) Polymer-based preparation of soil inoculants: applications to arbuscular mycorrhizal fungi. Rev Environ Sci Biotechnol 4:235–243

Vassilev N, Vassileva M, Azcon R, Medina A (2001) Preparation of gel-entrapped mycorrhizal inoculum in the presence or absence of Yarowia lipolytica. Biotechnol Lett 23:907–909

Vivas A, Vörös I, Biró B, Campos E, Barea JM, Azcón R (2003) Symbiotic efficiency of autochthonous arbuscular mycorrhizal fungus (*G. mosseae*) and *Brevibacillus* sp. isolated from cadmium polluted soil under increasing cadmium levels. Environ Pollut 126:179–189

Warner A (1985) US patent 4,551,165, November

Weber J, Ducousso M, Tham FY, Nourissier-Mountou S, Galiana A, Prin Y, Lee SK (2005) Co-inoculation of *Acacia mangium* with *Glomus intraradices* and *Bradyrhizobium* sp. in aeroponic culture. Biol Fertil Soils 41:233–239

Weber E, George E, Beck DP, Saxena MC, Marschner H (1992) Vesicular-arbuscular mycorrhiza and phosphorus uptake of chickpea grown in Northern Syria. Exp Agric 28:433–442

Wu FY, Bi YL, Wong MH (2009) Dual inoculation with an arbuscular mycorrhizal fungus and *Rhizobium* to facilitate the growth of Alfalfa on coal mine substrates. J Plant Nutr 32:755–771

Xavier LJC, Germida JJ (2002) Response of lentil under controlled conditions to co-inoculation with arbuscular mycorrhizal fungi and rhizobia varying in efficacy. Soil Biol Biochem 34:181–188

Yun-Jeong L, Eckhard G (2005) Development of a nutrient film technique culture system for arbuscular mycorrhizal plants. Hort Sci 40:378–380

Zobel RW, Dei Tredici P, Torren JG (1976) Method for growing plants aeroponically. Plant Physiol 57:344–346

Inoculation Effects of Associative Plant Growth-Promoting Rhizobacteria on the Performance of Legumes

11

Mohd. Saghir Khan, Almas Zaidi, Asfa Rizvi, and Saima Saif

Abstract

Constantly increasing human population requires that the crop production including those of legumes be enhanced rapidly to fulfill the food demands across the globe. In order to optimize pulse production, growers generally apply agrochemicals including fertilizers and pesticides. However, the excessive and uncontrolled use of such chemicals has resulted in reduced crop production besides their adverse impact on environment. In order to protect losses in soil fertility and to preserve environmental quality, the use of inexpensive and eco-friendly microbial preparations (biofertilizers) has been exploited in farming practices with remarkable success. Among various plant growth-promoting rhizobacteria (PGPR), the associative nitrogen-fixing PGPR, belonging to the genus *Azospirillum*, has long been employed as microbial inoculant worldwide to promote legume production. *Azospirillum*, when used as inoculant, increase the production of root hairs and root growth which in effect benefit plants with better absorption of water and nutrients. The inoculation of *Azospirillum* either alone or in combination with other beneficial PGPR has been found to increase N_2 fixation and concomitantly the grain yield of legumes. Considering the importance of *Azospirillum*, this chapter highlights the role of *Azospirillum* in the production of legumes in different agronomic setup.

M.S. Khan (✉) • A. Zaidi • A. Rizvi • S. Saif
Department of Agricultural Microbiology, Faculty of Agricultural Sciences, Aligarh Muslim University, Aligarh, UP 20202, India
e-mail: khanms17@rediffmail.com

© Springer International Publishing AG 2017
A. Zaidi et al. (eds.), *Microbes for Legume Improvement*,
DOI 10.1007/978-3-319-59174-2_11

11.1   Introduction

Among plant nutrients, nitrogen (N) is the most essential element whose deficiency restricts the crop production very severely. However, to fulfill the N demands and to optimize crop production, chemically synthesized nitrogenous fertilizers are applied in farming practices. The excessive and abrupt use of agrochemicals, however, destructs soil fertility and environment quality. Therefore, to prevent soil fertility losses and to preserve environment quality, there is urgent need to find some inexpensive and hazardous-free alternative for use in agricultural practices. In this regard, the microbial preparations, often called as biofertilizers, have provided an option to farmers to replace/minimize the use of chemical fertilizers. The application of microbial fertilizers for accelerating the growth and yield of crops (Verma et al. 2010; Upadhyay et al. 2012) including legumes (Pérez-Montaño et al. 2014; Fatnassi et al. 2015) has been practiced across different production systems. Besides symbiotic rhizobia, associative plant growth-promoting rhizobacteria (PGPR) especially the genus *Azospirillum* has been exploited with considerable success in the farming practices. *Azospirillum* as an associative nitrogen fixer and better plant colonizer (Bashan and Holguin 2004) is remarkably a versatile PGPR that facilitates the production of root hairs (Hadas and Okon 1987; Ribaudo et al. 2006) and increases root growth. As a result of well-developed root systems, the absorption of water and nutrients by plants is increased (Sarig et al. 1988; Bashan and Levanony 1990; Okon and Vanderleyden 1997). Besides nitrogen fixation (Dobereiner and Day 1976), *Azospirillum* also sequester Fe, can survive under unfavoring environmental conditions, and support other beneficial PGPR (Bashan and Holguin 2004; Chibeba et al. 2015). Also, *Azospirillum* benefit plants by mitigating adverse impact of some abiotic stresses (Ullah and Bano 2015). Due to these beneficial effects, *Azospirillum,* an outstanding PGPR, has long been used as microbial inoculant both singly (Cassán et al. 2009a) or jointly with other PGPR for enhancing the production and quality of many crops (Hungria et al. 2010) including legumes in different production systems (Galal 1997; Hungria et al. 2013; Servani et al. 2014). Accordingly, the beneficial impact of *Azospirillum* inoculation alone and/or with other PGPR has been reported for several legumes cultivated under greenhouse and field conditions (Tchebotar et al. 1998; Rodelas et al. 1999). For example, enhancement in root growth and nodulation of vegetable soybean (AGS190) following application of *A. brasilense* (Sp7) and *A. lipoferum* (CCM3863) co-inoculated with two strains (TAL102 and UPMR48) of *Bradyrhizobium japonicum* is reported (Molla et al. 2001). Sole application of *Azospirillum* and its association with *Bradyrhizobium* significantly enhanced root growth and nodulation on root systems of soybean plants. Additionally, the single and composite culture of *Azospirillum* and *Bradyrhizobium* substantially increased root morphogenesis (length, number, dry matter, and root hair development) and dry matter accumulation in shoots. The roots of soybean plants bacterized with mixture of *Azospirillum* and *Bradyrhizobium* had significantly greater number of nodules and biomass relative to other treatments. Among the two azospirilla used in this study, *A. brasilense* performed better in terms of root growth and nodule development in comparison to *A. lipoferum*. Conclusively, the enhancement in dry matter accumulation and in N content of *Azospirillum*-inoculated legumes was suggested due to improvement in root development, early nodulation, formation of more number of nodules, and efficient N_2-fixation rates (Volpin and Kapulnik 1994; Burdman et al. 1998). Similarly, response of 56 PGPR strains to pot-grown faba bean (*Vicia faba* var. Giza 429) maintained in greenhouse was variable (Abd El-Azeem et al. 2007). Application of PGPR stimulated the growth and yield of faba bean, and an average increases in biomass of straw, seeds, and total yields were 105.2, 31.9, and 56.8%, respectively, over uninoculated control. Also, the tested PGPR strains increased the number of nodules formed on the faba bean roots which ranged from 46.7% with *Micrococcus luteus* TK1 and *Xanthobacter autotrophicus* AM2 to 121.7% with *A. brasilense* AC1. Even though all PGPR enhanced nodule dry weight, only 23 isolates showed some significant increase (ranging from 35% with *A. brasilense* GO1 to 84% with *Serratia liquefaciens* GO2) over control. Here, the impact of associative nitrogen fixer especially the *Azospirillum* on the promotion of growth and yield of different legumes grown distinctively in different agroecological niches is highlighted.

11.2   Prevalence, Colonization, and General Characteristics

Among the associative nitrogen fixers, *Azospirillum* (α-subclass of proteobacteria) has been the most widely investigated organism and used as microbial inoculant in agronomic practices due largely to their ability to facilitate plant growth (Fages and Arsac 1981; Zaady et al. 1993; Cassan et al. 2008) including legumes (Vicario et al. 2016). Initially, it was isolated from sandy soil and was named *Spirillum lipoferum* (Beijerinck 1925) which subsequently was renamed as *Azospirillum lipoferum* (Tarrand et al. 1978). Since then many species have been identified (Xie and Yokota 2005; Peng et al. 2006; Mehnaz et al. 2007). They are found in rhizosphere (Bashan et al. 1995; Cecagno et al. 2015) and roots of many plants (Pedraza et al. 2007; Carvalho et al. 2014) and within the stem nodules and leaves and stems of other plants (Jhala et al. 2016). *Azospirillum* has been isolated from the major cereals like wheat (Rasool et al. 2015), maize (Noumavo et al. 2015), sorghum (Kanchanashri et al. 2014), and rice (Hossain et al. 2015) and several temperate climatic regions (Steenhoudt and Vanderleyden 2000). Azospirilla can survive even in the hostile environments due to their ability to form cyst (Bashan et al. 1991; Li et al. 2011) and flocs (Neyra et al. 1995; Joe et al. 2010), to produce melanin (Givaudan et al. 1991, 1993) and poly-β-hydroxybutyrate (PHB) (Okon and Itzigsohn 1992; Hou et al. 2014), and to synthesize polysaccharide (Del Gallo and Idaegi 1990; Fibach-Paldi et al. 2012). Azospirilla can also be protected from harsh environment while growing inside ectomycorrhizal fungal spores (Li and Catellano 1987). *Azospirillum* are Gram-negative and vibrio- or spirillum-shaped organism with peritrichous flagella. They are oxidase positive and exhibit acetylene reduction activity (ARA) under microaerophilic environment. They display a versatile C and N metabolism and, hence, can adapt to a competitive environment. They grow both under anaerobic and aerobic conditions but can also grow under microaerophilic (Pereg et al. 2016) environment.

11.3 Associative Interaction of *Azospirillum* with Plants

The interaction and subsequent colonization of azospirilla with host plants (Fallik et al. 1994; Rodrigues et al. 2015) begin chemotactically where bacterium *Azospirillum* is attracted toward the chemically variable exudates secreted by roots. Later on, *Azospirillum* attaches to the root surface by means of flagella and some glycocalyx compounds (Saikia et al. 2012). Following this, the bacteria aggregates onto the surface which is mediated by plant lectins (Alen'kina et al. 2014). After aggregation and firm attachment, exchange of communication occurs between plant and bacterium. Due to the signal exchange between the two associative partners, cellulose fibrils are produced by *Azospirillum*, which help bacteria to attach more tightly to the root surface leading eventually to the establishment of a viable and fully functional association. After a successful colonization and establishment, the azospirilla start secreting plant growth-promoting substances which in turn stimulate the production of plant hormones (Vacheron et al. 2013; Arshad and Frankenberger 1997).

11.4 Plant Growth Promotions by Associative Nitrogen Fixers: A General Perspective

Like many conventional free living or symbiotic PGPR, associative nitrogen fixers also facilitate plant growth either directly or indirectly. Of these, the direct growth-promoting activity of associative nitrogen fixers includes nitrogen fixation (Ramadan et al. 2016), solubilization of insoluble P (Rodriguez et al. 2004), and production of phytohormones (Bárbaro et al. 2008). In contrast, the indirect growth-promoting activity of associative nitrogen fixers include production of siderophores (Jha and Saraf 2015), antibiotics (Lenin and Jayanthi 2012), and induced systemic resistance (Kundan et al. 2015). Associative bacteria in general do not promote the growth of legumes since they do not form symbiosis as those formed by rhizobia. However, they promote the growth of legumes by secreting numerous growth-promoting substances (Table 11.1). For instance, *Azospirillum* have been shown to produce phytohormones like IAA (Crozier et al. 1988; Lambrecht et al. 2000; Ona et al. 2005; Spaepen et al. 2007; Masciarelli et al. 2013), gibberellins (Bashan and Bashan 2005), cytokinin or cytokinin-like substances (Stezelczyk et al. 1994; Tien et al. 1979), and ethylene (Stezelczyk et al. 1994). Broadly, these plant hormones affect the root morphogenesis of legumes (Hadas and Okon 1987; Baca et al. 1994; Pattern and Glick 1996; Ribaudo et al. 2006) and benefit plants with better absorption of water and nutrients (Bashan and Levanony 1990). Inoculation of *Azospirillum* on *Arabidopsis thaliana*, for example, increased the length of individual root hairs by at least twofolds (Dubrovsky et al. 1994). Also, IAA secreted by *Azospirillum* has been reported to stimulate symbiosis between legumes and nodule bacteria (Cassan et al. 2014). The other unique feature of associative nitrogen fixers is their ability to solubilize insoluble P (Krishnaraj and Dahale 2014) and making soluble P available to plants including legumes.

Table 11.1 Plant growth-promoting substances secreted by *Azospirillum* species

Name of organisms	Plant growth-promoting traits	References
Azospirillum sp.	IAA	Moghaddam et al. (2012)
Azospirillum sp.	Phosphate solubilization	Tahir et al. (2013)
Azospirillum sp.	Siderophores	Pedraza (2015)
Azospirillum sp.	ACC deaminase	Karunya and Reetha (2014)
Azospirillum sp.	HCN, ammonia	Sakthivel and Karthikeyan (2012)
Azospirillum sp.	Siderophores, biocontrol activity	
A. brasilense	IAA	Meza and de-Bashan (2015)
A. brasilense	Siderophores	Raja and Muthuselvam (2014)
A. lipoferum	Antagonistic activity	El-Hamshary et al. (2010)
A. brasilense	Cadaverine production	Cassan et al. (2009b)
A. lipoferum	IAA, gibberellic acid, siderophores	Lenin and Jayanthi (2012)
A. irakense	Nitrate reductase activity	Aliasgharzad et al. (2014)
A. brasilense	Hydrolytic enzymes	Radif and Hassan (2014)
Azospirillum sp.	Nitrogenase activity, β-galactosidase activity	

Furthermore, *Azospirillum* have shown the antibacterial activities which are due to their ability to synthesize bacteriocins (Tapia-Hernandez et al. 1990; Walker et al. 2011), siderophores (Tapia-Hernandez et al. 1990; Perrig et al. 2007; Rodrigues et al. 2015), and phenylacetic acid (PAA), an auxin-like molecule with antimicrobial activity (Somers et al. 2005; Naz et al. 2013). These and other related studies therefore suggest that species of *Azospirillum* could play an important role in the management of various pant diseases. Also, *Azospirillum* secretes numerous vitamins, like thiamin, niacin, pantothenic acid, riboflavin, etc. (Saikia et al. 2012; Rodelas et al. 1993; Dahm et al. 1993; Russel and Muszyski 1995).

11.5 Legume Responses to Associative Nitrogen Fixers

11.5.1 Single and Composite Inoculation Effects of *Azospirillum* on Legumes

Legumes are grown primarily for grain seeds, for livestock forage and silage, and as soil-enhancing green manure in many agricultural production systems around the world. They are the rich sources of protein, dietary fibers, carbohydrates, and dietary minerals and are used in vegetarian dietary systems. Legumes form symbiosis with nodule bacteria (rhizobia), and as a consequence of this multistep interaction, ammonium is produced which is utilized as N source by many plants. And hence, due to this, legumes are also used in many crop rotations. However, like rhizobia, associative nitrogen fixers do not form symbiosis with legumes, but they facilitate the growth of legumes when used alone (Hamaoui et al. 2001; Cassán et al. 2009a) or in combination with other PGPR (Marks et al. 2013; Hungria et al. 2015) by different mechanisms. Broadly, *Azospirillum* when used either alone or in combination can stimulate root hair formation and root growth and concurrently create more sites for early root infection and nodule formation by nitrogen-fixing rhizobia. Due to their growth-promoting activity, *Azospirillum* is considered one of the important PGPR and a *Rhizobium* helper which stimulates nodulation, nodule function, and possibly plant metabolism (Andreeva et al. 1993). For example, Chibeba et al. (2015) in a recent study determined the co-inoculation effects of *A. brasilense* and *Bradyrhizobium* spp. on nodulation, N_2 fixation, dry matter production, and accumulation of N in shoots of soybeans grown under greenhouse and field conditions. In greenhouse experiment, the co-inoculated plants showed a significant increase in nodulation, N_2 fixation, and concentration of N in shoots which increased even further under field conditions suggesting that the presence of *Azospirillum* might have helped plants to overcome environmental stresses.

While growing legumes, the combined use of rhizobia and *Azospirillum* can greatly be an effective strategy due largely to the nitrogen-fixing ability of rhizobia and phytohormone (Cacciari et al. 1989) production by *Azospirillum*. Considering this, *B. japonicum* and *A. brasilense* were used together to evaluate their effect on the morphophysiological development and nodulation of soybean (cv. BRS Favorita RR) cultivated in greenhouse at the Federal University of Lavras (UFLA) in the Field Crop sector of Brazil (Zuffo et al. 2015). Plant height, number of trifoliate leaves, shoot dry matter, root dry matter, nodule dry matter, root volume, leaf chlorophyll content, and leaf N content were measured at the beginning of flowering. *B. japonicum* at the rate of 3 mL kg^{-1} of seed exhibited the best morphophysiological performance and the maximum production of nodules on the root system of soybean plants. However, the sole or composite application of *A. brasilense* with *B. japonicum* did not show any significant effect on the measured parameters. Similarly, the composite application of *Azospirillum* and *Rhizobium* enhanced dry matter production and N content of legume plants. Also, *Azospirillum* had positive and favorable impact on nodule number, development, dry weight, and N_2 fixation in many legumes (Holguin and Bashan 1996; Burdman et al. 1997; Bashan and Holguin 2004; Chibeba et al. 2015). To further substantiate this, Yahalom et al. (1984) conducted an experiment under gnotobiotic condition and reported that *Azospirillum* inoculation stimulated the infection efficiency of *Rhizobium* and resulted in considerable nodule formation on *Medicago* plants. In a follow-up study, the combined application of *Azospirillum* sp. either with *B. japonicum* CB 1809 or USDA 110 increased shoot and root dry weight of soybean over non-inoculated control plants grown under pot house conditions. Also, inoculation of *gus*-marked USDA 110 singly or its co-inoculation gave 93.21–94.75% and 74.21–100% in nodule occupancy and 23.5–41.95% and 50.37–73.24% promotion in biomass dry weight over non-inoculated control in Myanmar and Thailand soil samples, respectively. *Azospirillum* sp. at 106, 107, and 108 cfu/mL increased nodulation in combination with USDA 110 with a corresponding increase in 73.8, 62.25, and 95.34% and 51.52, 62.38, and 79.46% over non-inoculated control in Myanmar and Thailand soil, respectively. It was suggested from this study that *Azospirillum* sp. could be used as co-culture with *B. japonicum* for soybean production (Aung et al. 2013).

Similarly, Hungria et al. (2015) assessed the co-inoculation effects of bradyrhizobia and azospirilla on soybean seeds grown under different soil and climate conditions in Brazil. The results revealed that co-inoculation was more efficient and beneficial to the soybean and promoted yield production without requiring any chemical N fertilizers even in soils harboring soybean bradyrhizobia. Peres et al. (2016) in a recent study assessed the inoculation efficiency of *R. tropici* and *A. brasilense* against common beans grown during winter season. The experiment was conducted in a randomized block design involving split-plot scheme with two irrigation depths in the plots (recommended for common beans and 75% of the recommended) and five forms of N supply in the split-plots (control non-inoculated with 40 kg ha^{-1} of N in top dressing, 80 kg ha^{-1} of N in top dressing, *A. brasilense* inoculation with 40 kg ha^{-1} of N in top dressing, *R. tropici* inoculation with 40 kg ha^{-1} of N in top dressing, and co-inoculation of *A. brasilense* and *R. tropici* with 40 kg ha^{-1} of N in top dressing). Each individual treatment was repeated four times. Co-inoculation of *R. tropici* and *A. brasilense* increased nodulation during the second year, but none of the treatments increased the grain yield compared to non-inoculated control with 40 kg N ha^{-1} applied as top dressing. Also, the use of 75% of recommended irrigation depth yielded grain yields similar to those recorded for recommended irrigation depth in common beans. In other experiment, the effects of single inoculation of *Azospirillum* and its combination with other PGPR belonging to genera *Azotobacter*, *Mesorhizobium*, and *Pseudomonas* on nutrient uptake, growth, and yield of chickpea plants cultivated under field conditions were variable (Rokhzadi and Toashih 2011). At flowering stage, the impact of sole or composite inoculation on nodulation and nutrient concentration in shoots varied significantly. The combined inoculation of *Azospirillum, A. chroococcum, M. ciceri* SWRI7, and *P. fluorescens* P21 was found superior and demonstrated maximum increase in dry weight of root nodules. However, all inoculants in general were found superior in terms of N accumulation in shoots over uninoculated control. Furthermore, the composite application of *Azospirillum* and *Azotobacter* significantly improved P concentration in shoots. Also, grain yield, whole dry biomass, and nutrient (N and P) uptake of grains were enhanced due to microbial inoculation compared to control plants. The development, yield components, and grain yield of beans grown during winter season both in the presence and absence of *A. brasilense* (strain AbV5 e AbV6) and N supply varied considerably (Gitti et al. 2012). Seed inoculation of *A. brasilense* showed higher leaf N content but did not affect significantly the development of plants, yield components, and grain yield of beans. At 60 kg urea ha^{-1}, there were more enhancements in plant growth, yield components, and grain yield of beans. This strategy of co-inoculation further consolidates the fact that mixture of both azospirilla and other PGPR could serve as an effective biotechnological tool to improve legume yield without employing any chemical N fertilizers.

On the contrary, the results obtained from combined inoculation in leguminous plants have also shown some conflicting results, i.e., they may both stimulate and inhibit the formation of nodules and root growth in a symbiotic system, varying as a function of the inoculum concentration level and of the inoculation type (Bárbaro et al. 2008). As an example, root growth of *Medicago polymorpha* was reduced when high level of *Azospirillum* inoculum was applied. This resulted in decrease in cell division in the apical meristem of the root which consequently decreased root potential for nodule production (Yahalom et al. 1991). Also, reduction in nodulation has been suggested due to competition between *Rhizobium* and *Azospirillum* in other legumes. This has been explained by the fact that *Azospirillum* colonizes root surface within few hours of inoculation and aggregates on the root surface. As a result of this aggregation, space is not available for rhizobia to colonize the same surface, and hence, no infection and subsequently no nodulation occur on legume roots. This concept is validated by the fact that when nodulation commenced, no root hair curling was noticed (Volpin and Kapulnik 1994). *Azospirillum* has also been used as co-culture with fungi against legume and cereal crops. For instance, the impact of sole and composite inoculation of *A. brasilense* and a P-solubilizing and biocontrol fungus, *Trichoderma harzianum*, on dry bean and wheat grown in pots and field was investigated by Mehmet et al. (2005). The results revealed that the sole application of *Azospirillum* and the dual inoculation had no significant effect on nodule numbers and nodule mass of pot-grown beans uprooted 45 days after sowing (DAS). The *Azospirillum* in the presence of supplementary P, however, significantly enhanced dry matter accumulation in nodules. Moreover, the single or composite inoculations did not differ significantly in terms of dry biomass and total plant N and P of bean (at 45 DAS) grown both in pot and fields. Also, the combined inoculation of *Azospirillum* and *T. harzianum* in the presence of rock phosphate (RP) significantly increased seed yield and total seed N and P of bean plants grown under field soil. This study therefore suggested that the mixture of *Azospirillum* and *T. harzianum* in the presence of RP was most effective and productive compared to other inoculations. Similar enhancement in seed germination, nodule formation, and development of soybean seedlings following composite application of *A. brasilense* and *B. japonicum* is reported (Cassán et al. 2009a). The increase in measured parameters was suggested due to the secretion of IAA and gibberellic acid (GA3) by both bacteria strains. The *Azospirillum*-inoculated garden pea maintained in greenhouse had significantly higher nodule numbers relative to control. Field inoculation of garden peas and chickpea in winter with *Azospirillum* 1 week after emergence produced a significant increase in seed yield, but did not affect plant dry matter yield. Also, the *Azospirillum* inoculation significantly increased dry matter yield, %N, N content, and acetylene reduction activity (ARA) of *Vicia sativa* L. (vetch) grown in greenhouse and under field for forage during season. The inoculation of *Rhizobium* (by the slurry method) and *Azospirillum* increased ARA of *Hedysarum coronarium* (sulla clover) above *Rhizobium*-inoculated control plants (Sarig et al. 1986). Vicario et al. (2016) evaluated and compared the effects of various single inoculation and co-inoculation treatments on growth parameters of peanut grown in greenhouse and field experiments. The co-inoculation with different *Bradyrhizobium* strains (native 15A and PC34 and recommended peanut inoculant C145) and *A. brasilense* strain Az39 generally increased the measured parameters of peanut raised in the greenhouse. In the field studies, 15A-Az39 co-inoculation had a greater promoting effect on measured growth parameters than did C145-Az39 co-inoculation.

11.5.2 Interactive Effect of *Azospirillum* and AM Fungi on Legumes

Krishnan and Sharavanan (2016) assessed the effects of single and dual inoculation of arbuscular mycorrhizal fungi (*Glomus mosseae*) and *Azospirillum* on black gram *Vigna mungo* (L.) Hepper. The composite culture of *G. mosseae* and *Azospirillum* produced highest plant biomass, shoot length, root length, number of leaves, number of root nodules, fresh and dry weight of black gram plants, and the biochemical content such as total chlorophyll, total sugar, and protein. Similarly, the inoculation of common bean, in Egypt, with mixture of AM fungi, *Rhizobium* sp., *Azospirillum* sp., and *Bacillus circulans*, showed enhancement in plant height, branching, nodulation, and plant biomass compared to control plants (Massoud et al. 2009). During interaction, AMF has been reported to enhance nodulation and N_2 fixation by legumes (Andre et al. 2005). Also mycorrhizal and other symbioses often act synergistically on infection rate, mineral nutrition, and plant growth (Amora-Lazcano et al. 1998). The positive fungal effect on plant P uptake is beneficial for the functioning of the nitrogenase enzyme of the microbial symbiont leading to a higher N_2 fixation and consequently to a better root growth and mycorrhizal development (Johansson et al. 2004). In a similar experiment, Ardakani et al. (2014) determined the possibility of improving the lentil performance using composite culture of AM fungi (*G. intraradices* and *G. mosseae*) and *Azospirillum* (*A. brasilense* under rainfed conditions, in Iran. The results revealed a substantial impact of AM fungi on grain protein, root colonization, and shoot dry weight. Among these, *G. intraradices* showed maximum increase in shoot dry weight, while *G. mosseae* induced highest root colonization and grain protein content in plants. Also, *Azospirillum* had a significant effect on dry matter accumulation in shoot and root colonization.

Conclusion

The impact of sole or composite application of *Azospirillum* on legume production under different soil situations is obvious. However, identification of more promising strains of *Azospirillum* and its co-inoculation with other plant growth promoting requires further understanding. The use of plant growth-promoting bacteria, such as *Azospirillum*, however, represents an economically viable strategy, besides the environmental benefits associated with the reduction in the use of fertilizers. There are good possibilities for increasing nodulation, nitrogen fixation, and crop yield of legumes in the field by inoculation with *Azospirillum*. This requires a series of research to be conducted in a controlled and homogeneous/field environment in order to evaluate root growth promotion and nodulation by *Azospirillum* co-inoculated with other PGPR.

References

Abd El-Azeem SA, Mehana TA, Shabayek AA (2007) Response of Faba bean (*Vicia faba* L.) to inoculation with plant growth-promoting rhizobacteria. CATRINA J 2:67–75

Alen'kina SA, Bogatyrev VA, Matora LY, Sokolova MK, Chernyshova MP, Trutneva KA, Nikitina VE (2014) Signal effects of the lectin from the associative nitrogen-fixing bacterium *Azospirillum brasilense* Sp7 in bacterial–plant root interactions. Plant Soil 381:337–349

Aliasgharzad N, Heydaryan Z, Sarikhani MR (2014) *Azospirillum* inoculation alters nitrate reductase activity and nitrogen uptake in wheat plant under water deficit conditions. Int J Sci Engg Tech 4(4):94–98

Amora-Lazcano E, Vazquez MM, Azcon R (1998) Response of nitrogen-transforming microorganisms to arbuscular mycorrhizal fungi. Biol Fertil Soils 27:65–70

Andre SA, Le GN, Roux C, Prin Y, Neyra M, Duponnois R (2005) Ectomycorrhizal symbiosis enhanced the efficiency of inoculation with two *Bradyrhizobium* strains and *Acacia holosericea* growth. Mycorrhiza 15:357–364

Andreeva I, Redkina T, Izmailov S (1993) The involvement of indole-acetic acid in the stimulation of *Rhizobium*-legume symbiosis by *Azospirillum brasilense*. Russ J Plant Physiol 40:780–780

Ardakani MR, Maleki S, Aghayri F, Rejali F, Faregh AH (2014) Tripartite symbiosis of lentil (Lens culinaris L.), Mycorrhiza and Azospirillum brasilense under rainfed condition. In: Rahmann G, Aksoy U (eds) Building organic bridges, Thuenen report, no. 20, vol 3. Johann Heinrich von Thünen-Institut, Braunschweig, pp 691–694

Arshad M, Frankenberger JR (1997) Plant growth-regulating substances in the rhizosphere: microbial production and functions. In: Donald LS (ed) Advances in agronomy. Academic, San Diego, pp 45–151

Aung TT, Tittabutr P, Boonkerd N, Herridge D, Teaumroong N (2013) Co-inoculation effects of *Bradyrhizobium japonicum* and *Azospirillum* sp. on competitive nodulation and rhizosphere eubacterial community structures of soybean under rhizobia-established soil conditions. Afr J Biotechnol 12:2850–2862

Baca BE, Soto-Urzua L, Xochihua-Corona YG, Cuervo-Garcia A (1994) Characterization of two aromatic amino acid aminotransferases and production of indoleacetic acid in *Azospirillum* strains. Soil Biol Biochem 26:57–63

Bárbaro IM, Brancalião SR, Ticelli M, Miguel FB, Silva JAA (2008) Técnica alternativa: co-inoculação de soja com Azospirillum e Bradyrhizobium visando incremento de produtividade. Artigo em Hypertexto. Disponível em: http://www.infobibos.com/Artigos/2008_4/coinoculacao/index.htm. Acessado em: 04 de janeiro de 2012

Bashan Y, Bashan LE (2005) Bacteria/plant growth-promoting. In: Hillel D (ed) Encyclopedia of soils in the environment. Elsevier, Oxford, pp 103–115

Bashan Y, Holguin G, De-Bashan LE (2004) *Azospirillum*-plant relationships: physiological, molecular, agricultural, and environmental advances (1997-2003). Can J Microbiol 50:521–577

Bashan Y, Levanony H (1990) Current status of *Azospirillum* inoculation technology: *Azospirillum* as a challenge for agriculture. Can J Microbiol 36:591–608

Bashan Y, Levanony H, Whitmoyer RE (1991) Root surface colonization of non-cereal crop plants by pleomorphic *Azospirillum brasilense* Cd. J Gen Microbiol 137:187–196

Bashan Y, Puente ME, Rodriquez-Mendoza MN, Toledo G, Holguin G, Ferrea-Cerrats R, Pedrin S (1995) Survival of *Azospirillum brasilense* in the bulk soil and rhizosphere of 23 soil types. Environ Microbiol 61:1938–1945

Beijerinck MW (1925) über ein Spirillum, welches frein stickto ff binden kann? Zentralbl Bakeriol Parasitenkd Infectionskr Abt 63:353–359

Burdman S, Kigel J, Okon Y (1997) Effects of *Azospirillum brasilense* on nodulation and growth of common bean (*Phaseolus vulgaris* L.) Soil Biol Biochem 29:923–929

Burdman S, Vedder D, German M, Itzigsohn R, Kigel J, Jurkevitch E, Okon Y (1998) Legume crop yield promotion by inoculation with *Azospirillum*. In: Elmerich C, Kondorosi A, Newton WE (eds) Biological nitrogen fixation for the 21st century. Kluwer Academic, Dordrecht, pp 609–612

Cacciari I, Lippi D, Pietrosanti T, Pietrosanti W (1989) Phytohormone-like substances produced by single and mixed diazotrophic cultures of *Azospirillum* and *Arthrobacter*. Plant Soil 115:151–153

Carvalho TLG, Balsemão-Pires E, Saraiva RM, Ferreira PCG, Hemerly AS (2014) Nitrogen signalling in plant interactions with associative and endophytic diazotrophic bacteria. J Exp Bot 65(19):5631–5642

Cassan F, Vanderleyden J, Spaepen S (2014) Physiological and agronomical aspects of phytohormone production by model plant growth promoting rhizobacteria (PGPR) belonging to the genus *Azospirillum*. J Plant Growth Regul 33:440–459

Cassán F, Perrig D, Sgroy V, Masciarelli O, Penna C, Luna V (2009a) *Azospirillum brasilense* Az39 and *Bradyrhizobium japonicum* E109 inoculated singly or in combination promote seed germination and early seedling growth in corn (*Zea mays* L.) and soybean (*Glycine max* L.) Eur J Soil Biol 45:28–35

Cassán F, Maiale S, Masciarelli O, Vidal A, Luna V, Ruiz O (2009b) Cadaverine production by *Azospirillum brasilense* and its possible role in plant growth promotion and osmotic stress mitigation. Eur J Soil Biol 45:12–19

Cassan F, Sgroy V, Perrig D, Masciarelli O, Luna V (2008) Phytohormone production by *Azospirillum* spp. physiological and technological aspects of plant growth promotion. In: *Azospirillum* spp. cell physiol plant interactions and Agronomic Research in Argentina, pp 61–86

Cecagno R, Fritsch TE, Schrank IS (2015) The plant growth-promoting bacteria Azospirillum amazonense: genomic versatility and phytohormone pathway. BioMed Res Int 2015:898592. doi:10.1155/2015/898592

Chibeba AM, Guimarães MF, Brito OR, Nogueira MA, Araujo RS, Hungria M (2015) Co-inoculation of soybean with *Bradyrhizobium* and *Azospirillum* promotes early nodulation. Am J Plant Sci 6:1641–1649

Crozier A, Arruda P, Jasmim JM, Monteiro AM, Sandberg G (1988) Analysis of indole-3-acetic acid and related indoles in culture medium from *Azospirillum lipoferum* and *Azospirillum brasilense*. Appl Environ Microbiol 54(5):2833–2837

Dahm H, Rózycki H, Strzelczyk E, Li CY (1993) Production of B-group vitamins by *Azospirillum* spp. grown in media of different pH at different temperatures. Zentralbl Mikrobiol 148:195–203

Del Gallo M, Idaegi A (1990) Characterization and quantification of exocellular polysaccharide in *Azospirillum brasilense* and *Azospirillum lipoferum*. Symbiosis 9:155–161

Dobereiner J, Day JM (1976) Association symbiosis in tropical grasses: characterization of microorganisms and dinitrogen fixing sites. In: Newton WE, Nyman CJ (eds) Proceedings of the first international symposium on nitrogen fixation, 2, Washington State University Press, Pullman, pp 518–538

Dubrovsky JG, Puente ME, Bashan Y (1994) *Arabidopsis thaliana* as a model system for the study of the effect of inoculation by *Azospirillum brasilense* Sp 245 on root hair growth. Soil Biol Biochem 26:1657–1664

El-Hamshary OIM, El-Gebally OG, Abou-El-Khier ZA, Arafa RA, Mousa Sh A (2010) Enhancement of the chitinolytic properties of *Azospirillum* strain against plant pathogens via transformation. J Am Sci 6(9):169–176

Fages J, Arsac JF (1981) Sunflower inoculation with *Azospirillum* and other plant growth promoting rhizobacterial. Plant Soil 137:87–90

Fallik E, Sarig S, Okon Y (1994) Morphology and physiology of plant roots associated with *Azospirillum*. In: Okon Y (ed) Azospirillum/plant associations. CRC Press, Boca Raton, pp 77–85

Fatnassi IC, Chiboub M, Saadani O, Jebara M, Jebara SH (2015) Impact of dual inoculation with *Rhizobium* and PGPR on growth and antioxidant status of *Vicia faba* L. under copper stress. C R Biol 338:241–254

Fibach-Paldi S, Burdman S, Okon Y (2012) Key physiological properties contributing to rhizosphere adaptation and plant growth promotion abilities of *Azospirillum brasilense*. FEMS Microbiol Lett 326:99–108

Galal YGM (1997) Dual inoculation with strains of *Bradyrhizobium japonicum* and *Azospirillum brasilense* to improve growth and biological nitrogen fixation of soybean (*Glycine max* L.) Biol Fertil Soils 24:317–322

Gitti DC, Arf O, Kaneko FH, Rodrigues RAF, Buzetti S, Portugal JR, Corsini DCDC (2012) Inoculation of *Azospirillum brasilense* cultivars of beans types in winter crop. Rev Agrarian 5:36–46

Givaudan A, Effosse A, Bally R (1991) Melanin production by *Azospirillum lipoferum* strains. In: Polsinelli M, Materassi R, Vincenzini M (eds) Nitrogen fixation: proceedings of the fifth international symposium on nitrogen fixation with non-legumes, Florence, 10–14 Sep 1990, pp 311–312

Givaudan A, Effosse A, Faure D, Potier P, Bouillant ML, Bally R (1993) Polyphenol oxidase in *Azospirillum lipoferum* isolated from rice rhizosphere: evidence for laccase activity in non-motile strains of *Azospirillum lipoferum*. FEMS Microbiol Ecol 108:205–210

Hadas R, Okon Y (1987) Effect of *Azospirillum brasilense* inoculation on root morphology and respiration in tomato seedlings. Biol Fertil Soils 5:241–247

Hamaoui B, Abbadi JM, Burdman S, Rashid A, Sarig S, Okn Y (2001) Effects of inoculation with *Azospirillum brasilense* on chickpeas (*Cicer arietinum*) and faba beans (*Vicia faba*) under different growth conditions. Agronomie, EDP. Sciences 21(6–7):553–560

Holguin G, Bashan Y (1996) Nitrogen-fixation by *Azospirillum brasilense* Cd is promoted when co-cultured with a mangrove rhizosphere bacterium (*Staphylococcus* sp.) Soil Biol Biochem 28:1651–1660

Hossain MM, Jahan I, Akter S, Rahman MN, Rahman SMB (2015) Effects of *Azospirillum* isolates from paddy fields on the growth of rice plants. Res Australas Biotechnol 6:15–22

Hou X, McMillan M, Coumans JVF, Poljak A, Raftery MJ (2014) Cellular responses during morphological transformation in *Azospirillum brasilense* and its flcA knockout mutant. PLoS One 9(12):e114435

Hungria M, Campo RJ, Souza EM, Pedrosa FO (2010) Inoculation with selected strains of *Azospirillum brasilense* and *A. lipoferum* improves yields of maize and wheat in Brazil. Plant Soil 331:413–425

Hungria M, Nogueira MA, Araujo RS (2013) Co-inoculation of soybeans and common beans with rhizobia and azospirilla: strategies to improve sustainability. Biol Fertil Soils 49:791–801

Hungria M, Nogueira MA, Araujo RS (2015) Soybean seed co-inoculation with *Bradyrhizobium* spp. and *Azospirillum brasilense*: a new biotechnological tool to improve yield and sustainability. Am J Plant Sci 6:811–817

Jha CK, Saraf M (2015) Plant growth promoting rhizobacteria (PGPR): a review. E3 J Agric Res Dev 5:0108–0119

Jhala YK, Shelat HN, Panpatte DG (2016) Efficacy testing of *Acetobacter* and *Azospirillum* isolates on maize cv. GM-3. J Fertil Pestic 7:164

Joe M, Karthikeyan MB, Sekar C, Deiveekasundaram M (2010) Optimization of biofloc production in *Azospirillum brasilense* (MTCC-125) and evaluation of its adherence with the roots of certain crops. Indian J Microbiol 50(Suppl 1):S21–S25

Johansson JF, Paul LR, Finlay RD (2004) Microbial interactions in the mycorrhizosphere and their significance for sustainable agriculture. FEMS Microbiol Ecol 48:1–13

Kanchanashri B, Gundappagol RC, Annu T, Mahadevaswamy, Santhosh GP (2014) Isolation and characterization of *Azospirillum* strains from rainfed areas of Raichur district of northern Karnataka, India. Bioinfolet 11(2a):295–299

Karunya SK, Reetha D (2014) Screening of plant growth promoting rhizobacteria for ACC deaminase activity. Int J Curr Res Chem Pharma Sci 1(3):65–70

Krishnan A, Sharavanan PS (2016) Effects of *Glomus Mosseae* and *Azospirillum* on the growth behavior of black gram *Vigna Mungo* (L.) Hepper. Ind J Res 5:454–457

Krishnaraj PU, Dahale S (2014) Mineral phosphate solubilization: concepts and prospects in sustainable agriculture. Proc Ind Natl Sci Acad 80:389–405

Kundan R, Pant G, Jadon N, Agrawal PK (2015) Plant growth promoting rhizobacteria: mechanism and current prospective. J Fertil Pestic 6:2. doi:10.4172/2471-2728.1000155

Lambrecht M, Okon Y, Vande Brook A, Vandereyden J (2000) Indole-3-acetic acid: a reciprocal signaling molecule in bacteria-plant interactions. Trends Microbiol 8:298–300

Lenin G, Jayanthi M (2012) Indole acetic acid, gibberellic acid and siderophore production by PGPR isolates from rhizospheric soils of *Catharanthus roseus*. Int J Pharm Biol Arch 3(4):933–938

Li CY, Catellano MA (1987) *Azospirillum* isolated from within sporocarp of the mycorrhizal fungi *Hebeloma crustuliniforme*, *Laccaria laccata* and *Rhizopogon vinicolor*. Trans Br Mycol Soc 88:563–566

Li H, Cui Y, Wu L, Tu R, Chen S (2011) cDNA-AFLP analysis of differential gene expression related to cell chemotactic and encystment of *Azospirillum brasilense*. Microbiol Res 166:595–605

Marks BB, Megías M, Nogueira MA, Hungria M (2013) Biotechnological potential of rhizobial metabolites to enhance the performance of *Bradyrhizobium* spp. and *Azospirillum brasilense* inoculants with soybean and maize. AMB Express 3:21

Masciarelli O, Urbani L, Reinoso H, Luna V (2013) Alternative mechanism for the evaluation of indole-3-acetic acid (IAA) production by *Azospirillum brasilense* strains and its effects on the germination and growth of maize seedlings. J Microbiol 51(5):590–597

Massoud ON, Morsy EM, El-Batanony NH (2009) Field response of snap bean (*Phaseolus vulgaris* L.) to N_2-fixers *Bacillus circulans* and arbuscular mycorrhizal fungi inoculation through accelerating rock phosphate and feldspar weathering. Aust J Basic Appl Sci 3:844–852

Mehmet Ö, Cevdet A, Oral D, Mehmet AS (2005) Single and double inoculation with *Azospirillum/Trichoderma*: the effects on dry bean and wheat. Biol Fertil Soils 41:262–272

Mehnaz S, Weselowski B, Lazarovits G (2007) *Azospirillum canadense* sp. nov., a nitrogen-fixing bacterium isolated from corn rhizosphere. Int J Syst Evol Microbiol 57:620–624

Meza B, de-Bashan LE, Bashan Y (2015) Involvement of indole-3-acetic acid produced by *Azospirillum brasilense* in accumulating intracellular ammonium in *Chlorella vulgaris*. Res Microbiol 166:72–83

Molla AH, Shamsuddin ZH, Halimi MS, Morziah M, Puteh AB (2001) Potential for enhancement of root growth and nodulation of soybean coinoculated with *Azospirillum* and *Bradyrhizobium* in laboratory systems. Soil Biol Biochem 33:457–463

Moghaddam MJM, Emtiazi G, Salehi Z (2012) Enhanced auxin production by Azospirillum pure cultures from plant root exudates. J Agric Sci Technol 14:985–994

Naz S, Cretenet M, Vernoux JP (2013) Current knowledge on antimicrobial metabolites produced from aromatic amino acid metabolism in fermented products. In: Méndez-Vilas A (ed) Microbial pathogens and strategies for combating them: science, technology and education. Formatex Research Centre, Badajoz, pp 337–346

Neyra CA, Atkinson A, Obubayi O (1995) Coaggregation of *Azospirillum* with other bacteria: basis for functional diversity. In: Fendrik I, Del Gallo M, Vanderleyden J, de Zamaroczy M (eds) Azospirillum VI and related microorganism, genetics—physiology—pecology, NATO ASI series, Series G. Springer, Berlin, pp 429–439

Noumavo PA, Agbodjato NA, Gachomo EW, Salami HA, Baba-Moussa F, Adjanohoun A, Kotchoni SO, Baba-Moussa L (2015) Metabolic and biofungicidal properties of maize rhizobacteria for growth promotion and plant disease resistance. Afr J Biotechnol 14(9):811–819

Okon Y, Itzigsohn R (1992) Poly-β-hydroxybutyrate metabolism in *Azospirillum brasilense* and the ecological role of PHB in the rhizosphere. FEMS Microbiol Lett 103:131–139

Okon Y, Vanderleyden J (1997) Root-associated *Azospirillum* species can stimulate plants. ASM News 63:366–370

Ona O, Van Impe J, Prinsen E, Vanderleyden J (2005) Growth and indole-3-acetic acid biosynthesis of *Azospirillum brasilense* Sp245 is environmentally controlled. FEMS Microbiol Lett 246:125–132

Pattern CL, Glick BR (1996) Bacterial biosynthesis of indole-3-acetic acid. Can J Microbiol 42:207–220

Pedraza R, Motok J, Tortora M, Salazar S, Díaz-Ricci J (2007) Natural occurrence of *Azospirillum brasilense* in strawberry plants. Plant Soil 295:169–178

Pedraza RO (2015) Siderophores production by Azospirillum: biological importance, assessing methods and biocontrol activity. In: Cassan FD et al (eds) Handbook for azospirillum. Springer, Cham, pp 251–262

Peng G, Wang H, Zhang G, Hou W, Liu Y, Wang ET, Tan Z (2006) *Azospirillum melinis* sp. nov., a group of diazotrophs isolated from tropical molasses grass. Int J Syst Evol Microbiol 56:1263–1271

Pereg L, de-Bashan LE, Bashan Y (2016) Assessment of affinity and specificity of *Azospirillum* for plants. Plant Soil 399:389–414

Peres AR, Rodrigues RAF, Arf O, Portugal JR, Corsini DCDC (2016) Co-inoculation of *Rhizobium tropici* and *Azospirillum brasilense* in common beans grown under two irrigation depths. Rev Ceres 63(2):198–207

Pérez-Montaño F, Alías-Villegas C, Bellogín RA, del Cerro P, Espuny MR, Jiménez-Guerrero I, López-Baena FJ, Ollero FJ, Cubo T (2014) Plant growth promotion in cereal and leguminous agricultural important plants: from microorganism capacities to crop production. Microbiol Res 169:325–336

Perrig D, Boiero ML, Masciarelli OA, Penna C, Ruiz OA, Cassán FD, Luna MV (2007) Plant-growth-promoting compounds produced by two agronomically important strains of *Azospirillum brasilense*, and implications for inoculant formulation. Appl Microbiol Biotechnol 75:1143–1150

Radif HM, Hassan SS (2014) Detection of hydrolytic enzymes produced by *Azospirillum brasiliense* isolated from root soil. World J Exp Biosci 2(2):36–40

Raja SB, Muthuselvam K (2014) Interstrain differences of Chilli *Azospirillum* isolates on their plant growth promoting traits under *in vitro* conditions. Int J Adv Res Biol Sci 1:248–253

Ramadan EM, AbdelHafez AA, Hassan EA, Saber FM (2016) Plant growth promoting rhizobacteria and their potential for biocontrol of phytopathogens. Afr J Microbiol Res 10(15):486–504

Rasool L, Asghar M, Jamil A, Rehman SU (2015) Identification of *Azospirillum* species from wheat rhizosphere. J Anim Plant Sci 25(4):1081–1086

Ribaudo CM, Krumpholz EM, Cassán FD, Bottini R, Cantore ML, Curá JA (2006) *Azospirillum* sp. promotes root hair development in tomato plants through a mechanism that involves ethylene. J Plant Growth Regul 25:175–185

Rodelas B, González-López J, Martínez-Toledo MV, Pozo C, Salmerón V (1999) Influence of Rhizobium/Azotobacter and *Rhizobium/Azospirillum* combined inoculation on mineral composition of faba bean (*Vicia faba* L.) Biol Fertil Soils 29:165–169

Rodelas B, Salmeron V, Martinez-Toledo MV, Gonzalez-Lopez J (1993) Production of vitamins by *Azospirillum brasilense* in chemically—defined media. Plant Soil 153:97–101

Rodrigues AC, Bonifacio A, Fernando de Araujo F, Lira Junior MA, Figueiredo MVB (2015) Azospirillum sp. as a challenge for agriculture. In: Maheshwari DK (ed) Bacterial metabolites in sustainable agroecosystem, sustainable development and biodiversity, vol 12. Springer, New York. doi:10.1007/978-3-319-24654-3_2

Rodriguez H, Gonzalez T, Goire I, Bashan Y (2004) Gluconic acid production and phosphate solubilization by the plant growth-promoting bacterium *Azospirillum* spp. Naturwissenschaften 91:552–555

Rokhzadi A, Toashih V (2011) Nutrient uptake and yield of chickpea (*Cicer arietinum* L.) inoculated with plant growth promoting rhizobacteria. AJCS 5(1):44–48

Russel S, Muszyski S (1995) Reduction of 4-choloronitrobenzene by *Azospirillum lipoferum*. In: NATO ASI Series G 37, pp 369–375

Saikia SP, Bora D, Goswami A, Mudoi KD, Gogoi A (2012) A review on the role of *Azospirillum* in the yield improvement of non leguminous crops. Afr J Microbiol Res 6(6):1085–1102

Sakthivel U, Karthikeyan B (2012) Isolation and characterization of plant growth promoting rhizobacteria (pgpr) from the rhizosphere of *Coleus forskohlii* grown soil. Int J Recent Sci Res 3(5):288–296

Sarig S, Kapulnik Y, Okon Y (1986) Effect of *Azospirillum* inoculation growth of several winter legumes. Plant Soil 90:335–342

Sarig S, Blum A, Okon Y (1988) Improvement of the water status and yield of field-grown grain sorghum (*Sorghum bicolor*) by inoculation with *Azospirillum brasilense*. J Agric Sci 110:271–277

Servani M, Mobasser HR, Ganjali HR (2014) Effect of bacterium *Azospirillum* phosphate fertil 2 on soybean. Int J Farm Alli Sci 3(3):324–327

Somers E, Ptacek D, Gysegom P, Srinivasan M, Vanderleyden J (2005) *Azospirillum brasilense* produces the auxin-like phenylacetic acid by using the key enzyme for indole-3-acetic acid biosynthesis. Appl Environ Microbiol 71:1803–1810

Spaepen S, Vanderleyen J, Remans R (2007) Indole-3-acetic acid in microbial and microorganism plant signaling. FEMS Microbiol Rev 31:425–448

Steenhoudt O, Vanderleyden J (2000) *Azospirillum* a free living nitrogen fixing bacterium closely associated with grasses: genetic, biochemical and ecological aspects. FEMS Microbiol Rev 24:487–506

Stezelczyk E, Kampert M, Li CY (1994) Cytokinin like substances and ethylene production by *Azospirillum* in media with different carbon sources. Microbiol Res 149:55–60

Tahir M, Mirza MS, Zaheer A, Dimitrov MR, Smidt H, Hameed S (2013) Isolation and identification of phosphate solubilizer *Azospirillum*, *Bacillus* and *Enterobacter* strains by 16SrRNA sequence analysis and their effect on growth of wheat (*Triticum aestivum* L.) Aust J Crop Sci 7:1284–1292

Tapia-Hernandez A, Mascarua-Esparza MA, Caballero Mellado J (1990) Production of bacteriocins and siderophore-like activity by *Azospirillum brasilense*. Microbios 64:73–83

Tarrand JJ, Kreig NR, Döbereiner J (1978) A taxonomic study of the *Azospirillum lipoferum* group, with a descriptions of a new genus, *Azospirillum* gen. nov. and two species, *Azospirillum lipoferum* (Beijerink) comb. nov. and *Azospirillum brasilense* sp. nov. Can J Microbiol 24:967–980

Tchebotar VK, Kang UG, Asis CA, Akao S (1998) The use of GUS-reporter gene to study the effect of *Azospirillum-Rhizobium* coinoculation on nodulation of white clover. Biol Fertil Soils 27:349–352

Tien TM, Gaskins MH, Hubbell DH (1979) Plant growth substances produced by *Azospirillum brasilense* and their effect on the growth of pearl millet (*Pennisetum americanum* L.) Appl Environ Microbiol 37:1016–1024

Ullah S, Bano A (2015) Isolation of plant-growth-promoting rhizobacteria from rhizospheric soil of halophytes and their impact on maize (Zea mays L.) under induced soil salinity. Can J Microbiol 61:307–313

Upadhyay SK, Singh JS, Saxena AK, Singh DP (2012) Impact of PGPR inoculation on growth and antioxidant status of wheat under saline conditions. Plant Biol (Stuttg) 14:605–611

Vacheron J, Desbrosses G, Bouffaud ML, Touraine B, Moënne-Loccoz Y, Muller D, Legendre L, Wisniewski-Dyé F, Prigent-Combaret C (2013) Plant growth promoting rhizobacteria and root system functioning. Front Plant Sci. doi:10.3389/fpls.2013.00356

Verma JP, Yadav J, Tiwari KN, Lavakush, Singh V (2010) Impact of plant growth promoting rhizobacteria on crop production. Int J Agric Res 5:954–983

Vicario JC, Primo ED, Dardanelli MS, Giordano W (2016) Promotion of peanut growth by co-inoculation with selected strains of *Bradyrhizobium* and *Azospirillum*. J Plant Growth Regul 35:413–419

Volpin H, Kapulnik Y (1994) Interaction of Azospirillum with beneficial soil microorganisms. In: Okon Y (ed) *Azospirillum*/plant associations. CRC Press, Boca Raton, pp 111–118

Walker V, Bertrand C, Bellvert F, Moënne-Loccoz Y, Bally R, Comte G (2011) Host plant secondary metabolite profiling shows a complex, strain-dependent response of maize to plant growth-promoting rhizobacteria of the genus Azospirillum. New Phytol 189(2):494–506

Xie CH, Yokota A (2005) *Azospirillum oryzae* sp. nov., a nitrogen-fixing bacterium isolated from the roots of the rice plant *Oryza sativa*. Int J Syst Evol Microbiol 55:1435–1438

Yahalom E, Dovrat A, Okon Y, Czosnek H (1991) Effect of inoculation with *Azospirillum brasilense* strain Cd and rhizobium on the root morphology of burr medic (*Medicago polymorpha* L.) Isr J Bot 40:155–164

Yahalom E, Kapulnik Y, Okon Y (1984) Response of *Setaria italica* to inoculation with *Azospirillum brasilense* as compared to *Azotobacter chroococcum*. Plant Soil 82:77–85

Zaady E, Perevolotsky A, Okon Y (1993) Promotion of plant growth by inoculum with aggregated and single cell suspension of *Azospirillum brasilense* Cd. Soil Biol Biochem 25:819–823

Zuffo AM, Rezende PM, Bruzi AT, Oliveira NT, Soares IO, Neto GFG, Cardillo BES, Silva LO (2015) Co-inoculation of *Bradyrhizobium japonicum* and *Azospirillum brasilense* in the soybean crop coinoculação de *Bradyrhizobium japonicum* e *Azospirillum brasilense* na cultura da soja. Revista de Ciências Agrárias 38:87–93

Perspectives of Using Endophytic Microbes for Legume Improvement

12

Muhammad Naveed, Muhammad Zahir Aziz,
and Muhammad Yaseen

Abstract

Plant growth-promoting rhizobacteria (PGPR) have long been used as inoculant for optimizing legume production, but their survival under hostile field conditions is conflicted. Endophytes among PGPR are the microorganisms that live inside different plant tissues for at least part of their life without harming their host. Beneficial endophytes facilitate plant growth by enhancing uptake of plant nutrients, protecting plants from phytopathogens and increasing tolerance against environmental stresses. Nevertheless, the cellular interactions between pulses and endophytes for improving legumes growth and yields are variable. The endophytic colonization and diversity, various growth promontory aspects, and recent advances in endophyte-legume interactions with consequential impact on legume production have been discussed comprehensively. Considering the importance of endophytic microorganisms, it is likely that their use in agricultural practices will play a pivotal role and offer environmentally friendly strategy for increasing legume productivity while decreasing chemical inputs.

12.1 Introduction

Sustainable agricultural practices are getting much attention and revitalizing the interest in legumes-endophytes interaction due largely to the nutritive value and fodder usage of legumes (Pablo et al. 2015; Subramanian et al. 2015). In soil plant root system, consistent interactions among indigenous/applied microorganisms and release of some signaling compounds by soil microbiota determine a healthy

M. Naveed (✉) • M.Z. Aziz • M. Yaseen
Institute of Soil and Environmental Sciences, University of Agriculture,
Faisalabad 38080, Pakistan
e-mail: mnaveeduaf@yahoo.com

relationship between different soil microbes involving plant growth-promoting rhizobacteria, mycorrhizae, and various plant genotypes including legumes. So, root systems apart from acting as a suitable habitat also provide nutrients to plants. Use of microbes particularly PGPR is considered one of the significant approaches for stable production of crops (Adesemoye et al. 2009; Berg 2009). The PGPR in soils may act as symbionts or as free-living organisms and establish association with host plants (Rivera-Cruz et al. 2008). Among microbes, there are certain groups of organisms that live inside the plant body without causing any damage to their host and enhance their growth, called plant growth-promoting endophytes (PGPE). Plant growth-promoting endophytes have advantages over conventional PGPR because their preferred niche is endosphere of root that protects them from competition and numerous other soil stresses (Naveed et al. 2014a). Endophytes promote the growth of plants by various mechanisms such as through phosphate solubilization (Suman et al. 2016), nonsymbiotic nitrogen fixation, production of phytohormones, nutrient uptake, and water absorption (Verma et al. 2010) both under conventional and stressed environments (Khan et al. 2017; Shahzad et al. 2017). A large number of endophytes have been recovered from different parts of the crop plant. Even though both endophytic and rhizospheric microbes have been found to enhance plant growth separately and mutually (Subramanian et al. 2015), the benefits from endophytic microbes are not well documented for legumes. In legume crops, endophytic microbes are recovered from roots, nodules, and different parts of the plant body (Dudeja et al. 2012; Saini et al. 2015). Some of the notable genera like *Enterobacter*, *Pseudomonas*, and *Bacillus* spp. are recovered from legumes and other cereal crops which have shown growth improvement in different production systems (De Meyer et al. 2015; Subramanian et al. 2015). The prevalence of endophytes, however, depends on host genotypes, growth stage of crops, inoculum concentration, and abiotic conditions. Microbes living in plant bodies are naturally selected with special importance to plants, and roots are highly susceptible to invasion by endophytic microbes. Varying genera and species of endophytes may be present in single plant species, but how they interact inside the plant tissue is not known. In a recent investigation, it is revealed that inside the plant body, endophytic microbes interact by employing quorum sensing (Kusari et al. 2015). In addition, the root exudates released in the rhizosphere by plants help in communication between plants and soil microbes (Long et al. 2008).

Legume crops are significant with respect to soil fertility. Also, endophyte-inoculated legumes have shown some promising effects on other crops in rotation. Therefore, it has become imperative to better understand the role of endophytic microbes in legumes improvement. Considering the importance of endophytes, ecological diversity of endophytes in legumes and mechanisms used by endophytic microbes for promoting legume growth are highlighted. Also, the plant growth-promoting endophyte (PGPE) tracking techniques and impact of single and composite endophytes on legumes have been discussed comprehensively.

12.2 Ecological Occurrence of Endophytic Microbes in Legumes

Nodule formation is a specific and major trait of legumes which acts as a novel natural habitat for nitrogen-fixing microbes. Different taxonomic and genetic studies have shown that the ability of microbes to induce and inhibit nodulation depends on genetic composition of microbes and environmental factors (Saini et al. 2015). The presence of endophytes in nodules in many studies has been reported (Xu et al. 2014; De Meyer et al. 2015). For example, 12 bacterial genera in different studies were found to produce nodules, of which nine genera belonged to α-proteobacteria and three to β-proteobacteria (Dudeja and Narula 2008; Weir 2011). Moreover, the large number of bacterial genera including 24 non-rhizobia was also isolated, most of which belonged to *Agrobacterium*, *Burkholderia*, *Enterobacter*, and *Erwinia* (Lei et al. 2008; Muresu et al. 2008; De Meyer et al. 2015). In another study, *Pantoea agglomerans* from foliage, the *Agrobacterium rhizogenes* from taproot, and *R. leguminosarum*, *R. phaseoli*, and *Mesorhizobium loti* from nodules have been isolated (Sturz and Christie 1995). Similarly, 12 non-rhizobia were found hosted in nodules of clover. De Meyer et al. (2015) in a recent study isolated 654 strains from 30 different legumes and characterized them by RFLP and 16S rRNA. The results revealed that 84% bacterial isolates from root nodules belonged to rhizobia and 16% were endophytic microbes. *Micromonospora* strains have recently been reported as natural endophytes of legume nodules (Martinez-Hidalgo et al. 2014). The ecological role of the interaction of the 15 selected representative *Micromonospora* strains was tested against *M. sativa*. Selected *Micromonospora* isolates increased nodulation and N nutrition of alfalfa by co-inoculation with *Ensifer meliloti* 1021. Stajković et al. (2009) recovered 15 endophytic bacteria from sterilized nodule surface of the alfalfa plants which were Gram positive and were classified as *Microbacterium trichothecenolyticum*, *Bacillus megaterium*, and *Brevibacillus choshinensis*. When these endophytes were reinoculated under sterilized conditions, no nodulation was observed; however, their co-inoculation with rhizobia promoted the nodule numbers and growth attributes of alfalfa plants compared to un-inoculated plants. Similarly, Pandya et al. (2013) isolated eight non-rhizobial bacterial genera, predominantly *Bacillus* spp. and *Paenibacillus* spp., from nodules of field-grown *V. radiata*. Xu et al. (2014) performed phylogenetic analysis of 201 isolates obtained from Qilian Mountain legumes. These isolates belonged to 35 different species; however, *Sinorhizobium meliloti* was more similar to rhizobia due to horizontal gene transfer: from rhizobia to non-rhizobial strains. Furthermore, the altitude and host species played more important role in endosymbiont separation. Endophytic bacteria were incompetent to persuade nodule development in peanut; however, their inoculation improved plant produce. When these endophytic isolates were co-inoculated with rhizobia, a number of nodules and biomass were increased (Ibanez et al. 2009). Hoque et al. (2011) isolated many non-rhizobial and rhizobial strains from different legumes in Australia. These legumes were infected and nodulated mainly by

Rhizobium, Ensifer, Mesorhizobium, Burkholderia, Phyllobacterium, and *Devosia.* Recently, Chimwamurombe et al. (2016) isolated 73 endophytic bacterial species belonging to 14 genera including *Proteobacteria* (*Rhizobium, Massilia, Kosakonia, Pseudorhodoferax, Caulobacter, Pantoea, Sphingomonas, Burkholderia, Methylobacterium*), *Firmicutes* (*Bacillus*), *Actinobacteria* (*Curtobacterium, Microbacterium*), and *Bacteroidetes* (*Mucilaginibacter, Chitinophaga*) from gnotobiotically grown Marama bean seedlings. Further screening revealed that the selected isolates showed production of IAA, ACC deaminase, siderophores, and nutrient solubilization activities. Besides these genera, many non-nodulating bacterial endophytes were also colonized in the root nodules. Briefly, these and other similar studies suggest that nodules act as a major habitat for widespread diversity of microorganisms.

12.3   Genetic Diversity of Nodule Endophytic Bacteria in Legumes

Phylogenetic analysis suggests that there is greater diversity among nodule endophytes which belongs to different ancestries (Table 12.1) but they have no capacity to induce nodule formation. However, according to current knowledge, diverse array of microbes are found inside nodules which could promote plant growth by various mechanisms, but nodulation by rhizobia is an exclusive property. The advent of 16S rRNA technology has dramatically improved the identification of nodule inducing strains. Using this technique, *Alphaproteobacteria* and *Betaproteobacteria,*

Table 12.1 Endophyte diversity in legume nodules

Legume crop	Endophyte microbes	References
Trifolium repens	*Epichloë* endophyte	Pablo et al. (2015)
Legumes	*Alphaproteobacteria, Betaproteobacteria, Gammaproteobacteria, Actinobacteria, Firmibacteria, Flavobacteria*	De Meyer et al. (2015)
C. jubata	*Pseudomonas* genera	Xu et al. (2014)
Chickpea	*Bacillus*	Kumar et al. (2013)
Medicago sativa	*Arthrobacter, Bacillus, Dyella, Microbacterium, Staphylococcus*	Palaniappan et al. (2010)
Arachis hypogaea L.	*Enterobacter, Klebsiella, Pseudomonas*	Ibanez et al. (2009)
Soybean	*Pantoea, Acinetobacter, Agrobacterium, Burkholderia*	Li et al. (2008)
Herbaceous legumes	*Agrobacterium, Enterobacteriaceae*	Kan et al. (2007)
Spontaneous legumes	*Phyllobacterium, Sphingomonas, Rhodopseudomonas, Pseudomonas*	Zakhia et al. (2006)

nonclassical rhizobial strains, were isolated which are capable for fixing nitrogen in legume crops (Rivas et al. 2009). Also, the genetic analysis of endophytes isolated from legumes in Flanders (Belgium) was performed by De Meyer et al. (2015). The 16S rRNA analysis revealed a large diversity from the classes *Betaproteobacteria*, *Alphaproteobacteria*, *Sphingobacteria*, *Gammaproteobacteria*, *Flavobacteria*, *Firmibacteria*, and *Actinobacteria*, isolated from nodules which are non-rhizobial microbes. Moreover, *nif*H gene was also detected in isolates, which indicate the endophytic nature of the bacteria. Legumes have intrinsic ability to accommodate both rhizobia and endophytes. Moreover, endophytic nodule infection depends on many factors including rhizobia and endophyte strain type, node factors, exopolysaccharides, and host plant genotypes (Zgadzaj et al. 2015). Endophytic bacterial diversity in nodule of *Astragalus* species analyzed by 16S rRNA revealed that 53% strains were non-rhizobial, while others were the nodulating strains of *Mesorhizobium* species which had the capacity of symbiosis (Chen et al. 2015).

12.4 How Endophytes Enter Their Host

Endophytic microbes prior to their entry into plant tissues first colonize the plant roots by chemotaxis and electrotaxis and through accidental encounter. Numerous microbes belonging to genera *Pseudomonas*, *Alcaligenes*, and *Azospirillum* adopt these mechanisms for their movement (You et al. 1995). However, microbial colonization on root surface involves three steps: (1) bacterial adsorption onto the root surface (rapid but weak interaction between bacteria and host plant), (2) anchoring on the root surface (strong interaction), and (3) colonization of the root surface with some cells entering into the root (Michiels et al. 1991; You et al. 1995). Many scientists have reported that extensive colonization of bacterial endophytes occurs on secondary root emergence site. This is because endophytes easily enter at the point of breakage of the epidermis, colonize at the cortex, and then spread into vascular tissue across the endodermis (Mahaffee et al. 1997). Also, plant wounding resulting from many factors including biotic and abiotic stresses enhances the probability of entry for microbes due to release of exudates which allow microbes to attach on roots successfully (Hallmann et al. 1997). Endophytes release the cell wall-degrading enzymes before colonizing the roots, but after entry into the intercellular spaces, enzyme release has not been reported, suggesting that endophytes only release enzymes for penetration into the plant body (Bell et al. 1995). Recently, Pandya et al. (2013) investigated the bacterial entry into root nodules of *Vigna radiata*. They co-inoculated fluorescently tagged *Pseudomonas fluorescens* and *Klebsiella pneumoniae* with host-nodulating *Ensifer adhaerens* to *V. radiata* seedlings and monitored root hair infection using confocal microscopy 5 days after inoculation. They observed that *P. fluorescens* and *K. pneumoniae* invaded the root hair only when co-inoculated with *E. adhaerens*. Recovery of inoculated tagged strains and confirmation through CLSM and 16S rRNA gene sequencing confirmed that the test bacteria occupied nodules.

12.4.1 Movement

Once endophytes enter into plant body, it moves rapidly by apoplastic pathway and through conducting elements, apoplast (Hallmann et al. 1997). Microbes move in plant body intercellular spaces up to 1–4 cm or up to 15 cells longer distance (Mahaffee et al. 1997). Systemic colonization of microbes in cortical tissue is, however, limited due to conviction that endodermis offer physical barrier through movement by apoplastic pathway in vascular soft tissue (Kloepper et al. 1992). This hypothesis is based upon the theory that endodermis contain the Casparian bands which regulate the solute movement by limiting the apoplastic movement. However, endophytes existing in the cortex do not break the Casparian bands, but it is broken down at the site of secondary root which provides the route for microbe entry into the vascular system (Peterson et al. 1981). Microbes can also enter into vascular tissues by undifferentiated cells of root. Yet, colonization of endophytes in internal plant tissues is affected by plant genetic character. Bacterial endophytes, for example, *Pseudomonas* spp., are reported to colonize the vascular and cortical tissues of common bean following seed application (Mahaffee et al. 1997).

12.4.2 Localization

Endophyte location in a specific part of the plant depends upon the strain type. Commonly, microbes reside in plant body (Mahaffee et al. 1997; Gao and Mendgen 2006) which is followed by an order of movement as (1) intercellular space, (2) root cortex, (3) steel, (4) endodermis, (5) parenchyma, and (6) vascular tissues. For example, diazotrophic endophyte *Azoarcus* sp. was found initially in intercellular spaces, after which it moved to cortex and steel within endodermis and finally reached to parenchyma cells where it proliferates (Hurek et al. 1994; Gao and Mendgen 2006). However, it enters into the xylem through penetration from pith. In contrast, endophyte *Azospirillum brasilense* existed only in intercellular spaces and never entered into vascular system (Schank et al. 1979). Some other endophytes remain in intercellular spaces and reside mainly in cell vacuole (Jacobs et al. 1985; Maougal et al. 2014). Further investigation has shown that these microbes also colonized the vascular tissues. Considering the present scenario, endophyte *Ralstonia solanacearum* did not multiply and colonized in vascular tissue as compared to the pathogen bacteria (Vasse et al. 1995). Also, the deliberate growing and restricted spread of *Erwinia stewartii* resulted in its conversion from virulent to avirulent, specifying the participation of EPS in this sensation (Braun 1990).

12.5 Mechanisms of Actions of Endophytic Bacteria

The plant growth-promoting bacteria (PGPB) including endophytes benefit plants through nitrogen fixation (BNF), phytohormone production, siderophore production, ACC-deaminase activity, nutrient solubilization/availability, vitamin production, and

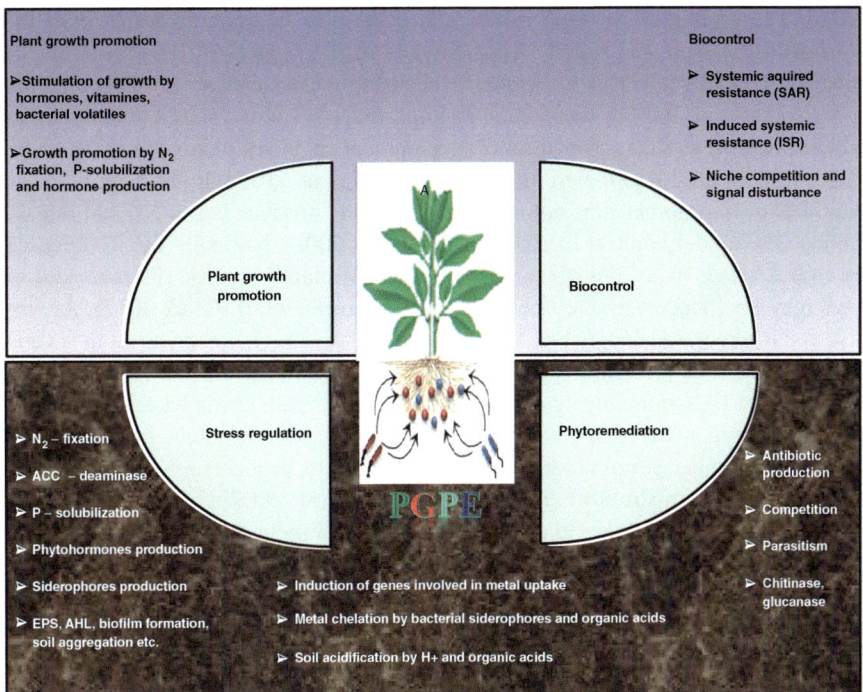

Fig. 12.1 Plant growth promotory activity of soil microbiota (modified from Naveed 2013)

suppression of phytopathogens (Muthukumar et al. 2010; Martinez-Viveros et al. 2010; Mitter et al. 2013). Also, several studies have shown that endophytic microbes improved plant growth, vigor, and yield, by abating abiotic stresses (drought, salinity, and temperature) (Wakelin et al. 2004; Naveed et al. 2014a, b, c). The endophytes like other free-living PGPR may adopt one or combination of these mechanisms for plant growth promotion (Fig. 12.1). However, under certain conditions, endophytes alter their behavior toward pathogen activity which affects the plant growth and development severely (Kobayashi and Palumbo 2000). The major mechanisms adopted by PGPB for growth promotion are discussed in the following sections.

12.5.1 Phytohormone Production

Phytohormones (also known as plant growth regulators) are versatile low molecular weight natural signaling molecules that act even at micromolar concentration and regulate essentially all physiological and developmental processes of plants (Chiwocha et al. 2003; Khan et al. 2012). Among phytohormones, auxins (IAA), gibberellins (GAs), cytokinins, abscisic acid (ABA), and ethylene are the best known and extensively investigated plant hormones secreted by many bacteria including endophytes (Khalid et al. 2006; Maheswari et al. 2013; Chimwamurombe et al.

2016). The production of gibberellins and cytokinins by endophytic bacteria like *Herbaspirillum* spp. and *Bacillus* spp. (Silva et al. 2015), auxins from *Pseudomonas* sp. isolated from ginger rhizome (Jasim et al. 2014), and abscisic acid by *Azospirillum* spp. (Cohen et al. 2009) is considered an important physiological trait of endophytes which positively affects growth and development of many plants (Shahzad et al. 2016; Halo et al. 2015) including legumes (Saini et al. 2015). It has been reported that 80% of microorganisms colonizing rhizosphere produce hormones and release auxins as secondary metabolites (Patten and Glick 2002). Normally, auxins released by PGPB affect many physiological processes of plants because internal pool of IAA may be changed by the release of IAA from microbes (Glick 2012). Auxins also act as a reciprocal signaling molecule which affects gene expression in several microorganisms. Furthermore, plant defense mechanisms are associated with down-regulation of IAA signaling against a number of phytopathogenic bacteria (Spaepen and Vanderleyden 2011; Grover et al. 2011). Broadly, auxin regulates cell extension, division, and tuber germination, controls processes of vegetative growth, and promotes resistance to stressful conditions (Santner et al. 2009; Grover et al. 2011). Auxin production by many plant-associated bacteria including *Klebsiella* (Celloto et al. 2012), *Enterobacter* (Ghosh and Sen 2015), and *Azospirillum* (Moghaddam et al. 2012) are reported. Upon inoculation, seed bacterization of auxin-generating *P. fluorescens* WCS365 did not increase the root or shoot weight of cucumber, sweet pepper, and tomato, but it significantly increased the root biomass of radish (Kravchenko et al. 2004). In other investigation, the length of canola roots was enhanced by 35–50% following wild-type GR12-2 inoculation over un-inoculated control which was due to IAA production (Patten and Glick 2002). In a recent experiment, Saini et al. (2015) isolated a total of 166 endophytic bacteria from roots and nodules of chickpea and roots of pea and lucerne and non-legumes like wheat and oat. Majority of the endophytes enhanced the root growth on agar plates in chickpea root growth promotion assay. However, endophytic bacteria isolated from chickpea nodule were found better root growth promoters relative to those recovered from roots. Furthermore, inoculation of endophytic bacteria in combination with *Mesorhizobium* enhanced growth, nodulation, and nitrogen-fixing parameters of chickpea. Dry matter accumulation and N contents in shoots of *Mesorhizobium*-inoculated chickpea plants were 665 and 1.54 mg plant^{-1}, respectively, compared to control plants (561 and 1.19 mg plant^{-1}, respectively). After inoculation with endophyte CNE 1036, both dry matter and N contents in shoots increased to 1532 and 8.15 mg plant^{-1}, respectively, which was followed by LRE3, isolated from the roots of lucerne. Of these, isolate CNE1036 was identified as *Bacillus subtilis*, while isolate LRE 3 was identified as *Bacillus amyloliquefaciens* by 16S rDNA analysis.

12.5.2 Nutrient Solubilization

12.5.2.1 Phosphorus Solubilization

Phosphorus (P) is another major nutrient required by plants. The solubilization of insoluble P and making it available to plants are yet other important traits of endophytes (Oteino et al. 2015). Several microbial species *Bacillus* and *Pseudomonas* have shown the potential to solubilize the insoluble P in soil (Banerjee et al. 2006; Tao et al. 2008; Grover et al. 2011). The endophytes release organic acids like 2-ketogluconic acid (Wagh et al. 2016), gluconic acid (Oteino et al. 2015; Wagh et al. 2016), etc. which lower the pH and ultimately solubilize the insoluble P (Halder and Chakrabarty 1993; Rodríguez et al. 2006; Naveed et al. 2014a). Numerous factors such as nutrients, soil composition, and environmental conditions affect P solubilizing capacity of endophytes (Nautiyal et al. 2000; Stephen and Jisha 2009).

12.5.2.2 Fe Uptake

Iron plays an important role in activation of several enzymes and is an essential micronutrient. Iron plays a vital role in photosynthesis, NO_2, SO_4 reduction, N_2 assimilation, and chlorophyll biosynthesis (Rashid 1996), while around 30% crops are iron deficient (Imsande 1998). In soil, total iron concentration is generally higher than those required by crops, but it is not present in available form. Under iron-deficient conditions, the leaves first become yellow followed by pale green and finally brown (Brittenham 1994). In soil solution, 10–18 M iron is present, which is not sufficient for sustaining plant and microbial growth. Several soil microbes including endophytes produce siderophores, a low molecular weight iron-loving compound (Prasad and Dagar 2014; Miliūtė and Buzaitė 2011) that holds Fe^{3+} and helps in Fe acquisition (Podile and Kishore 2006). Siderophores at pH less than six are less functional (Neilands and Nakamura 1991). Siderophores enhance crop growth by increasing iron availability (Abbamondi et al. 2016). Different kinds of siderophores such as ornibactin, ferrichrome, and desferrioxamine are produced by different species of endophytes (Verma et al. 2010). *Pseudomonas* sp. is the leading siderophore producer among PGPR which is used as inoculant in iron-deficient soils to fulfill the Fe requirements of crops (Sharma and Johri 2003).

12.5.3 Production of ACC-Deaminase

Ethylene (C_2H_4) is a potent phytohormone that can affect many stages of legume growth and development. Concentrations of ethylene as low as 6.25 ppm can evoke plant responses, while ethylene at 25 ppm is reported to decrease growth and development of plants (Zahir et al. 2009). To mitigate the effect of stress ethylene under stress conditions, PGPR produce 1-aminocyclopropane-1-carboxylate (ACC) deaminase (approximately 35–42 kDa) with pyridoxal 5-phosphate as an essential cofactor. Wide range of PGPR including species of *Pseudomonas* (Akhgar et al. 2014), *Bacillus* (Barnawal et al. 2013), *Azospirillum* (Li et al. 2005), *Serratia* (Carlos et al. 2016), *Arthrobacter* (Barnawal et al. 2014), *Enterobacter* (Li et al. 2016), *Pantoea* (Zhang et al. 2011), and *Burkholderia* (Jiang et al. 2008) produce ACC-deaminase (Saleem et al. 2007). Normally, microbes release IAA (Fig. 12.2) which activates ACC synthase which in turn converts SAM into ACC. ACC released into the rhizosphere is then converted to ammonia and α-ketobutyrate by ACC-deaminase secreted by PGPR (Fallik et al. 1994; Glick et al. 2007). ACC

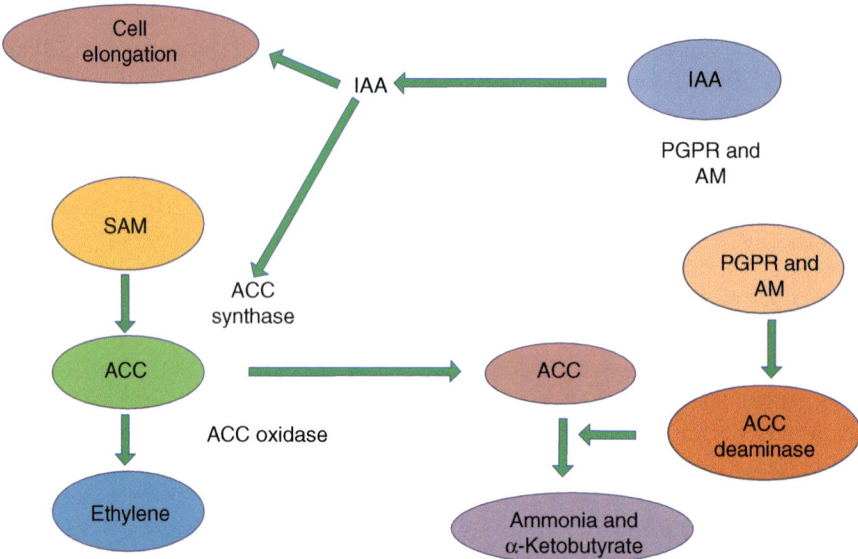

Fig. 12.2 Schematic model to explain how ACC deaminase secreted by PGPR lowers the ethylene concentration and thereby prevents ethylene inhibition of root elongation; *IAA* indole acetic acid, *ACC 1* aminocyclo-propane-1-carboxylic acid, *SAM* S-adenosyl methionine, *KB* α-ketobutyrate

deaminase-producing soil microbial populations stimulate plant growth by providing sole source of N. This mutual cooperation among plants and microbes plays important role in mitigation of stress-induced ethylene.

12.5.4 Production of Exopolysaccharides

Improving the soil structure is very important for healthy crop production because it influences physical, chemical, and biological composition of soils necessary for optimizing crop yields. In such a situation, the microbially produced exopolysaccharides (EPS) play an important role even under stressed conditions. The EPS consist of polysaccharide lipid (PL) complex and lipopolysaccharide protein (LP) complex. These compounds improve existence under abiotic stress (Konnova et al. 2001). Physiologically, the EPS secreted by microbes including endophytes (Orlandelli et al. 2016) protect the microbes against unfriendly environments. Also, EPS support microbes (Oades 1993) to colonize and permanently attach the roots. Furthermore, EPS lead to the development of microaggregates (<250 mm diameter) and macroaggregates (>250 mm diameter). EPS in soil are sorbed by clay surfaces by making cation bridges, van der Waals forces, hydrogen bonding, and anion adsorption. As a result, the soil aggregation and protection against weathering of aggregates are enhanced by making layer around soil aggregate (Tisdall and Oades 1982). Further investigation showed that EPS maintain high water potential in rhizosphere and in this way there could be more nutrient uptake by plants under abiotic stress conditions. And hence, due to enhanced water uptake, root and shoots are protected from desiccations, and consequently, the root and shoot growth is improved (Alami et al. 2000; Munns 2002).

12.5.5 Antioxidant Enzyme

Under abiotic and biotic stresses, the biochemical and physiological processes of plants are generally disturbed which ultimately result into the accumulation of reactive oxygen species (ROS). These reactive species are very toxic to living cells causing oxidative injury to DNA, phospholipids, and proteins. In addition, ROS perform as early signaling molecules for stress responses (Rodriguez et al. 2005; Gill and Tuteja 2010). Nevertheless, current studies have shown that reactive nitrogen species (RNS) and ROS are formed in both partners in many pathogenic and symbiotic systems (Tanaka et al. 2006; Jones et al. 2007; Gill and Tuteja 2010). However, ROS is not always damaging for the partners, but they also help in signaling the message between the two partners (Rouhier et al. 2008; Egamberdieva and Lugtenberg 2014). For example, during the formation of a mutualistic association among the grass *Lolium perenne* and *Epichloë festucae*, a fungal endophyte needs the production of hydrogen peroxide or superoxide through fungal NADPH oxidase, if it is not produced; it inactivates the gene expression which converts mutualistic interaction to antagonistic (Tanaka et al. 2006; White and Monica 2010). Low ROS concentrations are required for signaling and growth, while high concentration is detrimental to the cell and injures many macromolecules. Major molecules which produce ROS in cellular body are proteins (Rouhier et al. 2008; Rodrigues et al. 2008) and two main sites of ROS production including plastids and mitochondria (Navrot et al. 2007; White and Monica 2010). So, it is needed to maintain low level of ROS, while it is believed that antioxidant enzymes secreted by microbes play an important role in maintaining low level of ROS during symbiosis under abiotic stress tolerance (Rouhier et al. 2008; White and Monica 2010). The major antioxidants released by microbes include the low molecular weight compounds glutathione, tocopherol, and ascorbate and the enzymes catalases and superoxide dismutase (Rouhier et al. 2008; Zhang et al. 2014). These compounds and enzymes play a vital role in the removal of ROS indirectly through the formation of redox molecules in the cell, ascorbate, and glutathione and directly through superoxide dismutase, catalases, and ascorbate- or thiol-dependent peroxidases (Rouhier et al. 2008; Zhang et al. 2014). These compounds help plant-microbe symbiosis to withstand with abiotic stresses.

12.5.6 Biocontrol Activity

Plant diseases cause severe crop yield losses which are estimated to be more than 200 billion US dollar (Agrios 2005). Resistant plant varieties and chemicals are often used to control plant diseases. However, resistance does not exist against all diseases, and breeding of resistant plants takes many years (5–10 years). Likewise, acceptance of genetically engineered resistance is still a sensitive issue in some parts of the world. The use of agrochemicals (e.g., fungicides and insecticides) is not accepted by consumers and supermarket chains. To circumvent such problems, the adoption of biocontrol is advocated. Biocontrol is a mechanism wherein microorganisms are used to promote the growth of plants indirectly by inhibiting the growth of pathogens due to the secretion of secondary metabolites such as antibiotics (Raaijmakers et al. 2002), phenazines (Mavrodi et al. 2006), 2,4-diacetylphloroglucinol (Phl) (Dunne et al. 1998; Singh et al. 2011), pyoluteorin (Nowak-Thompson et al. 1999), pyrrolnitrin (Kirner et al. 1998), zwittermicin A (Emmert et al. 2004), kanosamine (Milner et al. 1996), and HCN (Chandra et al. 2007). Siderophore production in Fe stress conditions provides microorganism an additional advantage, resulting in the exclusion of pathogens due to Fe starvation (Arora et al. 2001). Generally, PGPB control phytopathogens via antagonism of the pathogen or by changing the host plant susceptibility. Plant growth-promoting endophytic bacteria can antagonize soil-borne pathogens through various mechanisms such as competition, antibiosis, and/or parasitism (Handelsman and Stabb 1996; Mehboob et al. 2009; Zhang et al. 2011; Mitter et al. 2013; Le Cocq et al. 2016).

12.5.7 Enhancement of Photosynthetic Activity

Photosynthesis is a process in which carbon dioxide (CO_2) is concerted into glucose by using the sunlight, but under stress condition, its activity is inhibited which leads to less crop production (Wahid and Rasul 2005). PGPB positively influence this vital reaction and maintain optimum photosynthesis under stress conditions. Shi et al. (2010) reported a significant increase in the photochemical efficiency and total chlorophyll content in leaves of sugar beet by inoculation with *Bacillus pumilus* and *Acinetobacter johnsonii*, respectively. Bacterization of grapevine with PsJN strain resulted in a 1.3 times higher CO_2 fixation rate and a 2.2 times higher O_2 evolution as compared to non-inoculated plants (Ait Barka et al. 2006). Xie et al. (2009) reported that *B. subtilis*-treated *Arabidopsis* plant enhances the volatile compounds because this is due to iron deficiency which occurs mainly under enhanced photosynthesis conditions. Likewise, Fernandez et al. (2012) assessed various photosynthesis parameters such as net photosynthesis, intercellular CO_2 concentration, stomatal conductance, activity of photosystem II, and total chlorophyll content in cold-stressed grapevine plantlets inoculated with *B. phytofirmans* PsJN as compared to non-bacterized controls. The authors clearly showed that the increase in plant photosynthetic activity was not due to a modulation of stomatal conductance in grapevine colonized by strain PsJN. Thus, the mechanism underlying the stimulation of plant photosynthesis by *B. phytofirmans* PsJN and other endophytic bacteria, however, remains elusive.

12.6   Single and Composite Effects of Endophytes on Legumes

There are different types of relationships (neutral, beneficial, and/or detrimental) that occur between host plant and endophytes. When applied as microbial inoculant, the sole application of endophyte has been found to promote plant growth via different mechanisms such as hormone production, P solubilization, siderophores, and production of organic acids (Zakhia et al. 2006; Li et al. 2008; Ibanez et al. 2009; Trujillo et al. 2010; Khalifa et al. 2016). However, the direct role of endophytes in nodule formation is not clear. Even though the co-inoculation of endophyte with other rhizosphere microbes has been found to enhance the growth and yield of legumes. For example, the co-inoculation of *Bradyrhizobium* sp. and endophyte *Enterobacter* sp. significantly enhanced nodules production in peanut. Meanwhile, it was found that the *Enterobacter* sp. produced IAA which promoted root growth and division. It is also speculated that legume co-inoculation with endophyte and rhizobia had more sites of specificity (Ibanez et al. 2009; Zgadzaj et al. 2015). Subramanian et al. (2015) in other experiment compared the effect of endophytic plant growth-promoting bacteria *B. megaterium* LNL6 and *Methylobacterium oryzae* CBMB20 on nodulation and effective rhizobial symbiosis (*B. japonicum* MN110) in soybean. The results revealed that co-inoculation exhibited an increase in nodule number, plant biomass, and N content compared to single inoculation of MN110. Similarly, co-inoculation of *Agrobacterium* sp. and *Sinorhizobium meliloti* favored the establishment of nodules in different legumes (Liu et al. 2010; Zhao et al. 2013). Annapurna et al. (2013) in yet other investigation observed stimulatory effect of *P. polymyxa* HKA-15 strain when used alone and in combination with *B. japonicum* DS-1 on soybean growth. They further observed that co-inoculation was more effective and increased the shoot by 7.2% and root dry weight by 14.5% when compared with *B. japonicum* inoculation alone. Endophyte microbe *Neotyphodium occultans*, *Bacillus megaterium*, *Bordetella avium*, and *Curtobacterium luteum* inoculated single or in combination with rhizobia improved the growth and nodulation of legumes (Eisenhauer 2012; Pablo et al. 2015). The isolated endophytic strain *Agrobacterium tumefaciens* from legume tissues showed improved growth when co-inoculated with *Rhizobium leguminosarum* under nitrogen-free environment. In another study, Stajković et al. (2009) described that single inoculation of endophyte enhanced growth of legumes under sterilized conditions, while the magnitude of growth promotion was more pronounced by co-inoculation of rhizobia and endophytic bacteria. Endophytes directly or indirectly interact with rhizobia by which they affect the growth and development of crops and nitrogen-fixing microbe's activity. Legumes have the ability to host both rhizobia and endophytes. Zgadzaj et al. (2015) found that infection of nodules with endophytes was established with the help of rhizobia toward nodule primordia where colonization occurred. They further suggested that endophyte-mediated nodule infection depends on strain type of rhizobia and endophyte, nod factors, EPS, and host genotypes.

12.7 Recent Advances in Endophyte and Legume Interaction Studies

There are different methods for determining plant-microbe interactions, and among these, autofluorescent protein (AFP) method is well known for biofilm formation (Larrainzar et al. 2005). These techniques are used for microbial detection and enumeration in situ on plant surface and in plant (Gage et al. 1996; Tombolini et al. 1997; Tombolini and Jansson 1998). For encoding of green fluorescent protein (GFP), AFP strategies are used as a marker. This technique has shown some promising results for detection of *Pseudomonas* monitoring in root tissues (Tombolini et al. 1997; Tombolini and Jansson 1998). GFP is a good AFP biomarker because it does not require any substrate or cofactor in order to fluoresce. GFP cassettes have been developed by scientists for integration of chromosome and GFP expression in a variety of bacteria (Tombolini et al. 1997; Tombolini and Jansson 1998; Xi et al. 1999). Bacterial cells with chromosomal integration of GFP can be identified by epi-fluorescence microscopy or confocal laser scanning microscopy (Villacieros et al. 2003). Germaine et al. (2004) studied GFP-labeled endophytes to check their colonization ability using different inoculation methods. A simple method of "stick dipping" showed efficient results. This method leads to identify colonization of the bacteria on specific plant tissues at levels of 10^2 to 10^4 CFU g^1 tissues depending on the strain, while pea plant inoculation was more efficient with imbibing method (Germaine et al. 2006; Germaine 2007). Annapurna et al. (2013) examined the colonization and spread of gfp-tagged *Paenibacillus polymyxa* alone and in combination with *Bradyrhizobium japonicum* in roots and root nodules of soybean plants grown under gnotobiotic conditions. They observed enhanced plant growth by co-inoculation, and in vivo visualization using confocal laser scanning microscopy (CLSM) showed the localization of the gfp-tagged *P. polymyxa* cells in the root nodules and its spread in the root tissue, both tap and lateral roots. Another method for visualization of colonies is the use of the β-glucuronidase (GUS) reporter system. James et al. (2002) inoculated a GUS-marked strain of *Herbaspirillum seropedicae* Z67 to field crop. GUS staining was most intense on coleoptiles, lateral roots, and also at some of the junctions of the main and lateral roots. They further showed that entry of endophytic bacteria in the roots occurred through cracks present at the point of lateral root emergence. Endophytic strain, *H. seropedicae*, colonized the root intercellular spaces, parenchyma, and cortical cells and enters the vascular tissue through stele. These bacteria also colonized the xylem vessel in leaves and stems. It is noteworthy that for successful colonization of endophytic bacteria a compatible host plant is required. Miche et al. (2006) observed the nitrogen-fixing strain BH72, which expresses nitrogenase (*nif*) genes inside plant roots. The proteomic analysis has revealed that plant defense system also restricts the colonization of endophytes in plant. Now, molecular based studies of endophyte and plant interaction in rhizosphere and endosphere were reported by using the in vivo expression technology (IVET) and recombination in vivo expression technology (Zhang et al. 2006) which provides the information about how bacteria enter and colonize the plant, suppress pathogens, and survive within the plant.

Conclusion

The endophytes have the ability to promote legume growth and yield due to their specific beneficial traits. These specific traits allow legumes to grow under abiotic stress conditions and also enhance their ability to promote plant growth under biotic stresses. The application of specific endophytic strains under a particular stressful environment could be effective for obtaining maximum benefits from microbial inoculation. It was also observed that in some cases dual inoculation enhanced the yield but this field is less explored. From the above discussion, it is also observed that microbes can improve plant growth by a number of mechanisms including nutrient fixation, solubilization, acquisition, and their uptake by the plants. The repertoire of their effects and functions in plant has not been comprehensively defined. The challenge and goal are to manage microbial communities to favor plant colonization by beneficial bacteria.

References

Abbamondi GR, Tommonaro G, Weyens N, Thijs S, Sillen W, Gkorezis P, Iodice C, Rangel WDM, Nicolaus B, Vangronsveld J (2016) Plant growth-promoting effects of rhizospheric and endophytic bacteria associated with different tomato cultivars and new tomato hybrids. Chem Biol Technol Agric 3:1–10

Adesemoye AO, Torbert HA, Kloepper JW (2009) Plant growth promoting rhizobacteria allow reduced application rates of chemical fertilizers. Microbial Ecol 58:921–929

Agrios GN (2005) Plant pathology, 4th edn. Academic, Amsterdam

Ait Barka E, Nowak J, Clement C (2006) Enhancement of chilling resistance of inoculated grapevine plantlets with a plant growth promoting rhizobacterium, *Burkholderia phytofirmans* PsJN. Appl Environ Microbiol 72:7246–7252

Akhgar M, Arzanlou R, Bakker PAHM, Hamidpour M (2014) Characterization of 1-Aminocyclopropane-1-carboxylate (ACC) deaminase-containing *Pseudomonas* spp. in the rhizosphere of salt-stressed canola. Pedosphere 24:461–468

Alami Y, Achouak W, Marol C, Heulin T (2000) Rhizosphere soil aggregation and plant growth promotion of sunflowers by exopolysaccharide producing *Rhizobium* sp. strain isolated from sunflower roots. Appl Environ Microbiol 66:3393–3398

Annapurna K, Ramadoss D, Bose P, Kumar LV (2013) In situ localization of *Paenibacillus polymyxa* HKA-15 in roots and root nodules of soybean (*Glycine max* L.) Plant Soil 373: 641–648

Arora NK, Kang SC, Maheshwari DK (2001) Isolation of siderophore-producing strains of *Rhizobium meliloti* and their biocontrol potential against *Macrophomina phaseolina* that causes charcoal rot of groundnut. Curr Sci 8:673–677

Banerjee MR, Yesmin L, Vessey JK (2006) Plant growth promoting rhizobacteria as biofertilizers and biopesticides. In: Rai MK (ed) Handbook of microbial biofertilizers. Haworth Press, New York

Barnawal D, Maji D, Bharti N, Chanotiya CS, Kalra A (2013) ACC deaminase-containing *Bacillus subtilis* reduces stress ethylene-induced damage and improves mycorrhizal colonization and rhizobial nodulation in *Trigonella foenum-graecum* under drought stress. J Plant Growth Regul 32:809–822

Barnawal D, Bharti N, Maji D, Chanotiya CS, Kalra A (2014) ACC deaminase-containing *Arthrobacter protophormiae* induces NaCl stress tolerance through reduced ACC oxidase activity and ethylene production resulting in improved nodulation and mycorrhization in *Pisum sativum*. J Plant Physiol 171:884–894

Bell CR, Dickie GA, Harvey WLG, Chan JWYF (1995) Endophytic bacteria in grapevine. Can J Microbiol 41:46–53

Berg G (2009) Plant–microbe interactions promoting plant growth and health: perspectives for controlled use of microorganisms in agriculture. Appl Microbiol Biotechnol 84:11–18

Braun EJ (1990) Colonization of resistant and susceptible maize plants by *Envinia stewartii* strains differing in exopolysaccharide production. Physiol Mol Plant Pathol 36:363–379

Brittenham GM (1994) New advances in iron metabolism, iron deficiency and iron overload. Curr Opin Hematol 1:549–556

Carlos MH, Stefani PV, Janette AM, Melani MS, Gabriela PO (2016) Assessing the effects of heavy metals in ACC deaminase and IAA production on plant growth-promoting bacteria. Microbiol Res 189:53–61

Celloto VR, Oliveira AJB, Gonçalves JE, Watanabe CSF, Matioli G, Gonçalves RAC (2012) Biosynthesis of indole-3-acetic acid by new *Klebsiella oxytoca* free and immobilized cells on inorganic matrices. Sci World J 2012:495970. doi:10.1100/2012/495970

Chandra S, Choure K, Dubey RC, Maheshwari DK (2007) Rhizosphere competent *Mesorhizobium loti* MP6 induced root hair curling, inhibits *Sclerotinia sclerotiorum* and enhance growth of Indian mustard (*Brassica compestris*). Braz J Microbiol 38:24–30

Chen W, Sun L, Lu J, Bi L, Wang L, Wei G (2015) Diverse nodule bacteria were associated with *Astragalus* species in arid region of northwestern China. J Basic Microbiol 55:121–128

Chimwamurombe PM, Grönemeyer JL, Reinhold-Hurek B (2016) Isolation and characterization of culturable seed-associated bacterial endophytes from gnotobiotically grown Marama bean seedlings. FEMS Microbiol Ecol 92(6):fiw083. doi:10.1093/femsec/fiw083

Chiwocha SD, Abrams SR, Ambrose SJ, Cutler AJ, Loewen M, Ross AR, Kermode AR (2003) A method for profiling classes of plant hormones and their metabolites using liquid chromatography-electrospray ionization tandem mass spectrometry: an analysis of hormone regulation of thermodormancy of lettuce (*Lactuca sativa* L.) seeds. Plant J 35:405–417

Cohen AC, Travaglia CN, Bottini R, Piccoli PN (2009) Participation of abscisic acid and gibberellins produced by endophytic *Azospirillum* in the alleviation of drought effects in maize. Botany 87:455–462

De Meyer SE, Beuf KD, Vekeman B, Willems A (2015) A large diversity of non-rhizobial endophytes found in legume root nodules in Flanders (Belgium). Soil Biol Biochem 83:1–11

Dudeja SS, Narula N (2008) Molecular diversity of root nodule forming bacteria. In: Khachatourians GG, Arora DK, Rajendran TP, Srivastava AK (eds) Agriculturally important microorganisms. Academic World International, Bhopal, pp 1–24

Dudeja SS, Giri R, Saini R, Suneja-Madan P, Kothe E (2012) Interaction of endophytic microbes with legumes. J Basic Microbiol 52:248–260

Dunne C, Moenne-Loccoz Y, McCarthy J, Higgins P, Powell J, Dowling DN, Gara F (1998) Combining proteolytic and phloroglucinol-producing bacteria for improved control of *Pythium*-mediated damping off of sugar beet. Plant Pathol 47:299–307

Egamberdieva D, Lugtenberg B (2014) Use of plant growth promoting rhizobacteria to alleviate salinity stress in plants. In: Miransari M (ed) Use of microbes for the alleviation of soil stresses, vol 1. Springer, New York, pp 73–96

Eisenhauer N (2012) Aboveground–belowground interactions as a source of complementarity effects in biodiversity experiments. Plant Soil 35:1–22

Emmert EA, Klimowicz AK, Thomas MG, Handelsman J (2004) Genetics of zwittermicin a production by *Bacillus cereus*. Appl Environ Microbiol 70:104–113

Fallik E, Sarig S, Okon Y (1994) Morphology and physiology of plant roots associated with *Azospirillum*. In: Okon Y (ed) *Azospirillum*/plant associations. CRC, London, pp 77–86

Fernandez O, Theocharis A, Bordiec S, Feil R, Jasquens L, Clement C, Fontaine F, Ait Barka E (2012) *Burkholderia phytofirmans* PsJN acclimates grapevine to cold by modulating carbohydrate metabolism. Mol Plant Microbe Interact 25:496–504

Gage DJ, Bobo T, Long SR (1996) Use of green fluorescent protein to visualize the early events of symbiosis between *Rhizobium meliloti* and alfafa (*Medicago satia*). J Bacteriol 178:7159–7166

Gao K, Mendgen K (2006) Seed-transmitted beneficial endophytic *Stagonospora* sp. can penetrate the walls of the root epidermis, but does not proliferate in the cortex, of *Phragmites australis*. Can J Bot 84:981–988

Germaine K (2007) Construction of endophytic xenobiotic degrader bacteria for improving the phytoremediation of organic pollutants. PhD thesis, Institute of Technology Carlow, Ireland

Germaine K, Keogh E, Garcia-Cabellos G, Borremans B, van Der Lelie D, Barac T, Oeyen L, Vangronsveld J, Moore FP, Moore ERB, Campbell CD, Ryan D, Dowling DN (2004) Colonisation of poplar trees by *gfp* expressing bacterial endophytes. FEMS Microbiol Ecol 48:109–118

Germaine K, Liu X, Cabellos G, Hogan J, Ryan D, Dowling DN (2006) Bacterial endophyte-enhanced phyto-remediation of the organochlorine herbicide 2,4-dichlorophenoxyacetic acid. FEMS Microbiol Ecol 57:302–310

Ghosh PK, Sen SK, Maiti TK (2015) Production and metabolism of IAA by *Enterobacter* spp. (*Gammaproteobacteria*) isolated from root nodules of a legume *Abrus precatorius* L. Biocatal Agric Biotechnol 3:296–303

Gill SS, Tuteja N (2010) Reactive oxygen species and antioxidant machinery in abiotic stress tolerance in crop plants. Plant Physiol Biochem 48:909–930

Glick BR (2012) Plant growth-promoting bacteria: mechanisms and applications. Scientifica 2012:963401. doi:10.6064/2012/963401

Glick BR, TodorovicB CJ, Cheng Z, Duan J, McConkey B (2007) Promotion of plant growth by bacterial ACC deaminase. Crit Rev Plant Sci 26:227–242

Grover WH, Bryan AK, diez-Silva M, Suresh S, Higgins JM, Manalis SR (2011) Measuring single-cell density. Proc Natl Acad Sci U S A 108:10992–10996

Halder AK, Chakrabarty PK (1993) Solubilization of inorganic phosphate by *Rhizobium*. Folia Microbiol 38:325–330

Hallmann J, Quadt-Hallmann A, Mahaffee WF, Kloepper JW (1997) Bacterial endophytes in agricultural crops. Can J Microbiol 43:895–914

Halo BA, Khan AL, Waqas M, Al-Harrasi A, Hussain J, Ali L, Adnan M, In-Jung L (2015) Endophytic bacteria (*Sphingomonas* sp. LK11) and gibberellin can improve *Solanum lycopersicum* growth and oxidative stress under salinity. J Plant Interact 10:117–125

Handelsman J, Stabb EV (1996) Biocontrol of soilborne plant pathogens. Plant Cell 8:1855–1869

Hoque MS, Broadhurst LM, Thrall PH (2011) Genetic characterisation of root nodule bacteria associated with *Acacia salicina* and *Acacia stenophylla* (*Mimosaceae*) across south eastern Australia. Int J Syst Evol Microbiol 61:299–309

Hurek T, Reinhold-Hurek B, Van Montagu M, Kellenberge E (1994) Root colonization and systemic spreading of *Azoarcus* sp. strain BH72 in grasses. J Bacteriol 176:1913–1923

Ibanez F, Angelini J, Taurian T, Tonelli ML, Fabra A (2009) Endophytic occupation of peanut nodules by opportunistic gamma proteobacteria. Syst Appl Microbiol 32:49–55

Imsande J (1998) Nitrogen de fi cit during soybean pod fill and increased plant biomass by vigorous N_2 fixation. Eur J Agron 8:1–11

Jacobs MJ, Bugbee WM, Gabrielson DA (1985) Enumeration, location, and characterization of endophytic bacteria within sugar beet roots. Can J Bot 63:1262–1265

James EK, Gyaneshwar P, Mathan N, Barraquio WL, Reddy PM, Iannetta PP, Olivares FL, Ladha JK (2002) Infection and colonization of rice seedlings by the plant growth-promoting bacterium *Herbaspirillum seropedicae* Z67. Mol Plant Microbe Interact 15:894–906

Jasim B, Joseph AA, Jimtha John C, Mathew J, Radhakrishnan EK (2014) Isolation and characterization of plant growth promoting endophytic bacteria from the rhizome of *Zingiber officinale*. 3 Biotech 4:197–204

Jiang C, Sheng X, Qian M, Wang Q (2008) Isolation and characterization of a heavy metal-resistant *Burkholderia* sp. from heavy metal-contaminated paddy field soil and its potential in promoting plant growth and heavy metal accumulation in metal-polluted soil. Chemosphere 72:157–164

Jones MPA, Cao J, O'Brien R, Murch SJ, Saxena PK (2007) The mode of action of thidiazuron: auxins, indoleamines, and ion channels in the regeneration of *Echinacea purpurea* L. Plant Cell 26:1481–1490

Kan FL, Chen ZY, Wang ET, Tian CF, Sui XH, Chen WX (2007) Characterization of symbiotic and endophytic bacteria isolated from root nodules of herbaceous legumes grown in Qinghai-Tibet plateau and in other zones of China. Arch Microbiol 188:103–115

Khalid A, Arshad M, Zahir ZA (2006) Phytohormones: microbial production and applications. In: Uphoff N, Ball AS, Fernandes E, Herren H, Husson O, Laing M, Palm C, Pretty J, Sanchez P, Sanginga N, Thies J (eds) Biological approaches to sustainable soil systems. Taylor & Francis/CRC, Boca Raton, FL, pp 207–220

Khalifa AYZ, Alsyeeh A, Almalki MA, Saleh FA (2016) Characterization of the plant growth promoting bacterium, *Enterobacter cloacae* MSR1, isolated from roots of non-nodulating *Medicago sativa*. Saudi J Biol Sci 23:79–86

Khan AL, Hamayun M, Kang SM, Kim YH, Jung HY, Lee JH, Lee IJ (2012) Endophytic fungal association via gibberellins and indole acetic acid can improve plant growth under abiotic stress: an example of *Paecilomyces formosus* LHL10. BMC Microbiol 12:3. doi:10.1186/1471-2180-12-3

Khan AL, Waqas M, Asaf S, Kamran M, Shahzad R, Bilal S, Khan MA, Sang-Mo K, Yoon-Ha K, Byung-Wook Y, Al-Rawahi A, Al-Harrasi A, In-Jung L (2017) Plant growth-promoting endophyte *Sphingomonas* sp. LK11 alleviates salinity stress in *Solanum pimpinellifolium*. Environ Exp Bot 133:58–69

Kirner S, Hammer PE, Hill DS, Altmann A, Fischer I, Weislo LJ, Lanahan M, van Pée KH, Ligon JM (1998) Functions encoded by pyrrolnitrin biosynthetic genes from *Pseudomonas fluorescens*. J Bacteriol 180:1939–1943

Kloepper JW, Wei G, Tuzun S (1992) Rhizosphere population dynamics and internal colonization of cucumber by plant growth-promoting rhizobacteria which induce systemic resistance to *Colletotrichurn orbiculare*. In: Tjamos ES (ed) Biological control of plant diseases. Plenum, New York, pp 185–191

Kobayashi DY, Palumbo JD (2000) Bacterial endophytes and their effects on plants and uses in agriculture. In: Bacon CW, White JF (eds) Microbial endophytes. Dekker, New York, pp 199–236

Konnova SA, Brykova OS, Sachkova OA, Egorenkova IV, Ignatov VV (2001) Protective role of the polysaccharide containing capsular components of *Azospirillum brasilense*. Microbiology 70:436–440

Kravchenko LV, Azarova TS, Makarova NM, Tikhonovich IA (2004) The effect of tryptophan present in plant root exudates on the phytostimulating activity of rhizobacteria. Microbiology 73:156–158

Kumar V, Pathak DV, Dudeja SS, Saini R, Giri R, Narula S, Anand RC (2013) Legume nodule endophytes more diverse than endophytes from roots of legumes or non-legumes in soils of Haryana, India. J Microbiol Biotech Res 3:83–92

Kusari P, Kusari S, Spiteller M, Kayser O (2015) Implications of endophyte-plant crosstalk in light of quorum responses for plant biotechnology. Appl Microbiol Biotechnol 99:5383–5390

Larrainzar E, Ogara F, Morrissey JP (2005) Application of autofluorescent proteins for insitu studies in microbial ecology 59:257–277

Le Cocq K, Gurr SJ, Hirsch PR, Mauchline TH (2016) Exploitation of endophytes for sustainable agricultural intensification. Mol Plant Pathol 18:469–473. doi:10.1111/mpp.12483

Lei X, Wang ET, Chen WF, Sui XH, Chen WX (2008) Diverse bacteria isolated from root nodules of wild *Vicia* species grown in temperate region of China. Arch Microbiol 190:657–671

Li Q, Saleh-Lakha S, Glick BR (2005) The effect of native and ACC deaminase containing *Azospirillum brasilense* Cd1843 on the rooting of carnation cuttings. Can J Microbiol 51:511–514

Li JH, Wang ET, Chen WF, Chen WX (2008) Genetic diversity and potential for promotion of plant growth detected in nodule endophytic bacteria of soybean grown in Heilongjiang province of China. Soil Biol Biochem 40:238–246

Li L, Nagai K, Yin F (2016) Progress in cold roll bonding of metals. Sci Technol Adv Mater 9:023001(11pp). doi:10.1088/1468-6996/9/2/023001

Liu J, Wang ET, da Ren W, Chen WX (2010) Mixture of endophytic *Agrobacterium* and *Sinorhizobium meliloti* strains could induce nonspecific nodulation on some woody legumes. Arch Microbiol 192:229–234

Long HH, Schmid DD, Baldwin IT (2008) Native bacterial endophytes promote host growth in a species specific manner, Phytohormone manipulations do not result in common growth responses. PLoS One 3:2702–2708

Mahaffee WF, Kloepper JW, Van Vuurde JWL, Van der Wolf JM, Van den Brink M (1997) Endophytic colonization of Phaseolus vulgaris by Pseudomonas fluorescens strain 89B-27 and Enterobacter asburiae strain JM22. In: Ryder MH, Stephens PM, Bowen GD (eds) Improving plant productivity in Rhizosphere bacteria. CSIRO, Melbourne, p 180

Maheswari TU, Anbukkarasi K, Hemalatha T, Chendrayan K (2013) Studies on phytohormone producing ability of indigenous endophytic bacteria isolated from tropical legume crops. Int J Curr Microbiol Appl Sci 2:127–136

Maougal RT, Bargaz A, Sahel C, Amenc L, Djekoun A, Plassard C, Drevon J (2014) Localization of the *Bacillus subtilis* beta-propeller phytase transcripts in nodulated roots of *Phaseolus vulgaris* supplied with phytate. Planta 239:901–908

Martinez-Hidalgo P, Galindo-Villardon P, Trujillo ME, Igual JM, Martýnez-Molina E (2014) *Micromonospora* from nitrogen fixing nodules of alfalfa (*Medicago sativa* L.). A new promising plant probiotic bacteria. Sci Rep 4:6389

Martinez-Viveros O, Jorquera MA, Crowley DE, Gajardo G, Mora ML (2010) Mechanisms and practical considerations involved in plant growth promotion by rhizobacteria. J Soil Sci Plant Nutr 10:293–319

Mavrodi DV, Blankenfeldt W, Thomashow LS (2006) Phenazine compounds in fluorescent *Pseudomonas* spp. biosynthesis and regulation. Annu Rev Phytopathol 44:417–445

Mehboob F, Junca H, Schraa G, Stams AJM (2009) Growth of *Pseudomonas chloritidismutans* AW-1T on n-alkanes with chlorate as electron acceptor. Appl Microbiol Biotechnol 83:739–747

Miche L, Battistoni F, Gemmer S, Belghazi M, Reinhold-Hurek B (2006) Up regulation of jasmonate-inducible defense proteins and differential colonization of roots of *Oryza sativa* cultivars with the endophyte *Azoarcus* sp. Mol Plant Microbe Interact 19:502–511

Michiels KW, Croes CL, Vanderleyden J (1991) Two different modes of attachment of *Azospirillum brasilense* sp7 to wheat roots. J Gen Microbiol 137:2241–2246

Miliūtė I, Buzaitė O (2011) IAA production and other plant growth promoting traits of endophytic bacteria from apple tree. Biologija 57:98–102

Milner J, Silo-Suh L, Lee JC, He H, Clardy J, Handelsman J (1996) Production of kanosamine by *Bacillus cereus* UW85. Appl Environ Microbiol 62:3061–3065

Mitter B, Brader G, Afzal M, Compant S, Naveed M, Trognitz F, Sessitsch A (2013) Advances in elucidating beneficial interactions between plants, soil and bacteria. Adv Agron 121:381–445

Moghaddam MJM, Emtiazi G, Salehi Z (2012) Enhanced auxin production by *Azospirillum* pure cultures from plant root exudates. J Agric Sci Technol 14:985–994

Munns R (2002) Comparative physiology of salt and water stress. Plant Cell Environ 25:239–250

Muresu R, Polone E, Sulas L, Baldan B, Tondello A, Delogu G, Cappuccinelli P, Alberghini S, Benhizia Y, Benhizia H, Benguedouar A, Mori B, Calamassi R, Dazzo FB, Squartini A (2008) Coexistence of predominantly nonculturable rhizobia with diverse, endophytic bacterial taxa within nodules of wild legumes. FEMS Microbiol Ecol 63:383–400

Muthukumar A, Bhaskaran R, Kumar SK (2010) Efficacy of endophytic *Pseudomonas fluorescens* (Trevisan) migula against chilli damping-off. J Biopest 3(105):109

Nautiyal CS, Bhadauria S, Kumar P, Lal H, Mondal R, Verma D (2000) Stress induced phosphate solubilization in bacteria isolated from alkaline soils. FEMS Microbiol Lett 182:291–296

Naveed M (2013) Maize endophytes–diversity, functionality and application potential. Ph.D. thesis, AIT–Austrian Institute of Technology/BOKU University, Vienna

Naveed M, Mitter B, Yousaf S, Pastar M, Afzal M, Sessitsch A (2014a) The endophyte *Enterobacter* sp. FD17: a maize growth enhancer selected based on rigorous testing of plant beneficial traits and colonization characteristics. Biol Fertil Soils 50:249–262

Naveed M, Mitter B, Reichenauer TG, Krzysztof W, Sessitsch A (2014b) Increased drought stress resilience of maize through endophytic colonization by *Burkholderia phytofirmans* PsJN and *Enterobacter* sp. FD17. Environ Exp Bot 97:30–39

Naveed M, Hussain MB, Zahir ZA, Mitter B, Sessitsch A (2014c) Drought stress amelioration in wheat through inoculation with *Burkholderia phytofirmans* strain PsJN. Plant Growth Regul 73:121–131

Navrot N, Rouhier N, Gelhaye E, Jacquot J (2007) Reactive oxygen species generation and anti-oxidant systems in plant mitochondria. Physiol Plantarum 129:185–195

Neilands JB, Nakamura K (1991) Detection, determination, isolation, characterization and regulation of microbial iron chelates. In: Winkelmann G (ed) Handbook of microbial iron chelates. CRC, London, pp 1–14

Nowak-Thompson B, Chaney N, Wing JS, Gould SJ, Loper JE (1999) Characterization of the pyoluteorin biosynthetic gene cluster of *Pseudomonas fluorescens* Pf-5. J Bacteriol 181:2166–2174

Oades JM (1993) The role of biology in the formation, stabilization and degradation of soil structure. Geoderma 56:182–186

Orlandelli RC, Vasconcelos AFD, Azevedo JL, Silva MLC, Pamphile JA (2016) Screening of endophytic sources of exopolysaccharides: preliminary characterization of crude exopolysaccharide produced by submerged culture of *Diaporthe* sp. JF766998 under different cultivation time. Biochimie Open 2:33–40

Oteino N, Lally RD, Kiwanuka S, Lloyd A, Ryan D, Germaine KJ, Dowling DN (2015) Plant growth promotion induced by phosphate solubilising endophytic *Pseudomonas* isolates. Front Microbiol 6:745

Pablo A, Parisi G, Lattanzi FA, Grimoldi AA, Omacini M (2015) Multi-symbiotic systems: functional implications of the coexistence of grass–endophyte and legume–rhizobia symbioses. Oikos 124:553–560

Palaniappan P, Chauhan PS, Saravanan VS, Anandham R, Sa T (2010) Isolation and characterization of plant growth promoting endophytic bacterial isolates from root nodule of Lespedeza sp. Biol Fertil Soils 46:807–816

Pandya M, Kumar GN, Rajkumar S (2013) Invasion of rhizobial infection thread by non-rhizobia for colonization of *Vigna radiata* root nodules. FEMS Microbiol Lett 348:58–65

Patten CL, Glick BR (2002) Regulation of indoleacetic acid production in *Pseudomonas putida* GR12-2 by tryptophan and the stationary-phase sigma factor RpoS. Can J Microbiol 48: 635–642

Peterson CA, Emanuel ME, Humphreys GB (1981) Pathway of movement of apoplastic fluorescent dye tracers through the endodermis at the site of secondary root formation in corn *(Zea mays)* and broad bean *(Vicia faba)*. Can J Bot 59:618–625

Podile AR, Kishore GK (2006) Plant growth promoting rhizobacteria. In: Gnanamanickam SS (ed) Plant associated bacteria. Springer, Amsterdam, pp 195–230

Prasad MP, Dagar S (2014) Identification and characterization of endophytic bacteria from fruits like Avacado and Black grapes. Int J Curr Microbiol Appl Sci 3:937–947

Raaijmakers JM, Vlami M, de Souza JT (2002) Antibiotic production by bacterial biocontrol agents. Antonie Van Leeuwenhoek 81:537–547

Rashid A (1996) Secondary and micronutrients. In: Saghir E, Bantel R (eds) Soil science, pp 341–379

Rivas R, García-Fraile P, Velazquez E (2009) Taxonomy of bacteria nodulating legumes. Microbiol Insights 2:51–69

Rivera-Cruz MC, Trujillo-Narcía A, Córdova-Ballona G, Kohler J, Caravaca F, Roldán A (2008) Poultry manure and banana wastes are effective biofertilizer carriers for promoting plant growth and soil sustainability in banana crops. Soil Biol Biochem 40:3092–3095

Rodrigues ML, Nimrichter L, Oliveira DL, Nosanchuk JD, Casadevall A (2008) Vesicular transcell wall transport in fungi: a mechanism for the delivery of virulence-associated macromolecules. Lipid Insights 2:27–40

Rodriguez GLJ, Valle R, Duran A, Roncero C (2005) Cell integrity signaling activation in response to hyperosmotic shock in yeast. FEBS Lett 579:6186–6190

Rodríguez JP, Beard TD, Bennett EM, Cumming GS, Cork S, Agard J, Dobson AP, Peterson GD (2006) Trade-offs across space, time and ecosystem services. Ecol Soc 11:28. http://www.ecologyandsociety.org/vol11/iss1/art28/

Rouhier N, Lemaire SD, Jacquot JP (2008) The role of glutathione in photosynthetic organisms: emerging functions for glutaredoxins and glutathionylation. Ann Rev Plant Biol 59:143–166

Saini R, Kumar V, Dudeja SS, Pathak DV (2015) Beneficial effects of inoculation of endophytic bacterial isolates from roots and nodules in chickpea. Int J Curr Microbiol Appl Sci 4:207–221

Saleem M, Arshad M, Hussain S, Bhatti AS (2007) Perspective of plant growth promoting rhizobacteria (PGPR) containing ACC deaminase in stress agriculture. J Ind Microbiol Biotechnol 34:635–648

Santner A, Calderon-Villalobos LIA, Estelle M (2009) Plant hormones are versatile chemical regulators of plant growth. Nat Chem Biol 5:301–307

Schank SC, Smith RL, Weiser GC, Zuberere DA, Bouton JH, Quesenberry KH, Tyler ME, Milam JR, Littell RC (1979) Fluorescent antibody technique to identify *Azospirillum brasilense* associated with roots of grasses. Soil Biol Biochem 11:287–295

Shahzad R, Waqas R, Khan AL, Asaf S, Khan MA, Sang-Mo K, Byung-Wook Y, In-Jung L (2016) Seed-borne endophytic *Bacillus amyloliquefaciens* RWL-1 produces gibberellins and regulates endogenous phytohormones of *Oryza sativa*. Plant Physiol Biochem 106:236–243

Shahzad R, Khan AL, Bilal S, Waqas M, Sang-Mo K, In-Jung L (2017) Inoculation of abscisic acid-producing endophytic bacteria enhances salinity stress tolerance in *Oryza sativa*. Environ Exp Bot 136:68–77. doi:10.1016/j.envexpbot.2017.01.010

Sharma A, Johri BN (2003) Growth promoting influence of siderophore producing *Pseudomonas* strains GRP3A and PRS9 in maize (*Zea mays* L.) under iron limiting conditions. Microbiol Res 158:243–248

Shi Y, Lou K, Li C (2010) Growth and photosynthetic efficiency promotion of sugar beet (*Beta vulgaris* L.) by endophytic bacteria. Photosynthesis Res 105:5–13

Silva JM, dos Santos TMC, de Albuquerque LS, Montaldo YC, de Oliveira JUL, da Silva SGM, Nascimento MS, Teixeira Rd RO (2015) Potential of the endophytic bacteria (*Herbaspirillum* spp. and *Bacillus* spp.) to promote sugarcane growth. Aust J Crop Sci 9:754–760

Singh SK, Strobel GA, Knighton B, Geary B, Sears J, Ezra D (2011) An endophytic *Phomopsis* sp. possessing bioactivity and fuel potential with its volatile organic compounds. Microbial Ecol 61:729–739

Spaepen S, Vanderleyden J (2011) Auxin and plant-microbe interactions. Cold Spring Harb Perspect Biol 3:a001438

Stajković O, Meyer SD, Miličić B, Willems A, Delić D (2009) Isolation and characterization of endophytic non-rhizobial bacteria from root nodules of alfalfa (*Medicago sativa* L.) Bot Serb 33:107–114

Stephen J, Jisha MS (2009) Buffering reduces phosphate solubilizing ability of selected strains of bacteria. World J Agric Sci 5:135–137

Sturz AV, Christie BR (1995) Endophytic bacterial systems governing red clover growth and development. Ann Appl Biol 126:285–290

Subramanian P, Kim K, Krishnamoorthy R, Sundaram S, Sa T (2015) Endophytic bacteria improve nodule function and plant nitrogen in soybean on co-inoculation with *Bradyrhizobium japonicum* MN110. Plant Growth Regul 76:327–332

Suman A, Yadav AN, Verma P (2016) Endophytic microbes in crops: diversity and beneficial impact for sustainable agriculture. In: Singh DP, Singh HB, Prabha R (eds), Microbial inoculants in sustainable agricultural productivity, vol 1, Springer, New Delhi, pp 117-143

Tanaka F, Ando A, Nakamura T, Takagi H, Shima J (2006) Functional genomic analysis of commercial baker's yeast during initial stages of model dough-fermentation. Food Microbiol 23:717–728

Tao G, Tian S, Cai M, Xie G (2008) Phosphate solubilizing and mineralizing abilities of bacteria isolated from soils. Pedosphere 18:515–523

Tisdall JM, Oades JM (1982) Organic matter and water stable aggregates in soils. J Soil Sci 33:141–163

Tombolini R, Jansson JK (1998) Monitoring of GFP-tagged bacterial cells. In: La Rossa RA (ed) Methods in molecular biology: bioluminescence methods and protocols. Humana Press, Totowa, pp 285–298

Tombolini R, Unge A, Davey ME, de Bruijn FJ, Jansson JK (1997) Flow cytometric and microscopic analysis of GFP tagged *Pseudomonas fluorescens* bacteria. FEMS Microbiol Ecol 22:17–28

Trujillo ME, Alonso-Vega P, Rodriguez R, Carro L, Cerda E, Alonso P, Martinez-Molina E (2010) The genus *Micromonospora* is widespread in legume root nodules: the example of *Lupinus angustifolius*. ISME J 4:1265–1281

Vasse J, Frey P, Trigalet A (1995) Microscopic studies of intercellular infection and protoxylem invasion of tomato roots by *Pseudomonas solanacearum*. Mol Plant Microbe Interact 8:241–251

Verma JP, Yadav J, Yiwari KN, Lavakush SV (2010) Impact of plant growth promoting rhizobacteria on crop production. Int J Agric Res 5:954–983

Villacieros M, Power B, Sánchez-Contreras M, Lloret J, Oruezabal RI, Martín M, Fernández-Piñas F, Bonilla I, Whelan C, Dowling DN, Rivilla R (2003) Colonization behaviour of *Pseudomonas fluorescens* and *Sinorhizobium meliloti* in the alfalfa (*Medicago sativa*) rhizosphere. Plant Soil 251:47–54

Wagh J, Chanchal K, Sonal S, Praveena B, Archana G, Kumar GN (2016) Inoculation of genetically modified endophytic *Herbaspirillum seropedicae* Z67 endowed with gluconic and 2-ketogluconic acid secretion, confers beneficial effects on rice (*Oriza sativa*) plants. Plant Soil 409:51–64

Wahid A, Rasul E (2005) Photosynthesis in leaf, stem, flower and fruit. In: Pessarakli M (ed) Handbook of photosynthesis, 3rd edn. CRC, Boca Raton, FL, pp 479–497

Wakelin SA, Warren RA, Harvey PR, Ryder MH (2004) Phosphate solubilization by *Penicillium* spp. closely associated with wheat roots. Biol Fertil Soils 40:36–43

Weir BS (2011) The current taxonomy of rhizobia New Zealand rhizobia. http://www.rhizobia.co.nz/taxonomy/rhizobia.html

White JF, Monica S (2010) Torres is plant endophyte-mediated defensive mutualism the result of oxidative stress protection? Physiol Plant 138:440–446

Xi C, Lambrecht M, Vanderleyden J, Michiels J (1999) Bi-functional gfp-and gusA-containing mini-Tn5 transposon derivatives for combined gene expression and bacterial localization studies. J Microbiol Methods 35:85–92

Xie X, Li Y, Liu Z, Haruta M, Shen W (2009) Low-temperature oxidation of CO catalysed by Co_3O_4 nanorods. Nature 458:746–749

Xu L, Zhang Y, Wang L, Chen W, Wei G (2014) Diversity of endophytic bacteria associated with nodules of two indigenous legumes at different altitudes of the Qilian Mountains in China. Syst Appl Microbiol 37:457–465

You CB, Lin M, Fang XJ, Song W (1995) Attachment of *Alcaligenes* to rice roots. Soil Biol Biochem 27:463–466

Zahir ZA, Ghani U, Naveed M, Nadeem SM, Arshad M (2009) Comparative effectiveness of *Pseudomonas* and *Serratia* sp. containing ACC-deaminase for improving growth and yield of wheat under salt-stressed conditions. Arch Microbiol 191:415–424

Zakhia F, Jeder H, Domergue O, Willems A, Cleyet-Marel JC, Gillis M, Dreyfus B, de Lajudie P (2006) Characterization of wild legume nodulating bacteria (LNB) in the infra-arid zone of Tunisia. Syst Appl Microbiol 27:380–395

Zgadzaj R, James EK, Kelly S, Kawaharada Y, de Jonge N, Jensen DB, Madsen LH, Radutoiu S (2015) A legume genetic framework controls infection of nodules by symbiotic and endophytic bacteria. PLoS Genet 11:1–21

Zhang XX, George A, Bailey MJ, Rainey PB (2006) The histidine utilization (hut) genes of *Pseudomonas fluorescens* SBW25 are active on plant surfaces, but are not required for competitive colonization of sugar beet seedlings. Microbiology 152:1867–1875

Zhang YF, He LY, Chen ZJ, Wang QY, Qian M, Sheng XF (2011) Characterization of ACC deaminase-producing endophytic bacteria isolated from copper-tolerant plants and their potential in promoting the growth and copper accumulation of *Brassica napus*. Chemosphere 83:57–62

Zhang HJ, Zhang N, Yang RC, Wang L, Sun QQ, Li DB, Guo YD (2014) Melatonin promotes seed germination under high salinity by regulating antioxidant systems, ABA and GA4 interaction in cucumber (*Cucumis sativus* L.) J Pineal Res 57:269–279

Zhao LF, YJ X, Ma ZQ, Deng ZS, Shan CJ, Wei GH (2013) Colonization and plant growth promoting characterization of endophytic *Pseudomonas chlororaphis* strain Zong1 isolated from *Sophora alopecuroides* root nodules. Braz J Microbiol 44:629–637

Legume-Microbe Interactions Under Stressed Environments

13

Hamdi H. Zahran

Abstract

Leguminous plants and their associated microbes are found in different hostile habitats. In order to adapt to the changing environments, microbes have evolved many mechanisms. Resilience among microbes to these changes is, however, essential for survival, which depends on swift and efficient control of genetic expression and metabolic responses. Legumes form mutual, antagonistic, and other beneficial associations with microbes, which face many unfavorable (stressed) environmental challenges. Stressed environments include deserts with arid climate, salinity, alkaline and acidic soils, and soils contaminated with toxic pollutants. To cope with such devastating environmental stress factors, microbes have traits that enable them to survive under undesirable conditions. Legumes, however, are stress sensitive, and only few of them can survive under stressed environments. Recent developments in molecular and genetic tools have helped to find some new stress-tolerant microbes and to reveal the regulatory mechanisms of stress tolerance in legume-microbe interactions. Recently, the need for using multi-microbial plant inoculants and the advantages of using crop-specific microbes have been recognized. Here, an attempt is made to highlight the importance of microbes in production of legumes. Therefore, future investigations have to consider the use of consortia of microbes in order to improve productivity of legumes in different agronomic setup.

H.H. Zahran
Department of Botany, Faculty of Science, University of Beni-Suef, Beni-Suef 62511, Egypt
e-mail: hhzahran1952@yahoo.co.uk

© Springer International Publishing AG 2017
A. Zaidi et al. (eds.), *Microbes for Legume Improvement*,
DOI 10.1007/978-3-319-59174-2_13

13.1   Introduction

Nitrogen (N) is one of the major liming nutrients for most crops and non-crop plant species. Biological nitrogen fixation (BNF) involves the conversion of atmospheric N_2 to ammonium, a form of N which can easily be utilized by plants. The BNF accounts for 65% of N used in agriculture worldwide (Yamal et al. 2016). Even though many diverse bacterial populations contribute to BNF in both terrestrial and aquatic systems, diazotrophs containing nitrogenase (an enzyme that catalyzes the conversion of gaseous N to the usable form) contribute immensely to BNF. The ability of a plant to supply all or part of its N requirements through BNF can be a great competitive advantage over non-N_2-fixing neighbors (Vessey et al. 2005; Yamal et al. 2016). An essential element of agricultural sustainability is the effective management of N in the environment. The biologically fixed N_2 is less susceptible to volatilization, denitrification, and leaching (Graham and Vance 2000). Symbiotic nitrogen fixation is therefore the main route for sustainable input of N into ecosystems (Lindström et al. 2010). Legume-rhizobia symbiosis is important in many aspects. For instance, it plays an important role in sustainable production of legumes and also helps to better understand symbiosis, evolution, and differentiation (Lindström et al. 2010). To achieve high rates of N_2 fixation, the legume host and *Rhizobium* must be closely matched, not only for infection but also for optimum development, nutrient exchange, and nitrogen fixation (Terpolilli et al. 2012). In most agricultural systems, the primary source of BNF (ca. 80%) occurs via the symbiotic interactions between legumes and rhizobia of the genera *Allorhizobium*, *Azorhizobium*, *Bradyrhizobium*, *Mesorhizobium*, *Rhizobium*, and *Sinorhizobium*. The actinorhizal (*Frankia*) and *Anabaena-Azolla* types of interactions on the contrary contribute the other 20% N to the plants. Legumes provide approximately 35% of the worldwide protein intake and that ca. 250 million ha of legumes are grown worldwide. For all legumes, there is great potential to increase N derived from N_2 fixation as well as to enhance the total N_2 fixed through improved management and genetic modification of the plant. Legume N_2 fixation is a variable but nonetheless valuable process in agriculture, contributing almost 20% of the N needed for world grain and oilseed production.

Legumes and their bacterial nodules evolved about 60 and 58 million years ago, respectively (Sprent 2006, 2008). Nodulation is one of the interesting characteristic features of legumes, but non-nodulation remains common in Caesalpinioideae, with smaller numbers in Mimosoideae and Papilionoideae. Legumes are within the order Fabales and represented by a single family, the Fabaceae (formerly the Leguminosae); however, most of the more than 650 genera in the family contain species that can form nodules (Vessey et al. 2005). Nodules are highly specialized organs formed by rhizobia on roots or stems of legume plants under N-limited conditions. Within nodules, rhizobia are transformed into an endosymbiotic form, the bacteroids, in which N_2 is reduced to ammonia.

Optimization of the symbiosis between the legumes and their respective microsymbionts (the rhizobia or non-rhizobial bacteria) requires a competitive, infective, and highly efficient N_2-fixing rhizobial strains in sufficient numbers to maximize nodulation. Over 150 species of rhizobia spread over 14 genera, varying in symbiotic and physiological characteristics, are now identified (Berrada and Fikri-Benbrahim 2014; Harun-or-Rashid et al. 2015; Peix et al. 2015). The infection and nodulation process in legume-rhizobia symbioses involve an intimate interaction of macro- and microsymbionts, mediated by bidirectional molecular communications between both symbionts. The rhizobia induce two types of nodules on legumes: determinate and indeterminate. The indeterminate nodules are formed most commonly on temperate legumes (e.g., pea, clover, alfalfa, etc.), inoculated with the fast-growing rhizobia, whereas determinate nodules are normally induced by bradyrhizobia on tropical legumes (e.g., soybean, common bean, etc.). Rhizobia infect host plants and induce root or stem nodules, using three fundamentally different mechanisms: via root hairs; via entry through wounds, cracks, or lesions; and via cavities located around primordia of adventitious roots.

Over the last two decades, advances in molecular biology and genetics have helped to identify a large number of genes having symbiotic functions. A comparative study of fully sequenced and annotated genomes of several rhizobial species of the order *Rhizobiales* has been recently used to investigate the feasibility of defining a core "symbiome," the essential genes required by all rhizobia for nodulation and nitrogen fixation (Black et al. 2012). However, no direct link between sequences of symbiotic genes and symbiotic effectiveness was found, and there were only a few examples of successful tracing of loci associated with some symbiotic traits in some legumes, e.g., pea *and* lotus (Zhukov et al. 2013). Therefore, there is a gap between understanding molecular basis of symbiosis development and effective functioning of the symbiotic systems. In the fast-growing species, symbiosis-related genes are clustered on one or several relatively large plasmids, whereas in the *Bradyrhizobium* and *Mesorhizobium*, these genes are chromosomally located, the symbiotic islands (Debelle et al. 2001; Gualtieri and Bisseling 2000; Laranjo et al. 2014). Symbiotic plasmids have been recently reported to be present in some *Bradyrhizobium* species (Okazaki et al. 2015). Symbiotic N_2 fixation requires the coordinated interaction of two major classes of genes, the *nif* and *fix* genes. The *nif* genes encode the molybdenum-based enzyme system having structural and functional relatedness to the N_2 fixation genes of *Klebsiella pneumoniae*. In most rhizobia, *nif* genes are plasmid borne but located on the chromosome in the bradyrhizobia (Lagares et al. 2014; Okazaki et al. 2015). Nitrogen fixation in symbiotic and free-living microbes is catalyzed by nitrogenase, an enzyme system encoded by the *nif*DK and *nif*H genes. Nitrogenase contains a molybdenum-iron protein (MoFe), called component 1, and an iron-containing protein (Fe), called component 2. Environmentally, *nif* gene expression is regulated by both O_2 and fixed N levels. Moreover, several other genes involved in exopolysaccharide secretion, hydrogen uptake, glutamine synthase formation, dicarboxylate transport, nodulation efficiency, and B-1,2-glucan and lipopolysaccharide synthesis influence N_2 fixation directly or indirectly (Sadowsky 2005). Legume species vary greatly in N_2 fixation ability, and the amount of fixed N under optimal conditions is several folds higher than the amount of N_2 usually fixed in the field. The major approaches for symbiotic N_2 fixation improvement are the selection and construction of effective rhizobial strains and breeding the symbiotically active plants (Zahran 2006a, b, 2009). The amount of N_2 fixed by legume-rhizobia symbioses may increase by 300% due to the crop breeding and management practices (Vance 1998).

To further understand the legume-microbe interactions, several model organisms have been chosen, which provide either genomic or expressed sequence tags (EST), a prerequisite for large-scale protein identification by peptide fingerprinting. Two model legumes include *Medicago truncatula* and *Lotus japonicus*, which have EST databases with about 180,000 and 32,000 entries, respectively, and whose genome is being sequenced (Rolfe et al. 2003). Proteomic analysis has mainly focused on *M. truncatula*, for which a proteome reference map has been established (Mathesius et al. 2001). On the other hand, the model symbiotic bacterium *S. meliloti*, able to infect both *M. truncatula* and its relative alfalfa (*M. sativa*), was chosen. *S. meliloti* genome consists of a 3.7 Mb chromosome and 2 megaplasmids of 1.4 and 1.7 Mb. The genome sequence contains 6294 protein-coding frames, which provide a better understanding of the possible functions of *S. meliloti* (Galibert et al. 2001). However, the gene sequence alone often reveals little about the function of the gene products. Thus, functional proteomics is beginning to play a role in the identification and analysis of gene networks at the level of protein expression. Among the grain crops, pulses or food legumes rank third after cereals and oilseeds in world production and represent an important dietary constituent for humans and animals. Grain legumes are mainly cultivated in developing countries accounting for 61.3 million ha in 2002, compared to 8.5 million ha in developed countries (Graham and Vance 2003). Grain legumes play a crucial role in sustainability of agricultural systems and in food protein supply in developing countries (Zahran 2006b). In this chapter, the impact of stressed environments on performance of legumes and their associated microbes is reviewed and discussed.

13.2 Arid and Saline Environments

Arid and semiarid areas characterized by high salinity and alkalinity occupy an increasing fraction of the Earth's surface (Pandey et al. 2016; Keshri et al. 2013). And more than 90% of arable land suffer from one or other kinds of stresses (Yamal et al. 2016). Salinity among stress is common in arid and semiarid areas, where evapotranspiration exceeds annual precipitation and where irrigation is therefore necessary to fulfill crop water demands (Manchanda and Garg 2008). Arid environments include desert areas which are generally water deficient due to low rainfall. This situation adversely affects the vegetation and concurrently degrades biological equilibrium of soils. Hence, the huge lands in arid and semiarid regions are barren. Under drying conditions, the soil water potential decreases and so does the soil hydraulic conductivity. It is more difficult for plants to extract water, and consequently, the plant water potential tends to decrease. This decrease may directly affect the physiological processes of plants. However, many plant species have been able to survive even under such unfavorable environments. Therefore, attempts have been made to rehabilitate such degraded lands, but such approaches have been limited to two possibilities (Tomar et al. 2003): first, the exploitation of plants native to arid environments and, second, devising efficient systems for using limited saline water resources either by preventing its unproductive evaporation loss to the dry environment or drainage below rooting zone.

Arid and semiarid environments offer optimal light and temperature for most crops, but insufficient precipitation requires extensive irrigation. Plants developing in the Mediterranean climate (hot and dry summer), for example, are periodically subjected to a combination of stresses which includes lack of water and high temperature, high evaporative demand and high light intensity, and limited supply of N, P, and other plant nutrients (Sánchez-Diaz 2001). Irrigated lands on the contrary are particularly prone to salinization, and salinity has profound effects on crop production. Irrigated agriculture is a major human activity, which often leads to secondary salinization of land and water resources in arid and semiarid conditions (Shrivastava and Kumar 2015; Yan et al. 2015). The salinization of lands has therefore become a major environmental issue and has been recognized as the most important economic, social, and environmental problem in many regions of the world (Egamberdieva and Mamedov 2015). Approximately 800 million ha of land throughout the world are salt affected, either by salinity (397 million ha) or the associated condition of sodicity (434 million ha); this is over 6% of the world's total land area (FAO 2005). Similar data have been reported, over 932 million hectares of land suffering salinization and alkalization in the world (Rengasamy 2006). Recovery of vegetation in these regions is closely related with the food insecurity, environmental health, and economic welfare (Rengasamy 2006).

Soil salinity is a major agronomic problem which restricts plant growth and biomass production and consequently reduces agricultural production (Yan et al. 2015). Apart from their direct impact on plant health, excess salt concentration negatively affects the physicochemical properties of soils leading eventually to an unsuitable environment for crop production. Soils having salts in the solution phase and/or sodium ions (Na^+) on the cation exchange sites exceeding the specified limits are called salt-affected soils (Yan et al. 2015). Major cations in salt-affected soils are Na^+, Ca^{2+}, Mg^{2+}, and to a lesser extent K^+. The major anions are Cl^-, SO_4^{2-}, HCO_3^-, CO_3^{2-}, and NO_3^-. These soils are generally divided into three broad categories: saline, sodic, and saline-sodic. A soil having electrical conductivity of saturated paste extract (EC_e) ≥ 4 dS/m and sodium adsorption ratio (SAR) <13 is called saline soil. Soils having EC_e < 4 dS/m and SAR ≥ 13 is designated as sodic soils. If a soil has $EC_e \geq 4$ dS/m and SAR > 13, it is categorized as a saline-sodic soil. About 23% of the 1.5×10^9 ha cultivated land considered as saline and about half of all the existing irrigation systems of the world (3×10^8 ha) are influenced by secondary salinization, alkalization, and waterlogging (Shrivastava and Kumar 2015). Further, about 10×10^6 ha of irrigated lands are abandoned each year because of the unfavorable effects of secondary salinization and alkalization (Dajic 2006). Because of the constantly increasing food demands and declining lands due to high salinity, research on plant responses to salinity has rapidly expanded in recent decades. Therefore, reducing the salinity and developing salt-tolerant high-yielding crops are a greater challenge before the scientists worldwide. In this regard, several strategies are adopted to ameliorate saline soils. For example, cropping in combination with leaching has been found as the most successful and suitable method for abatement of soil salinity (Hamdy 1990; Qadir et al. 2000). Moreover, the identification and use of plants adapted to saline environments are of increasing importance if such areas are to remain productive (Lebrazi and Benbrahim 2014). Another approach has been the use of microbes for enhancing the legume production under saline environment. Shrivastava and Kumar (2015) suggested that microorganisms could play a significant role in legume enhancement due to their unique multifarious properties such as tolerance to saline conditions, genetic diversity, synthesis of compatible solutes, production of plant growth-promoting hormones, biocontrol potential, and their interaction with crop plants.

Generally, the saline soils are deficient in N, and hence, such nutrient-deficient soils become unsuitable for cultivation. To circumvent such problems, salt-tolerant plants capable of fixing N through BNF are used (Zahran 1999, 2001; Lebrazi and Benbrahim 2014). Sadly, only a few salt-tolerant legume trees have been used in the remediation of salinized soils. For example, legume trees such as *Albizia lebbeck, Acacia nilotica, A. auriculiformis, A. farnesiana, A. tortilis, Cassia glauca, C. javanica, C. alata, Dalbergia sissoo, Gliricidia maculata, Prosopis juliflora,* and *Sesbania* spp. (Sharma et al. 2001; Zahran 2001; Giri et al. 2002; Tomar et al. 2003) have been used to remediate such degraded lands. Similarly, some herb legumes, like *M. intertexta* and *M. indicus,* are naturally growing in salt-affected soils (Al-Sherif et al. 2004; Zahran et al. 2007) or on seashores, while the halophytic herbs like *Canavalia rosea* (Chen et al. 2000) and licorice, *Glycyrrhiza uralensis* (Egamberdieva and Mamedov 2015), are also salt tolerant. Thus, rehabilitation/reclamation of arid soils with salt-tolerant legume trees and herbs would not only render these abandoned soils to be productive but would also ensure conservation and improvement of these lands.

13.3   Legume-Rhizobia Symbioses

13.3.1 The Rhizobial Bacteria in Stressful Environments

The rhizobial bacteria can live in different habitats: (1) soil, (2) the root-soil interface (rhizosphere), and (3) the root nodule. Within soil, rhizobia are found as free-living saprophytic heterotrophs or as host-specific N_2-fixing symbionts. These features provide rhizobial species multiple advantages with respect to survival and colonization over many other soil microbiota. A legume host is not always important for survival of rhizobia within soil because many of the rhizobia (bacteroids) released from nodules survive and persist in soil indefinitely as free-living, heterotrophic saprophytes until they colonize the specific legume host (Lindström et al. 1990). Rhizobia have traditionally been found as extremely stress-resistant organisms compared to their compatible host legumes (Abd-Alla et al. 2014; Abolhasani et al. 2010; Zahran et al. 2012; Laurette et al. 2015), and a few salt-tolerant rhizobia have been reported to form effective symbiosis with their host plants (Zahran et al.

2013). For example, salt-tolerant rhizobia like *Sinorhizobium* sp. were isolated from the halophytic herb *C. rosea*, grown at 3.5% NaCl (Chen et al. 2000), and from *M. sativa*, grown at 4.5% NaCl (Abolhasani et al. 2010). Other rhizobial species like *Mesorhizobium* strain CCNWGX035 showed higher tolerance to NaCl, pH, and temperature (Wei et al. 2008), and the halotolerant rhizobia that were recovered from seedlings of *A. gummifera* and *A. raddiana* grew at about 6% NaCl (Essendoubi et al. 2007). Similarly, *Bradyrhizobium* sp. isolated from lupine grew at 5% NaCl and survived under acidic (pH 4–5) and alkaline (pH 9–10) conditions (Raza et al. 2001). In a follow-up study, about 25% of lentil rhizobia showed higher tolerance to salinity while growing at electrical conductivity equals to 10 ds/m (Alikhani and Mohamadi 2010). Osmotolerant rhizobia strains, nodulating *P. vulgaris*, have been isolated from saline soil of Morocco and identified by analyzing core genes (*rrs*, *atpD*, *recA*) and symbiotic (*nodC*) genes (Faghire et al. 2012). The most abundant strains were closely related to *R. etli* and *R. phaseoli* followed by those related to *R. gallicum* and *R. tropici*.

Arid and semiarid lands harbor many microbial flora including rhizobia, which can be selected and used as inocula for plants grown in these habitats for rehabilitation purposes. Rhizobial strains belonging to the genera *Bradyrhizobium*, *Mesorhizobium*, and *Rhizobium* have been isolated from *Caragana* spp. grown in the alkaline sand land in the north of China (Li et al. 2012). Of these, *Mesorhizobium* spp. was universally predominant microsymbionts in these soils. Bacterial isolates which belong to *Mesorhizobium* species showed high salt tolerance and efficient symbiotic activity with *Lotus corniculatus*, and many of these isolates behaved like salt-dependent bacteria (Lorite et al. 2010). Rhizobia recovered from grain legumes in other investigation also expressed noticeable ability to grow under stress conditions (salt, heat, acid, etc.) that prevail in Egypt (Zahran et al. 2012). These rhizobia, which have been reported to belong to several *Rhizobium* and *Mesorhizobium* species, have been suggested to be a candidate to establish successful symbiosis with the compatible legume hosts under stress conditions (Zahran et al. 2013).

Abiotic stresses, such as salt, osmotic agents, and heat, alter the synthesis pattern of some essential cellular components (Laranjo and Oliveira 2011; Cardoso et al. 2015). Salt and heat altered protein and lipopolysaccharide profiles of the salt-tolerant rhizobia (Zahran et al. 1994), and the salt-tolerant *R. etli* strain (EBRI 26) formed 49 differentially expressed proteins at 4% NaCl (Shamseldin et al. 2006), of which 14 were overexpressed and 35 downregulated. Similarly, the tolerance of three *Mesorhizobium* species to temperature, pH, and salt stress has been investigated (Laranjo and Oliveira 2011). *Mesorhizobium* species exhibited a variable level of tolerance to temperature, pH, and salt stress. The highest growth level was found with *M. thiogangeticum* at 1% NaCl, with *M. ciceri* at pH 5, and with *M. plurifarium* at 37 °C. Furthermore, SDS-PAGE analysis revealed changes in the protein profiles, and a 60 kDa protein was overexpressed following heat stress. Similarly, overexpression of five proteins was identified in *M. plurifarium* and *M. thiogangeticum* under salt stress. Proteins induced in response to stress may have an important role in homeostasis and maintenance of vital cellular functions (Wankhade et al. 1996). A Tn5 mutant of *Rhizobium* sp., exopolysaccharide deficient (exo⁻), isolated from root nodules of pigeon pea (*Cajanus cajan*), showed a 50% growth inhibition at 350 mM NaCl (Unni and Rao 2001). Whole cell protein profiles of the wild type showed an overall increase in the levels of several proteins (22, 38, 68, >97 kDa) in the presence of NaCl, whereas in its exo⁻ mutant, certain low molecular weight outer membrane proteins (22 and 38 kDa) were decreased. Similarly, both the wild type and the exo⁻ mutant showed decreased levels of lipopolysaccharide (LPS) in the presence of NaCl. These observations suggested a possible involvement of the outer membrane components, along with other factors, during growth under salt stress (Unni and Rao 2001). Despite these findings, the understanding of mechanistic basis of salt tolerance in rhizobia remains unclear.

Increasing soil salinity has been shown to be a threat to plants and microbial life. Exposure of microorganisms to high-osmolality environments triggers rapid fluxes of cell water along the osmotic gradient out of the cell, thus causing a reduction in turgor and dehydration of the cytoplasm. Microorganisms have developed various adaptations to counteract the outflow of water (Paul 2012). The first response to osmotic upshifts and the resulting efflux of cellular water is uptake of K⁺, and cells start to accumulate compatible solutes. Yet another mechanism is by altering the cell envelope composition, resulting in changes in proteins, periplasmic glucans, and capsular, exo-, and lipopolysaccharides (Paul 2012). Rhizobia exposed to increased salinity can maintain osmotic equilibrium across the membrane by exclusion of salts and via intracellular accumulation of inorganic and/or organic solutes (Csonka 1991). For example, *R. meliloti* (currently *S. meliloti*) overcomes osmotic stress-induced growth inhibition by accumulating compatible solutes, such as K, glutamate, proline, glycine betaine, proline betaine, trehalose, and the dipeptide, N-acetylglutaminylglutamine amide (Boscari et al. 2002; Vriezen et al. 2007). Some compatible solutes are used as either N or C sources by rhizobia suggesting that their catabolism is regulated to prevent degradation during osmotic stress. However, the type of osmolytes and their concentrations depend on the level of osmotic stress, growth phase of the culture, C source, and the presence of osmolytes in the growth medium (Smith et al. 1994). Many bacteria are equipped with systems that facilitate the efficient transport of osmoprotectants under stressed conditions, and several of these osmoregulated systems have been identified (Wood et al. 2001). BetS, a system involved in the uptake of proline betaine (PB) in *S. meliloti*, is a Na⁺-coupled secondary transporter with a high affinity for glycine betaine and proline betaine (Boscari et al. 2004). This system is activated posttranslationally by osmotic stress and plays a crucial role in the rapid response to osmotic upshock. The salt tolerance of a salt-sensitive *B. japonicum* strain was improved after transformation with *bets* gene of *S. meliloti*. An increased tolerance of transformant cells to a moderate NaCl concentration (80 mM) was detected in the presence of glycine betaine or proline betaine, whereas the growth of the wild-type strain was totally abolished at 80 mM NaCl (Boscari et al. 2004). A genomic DNA clone library of the salt-tolerant (at 600 mM NaCl) and effective nitrogen fixer *Sinorhizobium* sp. strain BL3 has been constructed and transferred to the salt-sensitive *Rhizobium* sp. strain TAL1145 by conjugation, and the transconjugants were selected in a medium containing 100 mM NaCl (Payakapong et al. 2006).

Adaptation of rhizobia to salt is a complex and multilevel regulatory process involving many genes (Wei et al. 2004; Abd-Alla et al. 2014). As an example, Rüberg et al. (2003) found that the prolonged exposure of *S. meliloti* 1021 to 380 mM NaCl activated genes related to polysaccharide biosynthesis and transport of small biomolecules such as amino acids, amines, peptides, anions, and alcohols. In this bacterium, 137 identified genes showed significant changes in gene expression, resulting from the osmotic upshift; of these, 52 genes were induced and 85 were repressed. Similarly, sudden increase in external osmolarity of *S. meliloti* cultures, elicited by addition of either NaCl or sucrose stresses, induced large number of genes having unknown functions and in repression of many genes coding for proteins with known functions (Domínguez-Ferreras et al. (2006). Of the genes upregulated, 64% were located on plasmid (pSmbB), and 85% of the genes downregulated were chromosomal. This finding suggests the role of *S. meliloti* plasmid in osmoadaptation. Further, they reported that ribosomal genes and tricarboxylic acid cycle genes are repressed. Interestingly, 25% of all genes specifically downregulated by NaCl encode ribosomal proteins. Five salt-tolerant genes of *S. fredii* RT19 were identified by construction and screening of a Tn5-1063 library (Jiang et al. 2004). Na+ intracellular content measurements established that *phaA2*, *phaD2*, *phaF2*, and *phaG2* are mainly involved in the Na+ efflux in *S. fredii* RT19. Growth recovery of the *metH* mutants grown with different NaCl concentrations, obtained by addition of methionine, choline, and betaine, showed that the *metH* gene is probably involved in osmoregulation in *S. fredii* RT19 (Jiang et al. 2004). Nodulation factors or Nod factors (lipochitooligosaccharides) of rhizobia are communication signals with leguminous plants and are major host specificity determinants that trigger the nodulation program in a compatible legume host. Nod factor activities and cloning of genes required for their initiation lead to an understanding of the first steps in the signaling pathways and symbiotic interactions (Geurts et al. 2005; Chen et al. 2006). Nod factors, which possess hormone-like properties, stimulate the plant to produce more *nod* gene inducers to deform root hairs on their respective host plant and to initiate cell division in the root cortex. However, Nod factors from different *Rhizobium* species differ in the number of *N*-acetylglucosamine residues, the length and saturation of the acyl chain, and the nature of modifications on the basic backbone (e.g., sulfate, acetate, fucose, etc.). These differences define the host specificity observed in the symbiosis. The production of Nod factors and excretion of *nod* metabolites by *R. leguminosarum* bv. *trifolii* have been found to be disrupted by pH, temperature, and both P and N concentrations (McKay and Djordjevic 1993). For instance, *R. tropici* strain CIAT899 grown under acid conditions formed 52 Nod factors, 37 of which differ from the 29 formed under neutral conditions (Morón et al. 2005). Under salt stress conditions, 46 different Nod factors were identified in a *R. tropici* CIAT899 culture; 14 different new Nod factor structures identified were not produced under neutral or acid conditions. In the same strain (CIAT899), up to 38 different structures of Nod factors were detected, being higher under salt stress (Del Cerro et al. 2015). High concentration of sodium enhanced *nod* gene expression (using a *nodP::lacZ* fusion) and Nod factor biosynthesis (Estevéz et al. 2009). In the absence of flavonoid inducers, high concentrations of NaCl induced *nod* genes and the production of Nod factors in *R. tropici* (Guasch-Vidal et al. 2013). In other reports, salinity reduced Nod factor production in *R. etli*, *R. tropici*, and *S. arboris* (Lira Junior et al. 2015). Stimulation or suppression of Nod factors under stressed conditions might affect the legume-rhizobia symbioses. Furthermore, it has recently been reported that inhibition of cell surface molecules like glucan, LPS, and EPS, of some root-nodulating bacteria, from *Vigna radiata* and *Cicer arietinum*, under salt stress, affects the symbiotic performance of the sensitive bacteria (Singh et al. 2015). The tolerant isolates on the contrary produced all the active surface biomolecules and demonstrated high-quality symbiotic activity even under salt stress conditions.

13.3.2 Effects of Salt Stress

Stress conditions affect the host plant, the *Rhizobium*, and the interaction between the two partners. Among various stressors, salinity is one of the most destructive factors that severely affects crop production and agricultural sustainability in arid and semiarid regions of the world (Yamal et al. 2016). Fertility of soil is also greatly lost due to excessive accumulation of salt within soil ecosystems. Therefore, the introduction of plants capable of surviving under these conditions is worth interesting (Soussi et al. 1998). Plant responses to salt stress include an array of changes at the molecular, biochemical, and physiological levels (Manchanda and Garg 2008). Also, the salt-tolerant plants show a variable morphological and developmental process and physiological and biochemical processes (Ahmadian and Bayat 2016; Meng et al. 2016). Salinity disrupts cell function through the toxic effects of specific ions and by osmotic effects or both (Munns 2005). Specific ion effect results from a reduction in metabolic activity, due to the presence of excessive concentrations within cells, and causes plant death when a critical salinity level exceeds. Osmotic effects, however, are manifest by water deficit due to a reduction in cell turgor. The complexity of the plant response to salt stress partially was explained by the fact that salinity imposes salt toxicity in addition to osmotic stress (Hasegawa et al. 2000). Sodium is toxic to many organisms, except for halo-tolerant organisms like halobacteria and halophytes, which possess specific mechanisms that keep intracellular sodium concentrations low. Sodium accumulation in the cytoplasm is prevented by its uptake across the plasma membrane and by promoting its extrusion or sequestration in halophytes (Hasegawa et al. 2000). Therefore, a better understanding of physiological responses under salt conditions can be of value in programs conducted to breed salt-tolerant crop varieties. In the following section, plant responses to soil salinity are discussed with emphasis on molecular mechanisms of signal transduction and on the physiological consequences of altered gene expression.

Understanding the mechanisms by which plants perceive and transduce stress signals to initiate adaptive responses is essential for engineering stress-tolerant crop plants (Xiong and Zhu 2001). Thus, in addition to the existing salt-tolerant crop genotypes, research is needed to develop genotypes with increased tolerance to salinity (Qadir et al. 2000). Salt tolerance during germination and early seedling growth existed within *P. vulgaris* genotypes (Ahmadian and Bayat 2016). Genetic variability within a species offers a valuable tool for studying mechanisms of salt tolerance. One of these mechanisms depends on the capacity for osmotic adjustment. A general feature of many plants growing in a saline environment is that they decrease osmotic potential by accumulation of inorganic and/or compatible solutes in their cells. High soil salinity can limit legume productivity by adversely affecting the growth of the host plant, the development of root-nodule bacteria, and finally the N_2 fixation capacity (Zahran 1999; Latrach et al. 2014; El Sabagh et al. 2015). As an example, salt stress affected alfalfa growth directly by adversely affecting metabolism or indirectly by its effect on rhizobia capacity for symbiotic nitrogen fixation (Bertrand et al. 2015). Also, the cowpea rhizobia nodulation and N_2 fixation were reduced with the increase of soil salinity up to 9 dS/m (112 mM NaCl) as reported by Al-Saedi et al. (2016). Furthermore, high salinity causes suppression of photosynthesis; reduces the yield of dry mass of stems, roots, and nodules; decreases the survival of root-nodule bacteria in soil and rhizosphere; increases generation time; and disrupts the cell ultrastructure (Novikova and Gordienko 1999). The identification of tolerant genotypes that may sustain a reasonable yield in salt-affected soils has thus been a strategy adopted by scientists to overcome salinity. Variations in salt tolerance among Egyptian cultivars of soybean have been reported (El Sabagh et al. 2015). On the contrary, numerous reports are available that explain the formation of the symbiosis between root-nodule bacteria and various legume species at salinized soils (Zahran et al. 2003; Latrach et al. 2014). The root-nodule bacteria grown under saline conditions may have specific traits, which enable them to establish a symbiotic interaction under salt stress (Sobti et al. 2015). For example, some isolates of *S. meliloti* recovered from nodules of wild species of alfalfa, meliloti, and *Trigonella* preferably formed symbiosis with a salt-tolerant legume grown in both salinized and nonsalinized soils (Ibragimova et al. 2006). In *M. truncatula-S. meliloti* symbiosis, salinity modulated all the measured parameters like growth performance, nitrogen-fixing capacity (acetylene reduction assay), and nodule antioxidant system (Mhadhbi and Aouani 2008). The majority of *R. leguminosarum* bv. *trifolii* and *S. meliloti*, indigenous to Egyptian soil, were found halotolerant which grew at higher levels (4% NaCl) of salts (Abdel-Salam et al. 2010). These bacterial isolates were considered as an efficient candidate for plant inoculation in salt-affected soils. It appears that the efficiency of symbiotic interaction under salinized conditions depends on the symbiotic efficiency of the isolates under standard conditions but did not correlate with the source of nodule bacteria (soil or nodule) or their salt tolerance. The response of the symbiosis to a particular stress depends on a host factors, including legume genotype, cultivar, *Rhizobium* inoculant, climatic conditions, and the duration, timing, and severity of the stress (Chalk et al. 2010). Combining plant (legume) cultivar and rhizobial strains with superior salt tolerance is an effective strategy to improve legume productivity in salinity-affected areas (Lebrazi and Benbrahim 2014; Bertrand et al. 2015). Some symbiotic systems have been discovered that are tolerant to certain stress factors such as salinity, drought, extreme temperature, and metal toxicity (Monica et al. 2013). Legume trees such as

Acacia, Prosopis, and *Sesbania* and legume herbs such as *Melilotus* and *Medicago* are salt tolerant (Shamseldin and Werner 2005; Zahran et al. 2007). These legumes establish a symbiotic association with a wide range of rhizobia (*Rhizobium, Mesorhizobium,* and *Sinorhizobium*), well-adapted to the drastic conditions of arid climates (Marcar et al. 1991; Räsänen and Lindström 2003; Nguyen et al. 2004). In *A. ampliceps* seedlings, salinity levels (5–15 dS/m) decreased the number, size, and weight of the nodules/plant, though the seedlings were inoculated by a salt-tolerant (20 dS/m) *Rhizobium* (Rommi et al. 2002). The tree legumes *A. ehrenbergiana* and *A. tortilis* demonstrated a salt stress tolerance up to 2% NaCl in seedling growth parameters, root nodulation, and nitrogen fixation, and the harmful concentration of salt was 3% NaCl (Al-Shaharani and Shetta 2011). From this work, it has been suggested that inoculation of legume trees with salt-tolerant rhizobia strains in the nursery could establish an effective nitrogen-fixing symbiosis, which may be more salt tolerant after outplanting. The effects of seawater salinity (up to 50% concentration) on growth and nitrogen fixation of *V. faba* plants have been investigated (Fahmi et al. 2011). The nodulation, nitrogen content, nitrogenase activity, and chlorophyll a and b content of *V. faba* plants were decreased by increasing seawater salinity, while proline accumulation was increased under salt stress. Furthermore, salinity enhanced the occurrence of particular novel proteins in *V. faba* plants inoculated with a salt-tolerant rhizobial strain. In a similar study, soil salinity levels equal to or higher than 6.5 dS/m (about 80 mM NaCl) reduced the grain production of *V. faba* plants (Katerji et al. 2011). Higher salinity levels (10–40 mM NaCl) inhibited the population count of *R. tropici* CIAT899 and natural strains, growth, symbiotic efficiency, and total nitrogen content of dry bean (*P. vulgaris*), inoculated by these rhizobia (Uyanöz and Karaca 2006). Bacteria belonging to *R. leguminosarum* bv. *ciceri* and isolated from wild chickpeas exhibited varied salt tolerance (Ögütçğü et al. 2010). Inoculation of chickpea plants by *R. leguminosarum* bv. *ciceri* significantly increased growth, nodulation, and amounts of fixed and total nitrogen under saline (50 and 100 mM NaCl) conditions.

A best symbiotic N_2 fixation under saline conditions could be achieved if both symbionts and different stages of symbiosis such as recognition, root colonization, infection, nodulation, and nitrogen fixation are tolerant to the imposed stress factor. Salt stress effects on legume-rhizobia symbiosis can also be alleviated by inoculation of effective (N_2-fixing) and salt-tolerant rhizobia. For example, the soybean seedlings (*G. max*, salt tolerant, and *G. soja* salt sensitive) inoculated with *B. japonicum* and grown under 100 mM NaCl stress had improved morphological, anatomical, and physiological characters, suggesting the ameliorative effects of rhizobial inoculation on salt injury to soybean seedlings (Meng et al. 2016). *B. phymatum* GR01N inoculated *P. vulgaris* plants when grown under saline stressed soil had increased dry weight, greater nodule occupancy, and nitrogen fixation compared to those inoculated with *R. tropici* 899 (Talbi et al. 2013). However, the legume-*Rhizobium* (*V. unguiculata*) symbiosis has been reported to be decreased by salinity (1 dS/m), and a gradual shift in the spatial distribution of the nodules from the primary roots to the secondary roots under increased salinity is reported (Predeepa and Ravindran 2010). Similarly, inoculation of faba bean with salt-tolerant strains of

R. leguminosarum allowed the plant to grow and form effective (nitrogen-fixing) nodules under field and greenhouse conditions (Belal et al. 2013). However, in the absence of *Rhizobium*, salinity significantly reduced the height of alfalfa plants, dry biomass, and nodulation, in addition to other physiological parameters, e.g., relative water content, membrane permeability, stomatal conductance, etc. (Latrach et al. 2014). Under stress, plants maintain a low concentration of Na^+ and a high concentration of K^+ in the cytosol. However, Na^+ toxicity is not only due to toxic effects of Na^+ in the cytosol but also because K^+ homeostasis is disrupted possibly due to the ability of Na^+ competing for K^+ binding sites. Plants possess a number of mechanisms to prevent accumulation of Na^+ in the cytoplasm that include minimizing Na^+ influx, intracellular compartmentalization of Na^+, and maximizing Na^+ efflux as well as pre-circulation of Na^+ out of the shoot by the phloem (Ward et al. 2003; Bartels and Sunkar 2005). Salt tolerance correlates to an efficient Na^+ and Cl^- exclusion mechanism and to a better maintenance of leaf K^+ concentration at high levels of external NaCl (Sibole et al. 2003; Garthwaite et al. 2005). The tree legume, *A. nilotica*, however, had different mechanism of salt tolerance, being able to adjust osmotic potential by accumulation of Na^+, K^+, Cl^-, and proline under salt stress (Nabil and Coudret 1995). Salt-tolerant plants achieve the Na^+-K^+ balance in the cytosol by regulating the expression and activity of Na^+ and K^+ transporters and H^+ pumps that generate the driving force for transport (Zhu 2003). Na^+ transporters include the NHX and SOS families (salt overly sensitive) of Na^+/H^+ exchangers, HKT proteins, as well as components of the signaling pathway that regulate these transporters, such as SOS2 and SOS3 proteins (Horie and Schroeder 2004; Pardo et al. 2006). Proper regulation of ion flux is necessary for cells to maintain the concentrations of toxic ions low and to accumulate essential ions. The vacuolar sodium sequestration is mediated by a Na^+/H^+ antiporter at the tonoplast. Sequestration or compartmentalization of Na^+ into the vacuole through vacuolar Na^+/H^+ antiporters uses the proton motive force generated by the vacuolar H^+-translocating enzymes, H^+-adenosine triphosphatase (ATPase), and H^+-inorganic pyrophosphatase (PPiase), to couple the downhill movement of H^+ with the uphill movement of Na^+ against the electrochemical potential (Blumwald et al. 2000). The presence of Na^+/H^+ antiporter activities has been physiologically characterized in tonoplast vesicles and is molecularly represented by six *Arabidopsis* genes *AtNHX1–6* (Blumwald et al. 2000; Yokoi et al. 2002). The first Na^+/H^+ exchanger identified was *AtNHX1*, a member of a family of six genes (*AtNHX1–AtNHX6*) that show sequence homology to mammalian and yeast NHE or NHX exchangers, respectively (Yokoi et al. 2002). Many reports have indicated the existence of Na^+/H^+ antiporters in plant vacuoles (Blumwald et al. 2000; Zörb et al. 2005; Zahran et al. 2007). Several studies dealing with the occurrence, expression, and activity of Na^+/H^+ antiporters and NHX genes under salt stress were reviewed (Zahran et al. 2007). Overexpression of *NHX1* enhances salt tolerance in crop plants (e.g., tomato, rice, cotton, sugar beet, barley, sunflower, wheat, and maize) as well as some halophytic plants (*Atriplex*, *Suaeda*, and *Thellungiella*) and that this antiporter catalysis both Na^+/H^+ and K^+/H^+ exchange. Among legumes, a vacuolar antiporter (*MsNHX1*) was cloned from alfalfa whose gene was induced by NaCl and ABA treatments (Yang et al. 2005). The involvement of Na$^+$/H$^+$ transporters in *M. intertexta* and *M. indicus*)growing at salt-affected cultivated soils of Egypt, Zahran 1998; Al-Sherif et al. 2004) has been investigated (Zahran et al. 2007). NaCl induced gene expression of three genes in *M. intertexta* and one gene in *M. indicus* (Zahran et al. 2007). *NHX* gene triggered in *M. intertexta* plants to cope with tissue Na$^+$ accumulation, while in *M. indicus*, the absence of Na$^+$ accumulation and the lack of induction of *NHX* genes in response to NaCl indicate that this species relies on different mechanisms to cope with salt stress.

Under stress conditions, amendments by proline, glycine betaine, sucrose and mannitol, etc. protect major processes such as cell respiration, photosynthetic activity, nutrient transport, and N and C metabolism (Zhu 2002; Fernandez-Aunión et al. 2010; El Sabagh et al. 2015). Trehalose (a nonreducing disaccharide found in a wide variety of organisms, including bacteria and plants) plays an important role as an abiotic stress protectant, stabilizing dehydrated enzymes and membranes as well as protecting biological structures from desiccation damage (Benaroudj et al. 2001; Sampedro and Uribe 2004). Trehalose accumulation in *B. japonicum* enhanced the survival of this bacterium under salinity stress and played a role in the development of symbiotic nitrogen-fixing root nodules in soybean plants (Sugawara et al. 2010). Stress-induced trehalose biosynthesis in *B. japonicum* is due mainly to the OtsAB pathway and that the TreS pathway is likely involved in the degradation of trehalose to maltose. Similar accumulation of trehalose, mannitol, and alanine in *Burkholderia phymatum* (GR01N), in response to saline stress, is reported (Talbi et al. 2013). In alfalfa plants grown under salt stress, two major compatible sugars (sucrose and pinitol) involved in plant osmoregulation were found to increase in leaves, while starch was observed more in roots (Bertrand et al. 2015). In nodules, however, the above three solutes (sucrose, starch, and pinitol) increased under salt stress. These findings support the possible role of this disaccharide as an osmoprotectant against abiotic stress.

Cytokinins are a group of adenine derivatives that affects plant growth and development. Exogenous cytokinins induce cortical cell divisions in legume roots and the expression of several nodulin genes, thus enhancing legume-*Rhizobium* symbiosis (Mathesius et al. 2000; Gonzalez-Rizzo et al. 2006). Cytokinin levels decrease under adverse environmental conditions, but exogenous application of cytokinin counteracts the negative physiological effects of salt stress (Hare et al. 1997). In this regard, a new cytokinin receptor homologue (*MsHK1*) was induced in *M. sativa* seedlings by exogenous application of the cytokinin transzeatin (Coba De La Peña et al. 2008). *MsHK1* expressed in roots, leaves, and nodules of *M. sativa* under salt stress and transcript accumulation in the vascular bundles pointed to a putative role in osmosensing for *MsHK1* receptor homologue (Coba De La Peña et al. 2008). Similarly, exogenous abscisic acid (ABA) pretreatment to plants improved growth parameters and ameliorated the effects of salt on nodule weight and nitrogenase activity of a salt-sensitive cultivar of common bean (Mills et al. 2001; Khadri et al. 2007). ABA treatment seems to limit sodium translocation to shoot resulting in the maintenance of high K$^+$/Na$^+$ ratio in salt-stressed plants. Therefore, ABA may function as a stress signal and play an important role in the tolerance of plants to salinity. Likewise, homospermidine synthase, an enzyme involved in nodule organogenesis and salt tolerance, has been reported to be induced by salt stress in *R. tropici*-inoculated *P. vulgaris* plants (López-Gómez et al. 2016).

The role of the *rpoH2* gene of *Sinorhizobium* sp. BL3 in exopolysaccharide (EPS) synthesis, salt tolerance, and symbiosis with *Phaseolus lathyroides* has been determined (Tittabutr et al. 2006). Three *rpoH2* mutants of BL3 were constructed by transposon insertion mutagenesis. These mutants were not defective in EPS synthesis, nodulation, and nitrogen fixation, but they failed to grow in salt stress conditions. The results had indicated that *rpoH2* is required for salt tolerance in *Sinorhizobium* sp. BL3. The involvement of the plasmid (e.g., the megaplasmid SMb of *Sinorhizobium meliloti*) genes in the control of salt tolerance in some rhizobia has been shown (Roumiantseva and Muntyan 2015). The occurrence of insertion sequences (IS) and transposons in these megaplasmids gives evidence of their active involvement in horizontal gene transfer, which increases the possibility of introduction of foreign genes, in particular those affecting the stress tolerance of rhizobia. In *M. alhagi* (CCNWXJ12-2), grown under salt stress (0.4 M NaCl), a total of 1489 differential expressed genes were identified, and 933 genes were downregulated, while 916 genes were upregulated under these salt conditions (Liu et al. 2014). Notably, a gene encoding YadA domain-containing protein (yadA) was highly upregulated and involved in salt tolerance of *M. alhagi* (Liu et al. 2014). Isolates of *S. meliloti* from different sources, with reduced level of salt tolerance, had more diverse intergenic sequences of *rrn* operons (intergenic sequence, ITS) than salt-tolerant isolates related to the type strain of *S. meliloti* Rm1021 (Roumiantseva et al. 2014). Isolates with changed level of resistance to salt had more flexible intergenic sequences of *rrn* operons. The authors suggested the existence of chromosomal types different from that of Rm1021 and present in *S. meliloti* isolates with mainly altered level of salt tolerance. Thus, under the negative influence of salinity, the populations of bacterial symbionts undergo active microevolutionary processes, affecting the *rrn* operon structure, which seem to be linked to a change in the adaptation capacity of soil bacteria (Roumiantseva et al. 2014). The transcriptional analysis by northern hybridization of chaperone genes was performed using salt-sensitive and salt-tolerant isolates belonging to different *Mesorhizobium* species (Brígido et al. 2012). Upon salt shock, most isolates revealed a slight increase in the expression of the *dnaK* gene. No clear relationship was found between the chaperone gene induction and the level of salt tolerance of the isolates. In contrast, the transcriptional analysis by northern hybridization suggests a relationship between induction of major chaperone genes and higher tolerance to acid pH in the same *Mesorhizobium* species (Brígido and Oliveira 2013). Most acid-tolerant *Mesorhizobium* isolates displayed induction of the *dnaK* and *groESL* genes upon acid shock, while the sensitive ones showed repression. It has been suggested that bacterial ability to adapt to hyperosmotic salt stress conditions is also important for its nitrogen-fixing ability in legume root nodules (Dogra et al. 2013). *M. ciceri* mutants, with inability to survive in the presence of 0.1 M NaCl and obtained after Tn5 transposon insertion, formed symbiotic nodules with severely reduced nitrogenase activity (Dogra et al. 2013).

For plants, many genes encoding PR-5 proteins (proteins known to function as protein-based defensive system against abiotic and biotic stress) have been identified from a variety of plants, indicating that PR-5 is broadly distributed throughout higher plants. The involvement of PR-5 proteins in protection against abiotic stresses, such as osmotic imbalance, has been suggested (Kononowics et al. 1992). A novel soybean genes, *GmOLPA* and *GmOLPB* (*G. max* osmotin-like protein), encoding an acidic homologue of PR-5 protein (Onishi et al. 2006) and neutral homologue PR-5 protein (Tachi et al. 2009), respectively, were highly induced in the leaves of soybean plants under conditions of high salt stress. An alfalfa cDNA library was induced by salt stress constructed by suppression subtraction hybridization (SSH) technology (Jin et al. 2010). A total of 119 positive clones that were identified by reverse northern dot blotting resulted in 82 uni-ESTs, and most of the annotated sequences were homologous to genes involved in abiotic or biotic stress in plants. In addition, several ESTs, similar to genes from other plant species, closely involved in salt stress were isolated from alfalfa, such as aquaporin protein and glutathione peroxidase.

The root transcriptome of two genotypes of *M. truncatula*, having contrasting responses to salt stress (TN1.11, a salt-tolerant genotype and the reference Jemalong A17 genotype), has been analyzed (Zahaf et al. 2012). Transcriptomic analysis revealed specific gene clusters (those related to the auxin pathway and to changes in histone variant isoforms), and genes encoding transcription factors are differentially regulated between the two genotypes in response to salt. A novel Δ^1-pyroline-5-carboxylate synthetase gene (*MtP5CS3*) has been isolated from *M. truncatula*, which was strongly expressed under salinity and drought in shoots and nodulating roots (Kim and Nam 2013). Consistently, *MtP5cs3*, a loss-of-function mutant, accumulated much less proline, formed fewer nodules, and fixed nitrogen significantly less efficiently than the wild type under salinity (Kim and Nam 2013). Thus, *MtP5CS3* plays a critical role in regulating stress-induced proline accumulation during symbiotic nitrogen fixation. Recently, Long et al. (2015) identify novel and salt stress-regulated microRNAS from roots of *Medicago sativa* and *M. truncatula*. The salt stress-related conserved and novel miRNAs may have a large variety of target mRNA, some of which might play key roles in salt stress regulation of *Medicago* plants. To improve salt tolerance of alfalfa, a salt-tolerant *rstB* gene was introduced into alfalfa genome by *Agrobacterium*-mediated transformation (Zhang and Wang 2015). Significant enhancement of resistance to salt shock treatment was noted on the *rstB* transgenic plants. The effects of pre-incubation of *S. meliloti* strains, with the effective *nod* gene inducers (luteolin, methyl jasmonate, and genistein), on the growth and nitrogen fixation of alfalfa cultivars under salt stress, were determined (Ghasem et al. 2012). The *nod* gene inducers increased alfalfa growth and nitrogen fixation under normal as well as under salt stress conditions; luteolin was the most effective nod gene inducer under these conditions (Ghasem et al. 2012).

The production of reactive oxygen species (ROS) is yet another major damaging factor, which disrupts normal metabolism through oxidative damage of lipids and proteins in plants exposed to different environmental stresses. Plants with high concentrations of antioxidants (e.g., ascorbate peroxidase APOX, catalase CAT, peroxidase POD, and superoxide dismutase SOD), have greater resistance to these oxidative damages (Jiang and Zhang 2002). Nodules are particularly rich in both quantity and diversity of antioxidant defenses that may protect the nodule structures against high rates of nodule respiration as well as conserve the nitrogenase activity (Becana et al. 2000; Blokhina et al. 2003). Nitrogenase is O_2 sensitive; therefore, nodules have evolved mechanisms to downregulate their permeability to O_2 and maintain the infected cell O_2 concentration at approximately 5–50 nM compared to 250 μM for cells in equilibrium with air (Minchin 1997). Salinity induces the production of stress proteins or antioxidant enzymes in nodules to minimize damage caused by ROS such as, H_2O_2, O_2, and OH (Porcel et al. 2003). Salt stress (50 mM NaCl) or osmotic stress (50 mM mannitol) reduced plant growth, nitrogen fixation, and the activities of the antioxidant defense enzymes of common bean (*P. vulgaris*) nodules (Tejera et al. 2004; Jebara et al. 2005). Flavodoxins (electron carrier flavoproteins found in prokaryotes and some eukaryotic algae) are involved in the response to oxidative stress in bacteria and cyanobacteria. In *M. sativa* nodules, the decline in nitrogenase activity associated to salinity stress (100 mM NaCl) was significantly less in flavodoxin-expressing nodules than in wild-type nodules (Redondo et al. 2012). Furthermore, flavodoxin reduced salt-induced structural damage, which primarily affected young infected tissues and not fully differentiated bacteroids. Redondo et al. (2012) have further concluded that overexpression of flavidoxin in bacteroids has a protective effect on the function and structure of alfalfa nodules subjected to salinity stress. The production of ROS in root hairs is an important response to both symbiotic process and abiotic stress, and ROS are also involved in signaling and determination of the shape of root hair cells (Muñoz et al. 2012). Saline (150 mM NaCl), but not osmotic stress, markedly affects both apoplastic and intracellular ROS production, inhibiting root hair curling and inducing root hair death during *B. japonicum-G. max* symbiosis (Muñoz et al. 2012). The maintenance of sucrose synthase, together with isocitrate dehydrogenase, associated with a suitable antioxidant defense, may be relevant for osmotic tolerance in *P. vulgaris* N_2 fixation.

The performance and responses to osmotic stress (50 mM mannitol) have recently been evaluated in chickpea-*Mesorhizobium* symbiosis (Mhadhbi et al. 2008). Nodular POX and APOX activities were significantly enhanced in chickpea plants under osmotic stress. The increase of POX and APOX inversely correlated with the inhibition of aerial biomass production and nitrogen-fixing capacity, suggesting a protective role of these enzymes in nodules. In a similar report, salinity (75 mM NaCl) increased significantly the nodule conductance in four genotypes of *S. meliloti*-inoculated *M. truncatula* plants (Aydi et al. 2004). Thus, the sensitivity to salinity appears to be associated with an increase in nodule conductance that supports the increased respiration of N_2-fixing nodules under salinity. In contrast, salinity did not change the nodule conductance and nodule permeability of the salt-tolerant variety of chickpea (L'taief et al. 2007). The salt tolerance of this variety appears to be associated with stability in nodule conductance and the capacity to form nodules under salt constraint.

Nodule conductance to O_2 diffusion has been found as a major factor of the inhibition of N_2 fixation by salinity that severely reduces the production of legumes. In soybean, nitrogenase activity and leghemoglobin content were diminished, and ammonium content increased only under 200 mM NaCl (Zilli et al. 2008). Furthermore, heme oxygenase (HO) activity, protein synthesis, and gene expression were significantly increased under 100 mM salt treatment. This finding suggested that the upregulation of HO, as part of antioxidant defense system, could protect the nitrogen fixation and assimilation under saline stress conditions.

13.3.3 Effects of Water, Osmotic, and Desiccation Stresses

While growing, plants are often exposed to unpleasant environmental conditions, which have many harmful impacts on their growth, development, and overall performance (Lebrazi and Benbrahim 2014). Among various stressors, drought and salinity are considered as the most important abiotic factors limiting plant growth and yield in many areas of the world (Lebrazi and Benbrahim 2014; Shetta 2015). Osmotic stress refers to a situation where insufficient water availability limits growth and development of plants (Zhu et al. 1997). Soil water content directly influences the growth of rhizospheric microbes by decreasing water activity below critical tolerance limit and indirectly affects plant by altering their growth, nutrient concentration, root architecture, and exudates.

Microbial cells withstand lower water potentials better than most higher plant cells. Generally, the root-nodule bacteria (rhizobia) are more resistant to soil water deficit (drought) than their host plant, and hence, the impact of drought stress conditions on N_2 fixation might be due to direct influence on the macrosymbionts (Serraj et al. 1999; Hungria and Vargas 2000). Consequently, from the beginning of infection by rhizobia until the functioning of differentiated nodules, the most important factors limiting the fixation under water stress will probably depend on the host plant. The work made on different lucerne (*M. sativa*) cultivars suggests that those adapted to dry conditions are likely to show little water stress effects on N_2 fixation than those less adapted cultivars (Aguirreolea and Sánhez-Diaz 1989). Species of rhizobia, however, differ in their susceptibility to the detrimental effects of desiccation in natural soils. For example, slow-growing rhizobia generally thought to survive desiccation better than fast-growing rhizobia (Zahran 2001). Furthermore, indigenous rhizobia isolated from the tree legumes (*A. tortilis*, *A. saligana*, and *Leucaena leucocephala*), grown in Saudi Arabia, differed greatly in their responses to drought stress (Shetta 2015). Rhizobia from wild legumes of semiarid land displayed a variable growth pattern, production of extracellular polysaccharide (EPS), IAA, siderophores, and phosphate solubilization activity (Bhargava et al. 2016). So far, the effect of water stress on symbiosis is the concern; it affects nodule establishment, C and N metabolism, nodule O_2 permeability, nitrogenase activity, and total plant N_2 fixation ability (Zahran and Sprent 1986; Aguirreolea and Sánhez-Diaz 1989; Sadowsky 2005). However, N_2 fixation is widespread in arid land legumes (e.g., *Acacia* and *Prosopis* species), and drought-tolerant rhizobial strains have been reported for both tree and crop legume species (Nijiti and Galiana 1996). Similar variations in influence of drought or osmotic stress on other legumes, for example,

A. mangium (Galiana et al. 1998), *Gliricidia sepium* (Melchior-Marroquin et al. 1999), *Sesbania* (Rehman and Nautiyal 2002), *Albizzia adianthifolia* (Swaine et al. 2007), *Retama raetam* (Mahdhi and Mars 2006; Mahdhi et al. 2008), and *A. origena* (Shetta 2015), are reported.

Like bacterial partners, plants may also alleviate the impact of stress (e.g., osmotic), if grown with soil microorganisms, like PGPR and AM fungi (Valdenegro et al. 2001; Ruiz-Lozano 2003). The AM fungi have an improved ability for nutrient uptake and tolerance to biotic and abiotic stresses. Tree legumes form an association with AM fungi and rhizobia. This association could further be beneficial if they are used with PGPR. In this regard, *M. arborea*, a leguminous tree used for revegetation purposes under semiarid conditions, was inoculated either singly or in combination with microorganisms (three *Glomus* species, two strains—wild type and genetically modified *S. meliloti*—and PGPR (Valdenegro et al. 2001). Mycorrhizal fungi were effective in all cases, while PGPR inoculation was only effective when co-inoculated with specific mycorrhizal endophytes (*G. mosseae* plus wild-type rhizobia and *G. deserticola* plus genetically modified rhizobial strain). The effect of double inoculation with two species of AM fungi (*G. deserticola and G. intraradices*) and two strains of *S. meliloti* (wild type and its genetic variant) was examined in three *M. sativa* (*M. nolana*, *M. rigidula*, and *M. rotata*) plants. Nodulation and mycorrhizal dependency changed in each plant genotype in accordance with the *Sinorhizobium* strain and AM fungi involved. Plants inoculated with both the AM fungi and the genetically modified *S. meliloti* were better adapted to drought stress (Vázquez et al. 2001).

Arbuscular mycorrhizal fungal symbiosis can also alleviate drought-induced reductions in nodule activity and senescence. The most remarkable observation was the substantial reduction in oxidative damage to lipids and proteins in nodules of mycorrhizal plants subject to drought as compared to the nodules of non-mycorrhizal plants. Mycorrhizal protection against the oxidative stress caused by drought is perhaps one of the most important mechanisms by which the AM symbiosis increases the tolerance of plants against drought (Ruiz-Lozano et al. 2001). The AM symbiosis considerably increased the glutathione reductase activity (an important component of the ascorbate glutathione cycle) both in roots and nodules of soybean plants subject to drought stress (Porcel et al. 2003). The AM soybean plants respond to drought stress by downregulating the expression of two plasma membrane intrinsic protein (PIP) genes (Ruiz-Lozano et al. 2006); this is likely to be a mechanism to decrease membrane water permeability and to allow cellular water conservation. In another study, four *P. vulgaris* varieties were inoculated singly or jointly with two different AM fungi and/or two different *Rhizobium* strains (Franzini et al. 2010). The results revealed that one fungus and one *Rhizobium* when used together caused negative effects on plant growth. These effects were, however, found dependent on the *P. vulgaris* variety used and on the symbionts applied. Also, the AM symbiosis inhibited nodule development and N_2 fixation leading to poor plant growth. Conclusively, under moderate drought conditions, the dual symbiosis formed by AM fungi and *Rhizobium* can be deleterious to *P. vulgaris* growth.

The use of genetic engineering technology could also serve as an effective gene-based tool for improving crop tolerance to drought. Certain genes are expressed at elevated stress levels, and some specific proteins like water channel proteins, key enzymes for osmolyte biosynthesis, detoxification enzymes, and transport proteins (Vinocur and Altman 2005), are induced by abiotic stress. Therefore, a successful strategy may be to use genetic engineering to switch on a transcription factor regulating the expression of several genes related to abiotic stress (Bartels and Sunkar 2005; Chinnusamy et al. 2005). Cytokinins are classical phytohormones regulating cell growth, cell differentiation, apical dominance, and leaf senescence. In a recent study, two engineered *Sinorhizobium* strains, overproducing cytokinins, were constructed (Xu et al. 2012). Most of the alfalfa plants inoculated with the engineered strains survived, and the nitrogenase activity in their root nodules showed no apparent change, after being subjected to severe drought stress. It was suggested from this study that the genetically engineered *Sinorhizobium* strains, synthesizing more cytokinins, could improve the tolerance of alfalfa plants when grown under severe drought stress without affecting alfalfa nodulation or nitrogen fixation. In yet similar experiment, transgenic plants overexpressed the *P5CS* (Δ^1-pyroline-5-carboxylate synthetase) gene from *Vigna aconitifolia*; they accumulated high proline levels and were more tolerant to osmotic stress (Kishor et al. 2005). Two *P5CS* genes have been isolated from the model legume *M. truncatula*: *MtP5CS1* (encode a developmental "housekeeping" enzyme) and *MtP5CS2* (shoot-specific osmoregulated isoform). *M. truncatula* transformed with the *P5CS* gene from *Vigna aconitifolia* (Verdoy et al. 2006). Overexpression of *P5CS* genes accumulates high levels of proline in tissues of *M. truncatula*, which display enhanced osmotolerance (Verdoy et al. 2006). Transgenic legume models allow analysis of some biochemical and molecular mechanisms that are activated in the nodule in response to high osmotic stress and ascertain the essential role of proline in the maintenance of nitrogen-fixing activity under these conditions. A transcription factor DREB 1A from *Arabidopsis thaliana*, driven by the stress-inducible promoter from the *rd29A* gene, was introduced in a drought-sensitive peanut cultivar JL24 through *Agrobacterium tumefaciens*-mediated gene transfer. All transgenic events were able to maintain a transpiration rate equivalent to the well-watered control in soils dry enough to reduce transpiration rate in the wild type (Bhatnagar-Mathur et al. 2007). Recent molecular investigations thus indicate the active role of proline in alleviating the effects of osmotic stress on legume-*Rhizobium* symbiosis.

13.3.4 Effects of pH and Temperature

Acid soils limit agriculture production, and as much as 25% of crops suffer from soil acidity (Munns 1986). The ability of rhizobia to tolerate higher acidity is due to their ability to maintain internal pH approaching neutrality (Graham et al. 1994). Among rhizobia, bradyrhizobia, in general, are more acid tolerant than other rhizobia, although some strains of *R. tropici* have been found highly acid tolerant

(Graham et al. 1994) due to production of glutathione in order to grow in extreme acid stress conditions (Riccillo et al. 2000). Using Tn5 mutagenesis, acid-sensitive mutant of *S. meliloti* was isolated, and some genes involved in acid tolerance were characterized (Tiwari et al. 1996). Apart from rhizobia, acidity also affects the growth of the legumes and the infection process (Munns 1986). This effect is, in part, most likely due to both a disruption of signal exchange between macro- and microsymbionts (Hungria and Stacey 1997) and repression of nodulation genes and excretion of Nod factors in the rhizobia (Richardson et al. 1988). Soil acidity is reported to affect rhizobial persistence, nodulation efficiency, and N_2 fixation of some legumes (Graham and Vance 2000). Rhizobial strains nodulating *P. vulgaris* under arid conditions were analyzed for pH tolerance (Priefer et al. 2001). One strain (RP163) exhibiting high nodulation efficiency and a broad pH tolerance was mutagenized by Tn5, and the resulting mutants unable to grow on extreme pH media were isolated. In these mutants, a suitable well-characterized promoter is now available to drive expression of rhizobial stress-tolerant genes. In a similar approach, promoters and genes inducible under extreme pH values were identified in *R. leguminosarum* bv. *viceae* VF39 (Priefer et al. 2001); among them, *gabT* encodes the GABA (γ-aminobutyrate) transaminase which is induced under acidic conditions. Soil nutrients have a profound impact on legume-*Rhizobium* symbiosis. A nutrient stress is indirectly caused by changes in soil matric potential or acidity, which in turn limit nutrient bioavailability (Sadowsky 2005). Stress conditions apparently increase requirements for essential elements, such as Ca^{2+}, P, and N, in both plants and microbes. The presence of Ca^{2+} may offset the deleterious influence of low pH on root growth, while ion uptake increases *nod* gene induction and expression and concurrently affects the attachment of rhizobia to root hairs and nodule development (Richardson et al. 1988; Alva et al. 1990; Smit et al. 1992). Phosphorous (P) availability is another limiting factor for N_2 fixation and symbiotic interactions (Saxena and Rewari 1991), and about 33% of the arable land in the world is P deficient, especially in low pH soil (Graham and Vance 2000). There are marked differences in rhizobial and plant requirements for P, and the slow growers are more tolerant to low P than the fast-growing rhizobia (Beck and Munns 1985). High soil temperature has a marked influence on survival and persistence of rhizobial strains in temperate climate (Boumahdi et al. 2001). Relatively high root temperature has also been shown to influence infection, nitrogen fixation ability, and legume growth (Mohammadi et al. 2012). However, strains from naturally growing legumes in tropical regions survive better at higher temperatures (Zahran et al. 1994). The influence of temperature on rhizobia appears to be strain dependent. For example, *Bradyrhizobium* sp. (lupine) was less susceptible than *R. leguminosarum* bv. *trifolii* to high soil temperature (Sadowsky 2005). However, rhizobial strains at elevated temperatures lose infectivity (Segovia et al. 1991). Moreover, excessive temperature shock cures plasmids in fast-growing strains, and some strains, which were isolated from warm environments, had a Fix⁻ phenotype (Moawad and Beck 1991; Hungria and Franco 1993). Soil temperature greatly influences competition for nodulation (Triplett and Sadowsky 1992). However, some high-temperature (up to 40 °C)-tolerant rhizobia formed effective nitrogen-fixing nodules with *P. vulgaris* (Hungria et al. 1993; Michiels et al. 1994), *Prosopis* (Kulkarni and Nautiyal 1999), and *Acacia* (Zerhari et al. 2000). Each legume-*Rhizobium* combination has an optimum temperature relationship around 30–40 °C; exposure of both symbiotic partners to temperature extremes much above or below these critical temperatures impairs infection, nodulation, nodule development, and general nodule functioning as well as plant growth and productivity (Michiels et al. 1994). Elevated temperatures directly influence the production or release of *nod* gene inducers as reported for soybean and bean (Hungria and Stacey 1997) where it altered nodule functioning particularly leghemoglobin synthesis, nitrogenase activity, and H_2 evolution and, in addition, hastened nodule senescence (Hungria and Vargas 2000). Therefore, in order to obtain most competitive and effective bacterial strains, bacteria need to be isolated and screened from the pool of indigenous microbes that could adapt to a wide range of climatic conditions and hence increase growth and enhance nutrient uptake by plants in disturbed soils. Chaperone systems, such as *dnaK-dnaJ* and *groEL-groES*, are mostly known as important components of the heat shock response. A microarray study in *S. meliloti* revealed the upregulation of 169 genes after heat shock, including genes coding for chaperones and other heat shock proteins (Sauviac et al. 2007). The molecular bases of temperature stress tolerance in chickpea rhizobia (mesorhizobia) have been investigated, by comparing the expression of chaperone genes *dnaKJ* and *groESL* in thermotolerant and thermosensitive isolates (Alexandre and Oliveira 2011). Transcript levels of these genes increased with heat, and a higher induction of chaperone genes was detected in heat-tolerant isolates when compared with sensitive isolates of the same species. In a similar study, northern analysis revealed an increase in *groEL* expression in *M. huakuii* and *M. septentrionale* after heat shock; by contrast, a decrease was detected in *M. albiziae* and *M. thiogangeticum*, upon salt shock (Laranjo and Oliveira 2011).

13.4 Legume-Free Living PGPR Interactions Under Stressed Environments

Rhizobacteria have been reported to be beneficial to plants in many ways (Paul 2012). The sole/mixed application of bacterial cultures provides a more balanced nutrition and improves nutrient uptake by plants (Belimov et al. 1995; El-Komy 2005). The use of beneficial microbes in agricultural practices for enhancing the crop production was started about 60 years ago, and there is now increasing evidence that the use of microbes can also facilitate plant resistance to adverse environmental stresses (Sheng 2005). Recent studies have shown that several plant species require microbial associations for stress abatement and survival (De Zelicourt et al. 2013). However, very little is known about how microbes confer stress tolerance to plants. Among microbial communities colonizing different habitats, plant growth promoting rhizobacteria (PGPR) including both free living and symbiotic organisms promote growth and development of plants directly or indirectly (Khan et al. 2009). Direct stimulation includes providing plants with essential plant nutrients such as N (N_2 fixation) and P (P solubilization, PSM) and phytohormones (indole-3-acetic acid, zeatin, gibberellic acid, and abscisic acid). Indirect stimulation, on the contrary, involves iron sequestered by bacterial siderophores, phytopathogen management, allelopathy, antibiotic production, and competition with deleterious agents (Egambediyeva and Islam 2008). Apart from their conventional activity, PGPR have also been reported to increase crop yield under stressed environment (Pandey et al. 2016). For example, *Rhizobium* and *Pseudomonas*, applied either alone or as mixture, reduced the adverse effects of salinity on growth and physiology of mung bean plants (Ahmad et al. 2013; Egamberdieva et al. 2013b) and concurrently improved the ionic balance and P content and protein concentration in grain of mung bean. This study therefore suggested that this bacterial pairing could effectively be used to augment the growth, physiology, and quality of mung bean plants even at salt-affected areas. Additionally, when PGPR alleviate the salt stress experienced by the plant, more nodules might develop into nitrogen-fixing ones, thereby enabling the plant to obtain part of its N from the atmosphere (Egamberdieva et al. 2013b). The salt tolerance of *Galega officinalis* was clearly improved when this plant was inoculated with host-specific *R. galegae*, alone or with either of the two salt-tolerant *P. extremorientalis* or *P. trivialis* (Egamberdieva et al. 2013a). Other most significant PGPR are the genus *Azospirillum*, a free-living surface colonizing (sometimes living as endophyte) diazotroph. *Azospirillum* inoculation improves root development and enhanced water and mineral uptake due to the secretion of IAA (Spaepen et al. 2007). Many reports have focused on the ability of *Azospirillum* species to promote plant growth and increase agricultural productivity through certain mechanisms that act additively or synergistically with BNF in order to enhance the overall performance of plants. For example, *Azospirillum* significantly improved yield of legumes when co-inoculated with other effective, N_2-fixing bacteria. It has been shown that dual inoculation of *Azospirillum with Rhizobium* and other PGPR (e.g., *Azotobacter*) significantly increased nodulation and N_2 fixation of *Vicia faba* l (Rodelas et al. 1996, 1999). In yet further experiment, inoculation of *A. brasilense* significantly reduced the negative effects on growth and nodulation of chickpeas and faba bean caused by saline water, used for irrigation (Hamaoui et al. 2001). In a follow-up study, the effects of *A. brasilense* inoculation on growth, nodulation, and production of flavonoids and lipochitooligosaccharides (LCOs) have been reported for a *Rhizobium-P. vulgaris* interaction under salt stress (Dardanelli et al. 2008). *A. brasilense* promoted root branching in seedlings of *P. vulgaris* and increased secretion of *nod* gene-inducing flavonoid species. The negative effects detected under salt stress on gene expression and on Nod factor production were relieved in co-inoculated plants. Phosphorus (P) is yet another important nutrient that facilitates the growth of plants including legumes. However, the locked/complex insoluble P compounds in the rhizosphere are converted by many bacteria, often called as phosphate-solubilizing bacteria (PSB) into available P form for uptake by plants. Numerous bacterial genera including *Bacillus* and *Pseudomonas* (Karpagam and Nagalakshmi 2014; Oteino et al. 2015), *Sphingomonas and Burkholderia* (Panhwar et al. 2014; Song et al. 2008), *Achromobacter* (Ma et al. 2009*)*, *Acinetobacter* (Gulati et al. 2010), and *Rhizobia* (Kenasa et al. 2014;

Satyanandam et al. 2014) are involved in P solubilization. The effects of the P-solubilizing *P. putida* on the symbiosis between rhizobia and legumes (e.g., soybean and alfalfa), usually grown in arid climates, were investigated (Rosas et al. 2006). Modification of shoot and root system dry weights occurred in soybean but not in alfalfa in the presence of *Pseudomonas* strains.

A greater number of nodules and dry weight were recorded for soybean when co-inoculated with *P. putida* and *B. japonicum*. In addition to N_2 fixation and phytohormone biosynthesis, *A. brasilense* produces specific polyamines. Among polyamines, cadaverine (1,5-diaminobentane) has been identified in *A. brasilense* and some α-proteobacteria (Bohin et al. 2005; Perrig et al. 2007). Cadaverine correlates with root growth promotion and osmotic stress mitigation in some plant species, like *V. faba* (Liu et al. 2000), lettuce *Lactuca* (Barassi et al. 2006), and *Oryza sativa* (Cassán et al. 2009).

Certain other PGPR produce 1-aminocyclopropane-1-carboxylate (ACC) deaminase (Singh and Jha 2015; Magnucka and Pietr 2015), which regulates ethylene production by metabolizing ACC (an immediate precursor of ethylene biosynthesis in higher plants) into α-ketobutyrate and ammonia (Glick 2005; Glick et al. 2007). Bacterial strains containing ACC deaminase alleviate stress-induced ethylene-mediated negative impact on plants (Safronova et al. 2006; Zia-ul-Hassan et al. 2015). PGPR containing ACC deaminase activity sustain plant growth and development under stress conditions by reducing stress induced by ethylene production (Saleem et al. 2007). Some rhizobacteria (e.g., *Bacillus* species), associated with plants in saline soils, grew and fixed N_2 at 5% NaCl (Zahran et al. 1995; Egambediyeva and Islam 2008). Seed inoculation with the salt-tolerant bacteria, *B. japonicum*, *B. polymyxa*, *B. amyloliquefaciens*, *M. phlei*, and *P. alcaligenes*, significantly increased shoot growth; root length; uptake of N, P, and K; and yield of soybean, pea, and wheat as compared to the control (Egambediyeva and Hoflich 2004). PGPB species, e.g., *Halomonas variabilis* (HT1) and *Planococcus rifietoensis* (RT4), can be used to improve growth of *Cicer arietinum* plant, as they form biofilm and accumulate exopolysaccharides at increasing salt stress (Qurashi and Sabri 2012). Inoculation of both strains increased chickpea growth at elevated salt stress treatments (up to 200 mM NaCl).

13.5   Legume-Fungal Associations Under Stressed Environments

Besides N, P plays an important role in crop productivity. Among the natural bio-resources, arbuscular mycorrhizal fungi (AMF) that form beneficial symbiotic associations with majority of plants play a pivotal role in growth and development of plants growing under varied environmental conditions. Generally, they modify the root systems and enhance the mobilization and the uptake of several essential elements. Additionally, AM fungi stimulate plant stress tolerance by enhancing enzymatic and nonenzymatic antioxidant defense systems (Wu et al. 2014; Ahmad et al. 2015), lipid peroxidation (Abd-Allah et al. 2015), and phytohormone synthesis (Navarro et al. 2013) and defending roots against soil-borne diseases (Ghorbanli et al. 2004; Rabie 2005). Traditionally, the legume-rhizobia symbiotic performance of faba bean (Yinsuo et al. 2004), lucerne (Ardakani et al. 2009), lentil (Xavier and Germida 2002), and common bean (Tajini et al. 2012) has been reported to be enhanced following application of AMF. The AM fungi also protect plants against salt stress via better access to nutritional status and plant physiology modification (Rabie and Almadini 2005) and are considered as bio-ameliorators of saline soils (Yano-Melo et al. 2003; Tain et al. 2004). In saline environments (e.g., saline-alkali soils), vesicular AM plant root colonization is host dependent and is significantly affected by various amendments (e.g., PGPR amendments) employed to reclaim such soils. Double inoculation with rhizobia and an endomycorrhizal complex increased tolerance of *A. cyanophyla* plants to salinity (Hatimi 1999). The leguminous plants possessing high level of vesicular AM colonization (50–70%) in saline-alkali soil included *A. nilotica*, *A. lebbeck*, *and Dalbergia sissoo* (Raghuwanshi and Upadhyay 2004). Mycorrhizal seedlings of two species of *Sesbania* (*S. aegyptica and S. grandiflora*) had significantly higher root and shoot dry biomass, chlorophyll content, nodule number, and increased concentrations of P, N, and Mg^{++} but lower Na^+ concentration than non-mycorrhizal seedlings (Giri and Mukerji 2004). Mycorrhizal fungus (*G. fasciculatum*) alleviated the deleterious effects on growth of *A. nilotica* plants grown in saline soils that may be related to improved P nutrition (Giri et al. 2007). The reduction of Na^+ uptake, together with a concomitant increase in P, N, and Mg^{++} absorption and high chlorophyll content in mycorrhizal plants, may be important salt-alleviating mechanisms for plants growing in saline soil. Under saline conditions (150 mM NaCl), the halotolerant legume (*L. glaber*) colonized by *M. loti and G. intraradices* was more dichotomous, and the total biomass increased (Echeverria et al. 2008). The improved K^+/Na^+ ratios in root and shoot tissues of mycorrhizal *A. nilotica* plants may help in protecting disruption of K-mediated enzymatic processes under salt stress conditions. Exposure of pigeon pea plants to salinity stress (up to 8 dS/m) markedly decreased nodule mass, acetylene reduction activity (ARA), and leghemoglobin content (Garg and Manchanda 2008). However, AM fungi inoculation significantly improved nodulation, nitrogenase activity, and leghemoglobin content of salt-stressed pigeon pea plants. Under salt stress, soybean plants inoculated with salt-pretreated AM fungi showed increased superoxide dismutase (SOD) and peroxidase (POX) activity in shoots, relative to those inoculated with the non-pretreated AM fungi (Ghorbanli et al. 2004). Further, activities of enzymes involved in the detoxification of O_2^- radicals and H_2O_2 (SOD, catalase CAT, and POX) and enzymes of the ascorbate glutathione pathway responsible for the removal of H_2O_2 (glutathione reductase GR and ascorbate peroxidase APOX) increased markedly in AM salt-stressed plants (Garg and Manchanda 2008). In *Medicago-Rhizobium-Glomus* symbiosis, subjected to drought stress, nodule activity in infected plants was significantly higher than in noninfected plants (Peña et al. 1988). AM fungi may increase drought resistance of plants by several mechanisms including enhancing water uptake due to hyphal extraction of soil water and lowering leaf osmotic potential for greater turgor maintenance by regulating photosynthesis (Sánchez-Diaz et al. 1990; Ruiz-Lozano and Azcón 1995). However, this effect is independent of the P nutrition in plant tissues.

Conclusion

The legume-rhizobia symbiosis is of tremendous ecological and agronomic importance. Optimization of symbiosis between the legumes and their respective microsymbionts requires a competitive, infective, and highly efficient N_2-fixing rhizobial strain in sufficient numbers to optimize legume production. Soil salinity is one of the major destructive environmental factors limiting the productivity of legumes worldwide. And, hence, different adaptations and mitigation strategies are required to overcome salinity impact on plants. In this regard, efficient resource management and crop improvement strategies can help to circumvent salinity stress. However, such methods are expensive, and therefore, there is a pressing need to develop simple and low-cost biological strategies to cope with salinity stress. Plant growth promoting rhizobacteria including both free living and symbiotic/associative could play a pivotal role in the remediation of stressed soils, if functional properties of such organisms, for example, the ability to tolerate high salinity, genetic diversity, ability to synthesize bioactive compounds, production of phytohormones, biocontrol potential, etc., are exploited judiciously. The application of salt-tolerant PGPR including rhizobia endowed with multiple growth promoting activities should be encouraged to enhance legume production in stressed soils. Apart from these, cultivation of legumes capable of forming symbiosis with rhizobia, especially the nitrogen-fixing trees, could be used to rehabilitate arid saline soil. Furthermore, in order to better understand the mechanistic basis of such interactions, molecular tools can be employed. Considering the potential benefits of rhizobia and other PGPR in legume-rhizobia symbiosis under both conventional and stressed environment, it is suggested that extensive research be conducted and the outcome of such investigations be tested under field environments so that such organisms are recommended to farming communities for enhancing legume production in different agronomic regions of the world.

References

Abd-Alla MH, El-enany AE, Bagy MK, Bashandy SR (2014) Alleviating the inhibitory effect of salinity stress on *nod* gene expression in *Rhizobium tibeticum* – fenugreek (*Trigonella foenum graecum*) symbiosis by isoflavonoids treatment. J Plant Interact 9:275–284

Abd-Allah EF, Hashem A, Alqarawi AA, Hend A (2015) Alleviation of adverse impact of cadmium stress in sunflower (*Helianthus annuus* L.) by arbuscular mycorrhizal fungi. Pak J Bot 47:785–795

Abdel-Salam MS, Ibrahim SA, Abd-El-Halim MM, Badawy FM, Abo-Aba SEM (2010) Phenotypic characterization of indigenous Egyptian rhizobial strains for abiotic stresses performance. J Am Sci 6:498–503

Abolhasani M, Lakzian A, Tajabadipour A, Haghnia G (2010) The study salt and drought tolerance of *Sinorhizobium* bacteria to the adaptation to alkaline condition. Aust J Basic Appl Sci 4:882–886

Aguirreolea J, Sánhez-Diaz M (1989) CO_2 evolution by nodulated roots in *Medicago sativa* L. under water stress. J Plant Physiol 134:598–602

Ahmad M, Zahir ZA, Khalid M, Nazil F, Arshad M (2013) Efficacy of *Rhizobium* and *Pseudomonas* strains to improve physiology, ionic balance and quality of mung bean under salt-affected conditions in farmer's fields. Plant Physiol Biochem 63:170–176

Ahmad P, Hashem A, Abd-Allah EF, Alqarawi AA, John R, Egamberdieva D et al (2015) Role of *Trichoderma harzianum* in mitigating NaCl stress in Indian mustard (*Brassica juncea* L.) through antioxidative defense system. Front Plant Sci 6:868

Ahmadian S, Bayat F (2016) Morphological responses to salinity tolerance in common bean (*Phaseolus vulgaris* L.) Afr J Agric Res 11:1289–1298

Alexandre A, Oliveira S (2011) Most heat-tolerant rhizobia show high induction of major chaperone genes upon stress. FEMS Microbiol Ecol 75:28–36

Alikhani HA, Mohamadi L (2010) Assessing tolerance of rhizobial lentil symbiosis isolates to salinity and drought in dry land farming condition. In: 19th world congress of soil science, soil solutions for a changing world, 1–6 Aug 2010, Brisbane, Australia

Al-Saedi SA, Razaq IB, Ali NA (2016) Utilization of ^{15}N dilution analysis for measuring efficiency of biological nitrogen fixation under soil salinity stress. World J Pharm Pharm Sci 5:1468–1479

Al-Shaharani TS, Shetta ND (2011) Evaluation of growth, nodulation and nitrogen fixation of two *Acacia* species under salt stress. World Appl Sci J 13:256–265

Al-Sherif EA, Zahran HH, Atteya MA (2004) Nitrogen fixation and chemical composition of wild annual legumes at Beni-Suef Governorate, Egypt. Egypt J Biol 6:32–38

Alva AK, Assher CJ, Edwards DG (1990) Effect of solution pH, external calcium concentration, and aluminum activity on nodulation and early growth of cowpea. Aust J Agric Res 41:359–365

Ardakani MR, Pietsch G, Moghaddam A, Raza A, Friedel JK (2009) Response of root properties to tripartite symbiosis between lucerne (*Medicago sativa* L.), rhizobia and mycorrhiza under dry organic farming conditions. Am J Agric Biol Sci 4:26–277

Aydi S, Drevon J-J, Abdelly C (2004) Effect of salinity on root-nodule conductance to the oxygen diffusion in the *Medicago truncatula-Sinorhizobium meliloti* symbiosis. Plant Physiol Biochem 42:833–840

Barassi C, Ayrault G, Creus C, Sueldo R, Sobrero M (2006) Seed inoculation with *Azospirillum* mitigates NaCl effects on lettuce. Sci Hortic 109:8–14

Bartels D, Sunkar R (2005) Drought and salt tolerance in plants. Crit Rev Plant Sci 24:23–58

Becana M, Dalton DA, Moran JF, Iturbe-Ormaetxe I, Matamoros MA, Rubio MC (2000) Reactive oxygen species and antioxidants in legume nodules. Physiol Plant 109:372–381

Beck DP, Munns DN (1985) Effect of calcium on the phosphorous nutrition of *Rhizobium meliloti*. Soil Sci Soc Am J 49:334–337

Belal EB, Hassan MM, El-Ramady HR (2013) Phylogenetic and characterization of salt-tolerant rhizobial strain nodulating faba bean plants. Afr J Biotechnol 12:4324–4337

Belimov AA, Kojemiakov AP, Chuvarliyeva CV (1995) Interaction between barley and mixed cultures of nitrogen fixing and phosphate-solubilizing bacteria. Plant Soil 173:29–37

Benaroudj N, Lee DH, Glodberg A (2001) Trehalose accumulation during cellular stress protects cells and cellular proteins from damage by oxygen radicals. J Biol Biochem 276:24261–24267

Berrada H, Fikri-Benbrahim K (2014) Taxonomy of the rhizobia: current perspectives. Br Microbiol Res J 4:616–639

Bertrand A, Dhont C, Bipfubusa M, Chalifour F-P, Drouin P, Beauchamp CJ (2015) Improving salt stress responses of the symbiosis in alfalfa using salt-tolerant cultivar and rhizobial strain. Appl Soil Ecol 87:108–117

Bhargava Y, Murthy SR, Rajesh Kumar TV, Narayana Rao M (2016) Phenotypic, stress tolerance and plant growth promoting characteristics of rhizobial isolates from selected wild legumes of semiarid region, Tirupati, India. Adv Microbiol 6:1–12

Bhatnagar-Mathur P, Devi MJ, Reddy DS, Lavanya M, Vadez V, Serraj R, Yamaguchi-Shinozaki K, Sharma KK (2007) Stress-inducible expression of *At* DREB1A in transgenic peanut (*Arachis hypogaea* L.) increases transpiration efficiency under water-limiting conditions. Plant Cell Rep 26:2071–2082

Black M, Moolhuijzen P, Chapman B, Barrero R, Howieson J, Hungria M, Bellgard M (2012) The genetics of symbiotic nitrogen fixation: comparative genomics of 14 rhizobia strains by resolution of protein clusters. Genes 3:138–166

Blokhina O, Virolainen E, Fagerstedt KV (2003) Antioxidants, oxidative damage and oxygen deprivation stress; a review. Ann Bot 91:179–194

Blumwald E, Aharon GS, Apse MP (2000) Sodium transport in plant cells. Biochim Biophys Acta 1465:140–151

Bohin A, Bouchart F, Richet C, Kol O, Leroy V, Timmerman P, Huet G, Bohin J, Zane J (2005) GC/MS identification and quantification of constituents of bacterial lipids and glycoconjugates obtained after methanolysis as heptafuorobutyrate derivatives. Ann Biochem 340:231–244

Boscari A, Mandon K, Dupont L, Poggi M-C, Le Rudulier D (2002) BetS is a major glycine betaine/proline betaine transporter required for early osmotic adjustment in *Sinorhizobium meliloti*. J Bacteriol 184:2654–2663

Boscari A, Mandon K, Poggi M-C, Le Rudulier D (2004) Functional expression of *Sinorhizobium meliloti* BetS, a high-affinity betaine transporter, in *Bradyrhizobium japonicum* USDA110. Appl Environ Microbiol 70:5916–5922

Boumahdi M, Mary P, Hornez JP (2001) Changes in fatty acid composition and degree of unsaturation of (brady) rhizobia as a response to phases of growth, reduced water activities and mild desiccation. Anton Van Leeuwen 79:73–79

Brígido C, Alexandre A, Oliveira S (2012) Transcriptional analysis of major chaperone genes in salt-tolerant and salt-sensitive mesorhizobia. Microbiol Res 167:623–629

Brígido C, Oliveira S (2013) Most acid-tolerant chickpea mesorhizobia show induction of major chaperon genes upon acid shock. Microbial Ecol 65:145–153

Cardoso P, Freitas R, Figueira E (2015) Salt tolerance of rhizobial populations from contrasting environmental conditions: understanding the implications of climate change. Ecotoxicology 24:143–152

Cassán F, Maiale S, Masciarelli O, Vidal A (2009) Cadaverine production by *Azospirillum brasilense* and its possible role in plant growth promotion and osmotic stress mitigation. Eur J Soil Biol 45:12–19

Chalk PM, Alves BJR, Boddey RM, Urquiaga S (2010) Integrate effects of abiotic stresses on inoculant performance, legume growth and symbiotic dependence estimated by ^{15}N dilution. Plant Soil 328:1–16

Chen W-M, Lee TM, Lan C-C, Cheng C-P (2000) Characterization of halotolerant rhizobia isolated from root nodules of *Canavalia rosea* from seaside areas. FEMS Microbiol Ecol 34:9–16

Chen J-L, Lin S, Lin L-P (2006) Rhizobial surface biopolymers and their interaction with lectin measured by atomic force microscopy. World J Microbiol Biotechnol 22:565–570

Chinnusamy V, Jagendorf A, Zhu JK (2005) Understanding and improving salt tolerance in plants. Crop Sci 45:437–448

Coba De La Peña T, Cárcamo CB, Almonacid L, Zaballos A, Lucas MM, Balomenos D, Pueyo JJ (2008) A salt stress-responsive cytokinin receptor homologue isolated from *Medicago sativa* nodules. Planta 227:769–779

Csonka LN (1991) Prokaryotic osmoregulation: genetics and physiology. Annu Rev Microbiol 45:569–606

Dajic Z (2006) Salt stress. In: Madhava Rao KV, Raghavendra AS, Janardhan Reddy K (eds) Physiology and molecular biology of stress tolerance in plants. Springer, Netherlands, pp 41–99

Dardanelli MS, Fernández De Córdoba FJ, Espuny MR, Rodríguez Carvajal MA, Soria Díaz ME, Gil Serrano AM (2008) Effect of *Azospirillum brasilense* coinoculated with *Rhizobium* on *Phaseolus vulgaris* flavonoids and nod factor production under salt stress. Soil Biol Biochem 40:2713–2721

De Zelicourt A, Al-Yousif M, Hirt H (2013) Rhizosphere microbes as essential partners for plant stress tolerance. Mol Plant 6:242–245

Debelle F, Moulin L, Mangin B, Denarie J, Boivin C (2001) Nod genes and nod signals and the evolution of the *Rhizobium*-legume symbiosis. Acta Biochim Pol 48:359–365

Del Cerro P, Rolla-Santos AAP, Gomes DF et al (2015) Opening the "black box" of *nodD3*, *nodD4* and *nodD5* genes of *Rhizobium tropici* strain CIAT 899. BMC Genomics 16:864

Dogra T, Priyadarshini A, Kumar A, Singh NK (2013) Identification of genes involved in salt tolerance and symbiotic nitrogen fixation in chickpea rhizobium *Mesorhizobium ciceri* Ca181. Symbiosis 61:135–143

Domínguez-Ferreras A, Pérez-Arnedo R, Becker A, Olivares J, Soto MJ, Sanjuan J (2006) Transcriptome profiling reveals the importance of plasmid pSymB for osmoadaptation of *Sinorhizobium meliloti*. J Bacteriol 188:7617–7625

Echeverria M, Scambato AA, Sannazzaro AI, Maiale S, Ruiz OA, Menéndez AB (2008) Phenotypic plasticity with respect to salt stress response by *Lotus glaber*: the role of its AM fungal and rhizobial symbionts. Mycorrhiza 18:317–329

Egambediyeva D, Hoflich G (2004) Effect of plant growth promoting bacteria on growth and nutrient uptake of cotton and pea in a semi-arid region of Uzbekistan. J Arid Environ 56:293–301

Egambediyeva D, Islam KR (2008) Salt-tolerant rhizobacteria: plant growth promoting traits and physiological characterization within ecologically stressed environments. In: Ahmad I, Pichtel J, Hayat S (eds) Plant-bacteria interactions: strategies and techniques to promote plant growth. Wiley-VCH Verlag Gmbh & Co KGaA, Weinheim, pp 257–281

Egamberdieva D, Mamedov NA (2015) Potential use of Licorice in phytoremediation of salt affected soils. In: Oztürk M et al (eds) Plants, pollutants, and remediation. Springer Science+Business Media, Dordrecht, pp 309–318

Egamberdieva D, Berg G, Lindström K, Räsänen LA (2013a) Alleviation of salt stress of symbiotic *Galega officinalis* L. (Goat's Rue) by co-inoculation of Rhizobium with root colonizing Pseudomonas. Plant Soil 369:453–465

Egamberdieva D, Jabborova D, Wirth S (2013b) Alleviation of salt stress in legumes by co-inoculation with *Pseudomonas* and *Rhizobium*. In: Arora NK (ed) Plant microbe symbiosis: fundamentals and advances. Springer, India, pp 291–303

El Sabagh A, Omar A, Saneoka H, Barutcular C (2015) Comparative physiological study of soybean (*Glycine max* L.) cultivars under salt stress. YYU J AGR SCI 25:269–284

El-Komy HMA (2005) Coimmobilization of *A. lipoferum* and *B. megaterium* for plant nutrition. Food Technol Biotechnol 43:19–27

Essendoubi M, Brhada F, Eljamail JE, Filali-Maltouf A, Bonnassie S, Georgeault S, Blanco C, Jebbar M (2007) Osmoadaptive responses in the rhizobia nodulating *Acacia* isolated from south-eastern Moroccan Sahara. Environ Microbiol 9:603–611

Estevéz J, Soria-Díaz ME, De Córdoba FF, Morón B, Manyani H, Gil A, Thomas-Oates J, Van Brussel AAN, Dardanelli MS, Sousa C, Megías M (2009) Different and new Nod factors produced by *Rhizobium tropici* CIAT899 following Na⁺ stress. FEMS Microbiol Lett 293:220–231

Faghire M, Mandri B, Oufdou K, Bargaz A, Ghoulam C, Ramírez-Bahena MH, Velazquez E, Peix A (2012) Identification at the species and symbiovar levels of strains nodulating *Phaseolus vulgaris* in saline soils of the Marrrakech region (Morocco) and analysis of the *otsA* gene putatively involved in osmotolerance. Syst Appl Microbiol 35:156–164

Fahmi AI, Nagaty HH, Eissa RA, Hassan MM (2011) Effects of salt stress on some nitrogen fixation parameters in faba bean. Pak J Biol Sci 14:385–391

FAO (2005) Global network on integrated soil management for sustainable use of salt-affected soils. FAO Land and Plant Nutrition Management Service, Rome, Italy. http://www.fao.org. ag/agl/agll/spush

Fernandez-Aunión C, Ben Hamouda T, Iglesias-Guerra F, Argandoña M, Reina-Bueno M, Nieto JJ, Aouani ME, Vargas C (2010) Biosynthesis of compatible solutes in rhizobial strains isolated from *Phaseolus vulgaris* nodules in Tunisian fields. BMC Microbiol 10:192

Franzini VI, Azcón R, Mendes FL, Aroca R (2010) Interactions between Glomus species and *Rhizobium* strains affect the nutritional physiology of drought-stressed legume hosts. J Plant Physiol 167:614–619

Galiana A, Gnahoua GM, Chaumont J, Lesueur D, Prin Y, Mallet B (1998) Improvement of nitrogen fixation in *Acacia mangium* through inoculation with *Rhizobium*. Agroforest Syst 40:297–307

Galibert F, Finan TM, Long SR et al (2001) The composite gnome of the legume symbiont *Sinorhizobium meliloti*. Science 293:668–672

Garg N, Manchanda G (2008) Effect of mycorrhizal inoculation on salt-induced nodule senescence in *Cajanus cajan* (pigeonpea). J Plant Growth Regul 17:115–124

Garthwaite AJ, Von Bothmer R, Colmer TD (2005) Salt tolerance in wild *Hordeum* species is associated with restricted entry of Na^+ and Cl^- into the shoots. J Exp Bot 56:2365–2378

Geurts R, Fedorova E, Bisseling T (2005) Nod factor signaling genes and their function in the early stages of *Rhizobium* infection. Curr Opin Plant Biol 8:346–352

Ghasem F, Poustini K, Besharati H, Mohammadi VA, Abooei Mehrizi F, Goettfert M (2012) Pre-incubation of *Sinorhizobium meliloti* with luteolin, methyl jasmonate and genistein affecting alfalfa (*Medicago sativa* L.) growth, nodulation and nitrogen fixation under salt stress conditions. J Agric Sci Technol 14:1235–1264

Ghorbanli M, Ebrahimzadeh H, Sharifi M (2004) Effect of NaCl and mycorrhizal fungi on antioxidative enzymes in soybean. Biol Plant 48:575–581

Giri B, Mukerji KG (2004) Mycorrhizal inoculation alleviates salt stress in *Sesbania aegyptica* and *Sesbania grandiflora* under field conditions: evidence for reduced sodium and improved magnesium uptake. Mycorrhiza 14:307–312

Giri B, Kapoor R, Mukerji KG (2002) Va mycorrhizal techniques/VAM technology in establishment of plants under salinity stress conditions. In: Mukerji KG, Manorahari C, Singh J (eds) Techniques in mycorrhizal studies. Kluwer, Dordrecht, The Netherlands, pp 313–327

Giri B, Kapoor R, Mukerji KG (2007) Improved tolerance of *Acacia nilotica* to salt stress by arbuscular mycorrhiza, *Glomus fasciculatum* may be partly related to elevated K/Na ratios in root and shoot tissues. Microb Ecol 54:753–760

Glick BR (2005) Modulation of plant ethylene levels by the bacterial enzyme ACC deaminase. FEMS Microbiol Lett 251:1–7

Glick BR, Todorovic B, Czarny J, Cheng Z, Duan J, McConkey B (2007) Promotion of plant growth by bacterial ACC deaminase. Crit Rev Plant Sci 26:227–242

Gonzalez-Rizzo S, Crespi M, Frugier F (2006) The *Medicago truncatula* CRE1 cytokinin receptor regulates lateral root development and early symbiotic interaction with *Sinorhizobium meliloti*. Plant Cell 18:2680–2693

Graham PH, Vance CP (2000) Nitrogen fixation in perspective: an overview of research and extension needs. Field Crops Res 65:93–106

Graham PH, Vance CP (2003) Legumes: importance and constraints to greater use. Plant Physiol 131:872–877

Graham PH, Draeger KJ, Ferrey ML, Conroy MJ, Hammer BE, Martinez E, Arons SR, Quinto C (1994) Acid pH tolerance in strains of *Rhizobium* and *Bradyrhizobium*, and initial studies on the basis for pH tolerance of *Rhizobium tropici* UMR 1899. Can J Microbiol 40:198–207

Gualtieri G, Bisseling T (2000) The evolution of nodulation. Plant Mol Biol 42:181–194

Guasch-Vidal B, Estévez J, Dardanelli MS et al (2013) High NaCl concentrations induce the *nod* genes of *Rhizobium tropici* CIAT899 in the absence of flavonoid inducers. MPMI 26:451–460

Gulati A, Sharma N, Vyas P, Sood S, Rahi P, Pathania V, Prasad R (2010) Organic acid production and plant growth promotion as a function of phosphate solubilization by *Acinetobacter rhizosphaerae* strain BIHB 723 isolated from the cold deserts of the trans-Himalayas. Arch Microbiol 192:975–983

Hamaoui B, Abbadi JM, Burdman S, Rashid A, Sarig A, Okon Y (2001) Effects of inoculation of *Azospirillum brasilense* on chickpeas (*Cicer arietinum*) and faba bean (*Vicia faba*) under different growth conditions. Agronomie 21:553–560

Hamdy A (1990) Management practices under saline water irrigation. Acta Hortic 278:745–754

Hare PD, Cress WA, Van Staden J (1997) The involvement of cytokinins in plant responses to environmental stress. Plant Growth Regul 23:79–103

Harun-or-Rashid M, Krehenbrink M, Akhtar MS (2015) Nitrogen-fixing plant-microbe symbiosis. In: Lichtfouse E (ed) Sustainable agriculture reviews. Springer International Publishing, Switzerland, pp 193–234

Hasegawa PM, Bressan R, Zhu J-K, Bohmert H-J (2000) Plant cellular and molecular responses to high salinity. Annu Rev Plant Physiol Plant Mol Biol 51:463–499

Hatimi A (1999) Effect of salinity on the association between root symbionts and *Acacia cyanophyla* Lind: growth and nutrition. Plant Soil 216:93–101

Horie T, Schroeder JI (2004) Sodium transporters in plants. Diverse gene and physiological functions. Plant Physiol 136:2457–2462

Hungria M, Franco AA (1993) Effects of high temperature on nodulation and nitrogen fixation by *Phaseolus vulgaris* (L.) Plant Soil 149:95–102

Hungria M, Stacey G (1997) Molecular signals exchanged between host plants and rhizobia, basic aspects and potential application in agriculture. Soil Biol Biochem 29:519–530

Hungria M, Vargas MAT (2000) Environmental factors affecting N_2 fixation in grain legumes in the tropics, with emphasis on Brazil. Field Crops Res 65:151–164

Hungria M, Franco AA, Sprent JI (1993) New sources of high-temperature tolerant rhizobia for *Phaseolus vulgaris* (L.) Plant Soil 149:103–109

Ibragimova MV, Rumyantseva ML, Onishchuk OP, Belova VS, Kurchak ON, Andronov EE, Dzyubenko NI, Simarov BV (2006) Symbiosis between the root-nodule bacterium *Sinorhizobium meliloti* and alfalfa (*Medicago sativa*) under salinization conditions. Microbiology 75:77–81

Jebara S, Jebara M, Limam F, Aouani ME (2005) Changes in ascorbate peroxidase, catalase, guaiacol peroxidase and superoxide dismutase activities in common bean (*Phaseolus vulgaris*) nodules under salt stress. J Plant Physiol 162:929–936

Jiang M, Zhang J (2002) Water stress-induced abscisic acid accumulation triggers the increased generation of reactive oxygen species and up-regulates the activities of antioxidant enzymes in maize leaves. J Exp Bot 53:2401–2410

Jiang JQ, Wei W, Du BH, Li XH, Wang L, Yang SS (2004) Salt-tolerance genes involved in cation efflux and osmoregulation of *Sinorhizobium fredii* RT19 detected by isolation of Tn5 mutants. FEMS Microbiol Lett 239:139–146

Jin H, Sun Y, Yang Q, Chao Y, Kang J, Jin H, Li Y, Margaret G (2010) Screening of genes induced by salt stress from alfalfa. Mol Biol Rep 37:745–753

Karpagam T, Nagalakshmi PK (2014) Isolation and characterization of phosphate solubilizing microbes from agricultural soil. Int J Curr Microbiol Appl Sci 3:601–614

Katerji N, Mastrorilli M, Lahmer FZ, Maalouf F, Oweis T (2011) Faba bean productivity in saline-drought conditions. Eur J Agron 35:2–12

Kenasa G, Jida M, Assefa F (2014) Characterization of phosphate solubilizing faba bean (*Vicia faba L.*) nodulating rhizobia isolated from acidic soils of Wollega, Ethiopia. Sci Technol Arts Res J 3:11–17

Keshri J, Mishra A, Jha B (2013) Microbial population index and community structure in saline-alkaline soil using gene targeted metagenomics. Microbiol Res 168:165–173

Khadri M, Tejera NA, Lluch C (2007) Sodium chloride-ABA interaction in two common bean (*Phaseolus vulgaris*) cultivars differing in salinity tolerance. Environ Exp Bot 60:211–218

Khan MS, Zaidi A, Wani PA, Ahmad M, Oves M (2009) Functional diversity among plant growth-promoting rhizobacteria: current status. In: Khan MS, Zaidi A, Musarrat J (eds) Microbial strategies for crop improvement. Springer-Verlag, Germany, pp 105–132

Kim G-B, Nam Y-W (2013) A novel Δ1-pyroline-5-carboxylate synthetase gene of *Medicago truncatula* plays a predominant role in stress-induced proline accumulation during symbiotic nitrogen fixation. J Plant Physiol 170:291–302

Kishor PBK, Sangam S, Amrutha RN, Laxmi PS, Naidu KR, Rao KRSS, Rao S, Reddy KJ, Theriappan P, Sreenivasulu N (2005) Regulation of proline biosynthesis, degradation, uptake and transport in higher plants: its implications in plant growth and abiotic stress tolerance. Curr Sci 88:424–438

Kononowics AK, Nelson DE, Singh NK, Hasegawa PM, Bressan RA (1992) Regulation of the osmotin gene promoter. Plant Cell 4(513):524

Kulkarni S, Nautiyal CS (1999) Characterization of high-temperature tolerant rhizobia isolated from *Prosopis juliflora* grown in alkaline soil. J Gen Appl Microbiol 45:213–220

Lagares A, Sanjuán J, Pistorio M (2014) The plasmid mobilome of the model plant-symbiont *Sinorhizobium meliloti*: coming up with new questions and answers. Microbiol Spectr 2(5)

Laranjo M, Oliveira S (2011) Tolerance of *Mesorhizobium* type strains to different environmental stresses. Anton Van Leeuwen 99:651–662

Laranjo M, Alexandre A, Oliveira S (2014) Legume growth-promoting rhizobia: an overview on the *Mesorhizobium* genus. Microbiol Res 169:2–17

Latrach L, Farissi M, Mouradi M, Makoudi B, Bouizgaren A, Ghoulam C (2014) Growth and nodulation of alfalfa-rhizobia symbiosis under salinity: electrolyte leakage, stomatal conductance, and chlorophyll fluorescence. Turk J Agric For 38:320–326

Laurette NN, Maxémilienne NB, Henri F, Souleymanou A, Kamdem K, Albert N, Dieudonné N, Franços-Xavier E (2015) Isolation and screening of indigenous Bambara groundnut (*Vigna subterranea*) nodulating bacteria for their tolerance to some environmental stresses. Am J Microbiol Res 3:65–75

Lebrazi S, Benbrahim KF (2014) Environmental stress conditions affecting the N_2 fixing *Rhizobium*-legume symbiosis and adaptation mechanisms. Afr J Microbiol Res 8:4053–4061

Li M, Li Y, Chen WF, Sui XH, Li Y Jr, Li Y, Wang ET, Chen WX (2012) Genetic diversity, community structure and distribution of rhizobia in the root nodules of *Caragana* spp. from arid and semi-arid alkaline deserts, in the north of China. Syst Appl Microbiol 35:239–245

Lindström K, Lipsanen P, Kaijalainen S (1990) Stability of markers used for identification of two *Rhizobium galegae* inoculation strains after five years in the field. Appl Environ Microbiol 56:444–450

Lindström K, Murwira M, Willems A, Altier N (2010) The biodiversity of beneficial microbe-host mutualism: the case of rhizobia. Res Microbiol 161:453–463

Lira Junior MA, Nascimento LRS, Fracetto GGM (2015) Legume rhizobia signal exchange: promiscuity and environmental effects. Front Microbiol 6:945

Liu K, Fu H, Bei K, Luan S (2000) Inward potassium channel in guard cells as a target for polyamine regulation of stomatal movements. Plant Physiol 124:1315–1325

Liu X, Luo Y, Mohamed OA, Liu D, Wei G (2014) Global transcriptome analysis of Mesorhizobium alhagi CCNWXJ12-2 under salt stress. MBC Microbiol 14:319

Long R-C, Li M-N, Kang JM, Zhang TJ, Sun Y, Yang QC (2015) Small RNA deep sequencing identifies novel and salt stress-regulated microRNAS from roots of *Medicago sativa* and *Medicago truncatula*. Physiol Plant 154:13–27

López-Gómez M, Cobos-Porras L, Prell J, LLuch C (2016) Homospermidine synthase contributes to salt tolerance in free-living *Rhizobium tropici* and in symbiosis with *Phaseolus vulgaris*. Plant Soil 404:413–425

Lorite MJ, Muñoz S, Olivares J, Soto MJ, Sanjuán J (2010) Characterization of strains unlike *Mesorhizobium loti* that nodulate *Lotus* spp. in saline soils of Granada, Spain. Appl Environ Microbiol 76:4019–4026

L'taief B, Sifi B, Zaman-Allah M, Drevon J-J, Lachaâl M (2007) Effect of salinity on root-nodule conductance to the oxygen diffusion in the *Cicer arietinum-Mesorhizobium ciceri* symbiosis. J Plant Physiol 164:1028–1036

Ma Y, Rajkumar M, Freitas H (2009) Inoculation of plant growth promoting bacterium *Achromobacter xylosoxidans* strain Ax10 for the improvement of copper phytoextraction by *Brassica juncea*. J Environ Manag 90:831–837

Magnucka EG, Pietr SJ (2015) Various effects of fluorescent bacteria of the genus *Pseudomonas* containing ACC deaminase on wheat seedling growth. Microbiol Res 181:112–119

Mahdhi M, Mars M (2006) Genotypic diversity of rhizobia isolated from *Retama raetam* in arid regions of Tunisia. Ann Microbiol 56:305–311

Mahdhi M, Nzoué A, De Lajudie P, Mars M (2008) Characterization of root-nodulating bacteria on *Retama raetam* in arid Tunisian soils. Prog Nat Sci 18:43–49

Manchanda G, Garg N (2008) Salinity and its effects on the functional biology of legumes. Acta Physiol Plant 30:595–618

Marcar NE, Dart P, Sweeney C (1991) Effect of root zone salinity on growth and chemical composition of *Acacia ampliceps* B.R. Maslin, *A. auriculiformis*, *A. cunn*. ex. Benth. and *A. mangium* wild at two nitrogen levels. New Phytol 119:567–573

Mathesius U, Charon C, Rolfe BG, Kondorosi A, Crespi M (2000) Temporal and spatial order of events during the induction of cortical cell divisions in white clover by *Rhizobium leguminosarum* bv. *trifolii* inoculation or localized cytokinin addition. Mol Plant Microbe Interact 13:617–628

Mathesius U, Keijzers G, Natera SHA et al (2001) Establishment of a root proteome reference map for the model legume *Medicago truncatula* using the expressed sequence tag database for peptide mass fingerprinting. Proteomics 1:1424–1440

McKay IA, Djordjevic MA (1993) Production and excretion of nod metabolites by *Rhizobium leguminosarum* bv. *trifolii* are disrupted by the same environmental factors that reduce nodulation in the field. Appl Environ Microbiol 59:3385–3392

Melchior-Marroquin JI, Vargas-Hernandez JJ, Herrera-Cerrato R, Krishnamurthy L (1999) Screening *Rhizobium* spp. strains associated with *Gliricidia sepium* along an altidudinal transect in Veracruz, Mexico. Agroforest Syst 46:25–38

Meng N, Yu B-J, Guo J-S (2016) Ameliorative effects of inoculation with *Bradyrhizobium japonicum* on *Glycine max* and *Glycine soja* seedlings under salt stress. Plant Growth Regul. doi:10.1007/s10725-016-0150-6

Mhadhbi H, Aouani ME (2008) Growth and nitrogen-fixing performances of *Medicago truncatula-Sinorhizobium meliloti* symbiosis under salt stress (NaCl) stress: micro-and macrosymbiont contribution into symbiosis tolerance. In: Abdelly C, Öztürk M, Ashraf M, Grignon C (eds) Biosaline agriculture and high salinity tolerance. Birkhäuser Verlag, Switzerland, pp 91–98

Mhadhbi H, Jebara M, Zitoun A, Limam F, Aouani ME (2008) Symbiotic effectiveness and response to mannitol-mediated osmotic stress of various chickpea-rhizobia associations. World J Microbiol Biotechnol 24:1027–1035

Michiels J, Verreth C, Vnderleyden J (1994) Effects of temperature on bean-nodulating *Rhizobium* strains. Appl Environ Microbiol 60:1206–1212

Mills D, Zhang G, Benzioni A (2001) Effect of different salt and of ABA on growth and mineral uptake in jojoba shoots grown in vitro. J Plant Physiol 158:1031–1039

Minchin FR (1997) Regulation of oxygen diffusion in legume nodules. Soil Biol Biochem 29:881–888

Moawad H, Beck D (1991) Some characteristics of *Rhizobium leguminosarum* isolates from uninoculated field-grown lentil. Soil Biol Biochem 23:917–925

Mohammadi K, Sohrabi Y, Heidari G, Khalesro S, Majidi M (2012) Effective factors on biological nitrogen fixation. Afr J Agric Res 7:1782–1788

Monica N, Vidican R, Pop R, Rotar I (2013) Stress factors affecting symbiosis activity and nitrogen fixation by *Rhizobium* cultured *in vitro*. ProEnvironment 6:42–45

Morón B, Soria-Díaz ME, Ault J et al (2005) Low pH changes the profile of nodulation factors produced by *Rhizobium tropici* CIAT899. Chem Biol 12:1029–1040

Munns DN (1986) Acid soils tolerance in legumes and rhizobia. Adv Plant Nutr 2:63–91

Munns R (2005) Genes and salt tolerance: bringing them together. New Phytol 167:645–663

Muñoz N, Robert G, Melchiorre M, Racca R, Lascano R (2012) Saline and osmotic stress differentially affects apoplastic and intracellular reactive oxygen species production, curling and death of root hair during *Glycine max* L.-*Bradyrhizobium japonicum* interaction. Environ Exp Bot 78:76–83

Nabil M, Coudret A (1995) Effects of sodium chloride on growth, tissue elasticity and solute adjustment in two *Acacia nilotica* subspecies. Physiol Plant 93:217–224

Navarro JM, Perez-Tornero O, Morte A (2013) Alleviation of salt stress in citrus seedlings inoculated with arbuscular mycorrhizal fungi depends on the rootstock salt tolerance. J Plant Physiol 171:76–85

Nguyen NT, Moghaieb REA, Saneoka H, Fujita K (2004) RAPD markers associated with salt tolerance in *Acacia auriculiformis* and *Acacia mangium*. Plant Sci 167:797–805

Nijiti CF, Galiana A (1996) Symbiotic properties and *Rhizobium* requirements for effective nodulation of five tropical dry zone acacias. Agroforest Syst 34:265–275

Novikova TI, Gordienko NY (1999) Specific features of functioning of the symbiotic system *Rhizobium-Glycyrrhiza uralensis* under the conditions of chloride salinization. Sibirskii Ekolog Z 3:295–302

Ögütçü H, Kasimoğlu C, El Koca E (2010) Effects of rhizobium strains isolated from wild chickpeas on the growth and symbiotic performance of chickpeas (*Cicer arietinum* L.) under salt stress. Turk J Agric For 34:361–371

Okazaki S, Noisangiam R, Okubo T, Kaneko T, Oshima K et al (2015) Genome analysis of a novel *Bradyrhizobium* sp. DOA9 carrying a symbiotic plasmid. PLoS One 10(2):e0117392

Onishi M, Tachi H, Kojima T, Shiraiwa M, Takahara H (2006) Molecular cloning and characterization of a novel salt-inducible gene encoding an acidic isoform of PR-5 protein in soybean. Plant Physiol Biochem 44:574–580

Oteino N, Lally RD, Kiwanuka S, Lloyd A, Ryan D, Germaine KJ, Dowling DN (2015) Plant growth promotion induced by phosphate solubilizing endophytic *Pseudomonas* isolates. Front Microbiol 6:745. doi:10.3389/fmicb.2015.00745

Pandey S, Verma A, Chakraborty D (2016) Potential use of rhizobacteria as biofertilizer and its role in increasing tolerance to drought stress. In: Pati BR, Mandal SM (eds) Recent trends in biofertilizers. IK International Publishing House Pvt. Ltd., New Delhi, India, pp 115–140

Panhwar QA, Naher UA, Jusop S, Othman R, Latif MA, Ismail MR (2014) Biochemical and molecular characterization of potential phosphate solubilizing bacteria in acid sulphate soils and their beneficial effects on rice growth. PLoS One 9:e97241. PMC4186749

Pardo JM, Cubero B, Leidi EO, Quintero FJ (2006) Alkali cation exchangers: roles in cellular homeostasis and stress tolerance. J Exp Bot 57:1181–1199

Paul D (2012) Osmotic stress adaptations in rhizobacteria. J Basic Microbiol 52:1–10

Payakapong W, Tittabutr P, Teaumroong N, Boonkerd N, Singleton PW, Borthakur D (2006) Identification of two clusters of genes involved in salt tolerance in *Sinorhizobium* sp. strain BL3. Symbiosis 41:47–53

Peix A, Ramírez-Bahena MH, Velázquez E, Bedmar EJ (2015) Bacterial associations with legumes. Crit Rev Plant Sci 34:17–42

Peña JI, Sánchez-Diaz M, Aguirreolea J, Becana M (1988) Increased stress tolerance of nodule activity in the *Medicago-Rhizobium* Glomus symbiosis under drought. J Plant Physiol 133:79–83

Perrig D, Boiero L, Masciarelli O, Penna C, Ruíz O, Cassán F, Luna V (2007) Plant growth promoting compounds produced by two agronomically important strains of *Azospirillum brasilense*, and their implications for inoculant formulation. Appl Microbiol Biotechnol 75:1143–1150

Porcel R, Barea JM, Ruiz-Lozano JM (2003) Antioxidant activities in mycorrhizal soybean plants under drought stress and their possible relationship to the process of nodule senescence. New Phytol 157:135–143

Predeepa RJ, Ravindran DA (2010) Nodule formation, distribution and symbiotic efficacy of *Vigna unguiculata* L. under different soil salinity regimes. Emir J Food Agric 22:275–284

Priefer UB, Aurag J, Boesten B, Bouhmouch I, Defez R, Filali-Maltouf A, Miklis M, Moawad H, Mouhsine B, Prell J, Schlüter A, Senatore B (2001) Characterization of *Phaseolus* symbionts isolated from Mediterranean soils and analysis of genetic factors related to pH tolerance. J Biotechnol 91:223–236

Qadir M, Ghafoor A, Murtaza G (2000) Amelioration strategies for saline soils: a review. Land Degrad Dev 11:501–521

Qurashi AW, Sabri AN (2012) Bacterial exopolysaccharide and biofilm formation stimulate chickpea growth and soil aggregation under salt stress. Braz J Microbiol 43:1183–1191

Rabie GH (2005) Influence of VA-mycorrhizal fungi and kinetin on the response of mungbean plants to irrigation with seawater. Mycorrhiza 15:225–230

Rabie GH, Almadini AM (2005) Role of bioinoculants in development of salt-tolerance of *Vicia faba* plants under salinity stress. Afr J Biotechnol 4:210–222

Raghuwanshi R, Upadhyay RS (2004) Performance of vesicular-arbuscular mycorrhizae in saline-alkali soil in relation to various amendments. World J Microbiol Biotechnol 20:1–5

Räsänen LA, Lindström K (2003) Effects of biotic and a biotic constraints on the symbiosis between rhizobia and the tropical leguminous trees *Acacia* and *Prosopis*. Ind J Exp Biol 41:1142–1159

Raza S, Jørnsgård B, Abou-Taleb H, Christiansen JL (2001) Tolerance of *Bradyrhizobium* sp. (*Lupini*) strains to salinity, pH, CaCO₃ and antibiotics. Lett Appl Microbiol 32:379–383

Redondo FJ, De LA, Peña TC, Lucas MM, Pueyo JJ (2012) Alfalfa nodules elicited by a flavodoxin-overexpressing *Ensifer meliloti* strain display nitrogen-fixing activity with enhanced tolerance to salinity stress. Planta 236:1687–1700

Rehman A, Nautiyal CS (2002) Effect of drought on the growth and survival of the stress-tolerant bacterium *Rhizobium* sp. NBRI2505 sesbania and its drought sensitive trnsposon Tn5 mutant. Curr Microbiol 45:368–377

Rengasamy P (2006) World salinization with emphasis on Australia. J Exp Bot 57:1017–1023

Riccillo PM, Muglia CI, De Brujin FJ, Row AJ, Booth IR, Aguilar OM (2000) Glutathione is involved in environmental stress responses in *Rhizobium tropici*, including acid tolerance. J Bacteriol 182:1748–1753

Richardson AE, Simpson RJ, Djordjevic MA, Rolfe BG (1988) Expression of nodulation genes in *Rhizobium leguminosarum* bv. *trifolii* is affected by low pH and by Ca²⁺ and Al ions. Appl Environ Microbiol 54:2541–2548

Rodelas B, González-López J, Salmerón V, Pozo C, Martínez-Toledo MV (1996) Enhancement of nodulation, N₂ fixation and growth of faba bean (*Vicia faba* L.) by combined inoculation with *Rhizobium leguminosarum* bv. *viceae* and *Azospirillm brasilense*. Symbiosis 21:175–186

Rodelas B, González-López J, Martínez-Toledo MV, Pozo C, Salmerón V (1999) Influence of *Rhizobium/Azotobacter* and *Rhizobium/Azospirillum* combined inoculation on mineral composition of faba bean (*Vicia faba* L.) Biol Fertil Soils 29:165–169

Rolfe BG, Mathesius U, Djordjevic M, Weinman J, Hocart C, Weiller G, Bauer WD (2003) Proteomic analysis of legume-microbe interactions. Comp Funct Genome 4:225–228

Rommi DH, Ahmad R, Ismail S, Ghaffar A (2002) Effect of salt stress on *Rhizobium* and growth of *Acacia ampliceps*. In: Ahmad R, Malik KA (eds) Prospects for saline agriculture. Springer, Netherlands, pp 297–308

Rosas SB, Andrés JA, Rovera M, Correa NS (2006) Phosphate-solubilizing *Pseudomonas putida* can influence the rhizobia-legume symbiosis. Soil Biol Biochem 38:3502–3505

Roumiantseva ML, Muntyan VS (2015) Root nodule bacteria *Sinorhizobium meliloti*: tolerance to salinity and bacterial genetic determinants. Microbiology 84:263–280

Roumiantseva ML, Muntyan VS, Mengoni A, Simarov BV (2014) ITS-polymophism of salt-tolerant and salt-sensitive native isolates of *Sinorhizobium meliloti*-symbionts of alfalfa, clover and fenugreek plants. Russ J Genet 50:348–359

Rüberg S, Tian Z-X, Krol E, Linke B, Meyer F, Wang Y, Pühler A, Weidner S, Becker A (2003) Construction and validation of a *Sinorhizobium meliloti* whole genome DNA microarray: genome-wide profiling of osmoadaptive gene expression. J Biotechnol 106:255–268

Ruiz-Lozano JM (2003) Arbuscular mycorrhizal symbiosis and alleviation of osmotic stress. New perspectives for molecular studies. Mycorrhiza 13:309–317

Ruiz-Lozano JM, Azcón R (1995) Hyphal contribution to water uptake in mycorrhizal plants as affected by the fungal species and water status. Physiol Plant 95:472–478

Ruiz-Lozano JM, Collados C, Barea JM, Azcón R (2001) Arbuscular mycorrhizal symbiosis can alleviate drought-induced nodule senescence in soybean plants. New Phytol 151:493–502

Ruiz-Lozano JM, Porcel R, Aroca R (2006) Does the enhanced tolerance of arbuscular mycorrhizal plants to water deficit involve modulation of drought-induced plant genes? New Phytol 171:693–698

Sadowsky MJ (2005) Soil stress factors influencing symbiotic nitrogen fixation. In: Werner D, Newton WE (eds) Nitrogen fixation in agriculture, forestry, ecology, and the environment. Springer, The Netherlands, pp 89–112

Safronova VI, Stepanok VV, Engqvist GL, Alekseyev Y, Belimov AA (2006) Root-associated bacteria containing 1-aminocyclopropane-1-carboxylate deaminase improve growth and nutrient uptake by pea genotypes cultivated in cadmium supplemented soil. Biol Fertil Soils 42:267–272

Saleem M, Arshad M, Hussain S, Bhatti AS (2007) Perspective of plant growth promoting rhizobacteria (PGPR) containing ACC deaminase in stress agriculture. J Ind Microbiol Biotechnol 34:635–648

Sampedro JG, Uribe S (2004) Trehalose-enzyme interactions result in structure stabilization and activity inhibition. The role of viscosity. Mol Cell Biochem 256-257:127–319

Sánchez-Diaz M (2001) Adaptation of legumes to multiple stresses in Mediterranean-type environments. Options Mediterran 45:145–151

Sánchez-Diaz M, Pardo M, Antolin MC, Peña J, Aguirreolea J (1990) Effect of water stress on photosynthetic activity in the *Medicago-Rhizobium* Glomus symbiosis. Plant Sci 71:215–221

Satyanandam T, Babu K, Rosaiah G, Vijayalakshmi M (2014) Screening of *Rhizobium* strains isolated from the root nodules of *Vigna mungo* cultivated in rice fallows for their phosphate solubilizing ability and enzymatic activities. Br Microbiol Res J 4:996–1006

Sauviac L, Philippe H, Phok K, Bruand C (2007) An extracytoplasmic function sigma factor acts as a general stress response regulator in *Sinorhizobium meliloti*. J Bacteriol 189:4204–4216

Saxena AK, Rewari RB (1991) The influence of phosphate and zinc on growth, nodulation and mineral composition of chickpea (*Cicer arietinum* L.) World J Microbiol Biotechnol 7:202–205

Segovia L, Pinero D, Palacios R, Martinez-Romero E (1991) Genetic structure of a soil population of nonsymbiotic *Rhizobium leguminosarum*. Appl Environ Microbiol 57:426–430

Serraj R, Sinclair T, Purcell L (1999) Symbiotic N_2 fixation response to drought. J Exp Bot 50:143–155

Shamseldin A, Werner D (2005) High salt and high pH tolerance of new isolated *Rhizobium etli* strains from Egyptian soils. Curr Microbiol 50:11–16

Shamseldin A, Nyalwidhe J, Werner D (2006) A proteomic approach towards the analysis of salt tolerance in *Rhizobium etli* and *Sinorhizobium meliloti* strains. Curr Microbiol 52:333–339

Sharma MP, Bhatia NP, Adholeya A (2001) Mycorrhizal dependency and growth responses of *Acacia nilotica* and *Albizzia lebbeck* to inoculation by indigenous AM fungi as influenced by available soil P levels in a semi-arid Alfisol wasteland. New For 21:89–104

Sheng XF (2005) Growth promotion and increased potassium uptake of cotton and rape by a potassium releasing strain of *Bacillus edaphicus*. Soil Biol Biochem 37:1918–1922

Shetta ND (2015) Influence of drought stress on growth and nodulation of *Acacia origena* (Hunde) inoculated with indigenous *Rhizobium* isolates from Saudia Arabia. American-Eurasian J Agric Environ Sci 15:699–706

Shrivastava P, Kumar R (2015) Soil salinity: a serious environmental issue and plant growth promoting bacteria as one of the tools for its alleviation. Saudi J Biol Sci 22:123–131

Sibole JV, Cabot C, Poschenrieder C, Barceló J (2003) Efficient leaf ion partitioning, an overriding condition or abscisic acid-controlled stomatal and leaf growth responses to NaCl salinization in two legumes. J Exp Bot 54:2111–2119

Singh RP, Jha PN (2015) Plant growth promoting potential of ACC deaminase rhizospheric bacteria isolated from *Aerva javanica*: a plant adapted to saline environments. Int J Curr Microbiol Appl Sci 4:142–152

Singh AK, Singh G, Bhatt RP (2015) Effects of salt stress on cell surface properties and symbiotic performance of root nodulating bacteria. UK J Pharm Biosci 3:23–29

Smit G, Swart S, Lugtenberg BJJ, Kijne JW (1992) Molecular mechanisms of attachment of *Rhizobium* bacteria to plant roots. Mol Microbiol 6:2897–2903

Smith LT, Smith GM, D'Souza MR, Pocard J-A, Le Rudulier D, Madkour MA (1994) Osmoregulation in *Rhizobium meliloti*: mechanism and control by other environmental signals. J Exp Zool 268:162–165

Sobti S, Belhadj HA, Djaghoubi A (2015) Isolation and characterization of the native rhizobia under hyper-salt edaphic conditions in Ouargla (southeast Algeria). Energy Procedia 74:1434–1439

Song OR, Lee SJ, Lee YS, Lee SC, Kim KK, Choi YL (2008) Solubilization of insoluble inorganic phosphate by *Burkholderia cepacia* DA 23 isolated from cultivated soil. Braz J Microbiol 39:151–156

Soussi M, Ocaña A, Lluch C (1998) Effect of salt stress on growth, photosynthesis and nitrogen fixation in chick-pea (*Cicer arietinum* L.) J Exp Bot 49:1329–1337

Spaepen S, Vanderleyden J, Remans R (2007) Indole-3-acetic acid in microbial and microorganism-plant signaling. FEMS Microbiol Rev 31:425–448

Sprent JI (2006) Evolving ideas of legume evolution and diversity: a taxonomic perspective on the occurrence of nodulation. New Phytol 174:11–25

Sprent JI (2008) 60 Ma of legume nodulation: what's new? what's changing? J Exp Bot 59:1081–1084

Sugawara M, Cytryn EJ, Sadowsky MJ (2010) Functional role of *Bradyrhizobium japonicum* trehalose biosynthesis and metabolism genes during physiological stress and nodulation. Appl Environ Microbiol 76:1071–1081

Swaine EK, Swaine MD, Killham K (2007) Effect of drought on isolates of *Bradyrhizobium elkanii* cultured from *Albizzia adianthifolia* seedlings of different provenances. Agroforest Syst 69:135–145

Tachi H, Fukuda-Yamada K, Kojima T, Shiraiwa M, Takahara H (2009) Molecular characterization of a novel soybean gene encoding a neutral PR-5 protein induced by high-salt stress. Plant Physiol Biochem 47:73–79

Tain CY, Feng G, Li XL, Zhang PS (2004) Different effects of arbuscular mycorrhizal fungus isolates from saline or non-saline soil on salinity tolerance of plants. Appl Soil Ecol 26:143–148

Tajini F, Trabelsi M, Drevona JJ (2012) Combined inoculation with *Glomus intraradices* and *Rhizobium tropici* CIAT899 increases phosphorus use efficiency for symbiotic nitrogen fixation in common bean (*Phaseolus vulgaris* L.) Saudi J Biol Sci 19:157–163

Talbi C, Argandoña M, Salvador M, Allché JD, Vargas C, Bedmar EJ, Delgado MJ (2013) *Burkholderia phymatum* improves salt tolerance of symbiotic nitrogen fixation in *Phaseolus vulgaris*. Plant Soil 367:673–685

Tejera NA, Campos R, Sanjuan J, Lluch C (2004) Nitrogenase and antioxidant enzyme activities in *Phaseolus vulgaris* nodules formed by *Rhizobium tropici* isogenic strains with varying tolerance to salt stress. J Plant Physiol 161:329–338

Terpolilli JJ, Hood GA, Poole PS (2012) What determines the efficiency of N_2-fixing *Rhizobium*-legume symbioses? :–

Tittabutr P, Payakapong W, Teaumroong N, Boonkerd N, Singleton PW, Borthakur D (2006) The alternative sigma factor RpoH2 is required for salt tolerance in *Sinorhizobium* sp. strain BL3. Res Microbiol 157:811–818

Tiwari RP, Reeve WG, Dilworth MJ, Glenn AR (1996) Acid tolerance in *Rhizobium meliloti* strains WSM419 involved a two-component sensor-regulator system. Microbiology 142:1693–1704

Tomar OS, Minhas PS, Sharma VK, Singh YP, Gupta RK (2003) Performance of 31 tree species and soil conditions in a plantation established with saline irrigation. Forest Ecol Manag 177:333–346

Triplett EW, Sadowsky MJ (1992) Genetics of competition for nodulation. Annu Rev Microbiol 46:399–428

Unni S, Rao KK (2001) Protein and lipopolysaccharide profiles of a salt-sensitive *Rhizobium* sp. and its polysaccharide-deficient mutant. Soil Biol Biochem 33:111–115

Uyanöz R, Karaca Ü (2006) Effects of different salt concentrations and *Rhizobium* inoculation (native and *Rhizobium tropici* CIAT899) on growth of dry bean (*Phaseolus vulgaris* L.) Eur J Soil Biol 47:387–391

Valdenegro M, Barea JM, Azcón R (2001) Influence of arbuscular-mycorrhizal fungi, *Rhizobium meliloti* strains and PGPR inoculation on the growth of *Medicago arborea* used as model legume for re-vegetation and biological reactivation in a semi-arid mediterranean area. Plant Growth Regul 34:233–240

Vance CP (1998) Legume symbiotic nitrogen fixation: agronomic aspects. In: Spaink HP, Kondorosi A, Hooykaas PJJ (eds) The Rhizobiaceae. Kluwer Academic Publishers, Dordrecht, pp 509–530

Vázquez MM, Azcón R, Barea JM (2001) Compatibility of a wild type and its genetically modified *Sinorhizobium* strain with two mycorrhizal fungi on *Medicago* species as affected by drought stress. Plant Sci 161:347–358

Verdoy D, De La Peña TC, Redondo FJ, Lucas MM, Pueyo JJ (2006) Transgenic *Medicago truncatula* plants that accumulate proline display nitrogen-fixing activity with enhanced tolerance to osmotic stress. Plant Cell Environ 29:1913–1923

Vessey JK, Pawlowski K, Bergman B (2005) Root-based N_2-fixing symbiosis: legumes, actinorhizal plants, *Parasponia* sp. and cycads. Plant Soil 274:51–78

Vinocur B, Altman A (2005) Recent advances in engineering plant tolerance to abiotic stress: achievements and limitations. Curr Opin Biotechnol 16:123–132

Vriezen JAC, De Bruijn FJ, Nüsslein K (2007) Responses of rhizobia to desiccation in relation to osmotic stress, oxygen and temperature. Appl Environ Microbiol 73:3451–3459

Wankhade S, Apte SK, Rao KK (1996) Salinity and osmotic stress regulated proteins in cowpea *Rhizobium* 4a (peanut isolate). Biochem Mol Biol Int 39:621–628

Ward JM, Hirschi KD, Sze H (2003) Plants pass the salt. Trends Plant Sci 8:200–201

Wei W, Jiang J, Li X, Wang L, Yang SS (2004) Isolation of salt-sensitive mutants from *Sinorhizobium meliloti* and characterization of genes involved in salt tolerance. Lett Appl Microbiol 39:278–283

Wei G-H, Yang X-Y, Zhang Z-X, Yang Y-Z, Lindström K (2008) Strain *Mesorhizobium* sp. CCNWGX035: a stress-tolerant isolate from *Glycyrrhiza glabra* displaying a wide host range of nodulation. Pedosphere 18:102–112

Wood JM, Bremer E, Csonka IN, Kramer R, Poolman B, Van Der Heide T, Smith IT (2001) Osmosensing and osmoregulatory compatible solute accumulation by bacteria. Comp Biochem Physiol A Mol Integr Physiol 130:437–460

Wu QS, Ying-Ning Z, Abd-Allah EF (2014) Mycorrhizal association and ROS in plants. In: Ahmad P (ed) Oxidative damage to plants. Academic Press; Elsevier, New York, NY, pp 453–475

Xavier LJC, Germida JJ (2002) Response of lentil under controlled conditions to coinoculation with arbuscular mycorrhizal fungi and rhizobia varying in efficacy. Soil Biol Biochem 34:181–188

Xiong L, Zhu J-K (2001) Abiotic stress signal transduction in plants: molecular and genetic perspectives. Physiol Plant 112:152–166

Xu J, Li X-L, Luo L (2012) Effects of engineered *Sinorhizobium meliloti* on cytokinin synthesis and tolerance of alfalfa to extreme drought stress. Appl Environ Microbiol 78:8056–8061

Yamal G, Bidalia A, Vikram K, Rao KS (2016) An insight into the legume-*Rhizobium* interaction. In: Hakeem KR, Akhtar MS (eds) Plant, soil and microbes. Springer International Publishing, Switzerland, pp 359–384

Yan N, Marschner P, Cao W, Zuo C, Qin W (2015) Influence of salinity and water content on soil microorganisms. Int Soil Water Conserv Res 3:316–323

Yang Q, Wu M, Wang P, Kang J, Zhou X (2005) Cloning and expression analysis of a vacuolar Na^+/H^+ antiporter gene from alfalfa. DNA Seq 16:352–357

Yano-Melo AM, Saggin OJ, Maia LC (2003) Tolerance of mycorrhizal banana (*Musa* sp. cv. Pacovan) plantlets to saline stress. Agric Ecosyst Environ 95:343–348

Yinsuo J, Vincent MG, Colin JS (2004) The influence of *Rhizobium* and arbuscular mycorrhizal fungi on nitrogen and phosphorus accumulation by *Vicia faba*. Ann Bot 94:251–258

Yokoi S, Quintero FJ, Cubero B, Ruiz MT, Bressan RA, Hasegawa PM, Pardo JM (2002) Differential expression and function of *Arabidopsis thaliana* NHX Na^+/H^+ antiporters in the salt stress response. Plant J 30:529–539

Zahaf O, Blanchet S, Del Zélicourt A et al (2012) Comparative transcriptomic analysis of salt adaptation in roots of contrasting *Medicago truncatula* genotypes. Mol Plant 5:1068–1081

Zahran HH (1998) Structure of root nodules and nitrogen fixation in Egyptian wild herb legumes. Biol Plant 41:575–585

Zahran HH (1999) *Rhizobium*-legume symbiosis and nitrogen fixation under severe conditions and in an arid climate. Microbiol Mol Biol Rev 63:968–989

Zahran HH (2001) Rhizobia from wild legumes: diversity, taxonomy, ecology, nitrogen fixation and biotechnology. J Biotechnol 91:143–153

Zahran HH (2006a) Wild legume rhizobia: biodiversity and potential as biofertilizers. In: Rai MK (ed) Handbook of microbial biofertilizers. Haworth Press Inc., New York, pp 203–222

Zahran HH (2006b) Nitrogen (N₂) fixation in vegetable legumes: biotechnological perspectives. In: Ray RC, Ward OP (eds) Microbial biotechnology in horticulture, vol 1. Science Publishers, Inc., Enfield, USA, pp 49–82

Zahran HH (2009) Enhancement of rhizobia-legumes symbioses and nitrogen fixation for crops productivity improvement. In: Khan MS, Zaidi A, Musarrat J (eds) Microbial strategies for crop improvement. Springer-Verlag, Germany, pp 227–254

Zahran HH, Sprent JI (1986) Effects of sodium chloride and polyethylene glycol on root-hair infection and nodulation of *Vicia faba* L. plants by *Rhizobium leguminosarum*. Planta 167:303–309

Zahran HH, Räsanen LA, Karsisto M, Lindström K (1994) Alteration of lipopolysaccharide and protein profiles in SDS-PAGE of rhizobia by osmotic and heat stress. World J Microbiol Biotechnol 10:100–105

Zahran HH, Ahmad MS, Afkar E (1995) Isolation and characterization of nitrogen-fixing moderate halophilic bacteria from saline soils of Egypt. J Basic Microbiol 35:269–275

Zahran HH, Abdel-Fattah M, Ahmad MS, Zaki AY (2003) Polyphasic taxonomy of symbiotic rhizobia from wild leguminous plants growing in Egypt. Folia Microbiol 48:510–520

Zahran HH, Marin-Mansano MC, Sánchez-Raya AJ, Bedmar EJ, Venema K, Rodríguez-Rosales MP (2007) Effect of salt stress on the expression of NHX-type ion transporters in *Medicago intertexta* and *Melilotus indicus*. Physiol Plant 131:122–130

Zahran HH, Abdel-Fattah M, Yasser MM, Mahmoud AM, Bedmar EJ (2012) Diversity and environmental stress responses of rhizobial bacteria from Egyptian grain legumes. Aust J Basic Appl Sci 6:571–583

Zahran HH, Chahboune R, Moreno S, Bedmar EJ, Abdel-Fattah M, Yasser MM, Mahmoud AM (2013) Identification of rhizobial strains nodulating Egyptian grain legumes. Int Microbiol 16:157–163

Zerhari K, Aurag J, Khbaya B, Kharchaf D, Filali-Matlouf A (2000) Phenotypic characteristics of rhizobia isolates nodulating *Acacia* species in the arid and Saharan regions of Morocco. Lett Appl Microbiol 30:351–357

Zhang W-J, Wang T (2015) Enhanced salt tolerance of alfalfa (*Medicago sativa*) by *rstB* gene transformation. Plant Sci 234:110–118

Zhu J-K (2002) Salt and drought stress signal transduction in plants. Annu Rev Plant Boil 53:247–273

Zhu J-K (2003) Regulation of ion homeostasis under salt stress. Curr Opin Plant Biol 6:441–445

Zhu J-K, Hasegawa PM, Bressan R (1997) Molecular aspects of osmotic stress in plants. Crit Rev Plant Sci 16:253–277

Zhukov VA, Shtark Y, Borisov AY, Tikhonovich IA (2013) Breeding to improve symbiotic effectiveness of legumes. In: Andersen SB (ed) Plant breeding from laboratories to fields. InTech, Croatia, pp 167–207

Zia-ul-Hassan, Ansari TS, Shah AN, Jamro GM, Rajpar I (2015) Biopriming of wheat seeds with rhizobacteria containing ACC deaminase and phosphate solubilizing activities increases wheat growth and yield under phosphorus deficiency. Pak J Agri Agril Eng Vet Sci 31:24–32

Zilli CG, Balestrasse KB, Yannarelli GG, Polizio AH, Santa-Cruz DM, Tomaro ML (2008) Heme oxygenase up-regulation under salt stress protects nitrogen metabolism in nodules of soybean plants. Environ Exp Bot 64:83–89

Zörb C, Noll A, Karl S, Leib K, Yan F, Schubert S (2005) Molecular characterization of Na⁺/H⁺ antiporters (*ZmNHX*) of maize (*Zea mays* L.) and their expression under salt stress. J Plant Physiol 162:55–66

Rhizobial Amelioration of Drought Stress in Legumes

14

Muhammad Naveed, M. Baqir Hussain, Ijaz Mehboob, and Zahir Ahmad Zahir

Abstract

Rhizobia form a strong and effective symbiosis with legumes and make nitrogen available for uptake by plants. Biological nitrogen fixation (BNF) helps to maintain soil fertility at optimum level. However, among various environmental stress factors, drought stress is the most devastating factor destructing both rhizobial growth and *Rhizobium*-legume symbiosis. The establishment of a functional and efficient *Rhizobium*-legume interactions under these unfavorable environmental (arid/semiarid) conditions is therefore very critical. Considering these, understanding the responses of both rhizobia and legumes to drought is important for harnessing the maximum benefits of BNF. In this context, different strategies are adopted to overcome losses from drought stress. These strategies include germplasm screening, breeding drought-tolerant genotypes, transgenic approach, and biological approach. Also, rhizobia have been reported to adapt to severe water-deficit environment. Rhizobia participate in the regulation of plant's metabolite production (compatible solutes and antioxidant), molecular level responses (gene and protein expression), hormonal adjustment, and nutrient solubilization and uptake, to circumvent drought stress conditions. The advancement in omics has further provided an insight to identify specific proteins and metabolites which

M. Naveed (✉) • Z.A. Zahir
Institute of Soil and Environmental Sciences, University of Agriculture, Faisalabad 38040, Pakistan
e-mail: mnaveeduaf@yahoo.com

M.B. Hussain
Institute of Soil and Environmental Sciences, University of Agriculture, Faisalabad 38040, Pakistan

Department of Soil Science, Faculty of Agricultural Sciences and Technology, Bahauddin Zakariya University, Multan, Pakistan

I. Mehboob
District Fertility Lab, Kasur, Pakistan

© Springer International Publishing AG 2017
A. Zaidi et al. (eds.), *Microbes for Legume Improvement*,
DOI 10.1007/978-3-319-59174-2_14

341

could play pivotal roles in stress management and rhizobia-legume symbiosis. Here, in-depth insights into the impact of drought on rhizobia and legumes grown in drought affected areas are presented.

14.1   Introduction

A favorable environment is vital to legume production, but severe environmental conditions limit the growth and activity of both N_2-fixing microorganisms and host legumes. Among various stress factors, drought is considered an important environmental stress factor restricting legume production (Naresh et al. 2013). Drought situation also suppresses/inhibits growth of rhizobia and its symbiotic association with legumes. Therefore, to cope with drought stress, plants require changes at physiological, morphological, and biological levels. Many strategies have been developed to improve crop productivity under water-limiting conditions. Among different strategies, germplasm screening (Sharma et al. 2013), breeding activities (Fenta et al. 2014; Polania et al. 2016), and genomics (Varshney et al. 2013; Hossain and Komatsu 2014a, b) have led to some yield increase in many crop plants grown under drought environments. These methods are, however, time-consuming, labor-intensive, and expensive. Even some of these methods have environmental concerns also. Hence, an environmentally safe, fast, and cost-effective strategy for optimizing crop yield under drought stress conditions needs to be developed. Microbiological approach involving use of beneficial plant growth promoting rhizobacteria (PGPR) in this regard has been found environmentally safe and inexpensive (Naveed et al. 2014a, b). Among many useful rhizobacteria, rhizobia are group of symbiotic bacteria that transform atmospheric nitrogen (N) into usable form of N (Mehboob et al. 2009). Usually rhizobia are drought sensitive, but some of the physiologically efficient, effective, and compatible drought-tolerant strains have been found to induce drought tolerance and consequently increased the yield of associated plants by variety of mechanisms (Hussain et al. 2014a, b). The impact of drought stress on the physiology of leguminous plants and survival/adaptability of rhizobia under drought stress is highlighted. Also, various strategies used to improve drought endurance in leguminous plants and role of rhizobia in adaptation of plants to drought stress along with practical application of rhizobial inoculation are reviewed and discussed.

14.2   Severity of Drought Stress on the Physiology of Plants

Drought can be defined as a physiological condition where water potential and tissue turgor reach to a level that hampers normal growth, development, and yield of plants (Allahmoradi et al. 2011). Fundamental changes that occur due to dehydration include variation in physiological and biochemical processes (Sangtarash 2010; Abdullah et al. 2011) and disturbance in water relations (Gorai et al. 2010). Water deficit/drought also impairs normal turgor pressure leading to reduction in cell size and plant growth (Srivalli et al. 2003). Moreover, water stress increases root/shoot ratio, thickness of cell walls, and amount of cutinization and lignifications (Srivalli et al. 2003). Additionally, plants growing under water stress may have closed stomata, decreased leaf area index, variable plant metabolites (Jordan and Ritichie 2002; Bartels and Sunkar 2005), respiration, carbohydrates, ion uptake, nutrient uptake (Akinci and Losel 2010), and nutrient metabolism (Farooq et al. 2008). Drought limits flow of CO_2 into mesophyll tissue and impairs photosynthesis (Flexas et al. 2004; Zhou et al. 2007), alters photosynthetic pigments (Farooq et al. 2009) and carbohydrates (Jain et al. 2007), causes poor diffusion of CO_2 into leaves (Flexas et al. 2006), reduces photosynthetic assimilation rates (Zlatev and Lidon 2012), and accelerates the production of reactive oxygen species (ROS) (Montero-Tavera et al. 2008). At cellular level, water stress suppresses expansion and growth of cells (Jaleel et al. 2009), impairs structure and function of cell membrane and cell organelle (Gigon et al. 2004), changes ultrastructure of subcellular organelles (Yordanov et al. 2003) and endomembrane system and cell wall membrane integrity (Kacperska 2004), and disrupts homeostasis and ion distribution. Changes in synthesis of proteins (Bota et al. 2004; Barrera-Figueroa et al. 2007), proteases (Seki et al. 2002), and enzymes (Haupt-Herting and Fock 2000) are however required for detoxification and biosynthesis of various osmoprotectants (Iturriaga et al. 2009). Accumulation of ROSs (such as O_2^-, $^1O_2$, H_2O_2, OH^-) in plants has also been reported when plants were exposed to drought stress (Foyer and Noctor 2005). ROS in turn affects lipid peroxidation, membrane injuries, denaturation of functional and structural amino acids and protein, inactivation of enzymes, and DNA nicking (Sairam et al. 2005). Another effect of drought includes the activation of abscisic acid (ABA)-dependent signaling pathway (Kim et al. 2010). Drought stress triggers high levels of ABA (Raghavendra et al. 2010) produced in roots and leaves which are transported via ATP-binding cassette (ABC) transporters positioned in the plasma membrane into the guard cell (Kang et al. 2010). Ultimately, ROS is produced, because influx of Ca^{2+} across the membranes of plasma and vacuole starts. In turn, anion efflux activates and results in depolarization of the membrane which causes K^+ efflux across the membranes of both plasma and vacuole (Wasilewska et al. 2008) leading to reduced turgor pressure and induced stomata closure (Kim et al. 2010). Summarily, plants can adapt to water-deficit situations through changes in their molecular and physiological composition (Boutraa and Saders 2001). Severe deficit of water arrests photosynthesis; reduces turgor, water potential, and solute concentrations within the cytosol; increases extracellular matrices, etc. (Bhatt and Srinivasa Rao 2005). Also, continuous buildup of ABA and plant compatible osmolytes and excessive production of ROS result in wilting and ultimately cause the death of the plant (Jaleel et al. 2008). Above all, the severity of drought depends on its duration and intensity (Chaves et al. 2009), plant's sensitivity and capacity (Valladares et al. 2007), species and developmental stage of plants, and soil composition and climatic conditions (DaMatta and Ramalho 2006). As an example, the effect of water stress on mung bean varieties and its physiological responses to yield was assessed under field conditions using split-plot design with 20 treatments and three replications by

Naresh et al. (2013). The results revealed that drought-tolerant varieties maintained highest xylem water potential (XWP) and transpiration resistance (TR) and lowest leaf diffusive resistance (LDR) and canopy temperature minus air temperature $(T_c - T_a)$. In contrast, drought-sensitive varieties had lowest XWP and TR and highest LDR and $T_c - T_a$. Also, the rate of net photosynthesis measured at various stages of growth declined in all varieties. The free proline concentration increased from vegetative to active pod-filling stages in all the varieties. However, among the sensitive and resistant varieties, the resistant varieties in general had maximum level of proline under water stress.

14.3 Survival/Adaptability of Rhizobia Under Drought Stress

When rhizobial populations are restricted below 100 cells g^{-1} bulk soil, the environment is called stressful for rhizobia (Howieson and Ballard 2004). Within the soil, drought stress often affects survival and viability of *Rhizobium* (Issa and Wood 1995; Vanderlinde et al. 2010) and restricts their mobility (Gopalakrishnan et al. 2015), but the sensitivity/tolerance toward drought varies among rhizobial strains (Zeng 2003; Thiao et al. 2004; Moschetti et al. 2005). Water deficit may cause changes in hydration state of membrane proteins, cell turgor, and concentrations of intercellular ionic solute (Poolman et al. 2002) and induction and repression of genes (Dominguez-Ferreras et al. 2006). Alteration in the production of extracellular polysaccharides (EPS) and capsular lipopolysaccharides has also been recorded in rhizobia when exposed to drought stress (Zahran 1999). In order to cope drought stress, rhizobia have evolved different mechanisms (Billi and Potts 2002; Humann et al. 2009). For example, formation of cyst and floc, synthesis of polysaccharide and poly-β-hydroxybutyrate and melanin, etc. are some of the strategies adopted by rhizobia to counter drought stress (Rehman and Nautiyal 2002; Räsänen and Lindström 2003; Ben Romdhane et al. 2006; Vanderlinde et al. 2010). Generally, rhizobia use compatible solutes to maintain their cell turgor and survival under drought stress (Boncompagi et al. 1999) which can either be synthesized inside the cell when needed or they are collected from the surrounding, depending on the condition. Rhizobia can accumulate proline, choline, betaine, glycine betaine, trehalose, or quaternary amine compounds (QACs) as osmoprotectant (Gloux and Le Rudulier 1989) using *ProP* and *ProU* import system. However, rhizobial species use different solutes under different levels of stress (Miller and Wood 1996). On the other hand, glutamate, N-acetylglutaminylglutamine amide (NAGGN), pipecolic acid (PIP), and ectoine are the solutes synthesized within the rhizobial cells (Miller and Wood 1996) and have shown to enhance growth of rhizobial species under water stress (Gouffi et al. 2000). It has also been found that rhizobia activate mechanosensitive channels in the cell membrane to detect water tension and to escape solute and water stress (Poolman et al. 2002). Conclusively, most of the works have focused mainly on the occurrence and general behavior of rhizobia under drought stress, but little work is done to understand the role of proteins and enzymes in survival of rhizobia under water-deficit conditions.

14.4   Strategies for Improving Drought Stress Tolerance in Plants

Drought limits crop productivity leading to increase in desertification and food insecurity worldwide (FAO 2011). Hence, the advent and use of strategies capable of improving ability of plants to survive and proliferate under drought environment are urgently needed (Fig. 14.1). Some of the strategies which have been found highly effective in enhancing tolerance in plants against drought stress are discussed in the following section.

14.4.1 Germplasm Screening

Living materials bearing heredity from which new plants can grow are called germplasm (e.g. seeds, rootstock, or leaf plant tissues). In germplasm screening strategy, the superior germplasm is collected by selection of some lines with higher levels of

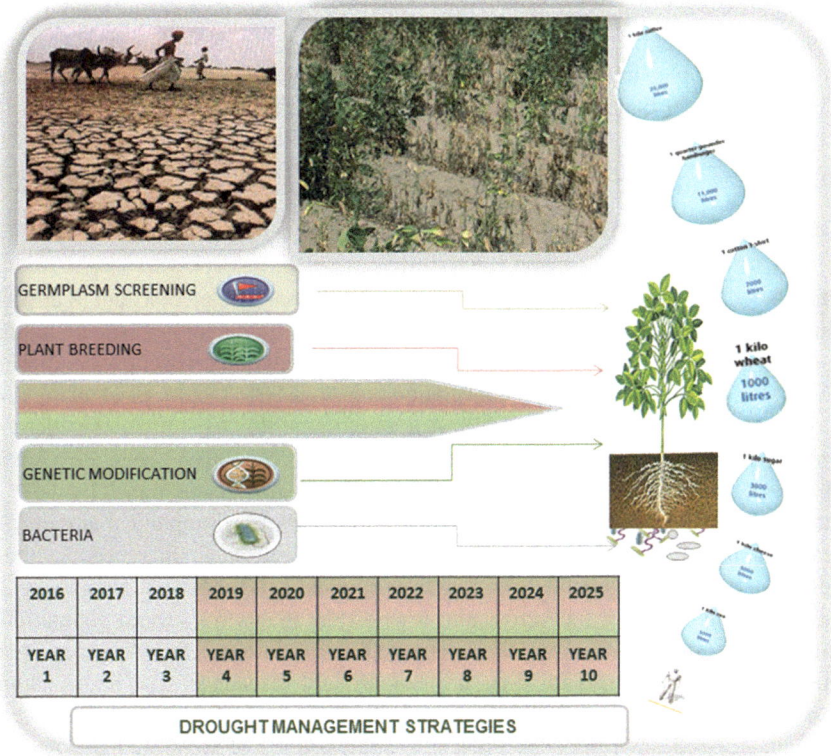

Fig. 14.1 Schematic view of different drought management strategies over their time periods (modified from Naveed 2013)

drought tolerance from a crop well adapted to harsh conditions. The wild relatives are important sources of genetic diversity for crop improvement. However, the exploitation of competitive genotypes limits crop improvement due to cross-incompatibility and linkages (Sharma et al. 2013). The selected germplasm is then screened several times under greenhouse and field conditions. The germplasm with higher levels of drought tolerance is evaluated further under greenhouse conditions to ascertain the levels of drought tolerance. The germplasm finally selected is propagated for multiplication and distribution to farming communities (Beebe et al. 2013). However, this germplasm strategy is a lengthy and time-consuming procedure (Sharma et al. 2013). Also, the huge number of germplasm in the gene bank poses a great problem during selection process. Despite these problems, more than a hundred thousand genotypes of chickpea (ICC 237, Yezin 1, Yelbey), pigeon pea [JK Sixer (JKPL 6), Guimu 4, Kamica], and groundnuts (Kalasin 2, JL24, Msandile) have been developed (Sharma et al. 2013). In another study, Acosta-Gallegos et al. (1995) and Sánchez-Valdez et al. (2004) developed mung bean cultivars Pinto Villa and Pinto Saltillo through germplasm screening in Mexico drought-stressed environments. Establishment of proper, practical, reliable, cheap, and fast selection methods and multiplication system is the prerequisite of this strategy. It requires large and expensive glasshouse and field trials to screen out drought-tolerant desired genotypes of plant materials. Multidisciplinary approaches to measure the impact of drought on the physiology, biochemistry, and morphology of the plants are required. Though increases in drought stress tolerance through this strategy are feasible but to a small extent.

14.4.2 Breeding Drought-Tolerant Genotypes

Plant breeding is another technique by which genotypes that are resistant to drought can be developed. It is considered a competent and profitable way of crop manipulation to enhance the capability of plants to grow efficiently under drought conditions. Historically, plant breeding has contributed enormously to tackle the challenges of global food security (Rajaram 2005). And employing this technique, considerable drought-tolerant cultivars/lines of important food crops have been developed. However, new crop varieties or lines with desirable traits are developed by deliberate crossing of related (closely or distantly) individuals. Traditionally, manipulation via plant breeding is done through controlled pollination. Plant breeders, firstly deliberately generate genetic diversity that would not exist in nature. Then to generate new plant varieties, they cross and re-cross the plants for several generations, followed by artificial selection of progeny with desired traits. The selected resistant plants are then evaluated for their level of drought resistance. The high-yielding resistant plants are then multiplied and distributed for field cultivation under drought-stressed environments. Presently, cultivar/lines of some of the popularly grown legumes in different agroclimatic regions, for example, chickpea (FLIP 87-59C), common bean (SEA5, NCB 280, NCB 226, SEN 56, SCR 2, SCR16, SMC 141, RCB 593, and BFS 67), peanut (ICGV 87354), and soybean (R01-416F,

Jackson, Prima 2000) (Sing et al. 1996; Reddy et al. 2001; Sing et al. 2001; Chen et al. 2007; Beebe et al. 2008, 2013; Fenta et al. 2014; Polania et al. 2016), have been developed through traditional breeding approach. However, in this method, only plants of same species are crossbred to introduce traits. And hence, the traditional breeding techniques have got limited success for improving drought tolerance ability in crops because drought tolerance is a multigenic trait. Also, it could result in inbreeding and can deteriorate the breed, and the plant may become more prone to disease or mutations (Araus et al. 2008). Undesirable traits can also be fixed through plant breeding unintentionally. It results in narrow genetic diversity and loss of some indigenous species and eliminates variation in population. Plant breeding through selection is however time-consuming, labor-intensive, and expensive and may not always be successful (Ashraf 2010).

14.4.3 Transgenic Approach

Plants developed by the insertion of specific/desired genes to express an intended/additional character are termed as transgenic plants. Plant breeders across the globe are using transgenic approaches to develop stress-tolerant varieties/cultivars/lines of various crops (Ashraf et al. 2008). Using this approach, the desired traits could be more carefully introduced from either a different variety of the same crop plants or a different plant species while excluding/reducing the undesirable plant characteristics. The prospects of developing drought-tolerant cultivars/varieties using transgenic approaches seem to be an attractive strategy where desired genes of specific character can be inserted and the transfer of undesirable gene can be restricted. Through transgenic approach it is possible to pyramid genes with similar effects (Gosal et al. 2009). Transgenic approach has advantage over traditional breeding that it can transform certain traits from other plant species which cannot be transferred through conventional breeding. Plants designed to resist drought via transgenic technology have the potential to withstand more strongly and could yield better under water-limited conditions. Legumes which have been improved for drought tolerance through successful incorporation of genes are soybean (Ronde et al. 2004), peanut (Bhatnagar-Mathur et al. 2009), and faba bean (Hanafy et al. 2013). However, the performance of transgenic cultivar/line developed so far has been tested mainly under controlled conditions, and hence, there is a need to test them under natural field conditions at global scale for ultimate transfer to end users.

14.4.3.1 Biological Approach

Biological approach involves the use of biological products or a substance which contains living microorganisms to manage stress in plants. Alleviation of drought-induced effects on plant growth following application of drought-tolerant plant beneficial bacteria is reported (Mnasri et al. 2007). For example, selected effective rhizobial strains enhanced the symbiotic nitrogen fixation under drought stress (Zahran 1999; Serraj et al. 1999) and consequently increased the yield of leguminous crops (Ballesteros-Almanza et al. 2010). When applied as inoculant, rhizobia improve the nutritional status of nodulated plants via BNF and also enhance drought tolerance of nodulated legumes by secreting compatible solutes (Rasanen et al. 2004). Similarly, rhizobial inoculation has been found as an economical strategy that could produce yield of legumes equal to or better than nitrogen fertilization under drought stress (Ben Romdhane et al. 2008). However, the use of tolerant and competitive rhizobial strains for better nodulation and yield of legumes under drought stress was suggested. Practically, even though biological approach is a new and emerging concept, it is an exciting, cost-effective, socially acceptable, and environment-friendly approach. It could improve plant growth and yield under water-deficit conditions without involving genetic engineering or traditional breeding of plants. Also, there are reports which suggest that naturally occurring rhizobia belonging to wild tree legumes in arid environments could possibly be higher drought tolerant (Zahran 2001; Ali et al. 2009) and could be an efficient approach for increasing N contents of plant under water-deficit environments (Mahmood and Athar 2008). Mechanistically, effective and tolerant microbes enhance plant growth under drought stress by synthesizing compatible solutes, antioxidants, and hormones, facilitating nutrient uptake, etc. Furthermore, numerous studies have suggested that along with stress-tolerant rhizobia selection of stress-tolerant cultivars could be a more rational strategy to increase legume yield under water-deficit conditions (Meuelenberg and Dakora 2007; Ben Romdhane et al. 2007; Mhadhbi et al. 2011; Yanni et al. 2016).

14.5 Role of *Rhizobium* in Adaptation of Legumes to Drought Stress

Water deficit reduces rhizobial viability (Hussain et al. 2014a, b), but legumes grown under arid and semiarid lands require drought-tolerant rhizobia to form effective symbiosis. In this context, the presence and performance of diverse rhizobial strains under drought stress have been reported (Upadyyay et al. 2009). Though rhizobial strains vary in effectiveness (Mhadhbi et al. 2008), rhizobia with better adaptive features could help legumes to cope with drought (Yang et al. 2009) (Fig. 14.2). Some of the active molecules synthesized by rhizobia and involved in protection against drought stress are described in the following section.

14.5.1 Metabolite Production/Adjustment

14.5.1.1 Compatible Solutes
Drought tolerance in plant is also regulated by compatible solutes (Ashraf and Foolad 2007). Rhizobia can influence the plant's response toward abiotic stresses through the intracellular accumulation of organic and inorganic solutes, often termed compatible solutes (Soliman et al. 2012). Some of the important compatible solutes which act as stress protectant are trehalose, glycine betaine, stachydrine,

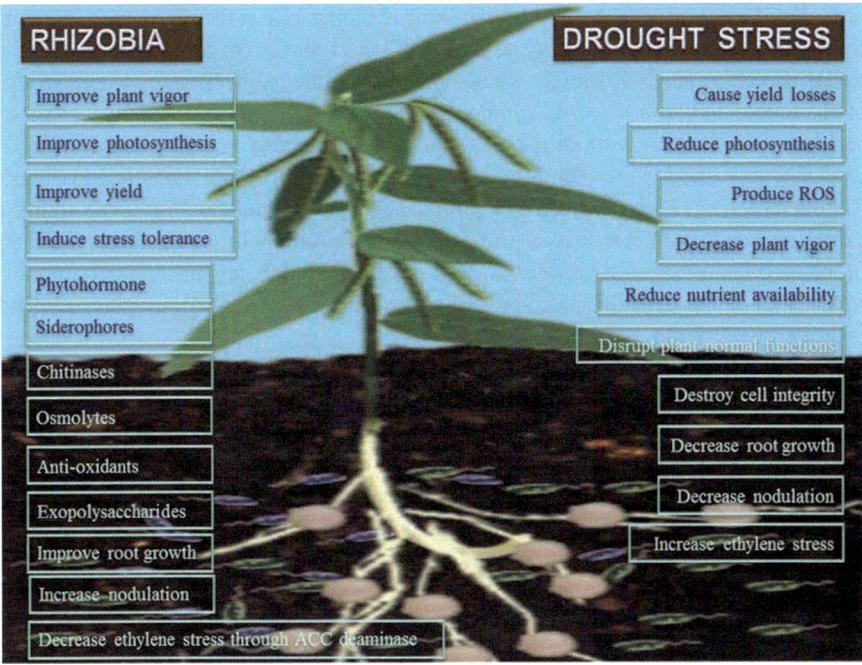

Fig. 14.2 Plant beneficial attributes of rhizobia and adverse impact of drought stress on legumes

polyamines, sucrose, cellobiose, maltose, turanose, Palatinose, gentiobiose, amino acids, and mostly proline, glutamate, ectoine, *N*-acetylglutaminylglutamine amide, and dimethylsulfoniopropionate (Pichereau et al. 1998; Boncompagi et al. 1999; Gouffi et al. 1999; Staudinger et al. 2016). Numerous studies have revealed that the rhizobial strains synthesize compatible solutes and when used as inoculant improve drought tolerance in higher plants (Sessitsch et al. 2002; Streeter 2003; Zacarias et al. 2004). These substances induce signal transduction and activation of stress tolerance in plants (Paul 2007). The plants may also take up these compatible solutes from exogenous sources including plant-rhizobia symbiotic association where rhizobia synthesize the solutes excessively (Lopez et al. 2008). Mhadhbi et al. (2011) in an experiment evaluated the effect of water stress on nodules of common bean inoculated with three rhizobial strains differing in stress tolerance ability. They observed that highly tolerant strain *Ensifer meliloti* 4H41 produced highest number of large-sized nodules. Furthermore, the drought tolerance response of the nodules was due to stimulated metabolic activities of the nodule triggered by strain 4H41. Similar increase in growth and productivity of legumes, for example, common bean (Zacarias et al. 2004), under water-deficit stress conditions due to inoculation with excessively trehalose producing rhizobia is reported (Rasanen et al. 2004; Suarez et al. 2008). In a follow-up study, accumulation of trehalose (up to 90 mg g^{-1}) in the nodules of common bean inoculated with rhizobia under drought stress was recorded. The overproduction of trehalose by rhizobia assists the legumes to recover from drought-like situation. Recently, Staudinger et al. (2016) revealed metabolite-based rhizobia-induced drought stress response strategy in *Medicago truncatula* plants inoculated with *Sinorhizobium medicae* WSM419 (NODe) or *Sinorhizobium meliloti* 2011 (NODi); both rhizobial strains differed greatly in their N fixation ability. They quantified leaf metabolites of the primary C and N metabolism, and out of 37 metabolites, 17 (the majority of amino acids and organic acids) were significantly enriched in NODe leaves. Increased concentrations of potassium and shifts in the carbon partitioning between starch and sugars and the enhanced allocation of reserves to osmolytes during drought were the possible reasons for enhanced tolerance against drought.

14.5.1.2 Antioxidant Production

Reactive oxygen species (ROS) are routinely generated in different cellular compartments as by-product of a variety of metabolic pathways (Bianco and Defez 2009). These ROS are toxic by-products and cause cellular damage and ultimately death (Matamoros et al. 2003). ROS also function as signaling molecules for the regulation of stomata opening and closing (Pie et al. 2000). They are also involved in programmed cell death (Gechev and Hille 2005). Usually, plant detoxifies these harmful species via antioxidants (Rouhier et al. 2006; Shao et al. 2008). Generally, antioxidants are enzymatic [such as a metalloenzyme, superoxide dismutases (SOD), ascorbate peroxidases (APX), peroxidases (POX), and catalases (CAT)] and nonenzymatic (such as carotenoids, ascorbic acid, and flavanones) (Comba et al. 1998; Jebara et al. 2010). ROS are also generated in plants as a result of drought stress response (Nanda et al. 2010). In order to adapt to prolonged stress, plant needs to increase its capacity to generate antioxidants. It has been recognized that rhizobia could modulate the growth and development of legume crops under stress via antioxidant secretion, if used as inoculant (Bianco and Defez 2009). As an example, Esfahani and Mostajeran (2011) evaluated the inoculation effect of *Mesorhizobium ciceri* on chickpea plants grown under drought stress and observed increased activity of nodular POX and APX. They suggested that improved tolerance of chickpea to drought stress was due to the excessive secretion of antioxidant enzymes by rhizobia under drought stress. In yet another investigation, the symbiotic performance and water stress response of six rhizobial strains with six chickpea cultivars in 36 symbiotic combinations varied considerably (Mhadhbi et al. 2008). Of the six rhizobial strains tested, the highest symbiotic efficiency was observed with local strains *M. ciceri* CMG6, *M. mediterraneum* C$_{11}$, and reference strain (835). Highest aerial biomass and nitrogen-fixing capacity were recorded for plants inoculated with reference strain (835) and *M. ciceri* (CMG6). Moreover, the enzymes APX, POX, and SOD were found to play a significant protective role in nodules under stressed environment. High tolerance of chickpea plants to drought stress when inoculated with *M. ciceri* strains C-15 and CP-36 has also been reported due to enhanced release of antioxidants such as APX and POX (Esfahani et al. 2010).

14.5.2 Molecular Level Adjustment

14.5.2.1 Gene Expression

Gene expression conferring to drought tolerance is a promising strategy (Jiang et al. 2012; Joshi et al. 2016). Inoculant/rhizobia could trigger higher expression of drought stress-adaptive genes in plants if used as inoculant (Farrar et al. 2014). Improvement in drought adaptation, growth, and yield in common bean inoculated with rhizobia over expressing *OtsA* genes (a trehalose synthase gene) has been observed by Suarez et al. (2008). They demonstrated that under drought stress, rhizobial inoculation significantly encouraged the higher expression of stress-adaptive genes in plants to eliminate free radicals and/or other toxic substances and assisted plant's recovery under drought.

14.5.2.2 Protein Expression

Protein expression has also been demonstrated as a plant's defense mechanism against drought stress (Asch and Padham 2005; Malamy 2005). Also, some studies correlate drought tolerance capacity of plants to its protein modulation as stress response mechanism. Under different stress intensities, various types of proteins, i.e., actin, tubulin, and xyloglucan endotransglucosylase (XET) expression, have been described to alter the plant growth rate (Sengupta et al. 2011). Rhizobial strains have been found to stimulate protein expression as a mechanism of ROS detoxification of nodules and improve nitrogen fixation (Redondo et al. 2009). However, studies at protein level on plants under water-deficit conditions are quite rare. Very recently, using a shotgun proteomics approach, Staudinger et al. (2016) studied the dynamic change of the leaf soluble proteome of *Medicago sativa* plants. Inoculation with NODe had increased abundance of pyruvate kinase, enzymes involved in ethylene and jasmonic acid metabolism, and of ribosomal proteins relative to their non-inoculated counterparts.

14.5.3 Hormone Level Adjustment

Under water-deficit conditions, plant hormones, for instance, abscisic acid (ABA), help plant to adapt to stress conditions by controlling stomatal closure (Wilkinson and Davies 2010). Production of phytohormone by rhizobia has been reported to improve growth, yield, and drought tolerance in legumes under water-deficit stress (Boiero et al. 2007; Marulanda et al. 2009; Mehboob et al. 2009, 2011) besides BNF (Ghosh and Basu 2006). Various researchers have revealed the phytohormone (such as auxin, gibberellic acid, cytokinins, abscisic acid, and ethylene)-producing ability of rhizobia isolated from stressed environments (Bhattacharyya and Basu 1997; Duhan et al. 1998; Datta and Basu 2000; Carrascoa et al. 2005; Mirza et al. 2007; Wani et al. 2008; Hussain et al. 2014a, b). While assessing the effect of cytokinin-producing strains of *Sinorhizobium* on severe drought tolerance potential of alfalfa plants, Xu et al. (2012) reported that *Sinorhizobium* strains capable to synthesize cytokinin could enhance the tolerance of alfalfa to severe drought stress without upsetting its nodulation or nitrogen fixation capacity. Now it is well recognized that plants respond to various environmental stresses through the biosynthesis of ethylene. So far, as drought stress is concerned, plant's response at cellular and molecular level has been reported (Bray 1997). But prolonged drought accelerates the production and accumulation of ethylene in plant tissues resulting in abnormal growth of a plant. Ma et al. (2003) observed that the inoculation with ACC deaminase expressing *Rhizobium leguminosarum* bv. *viciae* 128C53K modulated ethylene levels in plant roots and nodules of peas ultimately enhanced nodulation.

14.5.4 Nutrient Solubilization and Uptake

Rhizobia are recognized to enhance the bioavailability and uptake of nutrients like N, P, and K by legumes under drought conditions leading to increased pulse production. However, phosphorus (P) solubilization has been demonstrated as a potential characteristic of plant beneficial rhizobia (Ding et al. 2016; Hussain et al. 2016). Franzini et al. (2010) showed significant improvement in the nutritional physiology of four *Phaseolus vulgaris* cultivars due to the inoculation of rhizobia under mild drought conditions. Ding and coworkers (2016) observed significant increase in P and soluble sugar concentration in rhizobia-inoculated soybean grown under P-deficient conditions, whereas, Lazali et al. (2013) observed higher expression of phytase gene in mung bean due to rhizobial inoculation under P-deficient conditions. The P nutrition of chickpea was also increased by the inoculation of P-solubilizing *Mesorhizobium* strains (Imen et al. 2015). Rosas et al. (2006) revealed the improvement in rhizobia-legume symbiosis and increased shoot and root dry weight of soybean due to the inoculation of phosphate-solubilizing *Bradyrhizobium meliloti* 3DOh13. Similarly, Peix et al. (2001) observed significant increases in the concentration of Ca, Mg, K, P, and N of chickpea plants due to inoculation of phosphate-solubilizing *M. mediterraneum* strain PECA21. Yanni et al. (2016) conducted 16 field experiments in Nile Delta on mung bean and observed significant increase in N use efficiency due to the inoculation of indigenous rhizobial strains under water-deficit stress conditions. Similarly, Staudinger et al. (2016) demonstrated differential response of *Medicago truncatula* toward drought due to the inoculation of rhizobial strains where increased K concentration in leaves played a major role.

14.6   Recent Advances in *Rhizobium* Research at Omics Level

Advancement in technology has resulted in the development of new molecular techniques which can be used to better understand the response and regulation of plants to stress besides their role in exploring the symbiosis between rhizobia and legumes (Karmakar et al. 2015). Some of the recently developed omics like genomics (study the genes involved), transcriptomics (study the expression of genes in terms of messenger RNA), proteomics (study of proteins synthesized under certain conditions), and metabolomics (study of metabolites produced in varying conditions) are employed to study the finer details of response of plants toward specific stresses (Shulaev et al. 2008; Karmakar et al. 2015). These techniques assist in recognizing the specific genes and their expression and secretion of proteins and metabolites to reduce the impact of specific stress including drought. The proteomics and metabolomics profiling of model legumes like *Lotus japonicus*, *Medicago truncatula*, and *Glycine max* to elucidate their signaling pathways for stress management, biochemical pathways, cellular processes, and symbiosis have been studied (Ramalingam et al. 2015). For understanding the molecular basis of physiological and biochemical response of plants, the correlation between protein groups and metabolites produced during stress is used to identify specific responses (Weckwerth 2011; Rodziewicz et al. 2014). Furthermore, due to the recent development in the databases of legume genomics and proteomics, it has become possible to identify highly efficient and related stress proteins in legumes (Kosová et al. 2011; Varshney et al. 2013; Hossain and Komatsu 2014a, b). For example, Staudinger et al. (2016) in a study assessed the response of *M. truncatula* toward drought as influenced by rhizobial inoculation using ionomics, proteomics, and metabolomics techniques. They revealed a delayed leaf senescence in inoculated plants relative to uninoculated ones under drought. The inoculated plants accumulated more osmolytes under drought. Consequently, they suggested a mechanism following proteomics data that the phytohormone interaction and increased translational responses lead toward delayed senescence and higher NPK content under drought.

14.7 Practical Application/Agronomic Prospects of Rhizobial Inoculation

Even though rhizobia can inexpensively and easily be mass produced for commercial application as biofertilizer, the environmental conditions, soil characteristics, genetic attributes of symbionts and the host plant, the population of active and infective cell applied, the competence of indigenous microflora, and the method of inoculation are among the decisive factors that determine the success of rhizobial inoculation under field conditions (Grover et al. 2011). Despite these challenges, strains of rhizobia have been found to increase legume production under water-deficit stress (Mhadhbi et al. 2008; Marulanda et al. 2009; Elboutahiri et al. 2010; Abdel-Salam et al. 2011). And consequently, the usefulness and promising value of rhizobial strains as microbial inoculant to enhance the growth and yield of legumes by ameliorating the negative impact of drought (Ballesteros-Almanza et al. 2010; Mouradi et al. 2016) are reported (Table 14.1). For instance, Uma et al. (2013) in a study recovered 30 isolates of *B. japonicum* from soybean root nodules and evaluated their drought tolerance potential. Of these 30, only four isolates (SBJ-2, SBJ-10, SBJ-14 and SBJ-23) showed highest tolerance to drought stress. Of these four rhizobial cultures, *B. japonicum* strain SBJ-14 demonstrated maximum production of plant growth regulators

Table 14.1 Role of *Rhizobium* in adaptation of plants to drought stress

Rhizobial inoculants	Host legumes	Response	Reference
Rhizobial strains RhL9, RhL10	*Medicago sativa*	Rhizobial strains increased photosynthesis, growth, and yield attributes	Mouradi et al. (2016)
Rhizobium tropici and *Paenibacillus polymyxa*	Common/kidney bean	Altered stomatal conductance and phytohormone balance, improved nodule number and dry mass	Figueiredo et al. (2008)
Rhizobia sp.	Chickpea	Induced antioxidant production, improved biomass (up to 4 g plant^{-1}) and nitrogen fixation (up to 25 μmols h^{-1} plant^{-1})	Mhadhbi et al. (2008)
R. elti	Kidney bean	Increased synthesis of trehalose-6-phosphate, increased nodule number, higher nitrogenase activity	Suarez et al. (2008)
Rhizobia sp.	*Vigna mungo*	Capability to survive drought, gained more nitrogen content (4.2-fold) and total dry weight (1.9-fold)	Mahmood and Athar (2008)
M. mediterraneum LILM10	Chickpea	Increases in grain yield, nodule number, and shoot dry weight were also efficient to NaCl stress	Ben Romdhane et al. (2008)
Mesorhizobium ciceri	Chickpea	Antioxidant production (APX and POX), improved nodulation and nitrogenase activity	Esfahani et al. (2010)
Rhizobia sp.	Black locust	Improved plant survival up to 77% under drought	Ferrari et al. (2010)
Ensifer meliloti 4H41	*Phaseolus vulgaris*	Induced constructive adaptation for specific cortex structure and stress-adapted metabolic activities, improved nodule size and number	Mhadhbi et al. (2011)
Rhizobium gallicum 8a3	Common bean	Regulated water relations in plant, plant gained dry matter yield and leaf number	Sassi-Aydi et al. (2012)
Bradyrhizobium sp.	–	Exopolysaccharides and auxin production, enhanced nodule number, induced drought tolerance	Uma et al. (2013)

(IAA and EPS) and significantly enhanced nodulation, nodule ARA, and nodule N content in soybean plants. The enhancement in legume growth and yield by rhizobial species under drought situations has been attributed to the production of compatible solutes, antioxidants, phytohormones, and exopolysaccharides, increased solubilization and uptake of nutrients, and enhanced expression of protein and gene (Paul

2007; Suarez et al. 2008; Marulanda et al. 2009; Fernandez-Aunion et al. 2010; Grover et al. 2011; Staudinger et al. 2016).

Tajini et al. (2012) in a study investigated the comparative effect of inoculation of two rhizobial strains *R. etli* (12a3) and *R. tropici* (CIAT 899) on two genotypes of common bean CocoT and Flamingo under water stress. They observed that the strain 12a3 outcompeted and was too much proficient compared to CIAT 899 for Flamingo by expressing more values for number of nodules, chlorophyll content, N content, total leaf area, and plant dry mass. It was concluded that water stress had variable impact on symbiosis. Figueiredo et al. (2008) in another experiment conducted under greenhouse experiment evaluated the effect of *Rhizobium* when co-inoculated with other PGPR strains. They reported that *Rhizobium* in combination mitigated the impact of water-deficit stress in common bean plants and resulted in increased nodulation and plant N content. Dashadi et al. (2011) revealed the effect of *R. leguminosarum* bv. *viciae* (F46) inoculation on growth of faba bean under water-deficit stress condition. They reported increase in most of the growth parameters such as root dry weight, nodulation, number of nodule, total N content, day germination speed, mean day germination, and relative water content due to rhizobial inoculation. Finally, they recommended the inoculation of *R. leguminosarum* bv. *viciae* (F46) under water-deficit condition to improve growth and yield of faba bean. Ballesteros-Almanza et al. (2010) while assessing drought stress response of different bean genotypes inoculated with rhizobia reported that rhizobial inoculation improved production of nodule number (350 per plant), nodule dry weight (0.55 mg nodule^{-1}), plant leaf area (96 cm^2), total biomass (1.4 mg plant^{-1}), and drought tolerance (in terms of trehalose contents of nodules up to 90 mg g^{-1}) under water-deficit conditions compared to their well-watered counterparts. The impact of inoculation of rhizobial isolates on nodulation, total N contents, and dry matter production of *Vigna mungo* grown under arid conditions was variable (Mahmood and Athar 2008). In this study, the rhizobial inoculation increased significantly the number of nodules (23 per plant) and total N (3.3 mg plant^{-1}) and dry weight contents (298.6 mg plant^{-1}) of *V. mungo* plants as compared to un-inoculated control treatment. It was suggested from this investigation that inoculation of natural rhizobia of wild legumes having higher tolerance to drought might be useful for increasing N concentration of agriculturally important legumes under arid environment. Whereas when common bean plant was inoculated with overexpressing trehalose-6-phosphate synthase genes OtsA *R. etli,* the plant showed stress tolerance and produced 27% more number of nodules and 38% increase in nitrogenase activity and increased biomass and grain yield up to 57% as compared to wild-type *R. etli-*inoculated plants (Suarez et al. 2008). Moreover, they reported full recovery in 87% of plants (3 weeks old) under drought stress due to the inoculation of overexpressing trehalose-6-phosphate synthase genes *R. etli* in comparison to the plants inoculated with a wild-type *R. etli* in which only 7% plants were recovered. They suggested that inoculation of *R. etli* strain with overexpressing trehalose-6-phosphate synthase genes could enhance grain yield and tolerance to drought in common bean under drought stress. Ali et al. (2009) reported that ecological rehabilitation and increase in soil fertility of degraded soils could be achievable via effective, efficient, and compatible stress-tolerant rhizobial strains. Fitouri et al. (2012) studied the inoculation effect of strain *R. sulla* (Hc14) (moderately tolerant) and the strain Hc5 (most tolerant) on *Hedysarum coronarium* cv. Bikra 21 grown on a farmer's field without irrigation in a semiarid region and reported significant increases in the number of nodules, N contents, and shoot dry weight with respect to the un-inoculated plot. Barbosa et al. (2013) tested the effect of rhizobial inoculation on dry matter of *Vigna unguiculata* under water-deficit conditions and reported that rhizobial inoculation increased plant dry matter of tolerant cultivar compared to non-inoculated treatment. Gan et al. (2010) studied the water-use efficiency of chickpea upon inoculation with *Rhizobium* under water-scare conditions. They reported the inoculation significantly increased water-use efficiency in chickpea. Sanchez et al. (2012) also reported increased drought tolerance and biomass by *Clitoria ternatea* plants when inoculated with rhizobia. Abdel-Salam et al. (2011) while evaluating the growth performance and nodulation efficiencies of seven *R. meliloti* and 21 *R. leguminosarum* bv. *trifolii* strains against drought revealed that strains of *R. leguminosarum* were more drought tolerant than *R. meliloti* strains. Nodulation efficiency of the strains varied considerably, but a maximum of 16 nodules plant^{-1} was formed by Rt4 and Rt7 strains. All nodules produced by these strains were big or medium size. Likewise, Elboutahiri et al. (2010) revealed that the BNF in alfalfa could be improved with the inoculation of drought-tolerant rhizobia under drought conditions. Whereas, Yang et al. (2009) reported that rhizobia in combination with other PGPR could result in better plant productivity and improved drought tolerance. Esfahani et al. (2010) conducted an experiment to test antioxidant enzymes and nitrogenase activities in nodules of chickpea plant inoculated with two strains of *M. ciceri* CP-36 and C-15 under drought stress and unstressed controlled conditions. They reported significant increased activities of POX (38%) and APX (29.4%) in inoculated chickpea nodules of stressed plants compared to non-stressed plants. Furthermore, inoculated chickpea plants showed higher tolerance to drought. They suggested that possible mechanism for increased drought tolerance was possibly due to enhanced antioxidant enzymes activities as result of rhizobial inoculation under drought stress conditions.

14.8   Conclusion and Future Outlook

The adaptability and production of legumes are substantially restricted by many abiotic stresses, including drought. The identification of adaptive features among both rhizobia and legumes to drought is, therefore, of considerable importance. Considering these, there has been a dramatic increase in the use of rhizobia as microbial fertilizers (biofertilizer) to alleviate the effect of drought while growing legumes under stressed environment. It is an easy and inexpensive approach compared to other strategies. It offers the possibility to exploit rhizobia for enhancing legume production even under derelict environment. Efficient, effective, and drought-tolerant rhizobia could play remarkable roles in this context. Moreover, research at proteomics and metabolomics level is likely to pinpoint the exact mechanisms involved and the variants affecting the stress response pathways. However, more work is needed to discover and appraise competent and proficient combinations of microbial inocula and plant genotypes to expose the detailed mechanisms underlying their success.

References

Abdel-Salam MS, Ibrahim SA, Abd-El-Halim MM, Badawy FM, Abo-Aba SEM (2011) Phenotypic characterization of indigenous Egyptian Rhizobial strains for abiotic stresses performance. J Am Sci 7:168–174

Abdullah F, Hareri F, Naaesan M, Ammar MA, Kanbar OZ (2011) Effect of drought on different physiological characters and yield component in different varieties of Syrian Durum wheat. J Agric Sci 3:127–133

Acosta-Gallegos JA, Ochoa-Márquez R, Arrieta-Montiel MP, Ibarra-Pérez F, Pajarito-Ravelero A, Sánchez-Valdéz I (1995) Registration of "Pinto Villa" common bean. Crop Sci 35:1211

Akinci S, Losel DM (2010) The effects of water stress and recovery periods on soluble sugars and starch content in cucumber cultivars. Fresenius Environ Bull 19:164–171

Ali SF, Rawat LS, Meghvansi MK, Mahna SK (2009) Selection of stress-tolerant rhizobial isolates of wild legumes growing in dry regions of Rajasthan, India. ARPN J Agric Biol Sci 4:13–18

Allahmoradi P, Ghobadi M, Taherabadi S, Taherabadi S (2011) Physiological aspects of mungbean in response to drought stress. Int. Conf. Food Eng. Biotechnol. IPCBEE 9: IACSIT Press, Singapore

Araus JL, Slafer GA, Royo C, Serret MD (2008) Breeding for yield potential and stress adaptation in cereals. Crit Rev Plant Sci 27:377–412

Asch F, Padham JL (2005) Root associated bacteria suppress symptoms of iron toxicity in lowland rice. In: Tielkes E, Hulsebusch C, Hauser I, Deininger A, Becker K (eds) The global food and product chain-dynamics: innovations, conflicts, strategies. MDD GmbH, Stuttgart, p 276

Ashraf M (2010) Inducing drought tolerance in plants: recent advances. Biotechnol Adv 28:169–183

Ashraf M, Foolad MR (2007) Roles of glycine betaine and proline in improving plant abiotic stress tolerance. Environ Exp Bot 59:206–216

Ashraf M, Athar HR, Harris PJC, Kwon TR (2008) Some prospective strategies for improving crop salt tolerance. Adv Agron 97:45–110

Ballesteros-Almanza L, Altamirano-Hernandez J, Pena-Cabriales JJ, Santoyo G, Sanchez-Yanez JM, Valencia-Cantero E, Macias-Rodriguez L, Lopez-Bucio J, Cardenas-Navarro R, Farias-Rodriguez R (2010) Effect of co-inoculation with mycorrhiza and rhizobia on the nodule trehalose content of different bean genotypes. Open Microbiol J 4:83–92

Barbosa MAM, Lobato AKDS, Viana GDM, Coelho KNN, Barbosa JRS, Da Costa RCL, Filho BGDS, Neto CFDO (2013) Root contribution to water relations and shoot in two contrasting Vigna unguiculata cultivars subjected to water deficit and inoculation. Romanian Agric Res 30:155–162

Barrera-Figueroa B, Pena-Castro J, Acosta-Gallegos JA, Ruiz-Medrano R, Xoconostle-Cazares B (2007) Isolation of dehydration-responsive genes in a drought tolerant common bean cultivar and expression of a group 3 late embryogenesis abundant mRNA in tolerant and susceptible bean cultivars. Funct Plant Biol 34:368–381

Bartels D, Sunkar R (2005) Drought and salt tolerance in plants. Crit Rev Plant Sci 24:23–58

Beebe SE, Rao IM, Cajiao I, Grajales M (2008) Selection for drought resistance in common bean also improves yield in phosphorus limited and favorable environments. Crop Sci 48:582–592

Beebe SE, Rao IM, Blair MW, Acosta-Gallegos JA (2013) Phenotyping common beans for adaptation to drought. Front Physiol 4:1–20

Ben Romdhane S, Nasr H, Samba-Mbaye R, Neyra M, Ghorbal MH, Lajudie PD (2006) Genetic diversity of *Acacia tortilis ssp. raddiana* rhizobia in Tunisia assessed by 16S and 16S-23S rDNA genes analysis. J Appl Microbiol 100:436–445

Ben Romdhane S, Tajini F, Trabelsi M, Aouani ME, Mhamdi R (2007) Competition for nodule formation between introduced strains of *Mesorhizobium ciceri* and native populations of rhizobia nodulating chickpea (*Cicer aerietinum*) in Tunisia. World J Microbiol Biotechnol 23:1195–1201

Ben Romdhane S, Aouani ME, Trabelsi M, De Lajudie P, Mhamdi R (2008) Selection of high nitrogen-fixing rhizobia nodulating chickpea (*Cicer arietinum*) for semi-arid Tunisia. J Agron Crop Sci 194:413–420

Bhatnagar-Mathur P, Devi MJ, Vadez V, Sharma KK (2009) Differential antioxidative responses in transgenic peanut bear no relationship to their superior transpiration efficiency under drought stress. J Plant Physiol 166:1207–1017

Bhatt RM, Srinivasa Rao NK (2005) Influence of pod load response of okra to water stress. Indian J Plant Physiol 10:54–59

Bhattacharyya RN, Basu PS (1997) Bioproduction of indole acetic acid by a *Rhizobium* sp. from the root nodules of *Desmodium gangeticum* DC. Acta Microbiol Immunol Hung 44:109–118

Bianco C, Defez R (2009) *Medicago truncatula* improves salt tolerance when nodulated by an indole-3-acetic acid-overproducing *Sinorhizobium meliloti* strain. J Exp Bot 60:3097–3107

Billi D, Potts M (2002) Life and death of dried prokaryotes. Res Microbiol 153:7–12

Boiero L, Perrig D, Masciarelli O, Pena C, Cassan F, Luna V (2007) Phytohormone production by strains of *Bradyrhizobium japonicum* and possible physiological and technological implications. Appl Microbiol Biotechnol 74:874–880

Boncompagi E, Osteras M, Poggi MC, Ie Rudulier D (1999) Occurrence of choline and glycine betain uptake and metabolism in the family *Rhizobi*aceae and their roles in osmoprotection. Appl Environ Microbiol 65:2072–2077

Bota J, Medrano H, Flexas J (2004) Is photosynthesis limited by decreased Rubisco activity and RuBP content under progressive water stress? New Phytol 162:671–681

Boutraa T, Saders FE (2001) Influence of water stress on grain yield and vegetative growth of two cultivars of bean (*Phaseolus vulgaris* L.) J Agron Crop Sci 187:251–257

Bray EA (1997) Plant responses to water deficit. Trends Plant Sci 2:48–54

Carrascoa JA, Armarios P, Pajueloa E, Burgosa A, Caviedesc MA, Lopezb R, Chambera MA, Palomaresc AJ (2005) Isolation and characterization of symbiotically effective *Rhizobium* resistant to arsenic and heavy metals after the toxic spill at the Aznacollar pyrite mine. Soil Biol Biochem 37:1131–1140

Chaves MM, Flexas J, Pinheiro C (2009) Photosynthesis under drought and salt stress: regulation mechanisms from whole plant to cell. Ann Bot 103:551–560

Chen M, Wang QY, Cheng XG, ZS X, Li LC, Ye XG, Xia LQ, Ma YZ (2007) GmDREB2, a soybean DRE-binding transcription factor, conferred drought and high-salt tolerance in transgenic plants. Biochem Biophys Res Commun 353:299–305

Comba ME, Benavides MP, Tomaro ML (1998) Effect of salt stress on antioxidant defense system in soybean root nodules. Aust J Plant Physiol 25:665–671

DaMatta FM, Ramalho JDC (2006) Impacts of drought and temperature stress on coffee physiology and production: a review. Braz J Plant Physiol 18:55–81

Dashadi M, Khosravi H, Moezzi A, Nadian H, Heidari M, Radjabi R (2011) Co-inoculation *Rhizobium* and *Azotobacter* on growth indices of Faba bean under water stress in the green house condition. Adv Stud Biol 3:373–385

Datta C, Basu PS (2000) Indole acetic acid production by a *Rhizobium* species from root nodules of a leguminous shrub, *Cajanus cajan*. Microbiol Res 155:123–127

Ding X, Zhang S, Wang R, Li S, Liao X (2016) AM fungi and rhizobium regulate nodule growth, phosphorous (P) uptake, and soluble sugar concentration of soybeans experiencing P deficiency. J Plant Nutr 39:1915–1925

Dominguez-Ferreras A, Perez-Arnedo R, Becker A, Olivares J, Soto MJ, Sanjuan J (2006) Transcriptome profiling reveals the importance of plasmid pSymB for osmoadaptation of *Sinorhizobium meliloti*. J Bacteriol 188:7617–7625

Duhan JS, Dudeja SS, Khurana AL (1998) Siderophore production in relation to N₂ fixation and iron uptake in *Pigeon pea-Rhizobium* symbiosis. Folia Microbiol 43:421–426

Elboutahiri N, Thami-Alami I, Udupa SM (2010) Phenotypic and genetic diversity in *Sinorhizobium meliloti* and *S. medicae* from drought and salt affected regions of Morocco. BMC Microbiol 10:15

Esfahani MN, Mostajeran A (2011) Rhizobial strain involvement in symbiosis efficiency of chickpea-rhizobia under drought stress: plant growth; nitrogen fixation and antioxidant enzyme activities. Acta Physiol Plant 33:1075–1083

Esfahani MN, Mostajeran A, Emtiazi G (2010) The effect of drought stress on nitrogenase and antioxidant enzyme activities in nodules formed from symbiosis of chickpea with two strains of *Mesorhizobium ciceri*. World Appl Sci J 10:621–626

FAO (2011) The state of the world's land and water resources for food and agriculture the state of the world's land and water resources for food and agriculture. Managing systems at risk. http://www.fao.org/docrep/017/i1688e/i1688e.pdf

Farooq M, Basra SMA, Wahid A, Cheema ZA, Cheema MA, Khaliq A (2008) Physiological role of exogenously applied glycine betaine in improving drought tolerance of fine grain aromatic rice (*Oryza sativa* L.) J Agron Crop Sci 194:325–333

Farooq M, Wahid A, Kobayashi N, Fujita D, Basra SMA (2009) Plant drought stress: effects, mechanisms and management. Agron Sustain Dev 29:185–212

Farrar K, Bryant D, Cope-Selby N (2014) Understanding and engineering beneficial plant–microbe interactions: plant growth promotion in energy crops. Plant Biotechnol J 12:1193–1206

Fenta BA, Beebe SE, Kunert KJ, Burridge JD, Barlow KM, Lynch JP, Foyer CH (2014) Field phenotyping of soybean roots for drought stress tolerance. Agronomy 4:418–435

Fernandez-Aunion C, Hamouda TB, Iglesias-Guerra F, Argandona M, Reina-Bueno M, Nieto JJ, Aouani ME, Vargas C (2010) Biosynthesis of compatible solutes in rhizobial strains isolated from *Phaseolus vulgaris* nodules in Tunisian fields. BMC Microbiol 10:192

Ferrari AE, Esparrach CA, Galetti MA, Wall LG (2010) Afforestation of a desurfaced field with *Robinia pseudoacacia* inoculated with *Rhizobium* spp. and *Glomus deserticola*. Cienc Suelo 28:105–114

Figueiredo MVB, Burity HA, Martinez CR, Chanway CP (2008) Plant growth-promoting rhizobacteria for improving nodulation and nitrogen fixation in the common bean (*Phaseolus vulgaris* L.) World J Microbiol Biotechnol 24:1187–1193

Fitouri DS, Faysal BJ, Kais Z, Salah R, Ridha M (2012) Effect of inoculation with osmotolerant strain of *Rhizobium sullae* on growth and protein production of sulla (*Sulla coronarium* L.) under water deficit. J Appl Biosci 51:3642–3651

Flexas J, Bota J, Loreto F, Cornic G, Sharkey TD (2004) Diffusive and metabolic limitations to photosynthesis under drought and salinity in C3 plants. Plant Biol 6:269–279

Flexas J, Ribas-Carbo M, Bota J, Galmes J, Henkle M, Madrano S (2006) Decreased Rubisco activity during water stress is not induced by decreased relative water content but related to conditions of low stomatal conductance and chloroplast CO_2 concentration. New Phytol 172:73–82

Foyer CH, Noctor G (2005) Redox homeostasis and antioxidant signaling: a metabolic interface between stress perception and physiological responses. Plant Cell 17:1866–1875

Franzini VI, Azcón R, Mendes FL, Aroca R (2010) Interactions between Glomus species and Rhizobium strains affect the nutritional physiology of drought-stressed legume hosts. J Plant Physiol 167:614–619

Gan YT, Warkentin TD, Bing DJ, Stevenson FC, McDonal CL (2010) Chickpea water use efficiency in relation to cropping system, cultivar, soil nitrogen and rhizozial inoculation in semi-arid environment. Agric Water Manag 97:1375–1381

Gechev TS, Hille J (2005) Hydrogen peroxide as a signal controlling plant programmed cell death. J Cell Biol 168:17–20

Ghosh S, Basu PS (2006) Production and metabolism of indole acetic acid in roots and root nodules of *Phaseolus mungo*. Microbiol Res 161:362–366

Gigon A, Matos AR, Laffray D, Zuily-Fodil Y, Pham-Thi AT (2004) Effect of drought stress on lipid metabolism in the leaves of *Arabidopsis thaliana* (Ecotype Columbia). Ann Bot 94:345–351

Gloux K, Le Rudulier D (1989) Transport and catabolism of proline betaine in salt-stressed *Rhizobium meliloti*. Arch Microbiol 151:143–148

Gopalakrishnan S, Sathya A, Vijayabharathi R, Varshney RK, Gowda CLL, Krishnamurthy L (2015) Plant growth promoting rhizobia: challenges and opportunities. 3 Biotech 5:355–377

Gorai M, Hachef A, Neffati M (2010) Differential responses in growth and water relationship of *Medicago sativa* (L.) cv. Gabes and Astragalus gombiformis (Pom.) under water-limited conditions. Emir J Food Agric 22:01–12

Gosal SS, Wani SH, Kang MS (2009) Biotechnology and drought tolerance. J Crop Improv 23:19–54

Gouffi K, Pica N, Pichereau V, Blanco C (1999) Disaccharides as a new class of nonaccumulated osmoprotectants for *Sinorhizobium meliloti*. Appl Environ Microbiol 65:1491–1500

Gouffi K, Bernard T, Blanco C (2000) Osmoprotection by pipecolic acid in *Sinorhizobium meliloti*: specific effects of D and L isomers. Appl Environ Microbiol 66:2358–2364

Grover M, Ali SZ, Sandhya V, Rasul A, Venkateswarlu B (2011) Role of microorganisms in adaptation of agriculture crops to abiotic stresses. World J Microbiol Biotechnol 27:1231–1240

Hanafy MS, El-Banna A, Schumacher HM, Jacobsen HJ, Hassan FS (2013) Enhanced tolerance to drought and salt stresses in transgenic faba bean (*Vicia faba* L.) plants by heterologous expression of the PR10a gene from potato. Plant Cell Rep 32:663–674

Haupt-Herting S, Fock HP (2000) Exchange of oxygen and its role in energy dissipation during drought stress in tomato plants. Physiol Plant 110:489–495

Hossain Z, Komatsu S (2014a) Potentiality of soybean proteomics in untying the mechanism of flood and drought stress tolerance. Proteomes 2:107–127

Hossain Z, Komatsu S (2014b) Soybean proteomics. Methods Mol Biol 1072:315–331

Howieson J, Ballard R (2004) Optimizing the legume symbiosis in stressful and competitive environments within southern Australia-some contemporary thoughts. Soil Biol Biochem 36:1261–1273

Humann JL, Ziemkiewicz HT, Yurgel SN, Kahn ML (2009) Regulatory and DNA repair genes contribute to the desiccation resistance of *Sinorhizobium meliloti* Rm1021. Appl Environ Microbiol 75:446–453

Hussain MB, Zahir ZA, Asghar HN, Asgher M (2014a) Can catalase and EPS producing rhizobia ameliorate drought in wheat. Int J Agric Biol 16:3–13

Hussain MB, Zahir ZA, Asghar HN, Mahmood S (2014b) Scrutinizing rhizobia to rescue maize growth under reduced water conditions. Soil Sci Soc Am J 78:538–545

Hussain MB, Zahir ZA, Asghar HN, Mubaraka R, Naveed M (2016) Efficacy of rhizobia for improving photosynthesis, productivity and mineral nutrition of maize. Clean – Soil, Air, Water 44:1564–1571

Imen H, Neila A, Adnane B, Manel B, Mabrouk Y, Saidi M, Bouaziz S (2015) Inoculation with phosphate solubilizing *Mesorhizobium* strains improves the performance of chickpea (*Cicer aritenium* L.) under phosphorus deficiency. J Plant Nutr 38:1656–1671

Issa S, Wood M (1995) Multiplication and survival of chickpea and bean rhizobia in dry soils: the influence of strains, matric potential and soil texture. Soil Biol Biochem 27:785–798

Iturriaga G, Suarez R, Nova-Franco B (2009) Trehalose metabolism: from osmoprotection to signaling. Int J Mol Sci 10:3793–3810

Jain M, Prasad PVV, Boote KJ, Hartwell AL Jr, Chourey PS (2007) Effect of season-long high temperature growth conditions on sugar-to-starch metabolism in developing microspores of grain sorghum (*Sorghum bicolor* L. Moench). Planta 227:67–79

Jaleel CA, Manivannan P, Lakshmanan GMA, Gomathinayagam M, Panneerselvam R (2008) Alterations in morphological parameters and photosynthetic pigment responses of *Catharanthus roseus* under soil water deficits. Colloids Surf B: Biointerfaces 61:298–303

Jaleel CA, Manivannan P, Wahid A, Farooq M, Somasundaram R, Panneerselvam R (2009) Drought stress in plants: a review on morphological characteristics and pigments composition. Int J Agric Biol 11:100–105

Jebara S, Drevon JJ, Jebara M (2010) Modulation of symbiotic efficiency and nodular antioxidant enzyme activities in two *Phaseolus vulgaris* genotypes under salinity. Acta Physiol Plant 32:925–932

Jiang Y, Liang G, Yu D (2012) Activated expression of WRKY57 confers drought tolerance in Arabidopsis. Mol Plant 5:1375–1388

Jordan WR, Ritichie JT (2002) Influence of soil water stress on evaporation, root absorption and internal water status of cotton. Plant Physiol 48:783–788

Joshi R, Wani SH, Singh B, Bohra A, Dar ZA, Lone AA, Singla-Pareek SL (2016) Transcription factors and plants response to drought stress: current understanding and future directions. Front Plant Sci 7:1029. doi:10.3389/fpls.2016.01029

Kacperska A (2004) Sensor types in signal transduction pathways in plant cells responding to abiotic stressor: do they depend on stress intensity? Physiol Plant 122:159–168

Kang J, Hwang JU, Lee M, Kim YY, Assmann SM, Martinoia E, Lee Y (2010) PDR-type ABC transporter mediates cellular uptake of the phytohormone abscisic acid. Proc Natl Acad Sci USA 107:2355–2360

Karmakar K, Rana A, Rajwar A, Sahgal M, Johri BN (2015) Legume-rhizobia symbiosis under stress. In: Arora NK (ed) Plant microbe symbiosis-applied facets. Springer, New Delhi, pp 241–258

Kim TH, Bohmer M, HH H, Nishimura N, Schroeder JI (2010) Guard cell signal transduction network: advances in understanding abscisic acid, CO_2, and Ca^{2+} signaling. Annu Rev Plant Biol 61:561–591

Kosová K, Vítámvás P, Prášil IT, Renaut J (2011) Plant proteome changes under abiotic stress–contribution of proteomics studies to understanding plant stress response. J Proteome 74:1301–1322

Lazali M, Zaman-Allah M, Amenc L, Ounane G, Abadie J, Drevon JJ (2013) A phytase gene is overexpressed in root nodules cortex of *Phaseolus vulgaris*–rhizobia symbiosis under phosphorus deficiency. Planta 238:317–324

Lopez M, Herrera-Cervera JA, Tejera NA, Lluch C (2008) Growth and nitrogen fixation in *Lotus japonicus* and *Medicago truncatula* under NaCl stress: nodule carbon metabolism. J Plant Physiol 165:641–650

Ma WB, Guinel FC, Glick BR (2003) *Rhizobium leguminosarum biovar viciae* 1-aminocyclopropane-1-carboxylate deaminase promotes nodulation of pea plants. Appl Environ Microbiol 69:4396–4402

Mahmood A, Athar M (2008) Cross inoculation studies: response of *Vigna mungo* to inoculation with rhizobia from tree legumes growing under arid environment. Int J Environ Sci Technol 5:135–139

Malamy JE (2005) Intrinsic and environmental response pathways that regulate root system architecture. Plant Cell Environ 28:67–77

Marulanda A, Barea JM, Azcon R (2009) Stimulation of plant growth and drought tolerance by native microorganisms (AM fungi and bacteria) from dry environments: mechanisms related to bacterial effectiveness. J Plant Growth Regul 28:115–124

Matamoros MA, Dalton DA, Ramos J, Clemente MRE, Rubio MC, Becana M (2003) Biochemistry and molecular biology of antioxidants in the rhizobia-legume symbiosis. Plant Physiol 133:449–509

Mehboob I, Naveed M, Zahir ZA (2009) Rhizobial association with non-legumes: mechanisms and applications. Crit Rev Plant Sci 28:432–456

Mehboob I, Zahir ZA, Arshad M, Tanveer A, Farooq-e-Azam (2011) Growth promoting activities of different *Rhizobium* spp. in wheat. Pak J Bot 43:1643–1650

Meuelenberg F, Dakora FD (2007) Assessing the biological potential of N_2-fixing leguminosae in Botswana for increased crop yields and commercial exploitation. Afr J Biotechnol 6: 325–334

Mhadhbi H, Jebara M, Zitoun A, Limam F, Aouani ME (2008) Symbiotic effectiveness and response to mannitol-mediated osmotic stress of various chickpea-rhizobia association. World J Microbiol Biotechnol 24:1027–1035

Mhadhbi H, Chihaoui S, Mhamdi R, Mnasri B, Jebara M, Mhamdi R (2011) A highly osmotolerant rhizobial strain confers a better tolerance of nitrogen fixation and enhances protective activities to nodules of *Phaseolus vulgaris* under drought stress. Afr J Biotechnol 10:4555–4563

Miller KJ, Wood JM (1996) Osmoadaptation by rhizosphere bacteria. Annu Rev Microbiol 50:101–136

Mirza BS, Mirza MS, Bano A, Malik KA (2007) Coninoculation of chickpea with *Rhizobium* isolates from roots and nodules and phytohormone-producing *Enterobacter* strains. Aust J Exp Agric 47:1008–1015

Mnasri B, Mrabet M, Laguerre G, Aouani ME, Mhamdi R (2007) Salt tolerant rhizobia isolated from a Tunisian oasis that are highly-effective for N$_2$-fixation with *Phaseolus vulgaris* constitute a novel biovar (bv. Mediterranense) of *Sinorhizobium meliloti*. Arch Microbiol 187:79–85

Montero-Tavera V, Ruiz-Medrano R, Xoconostle-Cazares B (2008) Systemic nature of drought-tolerance in common bean. Plant Signal Behav 3:663–666

Moschetti G, Pelusoa A, Protopapa A, Anastasioa M, Pepe O, Defez R (2005) Use of nodulation pattern, stress tolerance, *nodC* gene amplification, RAPD-PCR and RFLP- 6S rDNA analysis to discriminate genotypes of *Rhizobium leguminosarum biovar viciae*. Syst Appl Microbiol 28:619–631

Mouradi M, Farissi M, Bouizgaren A, Makoudi B, Kabbadj A, Very AA, Sentenac H, Qaddourya A, Ghoulam C (2016) Effects of water deficit on growth, nodulation and physiological and biochemical processes in *Medicago sativa*-rhizobia symbiotic association. Arid Land Res Manag 30:193–208

Nanda AK, Andrio E, Marino D, Pauly N, Dunand C (2010) Reactive oxygen species during plant-microorganisms early interactions. J Integr Plant Biol 52:195–204

Naresh RK, Purushottam, Singh SP, Dwivedi A, Kumar V (2013) Effects of water stress on physiological processes and yield attributes of different mungbean (L.) varieties. Afr J Biochem Res 7:55–62

Naveed M (2013) Maize endophytes—diversity, functionality and application potential. Ph.D. Thesis, AIT—Austrian Institute of Technology/BOKU University, Tulln Campus, Vienna, Austria

Naveed M, Mitter B, Reichenauer TG, Krzysztof W, Sessitsc A (2014a) Increased drought stress resilience of maize through endophytic colonization by *Burkholderia phytofirmans* PsJN and *Enterobacter* sp. FD17. Environ Exp Bot 97:30–39

Naveed M, Hussain MB, Zahir ZA, Mitter B, Sessitsch A (2014b) Drought stress amelioration in wheat through inoculation with *Burkholderia phytofirmans* strain PsJN. Plant Growth Regul 73:121–131

Paul M (2007) Trehalose-6-phosphate. Curr Opin Plant Biol 10:303–309

Peix A, Rivas-Boyero AA, Mateos PF, Rodriguez-Barrueco C, Martinez-Molina E, Velazquez E (2001) Growth promotion of chickpea and barley by a phosphate solubilizing strain of *Mesorhizobium mediterraneum* under growth chamber conditions. Soil Biol Biochem 33:103–110

Pichereau V, Pocard JA, Hamelin J, Blanco C, Bemard T (1998) Different effects of dimethylsulfonipropionate, dimethylsulfonioacetate and other S-methylated compounds on the growth of *Sinorhizobium meliloti* at low and high osmolarities. Appl Environ Microbiol 64:1420–1429

Pie ZM, Murata Y, Benning G, Thomine S, Klusebner B, Allen GJ, Grill E, Schroeder JI (2000) Calcium channels activated by hydrogen peroxide mediate abscisic acid signaling in guard cells. Nature 406:731–734

Polania JA, Poschenrieder C, Beebe S, Rao IM (2016) Effective use of water and increased dry matter partitioned to grain contribute to yield of common bean improved for drought resistance. Front Plant Sci 7:1–10

Poolman B, Blount P, Folgering JHA, Friesn RHE, Moe PC, van der Heide T (2002) How do membrane proteins sense water stress? Mol Biol 44:899–902

Raghavendra AS, Gonugunta AK, Christmann A, Grill E (2010) ABA perception and signaling. Trends Plant Sci 15:395–401

Rajaram S (2005) Role of conventional plant breeding and biotechnology in future wheat production. Turk J Agric For 29:105–111

Ramalingam A, Kudapa H, Pazhamala LT, Weckwerth W, Varshney RK (2015) Proteomics and metabolomics: two emerging areas for legume improvement. Front Plant Sci 6:1116. doi:10.3389/fpls.2015.01116

Räsänen LA, Lindström K (2003) Effect of biotic and abiotic constraints on the symbiosis between rhizobia and the tropical leguminous trees Acacia and Prosopis. Indian J Exp Biol 41:1142–1159

Rasanen LA, Saijets S, Jokinen K, Lindstron K (2004) Evaluation of the roles of two compatible solutes, glycine betaine and trehalose, for the Acacia Senegal-*Sinorhizobium* symbiosis exposed to drought stress. Plant Soil 260:237–251

Reddy IJ, Nigam SN, Rao RCN, Reddy NS (2001) Registration of ICGV 87354 peanut germplasm with drought tolerance and rust resistance. Crop Sci 41:274–275

Redondo FJ, Coba de la Pena RT, Morcillo CN, Lucas MM, Pueyo JJ (2009) Overexpression of flavodoxin in bacteriods induces changes in antioxidant metabolism leading to delayed senescence and starch accumulation in alfalfa root nodules. Plant Physiol 49:1166–1178

Rehman A, Nautiyal CS (2002) Effect of drought on the growth and survival of the stress-tolerant bacterium *Rhizobium* sp. NBRI2505 sesbania and its drought-sensitive transposon Tn5 mutant. Curr Microbiol 45:368–377

Rodziewicz P, Swarcewicz B, Chmielewska K, Wojakowska A, Stobiecki M (2014) Influence of abiotic stresses on plant proteome and metabolome changes. Acta Physiol Plant 36:1–19

Ronde JAD, Cress WA, Krugerd GHJ, Strasserd RJ, Wan Staden J (2004) Photosynthetic response of transgenic soybean plants containing an Arabidopsis PSCR gene during heat and drought stress. J Plant Physiol 161:1211–1224

Rosas SB, Andres GA, Rovera M, Correa NS (2006) Phosphate solubilizing *Pseudomonas putida* can influence the rhizobia legume symbiosis. Soil Biol Biochem 38:3502–3505

Rouhier N, Santos CVD, Tarrago L, Rev P (2006) Plant methionine sulfoxide reductase A and B multigenic families. Photosynth Res 89:247–262

Sairam R, Srivastava G, Agarwal S, Meena R (2005) Differences in antioxidant activity in response to salinity stress in tolerant and susceptible wheat genotypes. Biol Plant 49:85–91

Sanchez DH, Schwabe F, Erban A, Udvardi MK, Kopka J (2012) Comparative metabolomics of drought acclimation in model and forage legume. Plant Cell Environ 35:136–149

Sánchez-Valdez I, Acosta-Gallegos JA, Ibarra-Pérez FJ, Rosales-Serna R, Singh SP (2004) Registration of Pinto Saltillo common bean. Crop Sci 44:1865–1866

Sangtarash MH (2010) Responses of different wheat genotypes to drought stress applied at different growth stages. Pak J Biol Sci 13:114–119

Sassi-Aydi S, Aydi S, Abdelly C (2012) Inoculation with the native rhizobium gallicum 8a3 improves osmotic stress tolerance in common bean drought-sensitive cultivar. Acta Agric Scand Sect B Soil Plant Sci 62:179–187

Seki M, Narusaka M, Ishida J, Nanjo T, Fujita M, Oono Y, Kamiya A, Nakajima M, Enju A, Sakurai T, Satou M, Akiyama K, Taji T, Yamaguchi-Shinozaki K, Carninci P, Kawai J, Hayashizaki Y, Shinozaki K (2002) Monitoring the expression profiles of 7000 Arabidopsis genes under drought, cold and high-salinity stresses using a full-length cDNA microarray. Plant J 31:279–292

Sengupta D, Kannan M, eddy AR (2011) A root proteomics-based insight reveals dynamic regulation of root proteins under progressive drought stress and recovery in *Vigna radiate* (L.) Wilczek. Planta 233:1111–1127

Serraj R, Sinclair TR, Purcell LC (1999) Symbiotic N_2 fixation response to drought. J Exp Bot 50:143–155

Sessitsch A, Howieson JG, Perret X, Antoun H, Martinez-Romero R (2002) Advances in *Rhizobium* research. Crit Rev Plant Sci 21:323–378

Shao HB, Chu LY, Jaleel CA, Zhao XA (2008) Water-deficit stress-induced anatomical changes in higher plants. C R Biol 331:215–225

Sharma S, Upadhyaya HD, Varshney RK, Gowda CLL (2013) Pre-breeding for diversification of primary gene pool and genetic enhancement of grain legumes. Front Plant Sci 4:1–14

Shulaev V, Cortes D, Miller G, Mittler R (2008) Metabolomics for plant stress response. Physiol Plant 132:199–208

Sing KB, Osaar M, Saxena MC, Johansen C (1996) Registration of FLIP 87-59C, a drought tolerant chickpea germplasm line. Crop Sci 36:1–2

Sing SP, Terah H, Gutierrez JA (2001) Registration of SEA 5 and SEA 13 drought tolerant dry bean germplas. Crop Sci 41:276–277

Soliman AS, Shanan NT, Massaoud ON, Swelim DM (2012) Improving salinity tolerance of *Acacia saligna* (Labill.) plant by *Arbuscular mycorrhizal* fungi and *Rhizobium* inoculation. Afr J Biotechnol 11:1259–1266

Srivalli B, Chinnusamy V, Chopra RK (2003) Antioxidant defense in response to abiotic stresses in plants. J Plant Biol 30:121–139

Staudinger C, Mehmeti-Tershani V, Gil-Quintana E, Gonzalez EM, Hofhansl F, Bachmann G, Wienkoop S (2016) Evidence for a rhizobia-induced drought stress response strategy in *Medicago truncatula*. J Proteome 136:202–213

Streeter JG (2003) Effect of trehalose on survival of *Bradyrhizobium japonicum* during desiccation. J Appl Microbiol 95:484–491

Suarez R, Wong A, Ramirez M, Barraz A, Orozco MDC, Cevallos MA, Lara M, Hernandez G, Iturriaga G (2008) Improvement of drought tolerance and grain yield in common bean by overexpressing trehalose-6-phosphate synthase in rhizobia. Mol Plant Microbe Interact 21:958–966

Tajini F, Trabelsi M, Drevon JJ (2012) Comparison between the reference *Rhizobium tropici* CIAT899 and the native *Rhizobium etli*12a3 for some nitrogen fixation parameters in common bean (*Phaseolus vulgaris* L.) under water stress. Afr J Microbiol Res 6:4058–4067

Thiao M, Neyra M, Isidore E, Sylla S, Lesueur D (2004) Diversity and effectiveness of *Rhizobium* from *Gliricidia sepium* native to Reunion Island, Kenya and New Caledonia. World J Microbiol Biotechnol 20:703–709

Uma C, Sivagurunathan P, Sangeetha D (2013) Performance of *Bradyrhizobial* isolates under drought conditions. Int J Curr Microbiol App Sci 2:228–232

Upadyyay SK, Singh DP, Saikia R (2009) Genetic diversity of plant growth promoting rhizobacteria from rhizospheric soil of wheat under saline conditions. Curr Microbiol 59:489–496

Valladares F, Gianoli E, Gomez JM (2007) Ecological limits to plant phenotypic plasticity. New Phytol 176:749–763

Vanderlinde EM, Harrison JJ, Muszynski A, Carlson RW, Tumer RJ, Yost CK (2010) Identification of a novel ABC transporter required for desiccation tolerance, and biofilm formation in *Rhizobium leguminosarum bv. viciae* 3841. FEMS Microbiol Ecol 71:327–340

Varshney RK, Mohan SM, Gaur PM, Gangarao NV, Pandey MK, Bohra A, Sawargaonkar SL, Chitikineni A, Kimurto PK, Janila P, Saxena KB, Fikre A, Sharma M, Rathore A, Pratap A, Tripathi S, Datta S, Chaturvedi SK, Mallikarjuna N, Anuradha G, Babbar A, Choudhary AK, Mhase MB, Bharadwaj CH, Mannur DM, Harer PN, Guo B, Liang X, Nadarajan N, Gowda CL (2013) Achievements and prospects of genomics-assisted breeding in three legume crops of the semi-arid tropics. Biotechnol Adv 31:1120–1134

Wani PA, Khan MS, Zaidi A (2008) Effect of metal-tolerant plant growth-promoting *Rhizobium* on the performance of pea grown in metal-amended soil. Arch Environ Contam Toxicol 55:33–42

Wasilewska A, Vlad F, Sirichandra C, Redko Y, Jammes F, Valon C, Frey NFD, Leung J (2008) An update on abscisic acid signaling in plants and more. Mol Plant 1:198–217

Weckwerth W (2011) Green systems biology-From single genomes, proteomes and metabolomes to ecosystems research and biotechnology. J Proteome 75:284–305

Wilkinson S, Davies W (2010) Drought, ozone, ABA and ethylene: new insights from cell to plant to community. Plant Cell Environ 33:510–525

Xu J, Li X, Luo L (2012) Effects of engineered *Sinorhizobium meliloti* on cytokinin synthesis and tolerance of alfalfa to extreme drought stress. Appl Environ Microbiol 78:8056–8061

Yang DG, Yang FP, Chen XQ, Zhang LQ, Zhang XD (2009) Obtainment of tranxformed maize with dehydration-responsive transcription factor CBF4 gene. Acta Agron Sin 35:1759–1763

Yanni Y, Zidan M, Dazzo F, Rizk R, Mehesen A, Abdelfattah F, Elsadany A (2016) Enhanced symbiotic performance and productivity of drought stressed common bean after inoculation with tolerant native rhizobia in extensive fields. Agric Ecosyst Environ 232:119–128

Yordanov I, Velikova V, Tsonev T (2003) Plant responses to drought and stress tolerance. Bulg J Plant Physiol (Special issue):187–206

Zacarias JJJ, Altamirano-Hernandez J, Cabriales JJP (2004) Nitrogenase activity and trehalose content of nodules of drought-stressed common bean infected with effective (Fix$^+$) and ineffective (Fix$^-$) rhizobia. Soil Biol Biochem 36:1975–1981

Zahran HH (1999) *Rhizobium*-legume symbiosis and nitrogen fixation under severe conditions and in an arid climate. Microbiol Mol Biol Rev 63:968–989

Zahran HH (2001) Rhizobia from wild legumes: diversity, taxonomy, ecology, nitrogen fixation and biotechnology. J Biotechnol 91:143–153

Zeng N (2003) Drought in the Sahel. Science 302:999–1000

Zhou Y, Lam HM, Zhang J (2007) Inhibition of photosynthesis and energy dissipation induced by water and high light stresses in rice. J Exp Bot 58:1207–1217

Zlatev Z, Lidon FC (2012) An overview on drought induced changes in plant growth, water relations and photosynthesis. Emir J Food Agric 24:57–72

Metal-Legume-Microbe Interactions: Toxicity and Remediation

15

Saima Saif, Almas Zaidi, Mohd. Saghir Khan, and Asfa Rizvi

Abstract

Heavy metals discharged from various sources accumulate within soils and disrupt ecosystems. The toxic metals are taken up by beneficial soil microbiota and growing plants and cause potential human risks via food chain. Also, heavy metals seriously affect the microbial compositions and their physiological functions. Among plant species, legumes play an important role in human dietary systems and supply nitrogen to legumes through symbiosis with rhizobia. Metals when present in legume habitat act as a devastating stress factor and restrict the growth of rhizobia, legumes, and legume-*Rhizobium* symbiosis. Several physical and chemical methods have been developed to remediate heavy metal-polluted soils, but these methods are unacceptable due to their high cost, and they are not environmentally friendly. Therefore, the use of metal-tolerant/metal-detoxifying microbes collectively called bioremediation offers a sustainable and low-cost option to clean up polluted soils. Besides remediation, the metal-tolerant microbes also promote plant growth by other direct or indirect means. Owing to the importance of legumes in maintaining soil fertility and human health, there is greater emphasis to identify the metal-resistant/metal-tolerant rhizobia and legume plants. The present chapter gives an in-depth insight into the impact of metals on rhizobia-legume symbiosis. Also, the role of metal-tolerant rhizobia in metal toxicity abatement is highlighted.

S. Saif (✉) • A. Zaidi • M.S. Khan • A. Rizvi
Department of Agricultural Microbiology, Faculty of Agricultural Sciences, Aligarh Muslim University, Aligarh 202002, Uttar Pradesh, India
e-mail: saima.saif3@gmail.com

© Springer International Publishing AG 2017
A. Zaidi et al. (eds.), *Microbes for Legume Improvement*,
DOI 10.1007/978-3-319-59174-2_15

15.1   Introduction

Heavy metals (HMs) discharged from different industrial operations cause a substantial threat to varied agroecosystems (Petrova et al. 2013; Qing et al. 2015). Agricultural soil may become contaminated with HMs emanating from a variety of anthropogenic sources such as smelters, mining, power station industries, application of metal-containing fertilizers, and sewage sludge (Li et al. 2014; Islam et al. 2015). The concentration of heavy metals deposited in soil, however, depends on the source of origin, transport to accumulation site, and their retention and fixation with soil constituents. Once accumulated in soil, heavy metals are adsorbed by soil materials and are redistributed into different chemical forms with varying bioavailability, mobility, and toxicity (Zhang et al. 2015; Alamgir 2016). This distribution is believed to be controlled by reactions of heavy metals in soils such as (1) mineral precipitation and dissolution; (2) ion exchange, adsorption, and desorption; (3) aqueous complication; (4) biological immobilization and mobilization; and (5) plant uptake (Mamindy-Pajany et al. 2014; Luo et al. 2016). At low concentrations, some metals like zinc, copper, nickel, and chromium can be nontoxic and are often involved in important enzyme functions. On the contrary, some metals like cadmium and mercury are nonessential and are highly toxic even at very low concentration (Hardiman et al. 1984). The effect of heavy metals on living constituents of environment, however, depends upon period of exposure, concentration, and species of metals used/available in the contaminated environment (Giller et al. 1998). The available fraction of HMs interact easily with soil microbiota (Xie et al. 2015) and growing vegetations (Rucińska-Sobkowiak 2016). Of these, bacteria use a reductive and chelating strategy for metal absorption (Kraepiel et al. 2009; Deicke et al. 2013; Zribi et al. 2015) which induce shift in microbial community (Klimek et al. 2016) and morphological, cellular, and physiological changes (Bajkic et al. 2013; Nahar et al. 2016; Xie et al. 2016). For example, HMs have been reported to cause changes in the taxonomic composition of *Anthyllis*-associated rhizobial populations, but such metals did not alter their symbiotic *nodA* diversity (Mohamad et al. 2017). Apart from their adverse impact on useful soil microflora, heavy metals also adversely affect the growth and development of legumes (Kandziora-Ciupa et al. 2016; Pireh et al. 2017) while growing in metal-polluted soils.

Legumes grown worldwide are the second most important crops for humans. Grain and forage legumes are grown in about 15% of the world's cultivated land and account for 27% of world's primary crop production. Legumes apart from acting as a natural nitrogenous fertilizer in association with rhizobia provide 33% of dietary nitrogen requirement (Graham and Vance 2003). Therefore, due to its nutritive importance, the study on the effect of heavy metals on legumes and associated microorganisms has become important. Most of the results have shown negative impacts of heavy metals on both the growth and nitrogen-fixing ability of symbiotic rhizobia (Marino et al. 2013). In other studies, the N_2-fixing rhizobia survived in metal-contaminated soils but failed to fix N with clover plants (Giller et al. 1989; Wakelin et al. 2016). Besides rhizobia, their symbiotic partners are also adversely affected by metal toxicity. Following uptake, metals are transported to different plant organs and accumulate within plant tissues. After accumulation to considerable levels, heavy metals affect various physiological processes of plants leading eventually to the reduction in yield of crops. Some of the important physiological processes that are adversely affected by heavy metals include photosynthesis and protein synthesis etc. (Oves et al. 2013; Rai et al. 2016). Also, the germination and biomass have been reported to be severely affected under chromium stress in chickpea (Velez et al. 2016). In another study, chromium treatment adversely affected nitrogenase, nitrate reductase, nitrite reductase, glutamine synthetase, and glutamate dehydrogenase in various plant organs of cluster beans at different growth stages, and specific activity of these enzymes decreased with an increase in chromium (VI) levels (Sangwan et al. 2014). These and other related data warrant that the strategies should be identified and developed to reduce or remove/detoxify metals from derelict soils in order to facilitate the growth and development of legumes even in metal-polluted soils. In this regard, several mechanical and chemical methods such as adsorption, ion exchange, membrane filtration, electrodialysis, reverse osmosis, ultrafiltration, and photocatalysis have been applied to remediate heavy metal-contaminated soils (Gunatilake 2015). Due to certain problems such as the unacceptability among masses and cost of operation of physicochemical methods (Segura and Ramos 2013), there is urgent need to find an alternative option. In this context, plants (phytoremediation) (Saadani et al. 2016) and free living/symbiotic PGPR alone (Delgadillo et al. 2015; Pajuelo et al. 2016; Yu et al. 2016a; Karthik et al. 2016a) or in combination with other endophytic and symbiotic bacteria have been widely explored for detoxification of metal-contaminated soils (Fatnassi et al. 2015b). In addition, among PGPR, rhizobia are special because apart from BNF, they possess several other plant growth-promoting potentials, for example, ability to synthesize (1) siderophores (Joshi 2016), (2) phytohormones (Imada et al. 2016), (3) ACC deaminase to lower ethylene levels (Duan et al. 2009), and (4) depression of plant diseases (Khan et al. 2002). As an example, the *Sinorhizobium meliloti* (isolated from a mining site) when used as inoculant against *Medicago sativa* grown under Cd stress (50 and/or 100 mg Cd kg^{-1} soil) hindered the occurrence of Cd-induced toxicity symptoms that appeared in the shoots of non-inoculated plants (Ghnaya et al. 2015). This beneficial effect of *S. meliloti* was accompanied by a considerable increase in biomass, nodulation, and improved nutrient acquisition relative to non-inoculated plants. The increase in plant biomass coupled with increase in Cd concentration in shoots of inoculated plants led to higher potential of Cd phytoextraction. At 50 mg Cd kg^{-1} soil, the amounts of Cd extracted in the shoots were 58 and 178 μg plant^{-1} in non-inoculated and inoculated plants, respectively. This study suggests that the *M. sativa-S. meliloti* interaction may be an efficient biological system to extract Cd from contaminated soils. Thus, metal-tolerant ability together with plant growth-promoting activity makes this group of bacteria very exciting and more practical in the production of legumes even in metal-contaminated soils (Teng et al. 2015).

15.2   Heavy Metal Toxicity to Rhizobial Diversity and Their Physiological Functions

Nitrogen-fixing bacteria forming symbiosis with legumes, collectively called rhizobia which include a range of genera, such as *Rhizobium*, *Bradyrhizobium*, *Sinorhizobium*, *Mesorhizobium*, *Allorhizobium*, and *Azorhizobium*, are physiologically versatile group of Gram-negative bacteria found in different habitats. Both rhizobia and legume plants in unison play an important role in maintaining soil fertility, but they also suffer heavily when exposed to varying concentrations of metals (Chaudri et al. 2000; Stan et al. 2011). Heavy metal contamination reduces microbial biomass, and even if they do not reduce their number, they reduce biodiversity or disturb the community structure (Xie et al. 2016) affecting their growth, morphology, and activities (Lakzian et al. 2002; Wang et al. 2007), including symbiotic N_2 fixation (McGrath et al. 1988; Lebeau et al. 2008). For example, the rate of motility of some substrains of soybean *Bradyrhizobium* strain USDA 143 was enhanced after exposure to Al, but the nodule number and nitrogen fixation were reduced (Octive et al. 1994). Infectiveness of clover *Rhizobium* strain RDG 2002 was also reduced. These results suggest that ecologically important traits in *Rhizobium* and *Bradyrhizobium* may be permanently affected by prior exposure of strains to Al and support the hypothesis that Al is potentially mutagenic. Toxicity of heavy metals to nodule bacteria (rhizobia) has, however, often been conflicting (Kinkle et al. 1994; Wani et al. 2008). For example, the effect of three heavy metals such as Al, Fe, and Mo on growth and symbiotic properties of two strains of rhizobia recovered from root nodules of two tropical legume species, *Mucuna pruriens* and *Trigonella foenum-graecum*, were variable (Paudyal et al. 2007). Among metals, Al at all concentrations showed detrimental effects under both in vitro and in vivo conditions, while iron supported bacterial growth and symbiotic functions (biomass production and nodulation) of rhizobia up to 25 μM. However, above 25 μM, iron had negative effect both on diversity and their associated activities. Molybdenum in contrast had no inhibitory effect on growth of both strains of rhizobia up to 75 μM concentration, while concentration beyond 20 μM of Mo inhibited nodulation and legume production. Similarly, different rates of seven heavy metals like Pb, Hg, Cd, Zn, Cu, Ni, and Cr had variable impact on the growth of 16 strains of *B. japonicum*, 15 strains of *S. meliloti*, 24 strains of *R. leguminosarum*, four strains of *R. loti*, and three strains of *R. galegae* (Miličić et al. 2006). The results further revealed that rhizobial strains differed considerably in intrinsic tolerance to the applied concentrations of heavy metals. In general, all rhizobial strains displayed the lowest intrinsic tolerance to Ni and Cu, while most strains exhibited highest intrinsic tolerance to Pb, Zn, and Hg. Among test metals, Cd and Cr at 50–75 μg mL^{-1} had maximum inhibitory effect on the growth of majority of rhizobial strains. Variation in sensitivity/resistance to different concentrations of heavy metals was found among different rhizobial species and even among rhizobia of identical species suggesting that rhizobial strains had variable genetic composition which resulted in varied sensitivity/resistance. In other experiment, *Rhizobium* strains capable of forming nodules on soybean were exposed to five metals like Fe,

Al, Mo, Co, and Hg. Of these metals, all concentrations of Co and Hg showed detrimental effects, while all concentrations of Fe, Mo, and Al supported rhizobial growth (Rane et al. 2014). The toxicity of Al to *R. leguminosarum* growth expressed as optical density (OD) was variable (Hosam et al. 2009). The results revealed that *Rhizobium* strain HB-3841str+ and E1012 strains could not grow at 25 μM $KAl(SO_4)_2$, but they grew at 25 μM $Al(NO_3)_3$. The results further revealed that the multiplication of majority of the *Rhizobium* strains was unaffected by 100 μM $Al_2(SO_4)_3$, while the growth of the rhizobial strains was adversely affected by 50 μM $AlCl_3$. The toxicity of Al compounds followed the order $Al(NO_3)_3 < Al_2(SO_4)_3 < KAl(SO_4)_2 < AlCl_3$. Moreover, it is suggested that there exists a relationship between the ability of rhizobia to tolerate heavy metals, concentration of heavy metals in soil, and alterations in protein pool. Considering these, alteration in expression of proteins in *R. radiobacter* VBCK1062 exposed to arsenate is reported (Deepika et al. 2016). Of the various proteins, one unique protein of approximately 21 kDa was highly expressed by *R. radiobacter* VBCK1062 grown in 5 mM arsenate; however, the same protein was downregulated in 10 mM arsenate. Realizing the deleterious impact of metals on diversity and physiological functions of rhizobia on one hand and a profound scientific and agricultural importance of legumes on the other hand, efforts should be directed to find metal-tolerant rhizobia for enhancing legume production in metal-contaminated soils.

15.3 Toxic Impact of Heavy Metals on Legume Production

Legumes belonging to the family Fabaceae are one of the most important pulse crops for human and animal consumption. Legumes are cultivated in different production systems primarily for grain seed called pulse, for livestock forage and silage, and as green manure. Legumes are a significant source of protein, dietary fiber, carbohydrates, and dietary minerals. However, when grown intentionally/ unintentionally in metal-contaminated soils, legumes show water and nutrient uptake problems, symptoms of injury, premature aging, retarded growth, decreased legume-*Rhizobium* symbiosis, decrease in fresh biomass of shoots and roots, and low yield and seed protein (Gramss and Voigt 2015; Rucińska-Sobkowiak 2016). In addition, after uptake by plants and translocation to various organs, metals can directly interact with cellular components and disrupt the metabolic activities causing cellular injuries and in some cases even may lead to the death of the plants (Fig. 15.1). Length, surface, and volume of roots also decrease under HM toxicity (Fahr et al. 2013), thereby leading to reduced root biomass and lesser exploration of soil volume. For example, lead has been found to retard the growth of pea when grown under lead-contaminated conditions. Also, lead toxicity causes abnormal enlargement and abnormal cell division in cortical cells. Due to the toxic effects of this metal, ectopic lignifications are also found in pith parenchyma cells. The ultimate effect of this metal was cell death (Chaudhari et al. 2016). In addition, Pb supply has been reported to increase leakage of K^+ ions which induces water stress along with the oxidative stress (Nautiyal and Sinha 2012; Reis et al. 2015). Heavy

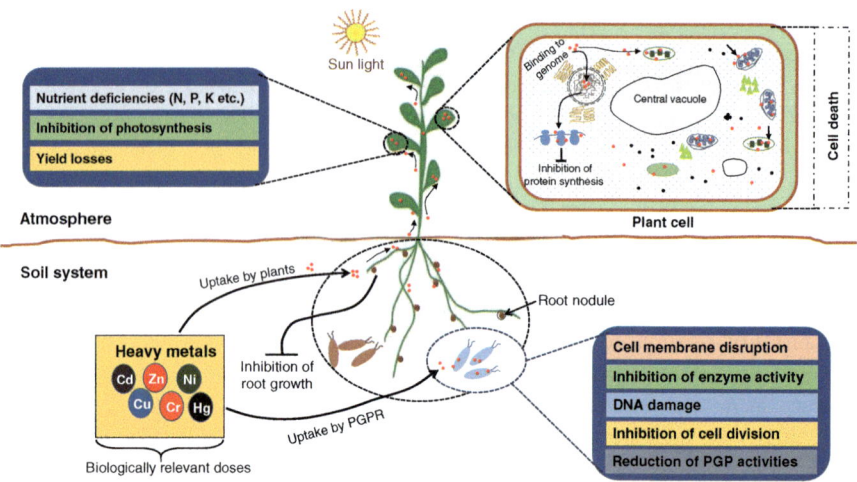

Fig. 15.1 Heavy metal toxicity to legumes grown in metal-contaminated soils

metals particularly Cd and Pb have been widely reported to reduce chlorophyll content and photosynthetic efficiency in plants (Parmar et al. 2013). Reduction in the chlorophyll content under As and Hg stress might be attributed to the inhibition of y-aminolevulinic acid dehydratase (ALAD) and protochlorophyllide reductase (Gupta et al. 2013; Mishra et al. 2016). Sulphydryl interaction of these enzymes was proposed as a mechanism for this inhibition. The destruction of photosynthetic pigments by metals could be due to impairment of an electron transport chain, replacement of Mg^{2+} ions associated with the tetrapyrrole ring of the chlorophyll molecules, inhibition of important enzymes associated with chlorophyll biosynthesis, or peroxidation processes in chloroplast membrane lipids by the ROS (Van Assche and Clijsters 1990; Sandalio et al. 2001; Dubey and Pandey 2011). Heavy metal stress can also cause changes in the protein content (Tamas et al. 2014).

The synergistic effect of Cd and Pb stresses resulted in higher increase of oxidative stress in pigeon pea plants which led to further increase of H_2O_2 content and lipid peroxidation (Garg and Aggarwal 2012). In case of legumes, Lafuente et al. (2010) reported that arsenic (As) reduces legume nodulation by affecting the first stages of the symbiotic interaction, which causes a 90% decrease in rhizobial infections using the model system *Medicago sativa-Sinorhizobium*. As affects the expression of nodulation genes that have been associated with processes that take place in the epidermis and the outer cortical cells and that the expression of genes associated with events that take place in the inner cortical cells is less affected. Similar to this Talano et al. (2013) determined the effect of As on soybean germination, development, and nodulation in soybean-*B. japonicum* E109 symbiosis and found significant reduction at 10 µM arsenic concentration. Despite that the microorganism tolerated the metalloid, the number of effective nodules was reduced for soybean seedlings inoculated with *B. japonicum*. The minor nodulation could be due to a reduced motility (swarming and swimming) of the microorganism in the presence of As. In another study, Chubukova et al. (2015) reported the effect of cadmium salts on the legume-*Rhizobium* symbiosis of the pea with *R. leguminosarum* and found that Cd inhibited the nodulation. Cadmium has an adverse effect on legume nodule metabolism even at low concentration. It has been widely reported that Cd impairs nodule functioning and nodule oxidative damage and causes chlorophyll depletion in soybean (Balestrasse et al. 2006), alfalfa (Shvaleva et al. 2010), white lupin (Carpena et al. 2003), and mung bean (Muneer et al. 2012). In a study with gene expression levels, Marino et al. (2013) observed that Lb and NifD declined locally in the nodules directly exposed to Cd, and the N_2 fixation inhibition provoked by Cd is due to a direct effect on nodules rather than a systemic effect through a control from the shoot. Panigrahi et al. (2013) observed significant reduction in nodule number, nitrogenase activity, and shoot N content of soybean when grown under arsenate stress (Reichman 2007) and in alfalfa plants grown in arsenic-contaminated Aznallcolár soil (El-Deeb and Al-Sheri 2005). Similarly, Neumann et al. (1998) have also reported 50% decline in nodule number in alfalfa plants at 5 μM As(V), while Kopitteke et al. (2007) have reported 10% reduction in nodule number in *Vigna unguiculata* at 0.2 μM Cu^{2+}. Similar reduction in N contents in pea and Egyptian clover in response to heavy metals of sewage water has also been reported (Chaudhary et al. 2004). Fatnassi et al. (2014) conducted a study to select appropriate legume-tolerant bacteria symbionts for specific metal contamination using four local legumes, *Vicia faba*, *Lens culinaris*, *Cicer arietinum*, and *Sulla coronaria*. Investigation of legume response to contamination showed that the greatest reduction in the shoot and root dry weights was observed in *Sulla coronaria* upon Cd contamination due to highly metal accumulation, while *Vicia faba* and *Lens culinaris* contained Cu and Pb respectively in their organs. Metal tolerance analyses showed that isolates from *Vicia faba* could grow with maximum Cu, Pb, and Cd levels of 2, 4, and 4.5 mM, respectively; however, isolates from the other tested legumes were more sensitive to heavy metals. Genetic characterization by PCR-RFLP of the 16S rDNA for 20% of the isolates revealed different species including *R. leguminosarum*, *R. phaseolus*, *R. etli*, and *Agrobacterium*.

15.4   How to Overcome Heavy Metal Stress?

Heavy metals are nondegradable and hence are difficult to remove from metal-contaminated soils. Among various metal-detoxifying/removal approaches (physical, chemical, and biological), bioremediation involving plants (phytoremediation) or microbes to remove/destroy or sequester hazardous substances from the environment is an inexpensive, most popular/widely acceptable, and environmentally sustainable option (Cunningham et al. 1995). Some of the notable phytoremediation strategies include phytoextraction, rhizofiltration, phytodegradation, phytovolatilization, and phytostabilization (Gallego et al. 2012; Bolan et al. 2014). In general, plants when used to clean up metal-polluted soils should exhibit two basic properties: (1) must be able to absorb and accumulate greater concentrations of metals and (2) be able to produce huge amounts of dry matters.

The other most striking and inexpensive approach to alleviate metal involves the use of plant growth-promoting bacteria (Khan et al. 2009; Hao et al. 2015; El Aafi et al. 2015). The use of PGPR in metal detoxification/removal offers several advantages. For example, the metabolites secreted by PGPR in the rhizosphere in situ are biodegradable and less toxic/harmful (Rajkumar et al. 2012). Also, there is no need for repeated inoculations of microbial agents in contaminated sites since once established, they compete well with indigenous organisms and express their full activity. Accordingly, metal-resistant PGPR have been widely investigated for their potential to improve plant growth, alleviate metal toxicity, and immobilize/mobilize/transform metals in soil, which may help to develop new microbe-assisted phytoremediation and restoration strategies. Due to these and other growth-promoting properties, the metal-resistant beneficial PGPR are often used as bioinoculants to enhance the establishment, growth, and development of plants in metal-contaminated soils. PGPR induce the growth of plants by acting as (1) biofertilizers, phytoavailability of minerals (N, P, K, Ca, and Fe); (2) phytostimulators, modulating phytohormones; (3) bioalleviators, reducing ethylene stress; (4) biocontrol agent, preventing deleterious effects of phytopathogens via production of antifungals/antibacterials and ISR; and (5) biomodifiers, modifying root biomass and morphology (Miransari 2011; Ullah et al. 2015; Ma et al. 2016). In a study, *Rhizobium* strain ND2 isolated from the root nodules of *Phaseolus vulgaris* grown in leather industrial effluent contaminated soil exhibited strong resistance to different heavy metals and reduced 30 and 50 μg mL^{-1} concentrations of Cr (VI) completely after 80 and 120 h of incubation, respectively. In addition, this strain produced 21.73 and 36.86 μg mL^{-1} of IAA at 50 and 100 μg mL^{-1} of tryptophan, respectively. Strain ND2 also secreted exopolysaccharide (EPS) ammonia, protease, and catalase and when used as inoculant stimulated root length of various crops grown under Cr(VI) stress (Karthik et al. 2016b). Some of the examples of growth-promoting biomolecules secreted by rhizobia are listed in Table 15.1.

In a recent study, Manohari and Yogalakshmi (2016), for example, observed copper tolerance and bioremediation potential in endophytic bacteria isolated from *Vigna unguiculata* root nodules. The results revealed that the endophytic bacteria were able to remove 82.8% of Cu (II) at pH 5, temperature 32.5 °C, and 600 mg Cu L^{-1} copper after 168 h incubation. The endophytic isolates also produced IAA and 1-aminocyclopropane-1-carboxylic acid (ACC) deaminase activity. The ACC deaminase induces physiological changes in plants by metabolizing ACC to ketobutyrate and ammonia and hence lowers the toxic effects of abnormally higher concentration of ethylene on plant, which otherwise inhibits plant growth (Belimov et al. 2002; Tittabutr et al. 2008; Singh et al. 2015; Han et al. 2015). In other study Deepika et al. (2016) explored the relationship between *Rhizobium* metal tolerance and its adaptations to metal-stressed environment using strain recovered from root nodules of *Vigna radiata*, based on viscous EPS production and arsenic-tolerant ability. The strain identified as *R. radiobacter* confirmed the role of EPS in arsenate sequestration. Interestingly, total arsenate uptake by strain VBCK1062 in whole-cell pellet and EPS were 0.045 mg and 0.068 mg g^{-1} of biomass, respectively. Thus, these results significantly contributed to better

Table 15.1 Bioactive molecules synthesized by rhizobial species affecting growth of legumes in metal-polluted environment

Rhizobial species	Heavy metals	Bioactive molecules	References
Rhizobium sullae	Cd	IAA, siderophore	Chiboub et al. (2016)
Sinorhizobium meliloti	Hg, As	EPS	Nocelli et al. (2016)
Rhizobium galegae *R. leguminosarum*	Pb, Zn	P solubilization, ammonia	Sbabou et al. (2016)
Rhizobium sp.	Fe, Pb	EPS and LPS	Singh and Singh (2015)
Sinorhizobium meliloti	Cu	ACC deaminase	Kong et al. (2015a)
Bradyrhizobium sp.	Cd, Cr, Ni, Pb, Zn, Cu	IAA, siderophore, NH_3, HCN	Wani and Khan (2014)
S. meliloti	Cu	IAA	Li et al. (2014)
Mesorhizobium sp.	Cd, Cr, Ni, Pb, Zn, Cu	IAA, NH_3, HCN, siderophore	Wani and Khan (2013a)
Rhizobium sp.	Cd, Co, Ni, Pb, Zn, Cu	IAA, siderophore	Yu et al. (2014)
Rhizobium sp.	Cd, Co, Cr, Hg, Pb, Zn, Cu	IAA, siderophore, NH_3, HCN	Singh et al. (2013)
R. leguminosarum sp.	Cd, Ni, Cr, Pb, Zn, Cu	IAA, siderophore, HCN, NH_3	Wani and Khan (2013b)
R. leguminosarum sp.	Pb	IAA, siderophore, HCN, NH_3	Wani and Khan (2012)
R. leguminosarum bv. trifolii	Co, Cr, Ni, Zn	P solubilization	Nonnoi et al. (2012)

understanding of plant-metal-microbe interactions, cellular-metabolic changes, and As-enhanced EPSs. Due to these properties, *R. radiobacter* was identified as a potential bioremediation agent for As-contaminated agroecosystems. In a study, the genetic manipulation of both symbiotic partners for Cu phytostabilization using composite *M. truncatula* plants expressing the metallothionein gene *mt4a* from *Arabidopsis thaliana* in roots was generated, in an attempt to increase the plant tolerance toward Cu. Also, an *Ensifer medicae* strain was genetically engineered by expressing the copper resistance genes *copAB* from *P. fluorescens*. The expression of *mt4a* in composite plants increases tolerance toward Cu and reduces oxidative stress caused by this pollutant. Lower levels of reactive oxygen species (ROS)-scavenging enzymes were found in *mt4a*-expressing plants besides improving nodulation, whereas inoculation with the genetically modified *Ensifer* has a synergistic effect; and the double symbiotic system enhances Cu accumulation in roots, without increasing metal translocation to shoots (Pajuelo et al. 2016). Considering all these, and based on studies in other bacteria, the metal resistance of rhizobia might be attributed to (1) changes in the metal efflux of microbial cell membranes, (2) intracellular chelation due to the production of metallothionein proteins (Nies 1995; Furukawa et al. 2015; Pérez-Palacios et al. 2017), and (3) the transformation of heavy metals to their less toxic oxidated forms through microbial metabolism (Nies 2003). The various strategies adopted by microbes to circumvent metal toxicity are presented in Fig. 15.2.

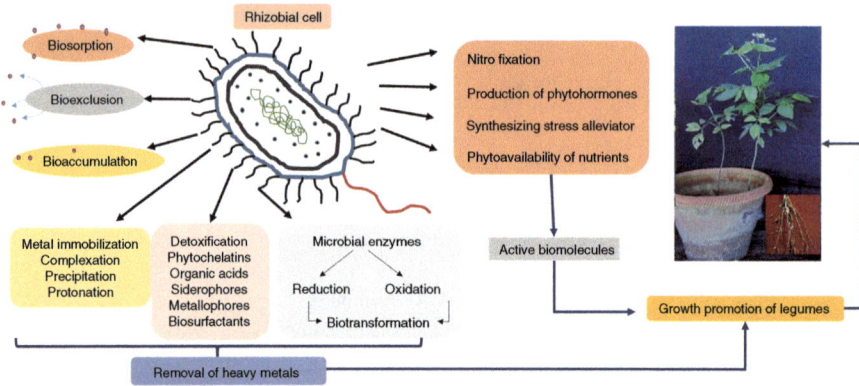

Fig. 15.2 Bioremediation strategies adopted by plant growth-promoting rhizobia in metal-stressed soils

15.5 Performance of Inoculated Legumes in Metal-Stressed Soils

Traditionally, nodule bacteria (rhizobia) have been used as inoculant by legume growers over the years as a viable, environmentally friendly and ecologically sound and inexpensive alternative to widely used chemical fertilizers in order to optimize pulse production in different ago-ecological regions. By forming symbiosis with legumes, the rhizobia (bacteroid) inhabiting a specialized organ, nodules, generally produced onto the root system, transform atmospheric nitrogen into ammonia via a process often referred to as biological nitrogen fixation (BNF). The ammonia is then taken up by plants. Of the two interacting partners, rhizobia in particular are reported to tolerate high levels of metals (Romaniuk et al. 2017; Lu et al. 2016) and hence could help to remediate heavy metal-polluted soils besides providing a good system to understand metal-microbe interactions (Joshi et al. 2014). The increase in growth and yields of legumes could be due to metal-reducing potential through adsorption/desorption mechanism of rhizobial strains (Bramhachari and Nagaraju 2017) besides their ability to fix N and synthesize growth regulators and ACC deaminase (Kong et al. 2015a; Nascimento et al. 2016). Thus, the potential of N_2-fixing bacteria in metal resistance/reduction and their ability to facilitate legume growth by several mechanisms other than nitrogen fixation in metal-stressed soil make them one of the most suitable choices for cleanup of the metal-contaminated sites and hence may further help in reducing toxicity problems to legumes when grown in derelict soils. Fatnassi et al. (2015a) in a study applied the consortium of bacteria containing *Rhizobium* sp. CCNWSX0481, *R. leguminosarum* bv. viciae, *Enterobacter cloacae*, and *Pseudomonas* sp. 2 (2010) as inoculant against *Vicia faba* L, cultivated in the vineyard of soil moderately contaminated with Cu. The results revealed a significant increase in nodulation where the number and the weight of nodules increased by 50%. Co-inoculation also influenced the growth and seed yield (number of seeds per pod and pods per plant) positively and enhanced the shoot and root weights by 33 and 26%, respectively. Moreover, the co-inoculation significantly reduced the accumulation of copper in roots by 35%. In a similar experiment conducted in pots maintained under controlled greenhouse conditions, four Cr(VI)-reducing bacterial strains (SUCR44, SUCR140, SUCR186, and SUCR188) were tested to evaluate their bioremediation potential against pea crops grown in Cr(VI)-contaminated soil (Soni et al. 2014). The effect of pretreatment of soil with chromate-reducing bacteria on plant growth, chromate uptake, bioaccumulation, nodulation, and population of *Rhizobium* was found to be directly influenced by the time interval between bacterial treatment and seed sowing. Pretreatment of soil with SUCR140 (*Microbacterium* sp.) 15 days before sowing showed a maximum increase in growth and biomass and hence enhanced the root length by 93%, plant height by 94%, dry root biomass by 99%, and dry shoot biomass by 99%. Co-inoculation of *Rhizobium* with SUCR140 further improved the measured parameters. The co-inoculation of SUCR140 and *Rhizobium* increased the root length and plant height by 117 and 116%, respectively, while dry matter accumulation in root and shoot was increased by 136 and 128%, respectively, over control. The bioavailability of Cr(VI) decreased significantly in soil by 61% and in SUCR140-inoculated plants by 36%. The populations of *Rhizobium* (126%) in soil and nodulation (146%) on *P. sativum* improved in the presence of SUCR140 resulting in greater N (54%) concentration in the plants. The biofuel leguminous tree *Pongamia pinnata* inoculated with *Bradyrhizobium liaoningense*, grown in pots, was used to remediate iron-vanadium-titanium oxide (V-Ti magnetite) mine tailing soil by Yu et al. (2016b). Inoculation with *B. liaoningense* PZHK1 increased the growth of *P. pinnata* both in V-Ti magnetite mine tailings and in Ni-contaminated soil. Furthermore, inoculation increased the metal accumulation capacity and superoxide dismutase activity of *P. pinnata*. The concentrations of Ni accumulated by inoculated plants were higher than the hyperaccumulator threshold. Inoculated *P. pinnata* accumulated high concentration of Fe, far exceeding the upper limit (1000 mg kg^{-1}) of Fe in plant tissue. Summarily, *P. pinnata-B. liaoningense* PZHK1 symbiosis showed potential to be applied as an effective phytoremediation technology for mine tailings and to produce biofuel feedstock on the marginal land. Kang et al. (2015) elucidated the role of free living *Enterobacter asburiae* KE17 in the growth and metabolism of soybean grown under copper (100 μm Cu) and zinc (100 μm Zn) stress. Plants grown under Cu and Zn stress exhibited significantly lower growth, but inoculation with *E. asburiae* KE17 increased growth rates of stressed plants. The concentrations of plant hormones (abscisic acid and salicylic acid) and rates of lipid peroxidation were higher in plants under heavy metal stress, while total chlorophyll, carotenoid content, and total polyphenol concentration were lower. The bacterial treatment reduced the abscisic acid and salicylic acid content and lipid peroxidation rate of Cu-stressed plants, whereas *E. asburiae* increased the concentration of photosynthetic pigments and total polyphenol. Moreover, the heavy metals induced increased accumulation of free amino acids such as aspartic acid, threonine, serine, glycine, alanine, leucine, isoleucine, tyrosine, proline, and gamma-aminobutyric acid, while *E. asburiae* KE17 significantly reduced concentrations of free amino acids in metal-affected plants. Co-treatment with *E. asburiae* KE17 regulated nutrient uptake by enhancing nitrogen content and inhibiting Cu and Zn accumulation in soybean plants. The results of this study suggest that *E. asburiae* KE17 mitigated the toxic effects of Cu and Zn by reprogramming plant metabolic processes.

Oxidative stress strongly affects BNF, while the antioxidant system of nodules reduces the damage caused by oxidizing compounds and therefore maintains functionality of BNF. To validate this further, de Jesus et al. (2016) in a study observed a substantial increase in the number of nodules, nodule dry mass, nitrogen fixation efficiency, and N content in *Bradyrhizobium* sp.-inoculated cowpea plants grown in tannery sludge enriched. The cowpea nodules had lower H_2O_2 levels, while leghemoglobin was maintained at the highest levels. The other antioxidative enzymes like catalase and phenol peroxidase were positively modulated in the nodules of the inoculated cowpea plants resulting in enhanced growth, N capture, and lower oxidative stress. In a similar experiment, *S. meliloti* CCNWSX0020 induced an increase in growth and N content of *M. lupulina* grown under copper stress. Furthermore, the total amount of Cu in inoculated plants significantly increased by 34 and 120.4% in shoots and roots, respectively, compared with non-inoculated plants. However, although the rhizobial symbiosis promoted Cu accumulation both in shoots and roots, the increase in roots was much higher than in shoots, thus decreasing the translocation factor and helping Cu phytostabilization. Additionally, rate of lipid peroxidation was decreased, but the secretion of antioxidant enzymes such as superoxide dismutase and ascorbate peroxidase was significantly elevated under Cu stress (Kong et al. 2015b). Broadly, considering these and other related studies, the toxicity of metals to each legume and remediating ability of rhizobia in general vary according to (1) legume genotype, (2) types/species and concentration of metals, and (3) intrinsic ability of nodule bacteria.

Conclusion

Considering the threat of heavy metals to legume production in metal-contaminated soils, the introduction of metal-tolerant rhizobial species into metal-polluted soils offers a sustainable and inexpensive option for enhancing the vegetative growth, nitrogen-fixing efficiency, yields, and grain quality of legumes. To achieve these goals, there is a need to isolate and select legume/metal-specific metal-tolerant rhizobia for restoration of metal-contaminated soils and consequently the production of legumes in metal-polluted soils. Also, metal resistance genes should be identified using some new molecular tools which subsequently can be transferred to other non-tolerant microbes used in remediation programs. Rhizobia due to their multiple plant growth-promoting activities, like ability to tolerate higher concentrations of varied metals, ability to synthesize plant growth-promoting substances in addition to their intrinsic property of fixing atmospheric nitrogen, and ability to detoxify heavy metals from contaminated sites, could therefore be used as an ideal and agronomically potential eco-friendly inoculant for raising the productivity of legumes in metal-polluted soils.

References

Alamgir M (2016) The effects of soil properties to the extent of soil contamination with metals. In: Environmental remediation technologies for metal-contaminated soils. Springer, Tokyo, pp 1–19

Bajkic S, Narancic T, Dokic L, Dordevic D, Nikodinovic-Runic J, Morić I, Vasiljević B (2013) Microbial diversity and isolation of multiple metal-tolerant bacteria from surface and underground pits within the copper mining and smelting complex bor. Arch Biol Sci 65:375–386

Balestrasse KB, Gallego SM, Tomaro ML (2006) Oxidation of the enzymes involved in nitrogen assimilation plays an important role in the cadmium-induced toxicity in soybean plants. Plant Soil 284:187–194

Belimov AA, Safroonova VI, Mimura T (2002) Response of spring rape to inoculation with plant growth promoting rhizobacteria containing 1-aminocyclopropane-1-carboxylate deaminase depends on nutrient status of the plant. Can J Microbiol 48:189–199

Bolan N, Kunhikrishnan A, Thangarajan R, Kumpiene J, Park J, Makino T, Kirkham MB, Scheckel K (2014) Remediation of heavy metal(loid)s contaminated soils–to mobilize or to immobilize? J Hazard Mater 266:141–166

Bramhachari PV, Nagaraju GP (2017) Extracellular polysaccharide production by bacteria as a mechanism of toxic heavy metal biosorption and biosequestration in the marine environment. In: Naik MM, Dubey SK (eds) Marine pollution and microbial remediation. Springer, Singapore, pp 67–85

Carpena RO, Vázquez S, Esteban E, Fernández-Pascual M, de Felipe MR, Zornoza P (2003) Cadmium-stress in white lupin: effects on nodule structure and functioning. Plant Physiol Biochem 41:911–919

Chaudhari J, Patel K, Patel V (2016) Exploring the toxic effects of Pb and Ni on stem anatomy of *Pisum Sativum* L. Int J Chem Environ Biol Sci 4:28–32

Chaudhary P, Dudeja SS, Kapoor KK (2004) Effectivity of host *Rhizobium leguminosarum* symbiosis in soils receiving sewage water containing heavy metals. Microbiol Res 159:121–127

Chaudri AM, Allain CMG, Barbosa-Jefferson VL, Nicholson FA, Chambers BJ, McGrath SP (2000) A study of the impacts of Zn and Cu on two rhizobial species in soils of a long-term field experiment. Plant Soil 221:167–179

Chiboub M, Saadani O, Fatnassi IC, Abdelkrim S, Abid G, Jebara M, Jebara SH (2016) Characterization of efficient plant-growth-promoting bacteria isolated from *Sulla coronaria* resistant to cadmium and to other heavy metals. C R Biol 339:391–398

Chubukova OV, Postrigan BN, Baimiev AK, Chemeris AV (2015) The effect of cadmium on the efficiency of development of legume-*Rhizobium* symbiosis. Biol Bull Russ Acad Sci 42:458–462

Cunningham SD, Berti WR, Huang JW (1995) Phytoremediation of contaminated soils. Trends Biotechnol 13:393–397

de Jesus MN, da Costa Neto VP, de Araújo ASF, Figueiredo MDVB, Bonifacio A, Rodrigues AC (2016) *Bradyrhizobium* sp. inoculation ameliorates oxidative protection in cowpea subjected to long-term composted tannery sludge amendment. Eur J Soil Biol 76:35–45

Deepika KV, Raghuram M, Kariali E, Bramhachari PV (2016) Biological responses of symbiotic *Rhizobium radiobacter* strain VBCK1062 to the arsenic contaminated rhizosphere soils of mung bean. Ecotoxicol Environ Saf 134:1–10

Deicke M, Bellenger JP, Wichard T (2013) Direct quantification of bacterial molybdenum and iron metallophores with ultra-high-performance liquid chromatography coupled to time-of-flight mass spectrometry. J Chromatogr A 1298:50–60

Delgadillo J, Lafuente A, Doukkali B, Redondo-Gómez S, Mateos-Naranjo E, Caviedes MA, Pajuelo E, Rodríguez-Llorente ID (2015) Improving legume nodulation and Cu rhizostabilization using a genetically modified rhizobia. Environ Technol 36:1237–1245

Duan J, Muller KM, Charles TC, Vesely S, Glick BR (2009) 1-aminocyclopropane-1-carboxylate (ACC) deaminase genes in rhizobia from southern Saskatchewan. Microb Ecol 57:421–422

Dubey D, Pandey A (2011) Effect of nickel (Ni) on chlorophyll, lipid peroxidation and antioxidant enzymes activities in black gram (*Vigna mungo*) leaves. Int J Sci Nat 2:395–401

El Aafi N, Saidi N, Maltouf AF, Perez-Palacios P, Dary M, Brhada F, Pajuelo E (2015) Prospecting metal-tolerant rhizobia for phytoremediation of mining soils from Morocco using *Anthyllis vulneraria* L. Environ Sci Pollut Res 22:4500–4512

El-Deeb SM, Al-Sheri FS (2005) Role of some chemical compounds on the detoxification of *Rhizobium leguminosarum* biovar *vicia* by some heavy metals. Pak J Biol Sci 8:1693–1698

Fahr M, Laplaze L, Bendaou N, Hocher V, El Mzibri M, Bogusz D, Smouni A (2013) Effect of lead on root growth. Front Plant Sci 4:175

Fatnassi IC, Chiboub M, Jebara M, Jebara SH (2014) Bacteria associated with different legume species grown in heavy-metal contaminated soils. Int J Agric Policy Res 2:460–467

Fatnassi IC, Chiboub M, Saadani O, Jebara M, Jebara SH (2015a) Phytostabilization of moderate copper contaminated soils using co-inoculation of *Vicia faba* with plant growth promoting bacteria. J Basic Microbiol 55:303–311

Fatnassi IC, Chiboub M, Saadani O, Jebara M, Jebara SH (2015b) Impact of dual inoculation with *Rhizobium* and PGPR on growth and antioxidant status of *Vicia faba* L. under copper stress. C R Biol 338:241–254

Furukawa K, Ramesh A, Zhou Z, Weinberg Z, Vallery T, Winkler WC, Breaker RR (2015) Bacterial riboswitches cooperatively bind Ni^{2+} or Co^{2+} ions and control expression of heavy metal transporters. Mol Cell 57:1088–1098

Gallego SM, Pena LB, Barcia RA, Azpilicueta CE, Iannone MF, Rosales EP, Zawoznik MS, Groppa MD, Benavides MP (2012) Unravelling cadmium toxicity and tolerance in plants: insight into regulatory mechanisms. Environ Exp Bot 83:33–46

Garg N, Aggarwal N (2012) Effect of mycorrhizal inoculations on heavy metal uptake and stress alleviation of *Cajanus cajan* (L.) Millsp. genotypes grown in cadmium and lead contaminated soils. Plant Growth Regul 66:9–26

Ghnaya T, Mnassri M, Ghabriche R, Wali M, Poschenrieder C, Lutts S, Abdelly C (2015) Nodulation by *Sinorhizobiummeliloti* originated from a mining soil alleviates Cd toxicity and increases Cd-phytoextraction in *Medicago sativa* L. Front Plant Sci 6:863

Giller KE, McGrath SP, Hirsch PR (1989) Absence of nitrogen fixation in clover grown on soil subject to long term contamination with heavy metals is due to survival of only ineffective *Rhizobium*. Soil Biol Biochem 21:841–848

Giller KE, Witter E, McGrath SP (1998) Toxicity of heavy metals to microorganisms and microbial processes in agricultural soils: a review. Soil Biol Biochem 30:1389–1414

Graham PH, Vance CP (2003) Legumes: importance and constraints to greater use. Plant Physiol 131:872–877

Gramss G, Voigt KD (2015) Regulation of the mineral concentrations in pea seeds from uranium mine and reference soils diverging extremely in their heavy metal load. Sci Hortic 194:255–266

Gunatilake SK (2015) Methods of removing heavy metals from industrial wastewater. J Multidiscip Eng Sci Stud (JMESS) 1:2912–1309

Gupta P, Jain M, Sarangthem J, Gadre R (2013) Inhibition of 5-aminolevulinic acid dehydratase by mercury in excised greening maize leaf segments. Plant Physiol Biochem 62:63–69

Han Y, Wang R, Yang Z, Zhan Y, Ma Y, Ping S, Zhang L, Lin M, Yan Y (2015) 1-aminocyclopropane-1-carboxylate deaminase from *Pseudomonas stutzeri* A1501 facilitates the growth of rice in the presence of salt or heavy metals. J Microbiol Biotechnol 25:1119–1128

Hao X, Xie P, Zhu YG, Taghavi S, Wei G, Rensing C (2015) Copper tolerance mechanisms of *Mesorhizobium amorphae* and its role in aiding phytostabilization by *Robinia pseudoacacia* in copper contaminated soil. Environ Sci Technol 49:2328–2340

Hardiman RT, Banin A, Jacoby B (1984) The effect of soil type and degree of metal contamination upon uptake of Cd, Pb and Cu in bush beans (*Phaseolus vulgaris* L.) Plant Soil 81:17–27

Hosam EAF, Hamuda B, Orosz E, Hamuda Y, Tóth N, Kecskés M (2009) *Vicia faba—Rhizobium leguminosarum* system symbiotic relationship under stress of soil pH and aluminium. Tájökológiai Lapok 7:301–318

Imada EL, de Oliveira ALM, Hungria M, Rodrigues EP (2016) Indole-3-acetic acid production via the indole-3-pyruvate pathway by plant growth promoter Rhizobium tropici CIAT 899 is strongly inhibited by ammonium. Microbiol Res 168:283–292

Islam MS, Ahmed MK, Raknuzzaman M, Habibullah-Al-Mamun M, Islam MK (2015) Heavy metal pollution in surface water and sediment: a preliminary assessment of an urban river in a developing country. Ecological Indicators 48:282–291

Joshi FR (2016) Studies on siderophore mediated iron uptake system in rhizobacteria and expression of fegA gene from *Bradyrhizobium japonicum* 61A152 in commercial rhizobial inocula impact on competitive survival in rhizosphere. Curr Microbiol 53:141–147

Joshi SR, Kalita D, Kumar R, Nongkhlaw M, Swer PB (2014) Metal–microbe interaction and bioremediation. In: Radionuclide contamination and remediation through plants. Springer International, Cham, pp 235–251

Kandziora-Ciupa M, Ciepał R, Nadgórska-Socha A, Barczyk G (2016) Accumulation of heavy metals and antioxidant responses in *Pinus sylvestris* L. needles in polluted and non-polluted sites. Ecotoxicology 25:970–981

Kang SM, Radhakrishnan R, You YH, Khan AL, Lee KE, Lee JD, Lee IJ (2015) *Enterobacter asburiae* KE17 association regulates physiological changes and mitigates the toxic effects of heavy metals in soybean. Plant Biol 17:1013–1022

Karthik C, Oves M, Thangabalu R, Sharma R, Santhosh SB, Arulselvi PI (2016a) *Cellulosimicrobium funkei*-like enhances the growth of *Phaseolus vulgaris* by modulating oxidative damage under Chromium (VI) toxicity. J Adv Res 7:839–850

Karthik C, Oves M, Sathya K, Sri Ramkumar V, Arulselvi PI (2016b). Isolation and characterization of multi-potential *Rhizobium* strain ND2 and its plant growth-promoting activities under Cr (VI) stress. Arch Agron Soil Sci 1–12

Khan MS, Zaidi A, Aamil M (2002) Biocontrol of fungal pathogens by the use of plant growth promoting rhizobacteria and nitrogen fixing microorganisms. J Ind Bot Soc 81:255–263

Khan MS, Zaidi A, Wani PA, Oves M (2009) Role of plant growth promoting rhizobacteria in the remediation of metal contaminated soils. Environ Chem Lett 7:1–19

Kinkle BK, Sadowsky MJ, Johnston K, Koskinen WC (1994) Tellurium and Selenium resistance in rhizobia and its potential use for direct isolation *R. meliloti* from soil. Appl Environ Microbiol 60:1674–1677

Klimek B, Sitarz A, Choczyński M, Niklińska M (2016) The effects of heavy metals and total petroleum hydrocarbons on soil bacterial activity and functional diversity in the upper silesia industrial region (Poland). Water Air Soil Pollut 227:1–9

Kong Z, Glick BR, Duan J, Ding S, Tian J, McConkey BJ, Wei G (2015a) Effects of 1-aminocyclopropane-1-carboxylate (ACC) deaminase-overproducing *Sinorhizobium meliloti* on plant growth and copper tolerance of *Medicago lupulina*. Plant Soil 391:383–398

Kong Z, Mohamad OA, Deng Z, Liu X, Glick BR, Wei G (2015b) Rhizobial symbiosis effect on the growth, metal uptake, and antioxidant responses of *Medicago lupulina* under copper stress. Environ Sci Pollut Res 22:12479–12489

Kopittke PM, Dart PJ, Menzies NW (2007) Toxic effects of low concentrations of Cu on nodulation of cowpea (Vigna unguiculata). Environ Pollut 145:309–315

Kraepiel AM, Bellenger JP, Wichard T, Morel FM (2009) Multiple roles of siderophores in free-living nitrogen-fixing bacteria. Biometals 22(4):573–581

Lafuente A, Pajuelo E, Caviedes MA, Rodríguez-Llorente ID (2010) Reduced nodulation in alfalfa induced by arsenic correlates with altered expression of early nodulins. J Plant Physiol 167:286–291

Lakzian A, Murphy P, Turner A, Beynon JL, Giller KE (2002) *Rhizobium leguminosarum* bv. viciae populations in soils with increasing heavy metal contamination: abundance, plasmid pro- files, diversity and metal tolerance. Soil Biol Biochem 34:519–529

Lebeau T, Braud A, Jézéquel K (2008) Performance of bioaugmentation-assisted phytoextraction applied to metal contaminated soils; a review. Environ Pollut 153:497–522

Li Z, Ma Z, Hao X, Rensing C, Wei G (2014) Genes conferring copper resistance in *Sinorhizobium meliloti* CCNWSX0020 also promote the growth of *Medicago lupulina* in copper-contaminated soil. Appl Environ Microbiol 80:1961–1971

Lu M, Li Z, Liang J, Wei Y, Rensing C, Wei G (2016) Zinc resistance mechanisms of P1B-type ATPases in *Sinorhizobium* meliloti CCNWSX0020. Sci Rep 6:29355

Luo L, Shen Y, Liu J, Zeng Y (2016) Investigation of Pb species in soils, celery and duckweed by synchrotron radiation X-ray absorption near-edge structure spectrometry. Spectrochim Acta B At Spectrosc 122:40–45

Ma Y, Oliveira RS, Freitas H, Zhang C (2016) Biochemical and molecular mechanisms of plant-microbe-metal interactions: relevance for phytoremediation. Front Plant Sci 7:918

Mamindy-Pajany Y, Sayen S, Mosselmans JFW, Guillon E (2014) Copper, nickel and zinc speciation in a biosolid-amended soil: pH adsorption edge, μ-XRF and μ-XANES investigations. Environ Sci Technol 48:7237–7244

Manohari R, Yogalakshmi KN (2016) Optimization of Copper (II) removal by response surface methodology using root nodule endophytic bacteria isolated from *Vigna unguiculata*. Water Air Soil Pollut 227:285

Marino D, Damiani I, Gucciardo S, Mijangos I, Pauly N, Puppo A (2013) Inhibition of nitrogen fixation in symbiotic *Medicago truncatula* upon Cd exposure is a local process involving leghemoglobin. J Exp Bot 64:5651–5660

McGrath SP, Brookes PC, Giller KE (1988) Effects of potentially toxic metals in soil derived from past applications of sewage sludge on nitrogen fixation by *Trifolium repens* L. Soil Biol Biochem 20:415–424

Miličić B, Delić D, Stajković O, Rasulić N, Kuzmanović Đ, Jošić D (2006) Effects of heavy metals on rhizobial growth. Roum Biotechnol Lett 11:2995–3003

Miransari M (2011) Soil microbes and plant fertilization. Appl Microbiol Biotechnol 92:875–885

Mishra S, Alfeld M, Sobotka R, Andresen E, Falkenberg G, Küpper H (2016) Analysis of sublethal arsenic toxicity to *Ceratophyllum demersum*: subcellular distribution of arsenic and inhibition of chlorophyll biosynthesis. J Exp Bot 67:4639–4646

Mohamad R, Maynaud G, Le Quéré A, Vidal C, Klonowska A, Yashiro E, Cleyet-Marel JC, Brunel B (2017) Ancient heavy metal contamination in soils as a driver of tolerant *Anthyllis vulneraria* rhizobial communities. Appl Environ Microbiol 83:1735–1716

Muneer S, Kim TH, Qureshi MI (2012) Fe modulates Cd-induced oxidative stress and the expression of stress responsive proteins in the nodules of *Vigna radiata*. Plant Growth Regul 68:421–433

Nahar K, Hasanuzzaman M, Alam MM, Rahman A, Suzuki T, Fujita M (2016) Polyamine and nitric oxide crosstalk: antagonistic effects on cadmium toxicity in mung bean plants through upregulating the metal detoxification, antioxidant defense and methylglyoxal detoxification systems. Ecotoxicol Environ Saf 126:245–255

Nascimento FX, Rossi MJ, Glick BR (2016) Role of ACC deaminase in stress control of leguminous plants. In: Plant growth promoting *Actinobacteria*. Springer, Singapore, pp 179–192

Nautiyal N, Sinha P (2012) Lead induced antioxidant defense system in pigeon pea and its impact on yield and quality of seeds. Acta Physiol Plant 34:977–983

Neumann H, Bode-Kirchhoff A, Madeheim A, Wetzel A (1998) Toxicity testing of heavy metals with the *Rhizobium*-legume symbiosis: high sensitivity to cadmium and arsenic compounds. Environ Sci Pollut Res 5:28–36

Nies DH (1995) The cobalt, zinc, and cadmium efflux system CzcABC from Alcaligenes eutrophus functions as a cation-proton antiporter in *Escherichia coli*. J Bacteriol 177:2707–2712

Nies DH (2003) Efflux-mediated heavy metal resistance in prokaryotes. FEMS Microbiol Rev 27:313–339

Nocelli N, Bogino PC, Banchio E, Giordano W (2016) Roles of extracellular polysaccharides and biofilm formation in heavy metal resistance of rhizobia. Materials 9:418

Nonnoi F, Chinnaswamy A, de la Torre VSG, de la Pena TC, Lucas MM, Pueyo JJ (2012) Metal tolerance of rhizobial strains isolated from nodules of herbaceous legumes (*Medicago* spp. and *Trifolium* spp.) growing in mercury-contaminated soils. Appl Soil Ecol 61:49–59

Octive JC, Johnson AC, Wood M (1994) Effects of previous aluminium exposure on motility and nodulation by *Rhizobium* and *Bradyrhizobium*. Soil Biol Biochem 26:1477–1482

Oves M, Khan MS, Zaidi A (2013) Chromium reducing and plant growth promoting novel strain *Pseudomonas aeruginosa* OSG41 enhance chickpea growth in chromium amended soils. Eur J Soil Biol 56:72–83

Pajuelo E, Pérez-Palacios P, Romero-Aguilar A, Delgadillo J, Doukkali B, Rodríguez-Llorente ID, Caviedes MA (2016) Improving legume–*rhizobium* symbiosis for copper phytostabilization through genetic manipulation of both symbionts. In: Biological nitrogen fixation and beneficial plant-microbe interaction. Springer International, Cham, pp 183–193

Panigrahi DP, Sagar A, Dalal S, Randhawa GS (2013) Arsenic resistance and symbiotic efficiencies of alfalfa and cowpea rhizobial strains isolated from arsenic free agricultural fields. Eur J Exp Biol 3:322–333

Parmar P, Kumari N, Sharma V (2013) Structural and functional alterations in photosynthetic apparatus of plants under cadmium stress. Bot Stud 54:45

Paudyal SP, Aryal RR, Chauhan SVS, Maheshwari DK (2007) Effect of heavy metals on growth of rhizobium strains and symbiotic efficiency of two species of tropical legumes. Sci World 5:27–32

Pérez-Palacios P, Agostini E, Ibáñez SG, Talano MA, Rodríguez-Llorente ID, Caviedes MA, Pajuelo E (2017) Removal of copper from aqueous solutions by rhizofiltration using genetically modified hairy roots expressing a bacterial Cu-binding protein. Environ Technol 1–12

Petrova S, Yurukova L, Velcheva I (2013) *Taraxacum officinale* as a biomonitor of metals and toxic elements (Plovdiv, Bulgaria). Bulg J Agric Sci 19:241–247

Pireh P, Yadavi A, Balouchi H (2017) Effect of cadmium chloride on soybean in presence of arbuscular mycorrhiza and vermicompost. Legum Res Int J 40:63–68

Qing X, Yutong Z, Shenggao L (2015) Assessment of heavy metal pollution and human health risk in urban soils of steel industrial city (Anshan), Liaoning, Northeast China. Ecotoxicol Environ Saf 120:377–385

Rai R, Agrawal M, Agrawal SB (2016) Impact of heavy metals on physiological processes of plants: with special reference to photosynthetic system. In: Plant responses to xenobiotics. Springer, Singapore, pp 127–140

Rajkumar M, Sandhya S, Prasad MNV, Freitas H (2012) Perspectives of plant-associated microbes in heavy metal phytoremediation. Biotechnol Adv 30:1562–1574

Rane MD, Shaikh EA, Malusare UG (2014) Effect of heavy metals on growth of *Rhizobium*. Int J Sci Eng Res 5:306–310

Reichman SM (2007) The potential use of the legume–*Rhizobium* symbiosis for the remediation of arsenic contaminated sites. Soil Biol Biochem 39:2587–2593

Reis GS, de Almeida AA, de Almeida NM, de Castro AV, Mangabeira PA, Pirovani CP (2015) Molecular, biochemical and ultrastructural changes induced by Pb toxicity in seedlings of *Theobroma cacao* L. PLoS One 10:e0129696

Romaniuk K, Dziewit L, Decewicz P, Mielnicki S, Radlinska M, Drewniak L (2017) Molecular characterization of the pSinB plasmid of the arsenite oxidizing, metallotolerant *Sinorhizobium* sp. M14–insight into the heavy metal resistome of sinorhizobial extrachromosomal replicons. FEMS Microbiol Ecol 93:fiw215

Rucińska-Sobkowiak R (2016) Water relations in plants subjected to heavy metal stresses. Acta Physiol Plant 38:257

Saadani O, FatnassiI C, Chiboub M, Abdelkrim S, Barhoumi F, Jebara M, Jebara SH (2016) *In situ* phytostabilisation capacity of three legumes and their associated plant growth promoting bacteria (PGPBs) in mine tailings of northern Tunisia. Ecotoxicol Environ Saf 130:263–269

Sandalio LM, Dalurzo HC, Gomez M, Romero-Puertas MC, Del Rio LA (2001) Cadmium-induced changes in the growth and oxidative metabolism of pea plants. J Exp Bot 52:2115–2126

Sangwan P, Kumar V, Joshi UN (2014) Effect of chromium (VI) toxicity on enzymes of nitrogen metabolism in clusterbean (*Cyamopsis tetragonoloba* L.) Enzyme Res 2014:1–9

Sbabou L, Idir Y, Bruneel O, Le Quere A, Aurag J (2016) Characterization of root-nodule bacteria isolated from *Hedysarum spinosissimum* L, growing in mining sites of Northeastern region of Morocco. SOJ Microbiol Infect Dis 4:1–8

Segura A, Ramos JL (2013) Plant–bacteria interactions in the removal of pollutants. Curr Opin Biotechnol 24:467–473

Shvaleva A, de la Peña TC, Rincón A, Morcillo CN, de la Torre VSG, Lucas MM, Pueyo JJ (2010) Flavodoxin over expression reduces cadmium-induced damage in alfalfa root nodules. Plant Soil 326:109–121

Singh AK, Singh G (2015) A study of multiple heavy metal tolerance in root nodulating bacteria. Int J Res Dev Pharm Life Sci 4:1713–1721

Singh Y, Ramteke PW, Shukla PK (2013) Characterization of *Rhizobium* isolates of pigeon pea rhizosphere from Allahabad soils and their potential PGPR characteristics. Int J Res Pure Appl Microbiol 3:4–7

Singh RP, Shelke GM, Kumar A, Jha PN (2015) Biochemistry and genetics of ACC deaminase: a weapon to "stress ethylene" produced in plants. Front Microbiol 6:937

Soni SK, Singh R, Singh M, Awasthi A, Wasnik K, Kalra A (2014) Pretreatment of Cr (VI)-amended soil with chromate-reducing rhizobacteria decreases plant toxicity and increases the yield of *Pisum sativum*. Arch Environ Contam Toxicol 66:616–627

Stan V, Gament E, Cornea CP, Voaides C, Dusa M, Plopeanu G (2011) Effects of heavy metal from polluted soils on the *Rhizobium* diversity. Not Bot Hort Agrobot Cluj 39:88–95

Talano MA, Cejas RB, González PS, Agostini E (2013) Arsenic effect on the model crop symbiosis *Bradyrhizobium*–soybean. Plant Physiol Biochem 63:8–14

Tamas MJ, Sharma SK, Ibstedt S, Jacobson T, Christen P (2014) Heavy metals and metalloids as a cause for protein misfolding and aggregation. Biomol Ther 4:252–267

Teng Y, Wang X, Li L, Li Z, Luo Y (2015) Rhizobia and their bio-partners as novel drivers for functional remediation in contaminated soils. Front Plant Sci 6:32

Tittabutr P, Awaya JD, Li QX, Borthakur D (2008) The cloned 1-aminocyclopropane-1-carboxylate (ACC) deaminase gene from *Sinorhizobium* sp. strain BL3 in *Rhizobium* sp. strain TAL1145 promotes nodulation and growth of *Leucaena leucocephala*. Syst Appl Microbiol 31:141–150

Ullah A, Heng S, Munis MFH, Fahad S, Yang X (2015) Phytoremediation of heavy metals assisted by plant growth promoting (PGP) bacteria: a review. Environ Exp Bot 117:28–40

Van Assche F, Clijsters H (1990) Effects of heavy metals on enzyme activity in plants. Plant Cell Environ 13:195–206

Velez PA, Talano MA, Paisio CE, Agostini E, González PS (2016) Synergistic effect of chick-pea plants and *Mesorhizobium* as a natural system for chromium phytoremediation. Environ Technol 1–9

Wakelin SA, Cavanagh JAE, Young S, Gray CW, van Ham RJC (2016) Cadmium in New Zealand pasture soils: toxicity to *Rhizobia* and white clover. N Z J Agric Res 59:65–78

Wang Y, Shi J, Wang H, Lin Q, Chen X, Chen Y (2007) The influence of soil heavy metals pollution on soil microbial biomass, enzyme activity, and community composition near a copper smelter. Ecotoxicol Environ Saf 67:75–81

Wani PA, Khan MS (2012) Bioremediaiton of lead by a plant growth promoting Rhizobium species RL9. J Bacteriol 2:66–78

Wani PA, Khan MS (2013a) Isolation of multiple metal and antibiotic resistant *mesorhizobium* species and their plant growth promoting activity. Res J Microbiol 8:25–35

Wani PA, Khan MS (2013b) Nickel detoxification and plant growth promotion by multi metal resistant plant growth promoting *Rhizobium* species RL9. Bull Environ Contam Toxicol 91:117–124

Wani PA, Khan MS (2014) Screening of multiple metal and antibiotic resistant isolates and their plant growth promoting activity. Pak J Biol Sci 17:206–212

Wani PA, Khan MS, Zaidi A (2008) Effects of heavy metal toxicity on growth, symbiosis, seed yield and metal uptake in pea grown in metal amended soil. Bull Environ Contam Toxicol 81:152–158

Xie P, Hao X, Herzberg M, Luo Y, Nies DH, Wei G (2015) Genomic analyses of metal resistance genes in three plant growth promoting bacteria of legume plants in Northwest mine tailings, China. J Environ Sci 27:179–187

Xie Y, Fan J, Zhu W, Amombo E, Lou Y, Chen L, Fu J (2016) Effect of heavy metals pollution on soil microbial diversity and Bermudagrass genetic variation. Front Plant Sci 7:755

Yu X, Li Y, Zhang C, Liu H, Liu J, Zheng W, Kang X, Leng X, Zhao K, Gu Y, Zhang X (2014) Culturable heavy metal-resistant and plant growth promoting bacteria in V-Ti magnetite mine tailing soil from Panzhihua, China. PLoS One 9:e106618

Yu X, Li Y, Cui Y, Liu R, Li Y, Chen Q, Gu Y, Zhao K, Xiang Q, Xu K, Zhang X (2016a) An indoleacetic acid producing *Ochrobactrum* sp. MGJ11 counteracts cadmium effect on soybean by promoting plant growth. J Appl Microbiol 122(4):987–996. doi:10.1111/jam.13379

Yu X, Li Y, Li Y, Xu C, Cui Y, Xiang Q, Gu Y, Zhao K, Zhang X, Penttinen P, Chen Q (2016b) *Pongamia pinnata* inoculated with *Bradyrhizobium liaoningense* PZHK1 shows potential for phytoremediation of mine tailings. Appl Microbiol Biotechnol 101:1739–1751

Zhang Y, Miró M, Kolev SD (2015) Hybrid flow system for automatic dynamic fractionation and speciation of inorganic arsenic in environmental solids. Environ Sci Technol 49:2733–2740

Zribi K, Nouairi I, Slama I, Talbi-Zribi O, Mhadhbi H (2015) *Medicago sativa*-Sinorhizobium meliloti symbiosis promotes the bioaccumulation of zinc in nodulated roots. Int J Phytoremediation 17:49–55

Nonsymbiotic and Symbiotic Bacteria Efficiency for Legume Growth Under Different Stress Conditions

16

Metin Turan, Nurgül Kitir, Erdal Elkoca, Deniz Uras,
Ceren Ünek, Emrah Nikerel, Bahar Soğutmaz Özdemir,
Leyla Tarhan, Ahmet Eşitken, Ertan Yildirim,
Negar Ebrahim Pour Mokhtari, Şefik Tüfenkçi,
M. Rüştü Karaman, and Adem Güneş

Abstract

In order to achieve maximum crop yields, excessive amounts of expensive fertilizers are applied in intensive farming practices. However, the biological nitrogen fixation via symbiotic and nonsymbiotic bacteria can play a significant role in increasing soil fertility and crop productivity, thereby reducing the need for

M. Turan (✉) • N. Kitir • D. Uras • C. Ünek • E. Nikerel • B. Soğutmaz Özdemir • L. Tarhan
Department of Genetics and Bioengineering, Faculty of Engineering, Yeditepe University,
Istanbul, Turkey
e-mail: m_turan25@hotmail.com

E. Elkoca
Department of Agronomy, Faculty of Agriculture, Ataturk University, Erzurum, Turkey

A. Eşitken
Horticulture and Viticulture Department, Faculty of Agriculture, Selcuk University,
Konya, Turkey

E. Yildirim
Department of Horticulture, Faculty of Agriculture, Ataturk University, Erzurum, Turkey

N.E.P. Mokhtari
Organic Farming Department, Islahiye Vocational School, Gaziantep University,
Gaziantep, Turkey

Ş. Tüfenkçi
Faculty of Agriculture, Department of Biosystem Enginering, Yüzüncü Yıl University,
Van, Turkey

M.R. Karaman
Vocational School of Medicals and Aromatics, Afyon Kocatepe University, Afyon, Turkey

A. Güneş
Agricultural Faculty, Soil and Plant Nutrition Science, Erciyes University, Kayseri, Turkey

© Springer International Publishing AG 2017
A. Zaidi et al. (eds.), *Microbes for Legume Improvement*,
DOI 10.1007/978-3-319-59174-2_16

chemical fertilizers. It is well known that a considerable number of bacterial species, mostly those associated with the plant rhizosphere, are able to exert a beneficial effect on plant growth. The use of those bacteria, often called plant growth-promoting rhizobacteria (PGPR), as biofertilizers in agriculture has been the focus of research for several years. The beneficial impact of PGPR is due to direct plant growth promotion by the production of growth regulators, enhanced access to soil nutrients, disease control, and associative nitrogen fixation. Legumes play a crucial role in agricultural production due to their capability to fix nitrogen in association with rhizobia. Inoculation with nodule bacteria called rhizobia has been found to increase plant growth and seed yields in many legume species such as chickpea, common bean, lentil, pea, soybean, and groundnut. However, both rhizobia and legumes suffer heavily and adversely from various abiotic factors. The impact of different stress factors on both PGPR and legume production is critically reviewed and discussed.

16.1   Introduction

Legumes are plants that belong to the family Fabaceae (approximately 700 genera and 18,000 species) and are categorized into two groups as cool season and warm or tropical season legumes (Toker and Yadav 2010; Miller et al. 2002). Broad bean (*Vicia faba*), lupins (*Lupinus* spp.), lentil (*Lens culinaris*), chickpea (*Cicer arietinum*), grass pea (*Lathyrus sativus*), common vetch (*Vicia sativa*), and dry pea (*Pisum sativum*) are placed in the cool season food legume group (FAOSTAT 2009; Andrews and Hodge 2010). In contrast, cowpea (*Vigna unguiculata*), soybean (*Glycine max* L.), mung bean (*Vigna radiata*), urd bean (*Vigna mungo*), and pigeon pea (*Cajanus cajan*) are included in the warm season food legume group (Latef and Ahmad 2015). Symbiotic relationship between legumes and rhizobia transforms atmospheric N into ammonium (Geurts et al. 2012) which is used as nutrient by legumes (Howard and Rees 1996) and other subsequent or intercropped crops (Liu et al. 2010). After cereals and oilseeds, legumes rank third in world production (Graham and Vance 2003). One-third (20–40%) of all dietary proteins are provided by legumes which are a primary source of amino acids (Zhu et al. 2005; Kudapa et al. 2013). Pulses are generally used as foods (Rebello et al. 2014). The low energy density and nutrient dense abilities make legumes a valuable food option to fulfill the requirement of undernourished or underserved populations (FAO 1994).

The frequent legume consumption reduces the risk of coronary heart disease by 22% and cardiovascular disease (CVD) risk by 11% (Flight and Clifton 2006). High intake of legumes protects from obesity and related disorders (Papanikolaou and Fulgoni 2008). The legumes also lower blood glucose and insulin responses (Mollard et al. 2012; Jenkins et al. 1980; Nestel et al. 2004) and increase sensitivity of insulin (Nestel et al. 2004). Since legumes are rich in sodium and low in potassium (Rebello et al. 2014), the probability of suffering from these disorders becomes low even when legume consumption is high. Phytochemicals, enzyme inhibitors, phytoestrogens, phytohemagglutinins (lectins), saponins, phenolic compounds, and oligosaccharides are also reported in the majority of legumes (Rebello et al. 2014). Legumes are low in fat content and rich in proteins (Campos-Vega et al. 2010) and complex carbohydrates (Kalogeropoulos et al. 2010) making legume an important and qualified food source. In addition, high content of fibers, polyunsaturated fatty acids, magnesium, and low glycemic index are other valuable properties of legumes (Bouchenak and Lamri-Senhadji 2013).

Biological nitrogen fixation (BNF) plays an important role in land improvement. Leguminous plants and rhizobia together form a symbiotic relationship (Freiberg et al. 1997; Zahran 2001) and have a great quantitative effect on the soil N pool (Ohyama et al. 2009; Abd-Alla et al. 2013). On the other hand, the N deficiency severely limits the plant growth. A prosperous BNF, however, increases agricultural productivity while minimizing soil loss and ameliorating adverse edaphic conditions. However, abiotic stresses have harmful impacts on plant development, including legumes (Singleton and Bohlool 1984; Subba Rao et al. 1999). Drought, salinity/alkalinity, unfavorable soil pH, nutrient deficiency, changes in temperature, inadequate or extreme soil moisture, and decreased photosynthetic activity conspire against a prosperous symbiotic process. In order to overcome these stress conditions, numerous inoculants have been developed to produce symbiotic legume-microbe formulations. In addition to this, experiments are performed in order to formulate new solutions supplemented with plant and microbe exudates which contain flavonoids, sugars, amino acids, and other low molecular weight molecules that are involved in microbe-plant interaction (Garg and Geetanjali 2009; Skorupska et al. 2010; Morel et al. 2012). By using these exudates, symbiotic relationships between bacteria and plants could be mimicked for plant development. Among symbiotic bacteria, rhizobia live in the rhizosphere of legumes and produce root nodules (Foth 1990; Abd-Alla et al. 2013). Structurally, rhizobia are small and rod-shaped Gram-negative bacteria which belong to *Rhizobiaceae* family (Long 1989) and spread over subclass *Alphaproteobacteria* and *Betaproteobacteria*. *Rhizobium*, *Mesorhizobium*, *Ensifer* (formerly *Sinorhizobium*), *Azorhizobium*, *Methylobacterium*, *Bradyrhizobium*, *Phyllobacterium*, *Devosia,* and *Ochrobactrum* are some of the notable genera. Briefly, the PGPR involving rhizobia promote the growth of legumes by stimulating the production of ACC deaminase and hormones as auxins, cytokinins, gibberellins, and certain volatiles; symbiotic nitrogen fixation; solubilization of mineral like phosphorus and other nutrients; etc. (Bashan and Holguin 1997; Ahmad et al. 2008). The growth of rhizobia and its nitrogen-fixing ability, however, are negatively impacted by several environmental factors (Singleton et al. 1982; Sherren et al. 1998; Abd-Alla et al. 2013). The effect of abiotic stresses on legume growth and nonsymbiotic/symbiotic bacteria efficiency and nodulation and nitrogen fixation is discussed in the following section.

16.2   Bacteria Involved in Legume Growth Under Stress Environment

16.2.1 Symbiotic Bacteria

Among symbiotic bacteria, rhizobia associate essentially with leguminous plants (Long 1989; Sprent 2001), and the other one *Frankia*, grouped in *Actinobacteria*, interacts with plants of eight different families (Huss-Danell 1997; Franchee et al. 2009). In land-based systems, symbiotic relationship between *Rhizobium* and legumes is the primary source of fixed N, and more than half of the biological N is supplied by BNF. Symbiotic bacteria infect the legume roots and form nodules (West et al. 2002). During preinjection stage, it is necessary for rhizobia to recognize the roots of the appropriate host in order to be able to colonize. During nodule formation, three root tissues (epidermis, pericycle, and cortex) must be transformed (Geurts et al. 2012). The roots secrete flavonoids, and when bacteria encounters flavonoids, bacterial nodulation genes (*nod/nol/noe*) are activated (Ovtsyna and Staehelin 2003). Nodulation genes in turn regulate the synthesis of nodulation factors which triggers the formation and deformation of root hairs, formation of nodule primordia, induction of early nodulin gene expression, ion flux changes, depolarization of membrane potential, and intra-extracellular alkalization (Broughton et al. 2000; Perret et al. 2000).

16.2.2 Nonsymbiotic Bacteria

The term "nonsymbiotic" could be defined as having an interdependent relationship. Nonsymbiotic bacteria also fix atmospheric nitrogen and in association with symbiotic bacteria increase plant growth. Nonsymbiotic nitrogen-fixing bacteria (free living, associative, and endophytes) are cyanobacteria, *Azospirillum*, *Azotobacter*, *Gluconacetobacter diazotrophicus* and *Azocarus*, etc. (Bhattacharyya and Jha 2012). Due to the inefficiency of suitable carbon and energy sources for free-living organisms, their role in nitrogen fixation is considered as minor (Wagner 2011). On the other hand, associative nitrogen fixer, *Azospirillum*, located predominantly on the root surface of the plant fixes remarkable amount of nitrogen within the rhizosphere of the host plants. Even if their nitrogen-fixing amount is outstanding, the level of the nitrogen fixation is determined by several factors. Soil temperature, low oxygen pressure, availability of photosynthates, efficiency of nitrogenase enzyme, and competitiveness of the bacteria are some of the factors that limit the nitrogen fixation process. *Azotobacter* is another aerobic bacterium with genomic content G-C of 63–67.5% and fixes nitrogen nonsymbiotically (Becking 2006). Soil, water, and sediments are the habitat of *Azotobacter* (Torres et al. 2004, 2005). *Azotobacter* facilitates plant growth by synthesizing IAA and other growth-promoting substances (Ahmad et al. 2005). Also, nodulation and nitrogen fixation in legumes have been found significantly increased following dual inoculation of *Rhizobium* and *Azospirillum* or another PGPR such as *Azotobacter* (Rodelas et al. 1996, 1999).

16.3   Impacts of Abiotic Stresses on PGPR and Legumes

16.3.1 Salinity Stress

Salinity is one of the biggest problems which decreases quality and productivity of crops worldwide. Approximately, 10% of the world's crop fields and 27% of irrigated lands are affected by the salinity stress. When precipitation is insufficient to leach the ions from the soil profile, salts accumulate and cause soil salinity (Blaylock 1994). In hot and dry climate conditions, the level of soil salinity is increased. Soil salinity has a negative impact on growth and yields of crops including legumes (Singleton et al. 1982; Kumari and Subbarao 1984). The level of salt toxicity, however, depends on plant species and concentration and composition of salts (Delgado et al. 1994). For soybean, it was reported that nodulation, total N content, and yields were reduced by soil salinity (Singleton and Bohlool 1984). Similarly, plant height of peanut (*Arachis hypogaea* cv. NC-7) decreased by 21.6% and fresh weight by 21.4% after application of 4 dS/m salinity levels, whereas root length decreased by 30% after 8 dS/m salinity levels (Aydınşakir et al. 2015). Salinity level also affects the net photosynthetic rate of plants. As an example, Stoeva and Kaymakanova (2008) revealed that the net photosynthetic rate (PN) of beans (*Phaseolus vulgaris*) measured on the seventh day of treatment was reduced to 65% at 50 mM NaCl, 56% at 50 mM Na_2SO_4, and 40% and 20% by 100 mM each of NaCl and Na_2SO_4, respectively. However, plants adapt to their environment to maintain their survival. In this regard, Moriuchi et al. (2016) found that *Medicago truncatula* plants merely adapted to the environment and removal of salinity stress led to lower growth potential for saline-adapted plants suggesting that adaptation to high salinity is inherited from parents to the offsprings.

Under salinity stress, legumes are not able to maintain their regular nitrogen fixation and nodulation abilities. In a study conducted with alfalfa cultivated in saline environment, it was observed that a number of active nodes and nitrogen fixation were decreased (Nabizadeh et al. 2011). Nodule structure is also affected by salinity stress. Serraj et al. (1995) found out that treatment with 100 mM NaCl had adverse impacts on the soybean nodules by turning nucleus into a lobed structure and with different chromatin distribution and enlarged periplasmic space after 2 h exposure. Changes in the nucleus lead to differences in gene expression that could be seen in phenotype as decreased nitrogen fixation activity. The sensitivity of nitrogen fixation process to saline conditions could be related with the tolerance level of the bacteria. Velagaleti and Marsh (1989) reported that salinity resulted in decreased rhizobia colonization and shrinkage of root formation, while salt-tolerant *Bradyrhizobium* symbiosis with soybean revealed lower inhibitory impact of salinity in N_2 fixation. Bacteria have evolved several mechanisms to counter salinity stress (Shrivastava and Kumar, 2015). And hence, symbiotic relationship of rhizobia and legume plants is helpful in adapting to the salinity stress. For example, *Rhizobium* and *Pseudomonas* when used as mixed inoculant enhanced the growth and nodulation of mung bean grown under salinity stress by providing auxin and ACC deaminase (Ahmad et al. 2012). Pro-betaine and proline are involved in salt stress tolerance in *Medicago sativa* (Trinchant et al. 2004). Under osmotic stress, nitrogen-fixing bacteria, *Sinorhizobium meliloti,* regulates the expression of BetS gene which has a role in Gly-betaine/Pro-betaine transporter (Boscari et al. 2002). Use of *S. meliloti* would be a useful method to overcome salinity stress. In another perspective, it is suggested that creating a symbiosis between a salt-tolerant plant genotype and a rhizobia maintains salt tolerance and effective nitrogen fixation activity (Zahran 1999; Keneni et al. 2010). Obtaining sucrose from phloem is significant for nodule nitrogen fixation (Gordon et al. 1987). However, the presence of C source on the roots of legumes is not enough for nitrogen fixation. Enzymatic activity is required to supply C to the bacteroides. López et al. (2008) detected more enzymatic activity of PEPC (phosphoenolpyruvate carboxylase), MDH (malate dehydrogenase), and ICDH (isocitrate dehydrogenase) in *Lotus japonicas*, nodulated by *M. loti*, than *Medicago truncatula*, nodulated by *S. meliloti*. *Lotu japonicus* nodule C metabolism was shown to be less sensitive to salinity than in *M. truncatula* since the enzymes that had a role in C supply could fuel the bacteroides for processing the nitrogen fixation. However, the nitrogenase activity in *L. japonicus* nodules was inhibited by salinity. Even some strategies have been developed to find better breeds, they are long drawn and cost intensive.

16.3.2 Cold Stress

Temperature is another important factor essentially required for proper growth and development of plants. However, if a plant is exposed to a colder temperature for a longer duration, it may suffer from cold stress which could lead to loss of flower, decrease in photosynthetic activity, reduced activity of conductive tissue and enzymatic activity, and slowing down of the growth rate. In order to avoid such harmful cold temperature effects, plants need to develop certain mechanisms. In this context, soluble sugar is even sensitive to abiotic stresses, but reserve of sugar has a role to fight against stress conditions. Sugar protects cells from damage by serving as osmoprotectant, nutrient, and primary messengers in signal transduction (Yuanyuan et al. 2009). Proline is yet another important biomolecule (an amino acid) that acts as osmoprotectant and protects plants from stress conditions and hence accelerates the plant recovery. For instance, proline content increased in roots and shoots of lentil grown under cold stress conditions (Oktem et al. 2008). The length and fresh weight of shoots were decreased significantly resulting in the loss of yield. Hekneby et al. (2001) exposed the 21-day-old *Medicago truncatula* plants to 20/15 °C or 10/5 °C (day/night temperatures) for 40 days. The results revealed a significant increase in root/shoot ratio of *M. truncatula* plants grown under cold environment while total dry matter, leaf area, and specific leaf area ratio did not differ between two temperature treatments showing the tolerance degree of *M. truncatula* to cold stress. Exposure of plants to cold stress can also affect *Rhizobium*-legume symbiosis resulting in poor nodulation and nitrogen fixation. As an example, Lidström et al. (1985) found out that population density of *Rhizobium* strains was decreased from 3×10^8 to 1×10^5/g after -5 °C in soil acidity conditions. Also, the nitrogen fixation was decreased which was attributed due to cold stress rather than soil acidity and caused by the reduction of bacterial numbers in soil after cold treatment. Lastly, molecular aspects of the cold stress response and adaptation to cold stress have also been reported for soybean (Zhang et al. 2014). As an example, molecular signal exchanges between rhizobia and the legume are affected by the temperature causing reduction in nodulation process. There are inter-organismal signaling between rhizobia and its symbiotic partners, and this could be inhibited by low temperature. Low temperatures inhibit biosynthesis and secretion of signal molecules so that the interaction between plant and bacterial symbiotic relationship is interrupted. For instance, genistein secretion from soybean roots, which is required for the induction of *nod* genes of *B. japonicum*, is retarded (Abd-Alla 2001, 2011).

16.3.3 Nutrient Deficiency Stress

Nutrients are required by plants to live, grow, and reproduce. Deficiency of nutrients restricts the growth of plants (Table 16.1). Plant nutrients are divided mainly into two groups: macronutrients (Ca, P, N, K, S, and Mg) and micronutrients (B, Cl, Mn, Fe, Zn, Cu, Mo, and Ni). The critical concentration of these nutrients required

Table 16.1 The effect of different nutrient deficiencies on plants and legume-rhizobia symbiosis

Element	Nutrient deficiency symptoms/damage	Importance in legume-rhizobia symbiosis
Nitrogen	Yellowing of older leaves while the rest of plant remain light green	Inhibits nodule formation and nitrogenase activity (Sprent et al. 1988)
Phosphorus	Leaf tips have a burnt look, older leaves turn dark green or reddish-purple	In case of deficiency, nitrogen fixation and symbiotic interactions are damaged (Weisany et al. 2013)
Potassium	Wilt of older leaves, interveinal chlorosis, and scorching inward from leaf margins	In case of deficiency, restrict rhizobial growth (Vincent et al. 1977)
Boron	Witches' broom formation and terminal buds die	Number of rhizobia infecting the host cell and number of infection thread are reduced during boron deficiency (Bolanos et al. 1996)
Molybdenum	Yellowing of older leaves (bottom of plant) while rest of the plant remain light green	Fe-Mo cofactor for most nitrogenases (Weisany et al. 2013)
Sulfur	Firstly, younger leaves turn yellow and sometimes this could be followed by older leaves	In case of deficiency, limited growth of rhizobia (O'Hara et al. 1987)
Calcium	Distorted or irregular shape of new leaves that are on the top of plant. It can cause blossom-end rot	In case of deficiency, nitrogen fixation in nodules is decreased (Banath et al. 1966) and nodulation and nodule development reduced (Banath et al. 1966)
Iron	Yellowing happens between the veins of young leaves	Fe-Mo cofactor for most nitrogenases (Weisany et al. 2013)

Modified from Guide to Symptoms of Plant Nutrient Deficiencies, Bradley and Hosier (1999)

for optimum growth of plants, however, varies from genotypes to genotypes and from organs to organs. The impact of nutrient deficiency on legumes is discussed in the following section.

16.3.3.1 Phosphorus Stress

Among plant nutrients, phosphorus (P) is an important element and is involved in numerous biochemical processes, particularly in energy acquisition, storage, and utilization (Epstein and Bloom 2005). N_2-fixing nodules have high requirement of P. Unlike N, the P resources are not renewable, and therefore, it is expected that high-grade rock phosphates (RP) will be depleted gradually. As a result, the production of legumes in P-deficient soil is likely to suffer heavily (Sulima et al. 2015). However, phosphate-solubilizing bacteria (PSB) belonging to genera *Bacillus*, *Pseudomonas*, *Achromobacter*, *Alcaligenes*, *Brevibacterium*, *Corynebacterium*, *Serratia*, *and Xanthomonas* can be useful in supplying soluble P to plants (Khan et al. 2007). The impact of PSB, however, differs from species to species when inoculated with symbiotic *Rhizobium* bacteria. Rosas et al. (2006), for example, designed an experiment to assess the impact of *Pseudomonas* when co-inoculated with *S. meliloti 3DOh13* against alfalfa and *B. japonicum TIIIB* against soybean. The results demonstrated no significant differences between *S. meliloti 3DOh13*-inoculated alfalfa plants and *S. meliloti 3DOh13+ Pseudomonas* co-inoculation. However, the number and dry weight of soybean nodules was greater for co-inoculation with *B. japonicum TIIIB* and *Pseudomonas* compared to the sole application of *B. japonicum TIIIB*. Considering these, it is suggested that PSB in combination with other PGPR including rhizobia could be useful for enhancing legume production.

16.3.3.2 Sulfur Stress

Sulfur (S) is yet another important nutrient element for plants. Sulfur plays an important role in development and functioning of nodules. However, the deficiency of S limits N_2 fixation. Sulfur-oxidizing bacteria, for example, *Beggiatoa*, *Chromatium*, *Chlorobium*, *Thiobacillus*, *Sulfolobus*, *Thiospira*, and *Thiomicrospira*, are used to fulfill the sulfate requirement of plants. Under sulfur-deficient conditions, these bacteria could be used to transform elemental S into sulfate that plants can utilize. For groundnut, Anandham et al. (2007) investigated the impact of co-inoculation of *Thiobacillus,* sulfur-oxidizing bacteria, and *Rhizobium* under S-deficient soil. The results indicated that the nodule number, nodule dry weight, and biomass were significantly increased, and pod yield was enhanced by 18%.

16.3.3.3 Iron Stress

Plants growing in calcareous soils suffer from iron deficiency. Some soil bacteria synthesize ferric chelate reductase (FC-R) enzyme and release organic acids that decrease apoplastic pH of root and leaf cells. Ferric chelate reductase reduces Fe^{3+} to available form (Donnini et al. 2009). Many experiments have shown that the increased FC-R activity helps plants to take up Fe while growing under Fe-deficient conditions (López-Millán et al. 2001; Manuel and Alcántara 2002). FC-R activity can be utilized for determination of Fe-chlorosis-tolerant rootstocks (Bavaresco et al. 1991; Romera et al. 1991). Furthermore, the other way for Fe acquisition from soil is releasing of organic acids such as citrate and malate (Jones 1998; Abadía et al. 2002). Many researches demonstrated that organic acid excretion makes iron available to plants under Fe-starved conditions (Jones et al. 1996; Abadía et al. 2002).

High lime in soil affects Fe nutrition detrimentally in many ways. At first, availability of Fe in soil is decreased under lime and high pH conditions. Fe is trapped in bicarbonate soils and becomes unavailable for uptake by plants. Due to increased bicarbonate concentration, Fe acquisition is deteriorated (Nikolic and Roemheld 2003). However, some treatments can help to alleviate lime-induced Fe deficiency of soils. Afterwards, Fe entered into root apoplast must be carried into xylem. However, some part of Fe^{3+} remains in the root apoplast under lime-contained soil conditions and cannot be carried into plant shoot as a result of high pH in root apoplast (Kosegarten and Koyro 2001; Molassiotis et al. 2005). It has been proposed that some part of Fe absorbed from soil remains in the root apoplast (Bienfait et al. 1983). In an experiment it was exhibited that chlorosis and root Fe content of chlorotic plants could be related to removing of root Fe into plant shoots. Iron (Fe^{3+} citrate)-loaded xylem must be distributed into the leaf from veins after removal from the leaf (Mengel 1995). There must be re-reduction of Fe^{3+} citrate into Fe^{2+} for distribution in leaves (Brüggemann et al. 1993; Mengel 1994; Toselli et al. 2000; Bohórquez et al. 2001). Iron present in leaf apoplast must enter cell in order to maintain distribution of Fe in the leaf vein to the leaf. Mengel (1994) reported that during Fe chlorosis in the leaves, active Fe concentration is lower than non-chlorosis plants, but total Fe concentration is the same in both plant leaves. Therefore, leaf FC-R enzyme possesses a remarkable importance for elevating Fe availability in the leaves.

Rhizobacteria lowers the rhizosphere pH by releasing organic acids which in turn increases FC-R activity. Many researchers have suggested that bacterial treatments cause a decrease in soil pH and an increase in nutrition availability in soil (Sharma and Johri 2003; Orhan et al. 2006; Karlidag et al. 2007; Zhang et al. 2009). Also, increase in root and leaf Fe concentration has been reported. Iron is available in soil complexes with many organic acids such as citrate and malate that increases availability of insoluble ferric oxyhydroxides (Jones et al. 1996). Thus, increase of active iron (Fe^{2+}) in soil may have increased Fe uptake by plant from soil. Root inoculations considerably influenced root FC-R activity. Fe^{2+} is returned into Fe^{3+} after loading to xylem and is transported to shoots as Fe^{3+}-citrate with complexing with citrate. Transportation type of Fe in xylem is mainly Fe^{3+} citrate complex. Therefore, increase of citrate in xylem helps Fe transportation from root to shoots. Therefore, distribution of Fe to leaves and regreening were maintained as a result of a decrease in leaf apoplastic pH. Leaf apoplastic pH may have been decreased with many treatments such as spraying diluted acid or citric acid (Tagliavini and Rombola 2001) to leaves or ammonium fertilizer application to soil; thus iron in veins can be distributed in leaves. In this regard, decrease in leaf apoplastic pH can be achieved by uptake and translocation of organic acids released by bacteria in rhizosphere.

16.3.4 Drought Stress

 The long exposure of plants to water-insufficient conditions, often called drought stress, has an adverse impact on plants (Zahran 1999) including legumes (Sangakkara and Hartwig 1996; Marino et al. 2007). Therefore, the assessment of drought impact on legume-*Rhizobium* symbiosis efficiency under abiotic stress conditions becomes highly critical. Ureides are nitrogenous compounds contributing to nitrogen recycling which accumulate in shoots and nodules of legumes under drought stress and consequently decline symbiotic nitrogen fixation (SNF) rapidly (Vadez et al. 2000). In addition, decreased transpiration rate diminishes N demand of shoot that lowers the rate of xylem translocation and reduces enzymatic activities which leads to decrease in nitrogen fixation rate (Valentine et al. 2011). Moreover, initiation of nodules, nodule growth, development, and function are affected by drought (Smith et al. 1988; Vadez et al. 2000; Streeter 2003). Drought situation also decreases photosynthetic activity which in turn adversely affects the SNF (Ladrera et al. 2007; Valentine et al. 2011).

In a study, Purcell et al. (1997) compared the nodulation patterns of two different soybeans: one tolerant to drought while the other was sensitive to drought. Drought-tolerant soybean was referred as "Jackson," while drought-sensitive one was referred as "SCE82-303." Even though the mass and number of nodules differed among two cultivars resulting, the nodule mass increased in "Jackson," while it decreased in "SCE82-303." Similarly, the impact of drought on SNF efficiency of *Rhizobium* was variable (Marino et al. 2007). For this, pea plants were grown in a split root system where one of the half was able to reach water, while the other half lacked water. Application of water-deficient conditions revealed decreased N_2 fixation. Furthermore, cell redox was imbalanced due to the reduction in the water potential of nodules. Besides, feedback signaling for systemic nitrogen did not work in the absence of water since the N_2 fixation was active and maintained at control values for half of the roots that were able to reach the water. This finding thus suggests that split root system controls the N_2 fixation at the local level rather by a systemic nitrogen signal. Considering these and other related studies, it becomes important to develop strategies that could protect both legumes and rhizobia from the negative impact of drought stress.

16.3.5 High Temperature and Heat Stress

Temperature is another important factor that affects N_2 fixation process among legumes. However, the temperature requirement of legumes varies from species to species or from cultivars to cultivars. For instance, the optimum temperature for N_2 fixation in clover and pea is 30 °C, while it is 35–40 °C for guar, soybean, peanut, and cowpea (Michiels et al. 1994). For beans, optimum temperature for nodule function is 25–30 °C, while 30–33°C temperature restricts nodule activity (Piha and Munnus 1987). However, nitrogen fixation by legumes is a main problem while growing at high temperatures in tropical and subtropical regions (Michiels et al. 1994). Infection of root hair, differentiations of bacteroides, structure of nodules, and legume root nodule function are affected by temperature (Zahran 1999). Additionally, photosynthetic rate, membrane stability, relations with water, and respiration are also impacted negatively by increased temperatures, which also regulate hormone levels and primary and secondary metabolite production. Heat stress also lowers the synthesis of ureides and decreases levels of nitrate reductase and glutamate synthase in legumes (Hungria and Vargas 2000; Christophe et al. 2011; Latef and Ahmad 2015). The decreased nitrogenase activity results in the reduction of N_2 fixation or accelerated nodule senescence leading to decreased nodule endurance (Bordeleau and Prevost 1994; Hungria and Vargas 2000; Christophe et al. 2011; Latef and Ahmad 2015). There are reports where increase in root temperatures has been found to adversely affect the bacterial infection and N_2 fixation of legumes, for example, soybean (Munevar and Wollum 1982), guar (Arayankoon et al. 1990), peanut (Kishinevsky et al. 1992), cowpea (Rainbird et al. 1983), and beans (Piha and Munnus 1987; Hungria et al. 1993). Plants have, however, evolved mechanisms to cope high temperature through heat-shock protein expression and other stress-related proteins and reactive oxygen species (ROS) production (Bhattacharya and Vijaylaxmi 2010; Hasanuzzaman et al. 2013).

16.3.6 Soil Acidity Stress

Globally, acidity covers nearly 40% of the lands that are available for farming (Valentine et al. 2011). An area which is larger than 1.5 Giga hectares is under acidity threat limiting the agricultural production (Graham and Vance 2000; Abd-Alla et al. 2014b). Soil acidity is increased by the impacts of global warming and agricultural applications that limit the legume crop productivity. However, alkalinity and acidity are the two extreme situations for any soil that may hamper growth, survival, and nitrogen fixation ability of rhizobia (Lapinskas 2007). During *Rhizobium*-legume symbiosis, *Rhizobium* was found more sensitive to acidic conditions than legumes. Virtually, since rhizobia are incapable of persisting and surviving under acidic conditions, this could reduce the effectiveness of symbiosis and concomitantly loss in legume productivity. Therefore, selection and application of acid-tolerant rhizobia become important for enhancing the production of legumes under acid stress environment. In this regard, mutants of *R. leguminosarum* that grew at pH as low as 4.5 (Chen et al. 1993) and *S. meliloti* which grew at pH level below 5.5 (Foster 2000) are reported. In addition, some rhizobial species can grow at a wide range of pH. For instance, *S. fredii* can grow at pH levels between 4 and 9.5 (Fujihara and Yoneyama 1993). Like acidity, alkalinity stress also destructs the growth of *Rhizobium* (Monica et al. 2013) and their symbiotic relationship with legumes (Zahran 1999). Therefore, it is also important to select *Rhizobium* isolates, which could survive the alkalinity stress and be capable of nitrogen fixation and, hence, the legume production (Abd-Alla et al. 2014a). Apart from rhizobia, yields and growth of legumes are also impacted by soil acidity (Ferguson et al. 2013). However, soil acidity helps to adjust the availability of mineral nutrients (e.g., phosphorus) and severity of some phytotoxic elements (e.g., aluminum, manganese, and iron) in natural/degraded ecosystems (Muthukumar et al. 2014).

Conclusion

Rhizobacteria including both symbiotic and nonsymbiotic bacteria are one of the important classes of soil microbiota which augment crop production including those of legumes in different agronomic regions. The application of PGPR provides a comparable yield and quality by supplying essential nutrients and hormones to legumes. Also, the application of PGPR helps to alleviate several stress conditions as drought, salinity, nutrient stress, and low/high temperature stress. The tolerance to high levels of stresses and the survival and persistence of PGPR in severe and harsh conditions make these bacteria a highly valuable organism to enhance legume production in extreme environmental conditions. However, further studies are needed to evaluate the performance of PGPR in different stressed conditions choosing a range of legume crops. The mineral nutrition and fertilization effects of N_2-fixing and other free-living PGPR should be examined regularly and carefully before they are recommended for application by farming communities.

References

Abadía J, López-Millán AF, Rombolà A, Abadía A (2002) Organic acids and Fe-deficiency: a review. Plant Soil 241:75–86

Abd-Alla MH (2001) Regulation of nodule formation in soybean-*Bradyrhizobium* symbiosis is controlled by shoot or/and root signals. Plant Growth Regul 34:241–250

Abd-Alla MH (2011) Nodulation and nitrogen fixation in interspecies grafts of soybean and common bean is controlled by isoflavonoid signal molecules translocated from shoot. Plant Soil Environ 57:453–458

Abd-Alla MH, El-enany AE, Bagy MK, Bashandy SR (2014) Alleviating the inhibitory effect of salinity stress on gene expression in ? fenugreek ( ) symbiosis by isoflavonoids treatment. J Plant Int 9(1):275–284

Abd-Alla MH, El-Enany AE, Nafady NA, Khalaf DM, Morsy FM (2014a) Synergistic interaction of *Rhizobium leguminosarum* bv. *viciae* and arbuscular mycorrhizal fungi as a plant growth promoting biofertilizers for faba bean (*Vicia faba* L.) in alkaline soil. Microbiol Res 169:49–58

Abd-Alla MH, Issa AA, Ohyama T (2014b) Impact of harsh environmental conditions on nodule formation and dinitrogen fixation of legumes. Agricultural and biological sciences "advances in biology and ecology of nitrogen fixation". ISBN: 978-953-51-1216-7

Ahmad F, Ahmad I, Khan MS (2005) Indole acetic acid production by the indigenous isolates of *Azotobacter* and fluorescent pseudomonas in the presence and absence of tryptophan. Turk J Biol 29:29–34

Ahmad F, Ahmad I, Khan MS (2008) Screening of free-living rhizospheric bacteria for their multiple plant growth promoting activities. Microbiol Res 163:173–181

Ahmad M, Zahir ZA, Asghar HN, Arshad M (2012) The combined application of rhizobial strains and plant growth promoting rhizobacteria improves growth and productivity of mung bean (*Vigna radiata* L.) under salt-stressed conditions. Ann Microbiol 62:1321–1330

Anandham R, Sridar R, Nalayini P, Poonguzhali S, Madhaiyan M, Sa T (2007) Potential for plant growth promotion in groundnut (*Arachis hypogaea* L.) cv. ALR-2 by co-inoculation of sulfur-oxidizing bacteria and Rhizobium. Microb Res 162:139–153

Andrews M, Hodge S (2010) Climate change, a challenge for cool season grain legume crop production. In: Climate change and management of cool season grain legume crops. Springer, Netherlands, pp 1–9

Arayankoon T, Schomberg HH (1990) Nodulation and N_2 fixation of guar at high room temperature. Plant Soil 126:209–213

Aydınşakir K, Büyüktaş D, Dinç N, Karaca C (2015) Impact of salinity stress on growing, seedling development and water consumption of peanut (*Arachis hypogaea* cv. NC-7). Akdeniz Univ Ziraat Fak Derg 28:77–84

Bashan Y, Holguin G (1997) Azospirillum–plant relationships: environmental and physio-logical advances (1990–1996). Can J Microbiol 43:103–121

Banath CL, Greenwood EAN, Loneragan JF (1966) Effects of calcium deficiency on symbiotic nitrogen fixation. Plant Physiol 41(5):760–763

Bavaresco L, Fregoni H, Fraschini P (1991) Investigations on iron uptake and reduction by excised roots of different grapevine rootstocks and a *V. vinifera* cultivar. In: Chen Y, Hadar Y (eds) Iron nutrition and interactions in plant. Kluwer Academic Publishers, Dordrecht, The Netherlands, pp 139–143

Becking J (2006) The family Azotobacteraceae. Prokaryotes 6:759–783

Bhattacharya A, Vijaylaxmi (2010) Physiological responses of grain legumes to stress environments. In: Yadav SS (ed) Chickpea breeding and management. CAB International, Wallingford, pp 35–86

Bhattacharyya PN, Jha DK (2012) Plant growth-promoting rhizobacteria (PGPR): emergence in agriculture. World J Microbiol Biotechnol 28:1327–1350

Bienfait HE, Bino RJ, Vander Blick AM, Duivenvoorden JF, Fontaine FM (1983) Characterization of ferric reducing activity in roots of fe-deficient *Phaseolus vulgaris*. Physiol Plant 59:196–202

Blaylock AD (1994) Soil salinity, salt tolerance and growth potential of horticultural and landscape plants. Co-operative Extension Service, University of Wyoming, Department of Plant, Soil and Insect Sciences, College of Agriculture, Laramie, Wyoming

Bohórquez JM, Romera FJ, Alcántara E (2001) Effect of Fe^{3+}, Zn^{2+} and Mn^{2+} on ferric reducing capacity and regreening process of the peach rootstock Nemaguard [*Prunus persica* (L.) Batsch]. Plant Soil 237:157–163

Bolanos L, Brewin NJ, Bonilla I (1996) Effects of boron on Rhizobium-legume cell-surface interactions and nodule development. Plant Physiol 110(4):1249–1256

Bordeleau LM, Prevost D (1994) Nodulation and nitrogen fixation in extreme environments. Plant Soil 161:115–125

Boscari A, Mandon K, Dupont L, Poggi MC, Le Rudulier D (2002) BetS Is a major glycine betaine/proline betaine transporter required for early osmotic adjustment in *Sinorhizobium meliloti*. J Bacteriol 184:2654–2663

Bouchenak M, Lamri-Senhadji M (2013) Nutritional quality of legumes, and their role in cardiometabolic risk prevention: a review. J Med Food 16:185–198

Bradley L, Hosier S (1999) Guide to Symptoms of Plant Nutrient Deficiencies ISO 690

Broughton WJ, Jabbouri S, Perret X (2000) Keys to symbiotic harmony. J Bacteriol 182:5641–5652

Bruggemann W, Maas-Kantel K, Moog PR (1993) Iron uptake by leaf mesophyll cells: The role of the plasma membrane-bound ferric-chelate reductase. Planta 190(2)

Campos-Vega R, Loarca-Pina G, Dave Oomah B (2010) Minor components of pulses and their potential impact on human health. Food Res Int 43:461–482

Chen H, Richardson AE, Rolfe BG (1993) Studies of the physiological and genetic basis of acid tolerance in *Rhizobium leguminosarum* biovar *trifolii*. Appl Environ Microbiol 59:1798–1804

Christophe S, Jean-Christophe A, Annabelle L, Alain O, Marion P, Anne-Sophie V (2011) Plant N fluxes and modulation by nitrogen, heat and water stresses: a review. Based on comparison of legumes and non legume plants. In: Shanker AK, Venkateswarlu B (eds) Abiotic stress in plants–mechanisms and adaptations. InTech, Croatia, pp 79–119

Delgado MJ, Ligero F, Lluch C (1994) Effects of salt stress on growth and nitrogen fixation by pea, faba-bean, common bean and soybean plants. Soil Biol Biochem 26:371–376

Donnini S, Castagna A, Ranieri A, Zocchi G (2009) Differential responses in pear and quince genotypes induced by Fe deficiency and bicarbonate. J Plant Physiol 166:1181–1193

Epstein E, Bloom AJ (2005) Mineral nutrition of plants: principles and perspectives, 2nd edn. Sinauer Associates, Sunderland, MA

FAO (1994) Pulses and derived products

FAOSTAT (2009) Food and Agriculture Organization of the United Nations, Rome

Ferguson BJ, Lin MH, Gresshoff PM (2013) Regulation of legume nodulation by acidic growth conditions. Plant Signal Behav 8:e23426

Flight I, Clifton P (2006) Cereal grains and legumes in the prevention of coronary heart disease and stroke: a review of the literature. Eur J Clin Nutr 60:1145–1159

Foster JW (2000) Microbial responses to acid stress. In: Storz G, Hengge-Aronis R (eds) Bacterial stress response. ASM Press, Washington, DC, pp 99–115

Foth HD (1990) Fundamentals of soil science. Wiley, New York

Franche C, Lindström K, Elmerich C (2009) Nitrogen-fixing bacteria associated with leguminous and non-leguminous plants. Plant Soil 321(1–2):35–59

Freiberg C, Fellay R, Bairoch A, Broughton WJ, Rosenthal A, Perret X (1997) Molecular basis of symbiosis between Rhizobium and legumes. Nature 387:394–401

Fujihara S, Yoneyama T (1993) Effects of pH and osmotic stress on cellular polyamine contents in the soybean *Rhizobia fredii* P220 and *Bradyrhizobium japonicum* A1017. Appl Environ Microbiol 59:1104–1109

Garg N, Geetanjali G (2009) Symbiotic nitrogen fixation in legume nodules: process and signaling: a review. In: Lichtfouse E, Navarette M, Véronique S, Alberola C (eds) Sustainable agriculture. Springer, Netherlands, pp 519–531

Geurts R, Lillo A, Bisseling T (2012) Exploiting an ancient signalling machinery to enjoy a nitrogen fixing symbiosis. Curr Opin Plant Biol 15:1–6

Gordon AJ, Mitchel DF, Ryle GJA, Powell CE (1987) Diurnal production and utilization of photosynthate in nodulated white clover. J Exp Bot 38:84–98

Graham PH, Vance CP (2000) Nitrogen fixation in perspective: an overview of research and extension needs. Field Crops Res 65:93–106

Graham PH, Vance CP (2003) Legumes: importance and constraints to greater use. Plant Physiol 13:872–877

Hasanuzzaman M, Gill SS, Fujita M (2013) Physiological role of nitric oxide in plants grown under adverse environmental conditions. In: Tuteja N, Gill SS (eds) Plant acclimation to environmental stress. Springer Science + Business Media, New York, pp 269–322

Hekneby M, Antolin MC, Sanchez-Diaz M (2001) Cold response of annual mediterranean pasture legumes. In: Delgado I, Lloveras J (eds) Quality in lucerne and medics for animal production. Options Mediterraneennes 45. CIHEAM, Zaragoza, pp 157–161

Howard JB, Rees DC (1996) Structural Basis of Biological Nitrogen Fixation. Chem Rev 96(7):2965–2982

Hungria M, Franco AA, Sprent JI (1993) New sources of high-temperature tolerant rhizobia for Phaseolus vulgaris L. Plant Soil 149(1):103–109

Hungria M, Vargas MAT (2000) Environmental factors affecting N_2 fixation in grain legumes in the tropics, with an emphasis on Brazil. Field Crops Res 65:151–164

Huss-Danell K (1997) Actinorhizal symbioses and their N_2 fixation. New Phytol 136:375–405

Jenkins DJ, Wolever TM, Taylor RH, Barker HM, Fielden H (1980) Exceptionally low blood glucose response to dried beans: comparison with other carbohydrate foods. Br Med J 281:578–580

Jones DL, Darrah PR, Kochian LV (1996) Critical evaluation of organic acid mediated iron dissolution in the rhizosphere and its potential role in root iron uptake. Plant Soil 180:57–66

Kalogeropoulos N, Chiou A, Ioannou M, Karathanos VT, Hassapidou M, Nikolaos K, Andrikopoulos NK (2010) Nutritional evaluation and bioactive microconstituents (phytosterols, tocopherols, polyphenols, triterpenic acids) in cooked dry legumes usually consumed in the Mediterranean countries. Food Chem 121:682–690

Karlidag H, Esitken A, Turan M, Sahin F (2007) Effects of root inoculation of plant growth promoting rhizobacteria (PGPR) on yield, growth and nutrient element contents of leaves of apple. Sci Hortic 114(1):16–20

Keneni AF, Assefa PC, Prabu (2010) Characterization of acid and salt-tolerant rhizobial strains isolated from faba bean fields of Wollo Northern Ethiopia. J Agric Sci Technol 12:365–376

Khan MS, Zaidi A, Wani PA (2007) Role of phosphate solubilizing microorganisms in sustainable agriculture – a review. Agron Sustain Dev 27:29–43

Kishinevsky BD, Sen D, Weaver RW (1992) Effect of high root temperature on Bradyrhizobium-peanut symbiosis. Plant Soil 143:275–282

Kosegarten H, Koyro HW (2001) Apoplastic accumulation of iron in the epidermis of maize (*Zea mays*) roots grown in calcareous soil. Physiol Plant 113:515–522

Kudapa H, Ramalingam A, Nayakoti S, Chen X, Zhuang WJ, Liang X, Varshney RK (2013) Functional genomics to study stress responses in crop legumes: progress and prospects. Func Plant Biol 40:1221–1233

Kumari MSL, Subbarao NS (1984) Root hair infection and nodulation of lucerne as influenced by salinity and alkalinity. Plant Soil 40:261–268

Ladrera R, Marino D, Larrainzar E, González EM, Arrese-Igor C (2007) Reduced carbon availability to bacteroids and elevated ureides in nodules, but not in shoots, are involved in nitrogen fixation response to early drought in soybean. Plant Physiol 145:539–546

Lapinskas EB (2007) The effect of acidity on the distribution and symbiotic efficiency of rhizobia in Lithuanian soils. Eurasian Soil Sci 40:419–425

Latef AAHA, Ahmad P (2015) Legumes and breeding under abiotic stress: an overview. In: Legumes under environmental stress yield, improvement and adaptations. John Wiley & Sons Ltd, Hoboken, NJ, pp 1–20

Lindström K, Sorsa M, Polkunen J, Kansaner P (1985) Symbiotic nitrogen fixation of *Rhizobium* (Galega) in acid soils, and its survival in soil under acid and cold stress. Plant Soil 87:293–302

Liu J-Q, Allan DL, Vance CP (2010) Systemic signaling and local sensing of phosphate in common bean: Cross-Talk between photosynthate and microRNA399. Mole Plant 3(2):428–437

Long SR (1989) *Rhizobium* genetics. Annu Rev Genet 23:483–506

López M, Herrera-Cervera JA, Iribarne C, Tejera NA, Lluch C (2008) Growth and nitrogen fixation in *Lotus japonicus* and *Medicago truncatula* under NaCl stress: nodule carbon metabolism. J Plant Physiol 165:641–650

López-Millán AF, Morales F, Abadía A, Abadía J (2001) Iron deficiency-associated changes in the composition of the leaf apoplastic fluid from field-grown pear (*Pyrus communis* L.) trees. J Exp Bot 52:1489–1498

Manuel D, Alcántara E (2002) A comparison of ferric-chelate reductase and chlorophyll and growth ratios as indices of selection of quince, pear and olive genotypes under iron deficiency stress. Plant Soil 241:49–56

Marino D, Frendo P, Ladrera R, Zabalza A, Puppo A, Arrese-Igor C, González EM (2007) Nitrogen fixation control under drought stress. Localized or systemic? Plant Physiol 144:1233

Mengel K (1994) Iron availability in plant tissues—iron chlorosis on calcareous soils. Plant Soil 165:275–283

Mengel K (1995) Iron availability in plant tissues—iron chlorosis on calcareous soils. In: Abadía J (ed) Iron nutrition in soils and plants. Kluwer Academic Publishers, Dordrecht, The Netherlands, pp 389–397

Michiels J, Verreth C, Vanderleyden J (1994) Effects of temperature stress on bean nodulating Rhizobium strains. Appl Environ Microbiol 60:1206–1212

Miller PR, McConkey BG, Clayton GW, Brandt SA, Staricka JA, Johnston AM, Neill KE (2002) Pulse crop adaptation in the northern Great Plains. Agron J 94:261–272

Molassiotis AN, Diamantidis GC, Therios IN, Tsirakoglou V, Dimassi KN (2005) Oxidative stress, antioxidant activity and Fe (III)-chelate reductase activity of five Prunus rootstocks explants in response to Fe deficiency. Plant Growth Regul 46:69–78

Mollard RC, Zykus A, Luhovyy BL, Nunez MF, Wong CL, Anderson GH (2012) The acute effects of a pulse-containing meal on glycaemic responses and measures of satiety and satiation within and at a later meal. Br J Nutr 108:509–517

Monica NISTE, Roxana VIDICAN, Ioan ROTAR, Rodica POP (2013) The effect of pH stress on the survival of *Rhizobium trifolii* and *Sinorhizobium meliloti* in vitro. Bull UASMV Ser Agric 70(2):449–450

Morel MA, Braña V, Castro-Sowinski S (2012) Legume crops, importance and use of bacterial inoculation to increase production. In: Goyal A (ed) Crop plant. InTech, Croatia. doi:10.5772/37413

Moriuchi KS, Friesen ML, Cordeiro MA, Badri M, Vu WT, Main BJ et al (2016) Salinity adaptation and the contribution of parental environmental effects in *Medicago truncatula*. PLoS One 11(3):e0150350

Munevar F, Wollum AG (1982) Response of soybean plants to high root temperature as affected by plant cultivar and Rhizobium strain. Agron J 74:138–142

Muthukumar T, Priyadharsini P, Uma E, Jaison S, Pandey RR (2014) Role of arbuscular mycorrhizal fungi in alleviation of acidity stress on plant growth. In: Miransari M (ed) Use of microbes for the alleviation of soil stresses. Springer Science + Business Media, New York, pp 43–71

Nabizadeh E, Jalilnejad N, Armakani M (2011) Effect of salinity on growth and nitrogen fixation of Alfalfa (*Medicago sativa*). World Appl Sci J 13:1895–1900

Nestel P, Cehun M, Chronopoulos A (2004) Effects of long-term consumption and single meals of chickpeas on plasma glucose, insulin, and triacylglycerol concentrations. Am J Clin Nutr 79:390–395

Nikolic M, Roemheld V (2003) Nitrate does not result in iron inactivation in the apoplast of sunflower leaves. Plant Physiol 132:1303–1314

O'Hara GW, Franklin M, Dilworth MJ (1987) Effect of sulfur supply on sulfate uptake, and alkaline sulfatase activity in free-living and symbiotic bradyrhizobia. Arch Microbiol 149(2):163–167

Ohyama T, Ohtake N, Sueyoshi K, Tewari K, Takahashi Y, Ito S, Nishiwaki T, Nagumo Y, Ishii S, Sato T (2009) Nitrogen fixation and metabolism in soybean plants. Nova Science Publishers, Inc., New York

Oktem HA, Eyidoan F, Demirba D et al (2008) Antioxidant responses of lentil to cold and drought stress. J Plant Biochem Biotechnol 17:15–21

Orhan E, Esitken A, Ercisli S, Turan M, Sahin F (2006) Effects of plant growth promoting rhizobacteria (PGPR) on yield, growth and nutrient contents in organically growing raspberry. Sci Hortic 111:38–43

Ovtsyna AO, Staehelin C (2003) Bacterial signals required for the Rhizobium-legume symbiosis. In: Pandalai SG (ed) Recent research developments in microbiology, Part II, vol 7. Research Signpost, Trivandrum, India, pp 631–648

Papanikolaou Y, Fulgoni VL III (2008) Bean consumption is associated with greater nutrient intake, reduced systolic blood pressure, lower body weight, and a smaller waist circumference in adults: results from the National Health and Nutrition Examination Survey 1999–2002. J Am Coll Nutr 27:569–576

Perret X, Staehelin C, Broughton WJ (2000) Molecular basis of symbiotic promiscuity. Microbiol Mol Biol Rev 64:180–201

Piha MI, Munnus DN (1987) Sensitivity of the common bean (*Phaseolus vulgaris* L.) symbiosis to high soil temperature. Plant Soil 98:183–194

Purcell LC, Silva M, King CA, Kim WH (1997) Biomass accumulation and allocation in soybean associated with genotypic differences in tolerance of nitrogen fixation to water deficits. Plant Soil 196:101–103

Rainbird RM, Akins CA, Pate JJS (1983) Effect of temperature on nitrogenase functioning in cowpea nodules. Plant Physiol 73:392–394

Rebello CJ, Greenway FL, Finley JW (2014) A review of the nutritional value of legumes and their effects on obesity and its related co-morbidities. Obes Rev 15:392–407

Rodelas B, González-López J, Salmerón V, Pozo C, Martínez-Toledo MV (1996) Enhancement of nodulation, N_2 fixation and growth of faba bean (*Vicia faba* L.) by combined inoculation with *Rhizobium leguminosarum* bv. *viciae* and *Azospirillum brasilense*. Symbiosis 21:175–186

Rodelas B, González-López J, Martínez-Toledo MV, Pozo C, Salmeró NV (1999) Influence of *Rhizobium/Azotobacter* and *Rhizobium/Azospirillum* combined inoculation on mineral composition of faba bean (*Vicia faba* L.) Biol Fertil Soils 29:165–169

Romera FJ, Alcantara E, De La MDG (1991) Characterization of the tolerance to iron chlorosis in different peach rootstocks grown in nutrient solution. Plant Soil 130:115–125

Rosas S, Andres J, Rovera M, Correa N (2006) Phosphate-solubilizing *Pseudomonas putida* can influence the rhizobia–legume symbiosis. Soil Biol Biochem 38:3502–3505

Sangakkara UR, Hartwig UA (1996) Soil moisture and potassium affect the performance of symbiotic nitrogen fixation in faba bean and common bean. Plant Soil 184:123–130

Serraj R, Fleurat-Lessard P, Jaillard B, Drevon JJ (1995) Structural changes in the innercortex cells of soybean root nodules are induced by short-term exposure to high salt or oxygen concentrations. Plant Cell Environ 18(4):455–462

Sharma A, Johri BN (2003) Growth promoting influence of siderophore-producing *Pseudomonas* strains GRP3A and PRS 9 in maize (*Zea mays* L.) under iron limiting conditions. Microbiol Res 158:243–248

Sherren A, Ansari R, Naqvi SSM, Soomaro AQ (1998) Effect of salinity on *Rhizobium* species, nodulation and growth of soybean. Pak J Bot 1:75–81

Singleton PW, Bohlool BB (1984) Effect of salinity on nodule formation by soybean. Plant Physiol 74:72–76

Singleton PW, Swaify SA, Bohlool BB (1982) Effect of salinity on *Rhizobium* growth and survival. Appl Environ Microbiol 44:884–890

Skorupska A, Wielbo J, Kidaj D, Marek-Kozaczuk M (2010) Enhancing Rhizobium legume symbiosis using signaling factors. In: Khan MS, Zaidi A, Musarrat J (eds) Microbes for legume improvement. Springer-Verlag, New York, pp 27–54. ISBN: 978-3-211

Smith DL, Dijak M, Hume DJ (1988) The effect of water deficit on N_2 (C_2H_2) fixation by white bean and soybean. Can J Plant Sci 68:957–967

Sprent JI, Stephens JH, Rupela OP (1988) Environmental effects on nitrogen fixation. In World crops: cool season food legumes. Springer Netherlands, pp. 801–810

Sprent JI (2001) Nodulation in Legumes. Cromwell Press, Royal Botanical Gardens, Kew

Stoeva N, Kaymakanova M (2008) Effect of salt stress on the growth and photosynthesis rate of bean plants. J Cent Eur Agric 9:385–392

Streeter JG (2003) Effects of drought on nitrogen fixation in soybean root nodules. Plant Cell Environ 26:1199–1204

Subba Rao GV, Johnseng C, Kumarrao JVDK, Jana MK (1999) Response of the Pigeon pea *Rhizobium* symbiosis to salinity stress: variation among *Rhizobium* strain in symbiotic ability. Biol Fertil 9:49–53

Sulima AS, Zhukov VA, Shtark OY, Borisov AY, Tikhonovich IA (2015) Nod-factor signaling in legume-rhizobial symbiosis. In: El-Shemy H (ed) Plants for the future. InTech, Croatia. doi:10.5772/61165

Tagliavini M, Rombola AD (2001) Iron deficiency and chlorosis in orchard and vineyard ecosystems. Eur J Agron 15:71–92

Tejera N, Lluch C, Martínez M, González J (2005) Isolation and characterization of *Azotobacter* and *Azospirillum* strains from the sugarcane rhizosphere. Plant Soil 27:223–232

Toker C, Yadav SS (2010) Legumes cultivars for stress environments. In: Climate change and management of cool season grain legume crops. Springer, Netherlands, pp 351–376

Torres M, Valencia S, Bernal J, Martínez P (2004) Isolation of *Enterobacteria, Azotobacter* sp. and *Pseudomonas* sp., producers of indole-3-acetic acid and siderophores, from Colombian rice rhizosphere. Rev Latin Microbiol 42:171–176

Toselli M, Marangoni B, Tagliavini M (2000) Iron content in vegetative and reproductive organs of nectarine trees in calcareous soils during the development of chlorosis. Eur J Agron 13(4):279–286

Trinchant JC, Boscari A, Spennato G, van de Sype G, le Rudulier D (2004) Proline betaine accumulation and metabolism in alfalfa plants under sodium chloride stress. Exploring its compartmentalization in nodules. Plant Physiol 135:1583–1594

Vadez V, Sinclair TR, Serraj R (2000) Asparagine and ureide accumulation in nodules and shoot as feedback inhibitors of N_2 fixation in soybean. Physiol Plant 110:215–223

Valentine AJ, Benedito VA, Kang Y (2011) Legume nitrogen fixation and soil abiotic stress: from physiology to genomic and beyond. Annu Plant Rev 42:207–248

Velagaleti RR, Marsh S (1989) Influence of host cultivars and *Bradyrhizobium* strains on the growth and symbiotic performance of soybean under salt stress. Plant Soil 119:133–138

Vincent JM (1977) Rhizobium: General microbiology. In R.W.F. Hardy and W.S. Silver (eds.) A Treatise on Dinitrogen Fixation Section III Biology. John Wiley & Sons, New York, p 277366

Wagner SC (2011) Biological nitrogen fixation. Nat Educ Knowl 3:15

Weisany W, Raei Y, Allahverdipoor KH (2013) Role of some of mineral nutrients in biological nitrogen fixation. Bull Env Pharmacol Life Sci 2(4):77–84

West SA, Kiers ET, Pen I, Denison RF (2002) Sanctions and mutualism stability: when should less beneficial mutualists be tolerated? J Evol Biol 15:830–837

Yuanyuan M, Yali Z, Jiang L, Hongbo S (2009). Roles of plant soluble sugars and their responses to plant cold stress. African J Biotechnol 8(10)

Zahran HH (1999) *Rhizobium*-legume symbiosis and nitrogen fixation under severe conditions and in an arid climate. Microbiol Mol Biol Rev 63:968–989

Zahran HH (2001) Rhizobia from wild legumes: diversity, taxonomy, ecology, nitrogen fixation and biotechnology. J Biotechnol 91:143–153

Zhang H, Sun Y, Xie X, Kim MS, Dowd SE, Paré PW (2009) A soil bacterium regulates plant acquisition of iron via deficiency-inducible mechanisms. Plant J 58:568–577

Zhang XN, Li X, Liu JH (2014) Identification of conserved and novel cold-responsive microRNAs in trifoliate orange (*Poncirus trifoliata* (L.) Raf.) using high-throughput sequencing. Plant Mol Biol Rep 32:328–341

Zhu H, Choi HK, Cook DR, Shoemaker RC (2005) Bridging model and crop legumes through comparative genomics. Plant Physiol 137:1189–1196

Index